AN INTRODUCTION TO AMERICAN POLICING

Dennis J. Stevens, PhD
Research Coordinator
The Justice Research Association
Jupiter, Florida

JONES AND BARTLETT PUBLISHERS
Sudbury, Massachusetts
BOSTON TORONTO LONDON SINGAPORE

World Headquarters
Jones and Bartlett Publishers
40 Tall Pine Drive
Sudbury, MA 01776
978-443-5000
info@jbpub.com
www.jbpub.com

Jones and Bartlett Publishers Canada
6339 Ormindale Way
Mississauga, Ontario L5V 1J2
Canada

Jones and Bartlett Publishers International
Barb House, Barb Mews
London W6 7PA
United Kingdom

Jones and Bartlett's books and products are available through most bookstores and online booksellers. To contact Jones and Bartlett Publishers directly, call 800-832-0034, fax 978-443-8000, or visit our website www.jbpub.com.

Library of Congress Cataloging-in-Publication Data
Stevens, Dennis J.
An introduction to American policing / by Dennis J. Stevens.
p. cm.
Includes bibliographical references and index.
ISBN 978-0-7637-4893-7 (pbk.)
1. Police—United States. 2. Police administration—United States.
I. Title.
HV8139.S84 2009
363.20973—dc22
2008022733

6048

Production Credits
Acquisitions Editor: Jeremy Spiegel
Editorial Assistant: Maro Asadoorian
Production Director: Amy Rose
Senior Production Editor: Renée Sekerak
Production Assistant: Jill Morton
Associate Marketing Manager: Lisa Gordon
Manufacturing and Inventory Control Supervisor: Amy Bacus
Composition: Northeast Compositors, Inc.
Interior Design: Anne Spencer
Cover Design: Brian Moore
Cover Image: © LiquidLibrary
Chapter Opener Image: © Stephen Sweet/ShutterStock, Inc.
Photo Research Manager and Photographer: Kimberly Potvin
Photo Researcher: Lee Michelsen
Assistant Photo Researcher: Jessica Elias
Text Printing and Binding: Malloy Incorporated
Cover Printing: Malloy Incorporated

Printed in the United States of America
12 11 10 09 08 10 9 8 7 6 5 4 3 2 1

Dedicated to my loving children:
David D. Stevens, Mark A. Stevens, and Alyssa P. Stevens

Brief Contents

Contents

About the Author

Dr. Dennis J. Stevens received a PhD from Loyola University of Chicago in 1991. Currently, he is a research coordinator at The Justice Research Association in Jupiter, Florida. Formerly, he worked for the University of Southern Mississippi as director of its Criminal Justice graduate program (Gulf Coast Campus) and PhD director (CJ Department, Hattiesburg and Gulf Coast) while his research and community outreach to the New Orleans Police Department afforded him opportunities to serve as an adjunct professor at Loyola University of New Orleans. Dr. Stevens was also employed by the Commonwealth of Massachusetts and instructed classes at the University of Massachusetts–Boston and Salem State College. In addition to teaching traditional students, he has taught and counseled law enforcement and correctional personnel at law academies

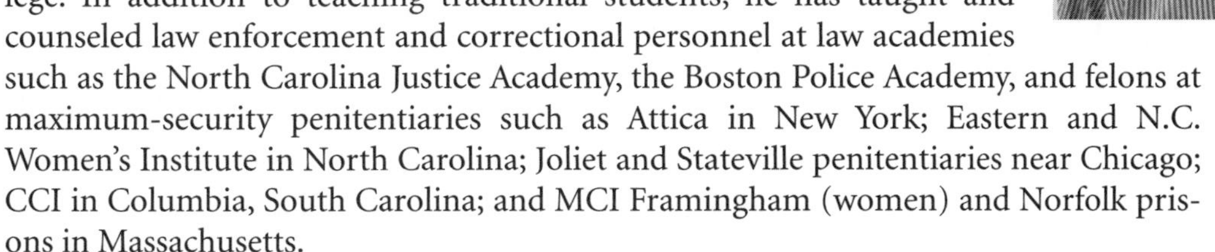

such as the North Carolina Justice Academy, the Boston Police Academy, and felons at maximum-security penitentiaries such as Attica in New York; Eastern and N.C. Women's Institute in North Carolina; Joliet and Stateville penitentiaries near Chicago; CCI in Columbia, South Carolina; and MCI Framingham (women) and Norfolk prisons in Massachusetts.

Dr. Stevens has published several books and almost 100 scholarly and popular literature articles on policing, corrections, and criminally violent offenders such as pedophiles and sexual predators (many of whom were instructed or counseled by Stevens). As a Criminal Justice/Sociology Department program director at Mount Olive College in North Carolina, he created and implemented a criminal justice lockstep curriculum (in collaboration with the Academic Vice President, Dorothy Whitley); four years after the curriculum was developed, more than 400 students were enrolled at five different locations across the state, including the North Carolina Justice Academy.

Dr. Stevens has been consulted by private agencies such as Wachenhut to write their "use of force" protocols for enforcement personnel subcontracted to the U.S. Department of Energy, by state legislators to examine the flow of drugs into prisons and the recidivism rates among student-incarcerated inmates, by federal and state agencies to study corruption among narcotic officers, by the American military related to cases on military bases, and by the governments of foreign countries (e.g., the

Provincial Government of Canada) to aid in research linked to incarcerated inmates and educational programs.

As a volunteer, Dr. Stevens has guided many sexually abused children and their families through church-affiliated programs in New York, North Carolina, and South Carolina and has led group crisis sessions among various police and correctional agencies. For example, he provided crisis intervention or debriefing sessions for officers of the New Orleans Police Department and deputies of Jefferson Parish for two years after Hurricane Katrina.

Foreword

On a November night in 2006, Sean Bell was holding his bachelor party at Club Kalua in the Jamaica Queens section of New York City. Unknown to the revelers, the strip club was under investigation by a New York Police Department (NYPD) undercover unit for suspicion of prostitution and complaints of guns and drugs. An altercation took place between the Bell party and other patrons, and an undercover police officer stationed inside the club thought he heard one of Bell's friends saying, "Yo, get my gun," as they headed outside to Bell's car. Fearing a shooting might occur, the detective followed the men to their car and alerted his backup team. When the team confronted Bell and ordered him to raise his hands after getting in his car, he instead accelerated the vehicle, hitting a police officer and an unmarked police minivan. The police team then shot Sean Bell 50 times, killing him, and wounded two of his companions.

Subsequent investigations showed that the incident was precipitated by mistaken perceptions on both sides: As it turned out, all three men were unarmed. Tried on charges ranging from manslaughter to reckless endangerment, on April 25, 2008, all three officers were cleared of any wrongdoing. In response to the verdict, on May 7, 2008, the Reverend Al Sharpton led a "slowdown" protest in the streets of New York City and 200 people were arrested as a result of the protest.

The Bell case aptly illustrates the exigencies facing police in contemporary society—challenges that make Dennis J. Stevens' *An Introduction to American Policing* a welcome addition to the police literature. This book arrives at a time when police are expected to confront a myriad of new problems ranging from cybercrime to terrorism. Never before have police officers been expected to be proficient in such esoteric skills as crime mapping, information technology, and biometric identification while at the same time dealing with more traditional issues such as gang control, community safety, and effective patrol techniques. At the same time, police agencies need to be sensitive to the demands of the populations they serve. While members of the general public may applaud police efforts and credit officers with having helped bring the crime rate down, they remain concerned—and rightfully so—about the power police officers have to monitor their behavior and control their lives. Even when most community members believe that police officers are competent and dependable, they are still worried about police's use of force and willingness to respect the rights of suspects.

The Bell case raises another key concern: Are police racially and ethnically biased, and do they use racial profiling to harass members of the minority community? In the example cited earlier, was Sean Bell killed because he was an African American man in the wrong place at the wrong time? By focusing its attention on cases like the Bell shooting, the media have given the general public the idea that some police officers actually believe "driving while black" is an enforceable offense, while ignoring the substantial efforts police departments have made to improve their officers' behavior, deal with personnel problems, and improve their relationships with the community they serve. The media-fueled complaints are not lost on police officers, who may become frustrated because they feel that they get little credit when they do a good job but get slammed when things go awry. Their misgivings are not misplaced, considering that any misstep they may make is apt to be recorded on a cell phone, posted on YouTube within the hour, and broadcast on CNN an hour later.

Even as they grapple with these ongoing problems, police and law enforcement agents are being challenged to confront new and emerging social problems. Concern about immigration has galvanized police agencies in border jurisdictions across the United States. Immigration and Customs Enforcement (ICE) has hired more agents, as has the Border Patrol. National Guard troops have even been stationed at the border, ground sensors have been installed, and funds have been authorized for continued construction of fencing.

Immigration is not the only emerging issue, of course. Since the events of September 11, 2001, a number of local policing agencies have responded to the threat of terrorism by creating special antiterror programs. New York City, for example, established a Counterterrorism Bureau whose teams have been trained to examine potential targets in the city. Other local departments are monitoring the Internet for sexual predators, employing sophisticated mapping techniques, and using newly emerging information technology (IT) programs to effectively control common crimes such as burglary and rape. And even if they are too late to prevent or deter crime, police must be familiar with sophisticated identification methods ranging from DNA testing to high-tech equipment that can create virtual crime scenes for sophisticated analysis.

Clearly, police officers today face challenges that officers of an earlier era could hardly imagine. All of these issues, challenges, and events make Stevens' *An Introduction to American Policing* especially timely. Rather than rely on traditional methods to get his points across, the author uses a combination of the practical and theoretical to engage the reader in contemporary police issues. This text does not depend on musings from the Ivory Tower that are not grounded in the day-to-day experience of police officers. Instead, *An Introduction to American Policing* presents students with the actual words, deeds, and experiences of people currently on the front line of contemporary policing—an invaluable teaching tool. And, of course, the book covers all of the important issues, from search and seizure to stings, from suicide by cop to police officer stress. I have learned a lot from Stevens' work and I am sure you will, too.

Larry Siegel
University of Massachusetts–Lowell

Preface

Toss yesterday's police textbooks out the window: Terrorism, immigration, and a civil rights resurgence have changed policing so much that it's unlikely that anyone can recognize their daddy's job. Today, street cops spend most of their time attending to quality-of-life policing standards and less than 15 percent of their time engaged in law enforcement initiatives, such as responding to burglary calls or trying to find a stolen car.[1] It's time to cruise with gang units in Central Los Angeles; do the electric slide with New Orleans' finest, who jam suspects into the back of cruisers because jails have lost electrical power; discover the perspective inspired by a commuter on a crowded New York City subway that revolutionized police initiatives; and take a look at the street gang leader who sued the Chicago police and won. But policing won't be outsmarted: COPLINK and other artificial intelligence systems, forensic technology, gang injunction and restraining orders, and approval of high-speed pursuit and no-knock searches by the U.S. Supreme Court has leveled the playing field.

In democracies the world over, criminal justice works hard to protect the idea of justice—that unyielding principle of freedom through the architecture of courtrooms, where it can be shaped by "rules of discovery, hearsay, visible accusers, and cross-examination that have been built through trials and lots of errors"[2] over centuries. The adversarial process is under attack across the world, because at times it is messy. This process seems to always start with the gatekeepers—police officers. Its centerpiece is freedom of choice, which means that it can be both unpredictable and combative. Of course, those are the very reasons why democracy is revered and cherished.

Americans want to be safe and feel secure in their individual everyday choices without fear of arbitrary government intrusion or the intrusion of others. They need police officers to effectively deal with the intruders and protect their freedom at the same time. Herein lies the incredible challenge of policing—balancing apprehension versus civil rights.

Every day, police face new legal, economic, and managerial challenges. Many labor under the burden of inadequate budgets and ever-increasing demand for quality police services, including a host of new responsibilities related to homeland security and immigration. Americans insist that police shield them from crime and the collateral damage of terrorism, natural calamities, and epidemics, yet both law makers and lawbreakers demand that police safeguard their rights as provided by the U.S. Constitution. When individuals believe that their rights have been violated, public sentiment often rushes them into the streets and the courtrooms.

One way to learn about the quest to maintain both public safety and freedom is by reading *An Introduction to American Policing*. This textbook provides a balanced investigation of police history and theory linked to police practice, thereby enabling students to make informed decisions about the issues.

It is clear that when human rights are safeguarded through professional police initiatives, there is less anarchy and government corruption.[3] The validity of this compelling argument is readily apparent among Americans, who are generally enjoying a steady rise in their quality-of-life experiences by influencing the decision-making routines of police behavior through leadership, practical police knowledge, evidence-based policing, litigation, and oversight activities. Support for this process is a benchmark of a democracy, as is police accountability.

This textbook also explains policing from the inside—that is, through the personal experiences of officers—and links those experiences to theories proposed in scholarly research. However, as you proceed through the pages ahead, be mindful that contemporary history is the hardest to write, but the easiest to criticize.

An Introduction to American Policing describes the criminal justice system, criminology, and law enforcement knowledge, using them to connect the dots linked to the progress of the police community. It will satisfy both scholars, who lean toward theoretical perspectives, and practitioners, who favor practical recommendations, because it includes investigative reports about police organizations, case studies, narratives from violators, and current research and impressions from colleagues, officers, and researchers. This technique results in a blend of theory and practice that prompts reader participation by probing the way some believe policing "should be" while examining evidence about "the way it is." After reading this book, readers will recognize the central theories and practical realities of U.S. law enforcement and quality-of-life practices. In addition, in recognition of the frequent partnerships being forged between local/state police organizations and federal agencies,[4] *An Introduction to American Policing* provides an up-to-date description of those partnerships and of the policing efforts of federal law enforcement agencies, including the Department of Homeland Security (DHS).

The goal of this book is to help the reader think for himself or herself, to understand, to imagine, to reflect, and to exercise good judgment when thinking about the principal issues related to American policing. In summary, the book is intended to provide relevant theoretical scholarly perspectives and practical police experiences to help a reader make an informed decision about policing in this new millennium.

Dennis J. Stevens
The Justice Research Association
Jupiter, Florida

Notes

1. Peter K. Manning, "The Police: Mandate, Strategies, and Appearances," in *The Police and Society: Touchstone Readings,* ed. Victor E. Kappeler (Mt. Prospect, IL: Waveland Press, 1995), 97–126.
2. Ron Suskind, "The Unofficial Story of the al-Qaeda 14," *Time* September 18, 2006, p. 34.
3. This perspective is advanced in Dennis J. Stevens, *Police Officer Stress: Sources and Solutions* (Upper Saddle River, NJ: Prentice Hall, 2008).
4. Robert Michael Goldman, *A Free Ballot and a Fair Count: The Department of Justice and the Enforcement of Voting Rights* (New York: Fordham University Press, 2001).

Acknowledgments

As with most publications, numerous individuals supported this project, although the opinions offered, alas, belong to the writer. The author wishes to acknowledge the efforts of my graduate assistants who make my job easier: Kim E. Cox (currently Special Agent, Homeland Security), Lacey Cochran Stewart (currently Makenzy's mom), Linda Moss (PhD candidate), Sergeant Luke Thompson (Gulfport Police Department, Mississippi), and Sergeant Shane Steel (Biloxi Police Department, Mississippi, on a leave of absence while attending law school in Los Angeles).

Larry Siegel, University of Massachusetts–Lowell, diligently provided a fantastic foreword, and Chief Theron Bowman, PhD, Arlington Texas Police Department wrote an excellent introduction, among his other contributions to this book. Thanks are also owed to Sergeant John Wiggins, Investigations and Internal Affairs and an instructor at N.C. Justice Academy; Timothy Bakken, Professor of Law, U.S. Military Academy at West Point; Lafayette, Louisiana, Police Officer Jennifer Taylor (currently a PhD candidate at the University of Southern Mississippi); and Sergeant Steven Dickenson, Palm Beach County Sheriff's Department, Florida.

Municipal officers (and former students) in the Commonwealth of Massachusetts also contributed to this work. At the Boston Police Department, thanks go to Officers Debra Blandin, Maria Gonzales, Kenisha Stewart, and Jennifer Costa; Sergeant Roy Chambers; Detectives Joseph Fiandaca, George Kelley, and Daniel O'Neill; and Captain Robert Dunford (superintendent of uniforms Boston, who was kind enough to take me under his wing). Other contributors included Massachusetts State Troopers (and former students) Lisa Cisso and Jason Powers; Salem Officer Christian Hanson; Nantucket Officer Nadya Marino; Rockport Officers Michael Marino and Paul Vansteenberg; and Essex County Sheriff Frank G. Cousins, Jr.

I am also grateful for the assistance and comments of District Police Training, Southern Counties, Angela Rogers; and Department of the Army, Criminal Investigation Command, Forensic Latent Print Examiner, Rooney A. Schenck. Others who contributed include Chief Harley Schinker, Long Beach, Mississippi; University of Southern Mississippi students: Lieutenant William Seal, Long Beach; Sergeant Vince Myrick, on leave from Biloxi Police Department, Mississippi, owing to special agent assignment with a federal law enforcement agency; Special Agent Brandon Hendry, Mississippi Bureau of Narcotics; Chief Charles Ramsey, Metropolitan Washington, DC, Police; New Orleans Officer David J. Lapene; Jefferson Parish Deputy Wayne A. Heims;

Chief Francis D'Ambra, Manteo, North Carolina; Police Superintendent Phillip Cline's office, Chicago Police Department; Sergeant Mike McCleery, Criminal Investigations, New Orleans Police Department; Sergeant Steven Morrison, Houston Police Department; Constable Ron Hickman, Precinct 4 Harris County–Houston, Texas; Dina Richardson, Tucson Police Department; and Sergeant Chad Gann, Arlington, Texas, Police Department.

My gratitude also extends to the following people: Bessey Hutchinson (PhD candidate at the University of Southern Mississippi); Supervisor Peter Cartmell, Crime Analysis Unit, Fort Lauderdale; my colleague at Southern Miss, Dr. Thomas Payne; Dr. James McCabe (former Division Commander, Captain, New York Police Department), and Dr. Patrick Morris (former Sergeant, Norwalk, Connecticut, Police Department), Sacred Heart University, Fairfield, Connecticut. Also, Dr. Mark L. Dantzker, University of Texas Pan-American, encouraged the writing of this textbook (and other works).

I must also acknowledge two prominent individuals who provided inspiration for me to write books and articles: Dr. Thomas B. Priest, a mentor and first my academic boss, who guided me through my first published articles and my dissertation at Johnson C. Smith University, and my friend, colleague, and mentor, Frank Schmalleger, who over the years provided inspiration to this writer.

Regardless how prolific we might think we are, without those patient folks working in information technology who go out of their way for writers and others, nary a line in a book would be written. My thanks go to Rob Smith, Annette Copeland, and David Siccone at the I-Tech Department at University of Southern Mississippi, Gulfport campus, who resurrected my computer from the dead (twice!) as I wrote this book.

Others who also aided in the development of this work include the very hard working and bright personnel at Jones and Bartlett including Jeremy Spiegel, Renée Sekerak, and Maro N. Asadoorian.

Finally acknowledgment goes to the reviewers of this work:

Michael Costelloe, Northern Arizona University
Paul C. Leccese, Old Dominion University
Charles A. Loftus, Arizona State University
Kirk Miller, Northern Illinois University
Nicole Romeiser, University of Maryland

Introduction

Welcome to an exciting time in policing. As police philosophies have evolved, so, too, have policing innovations and advances in technology, equipment, and forensic sciences. These tools—both theoretical and practical—offer today's officers more options and, in many instances, better outcomes. But how did we get here, and are some approaches better than others? By whose standards do we measure success: our own or those of the public? In his latest examination of these issues, *An Introduction to American Policing,* Dr. Dennis J. Stevens delves into just those questions.

Dr. Stevens' in-depth study of policing history includes current trends and practices. This text takes a comprehensive look at the policing world from the inside out, complete with research findings and perspectives from front-line workers on the job.

Perhaps never in its history has policing enjoyed (entertained? suffered?) such high-profile visibility with the public. Police officers are the focal point of endless television shows and movies, and the basis for countless novels and plays. And within this past decade, police have been forced to deal with a new and perhaps the most pervasive (invasive?) factor: the Internet. The prevalence of camera cell phones means everything police officers say or do is subject to be instantly recorded and broadcast to the world via the Internet in a matter of minutes. It seems as if *everyone* is interested in what the police are doing. Dr. Stevens explores policing's rising visibility and expectations. Today's law enforcement professionals are managing not just investigations but also expectations, both internally and with their outside "customers." How do officers balance the many demands placed on them and still return sanely and safely to their loved ones at home? This book offers some thoughts on those topics.

In the post–September 11 environment, there has been a renewed sense of respect for the heroes in blue. At the same time, recent upswings in violent crime across the United States have brought more intense scrutiny of police and greater demand for their services. Law enforcement personnel rely on terms such as "community ownership," "stakeholder participation," and "partnerships" to describe the cooperation needed to provide effective policing, but are expected to offer unparalleled service at all times—even when such cooperation is largely absent. How this is accomplished, and how it sometimes fails, are inspected in this book from several angles. The variety of views offered is sure to prompt discussion and debate, both of which are necessary if the noble profession of policing is to progress even further and continue to improve.

Chief Theron Bowman, PhD
Arlington Police Department
Arlington, Texas

Fundamentals of Policing

29
28

Police: The Essentials

"We must all work to create a culture of respect for the rule of law at home and abroad, where all members of the community respect the law and are protected and empowered."

—Secretary General of the United Nations

LEARNING OBJECTIVES

When you finish reading this chapter, you will be better prepared to

- Describe American policing
- Describe in what way the American police officer has changed
- Characterize the rule of law
- Describe the law of criminal procedure
- Characterize police discretion
- Describe mandatory arrest and its purpose
- Characterize clearance rates and their implications
- Describe street justice
- Characterize the "broken windows" perspective
- Define order maintenance as a police initiative
- Describe the stereotypes about policing held by the public and the police
- Explain the police–population ratio
- Identify what Americans want and don't want from the police[1]

CASE STUDY: YOU ARE THE POLICE OFFICER

You're at your first police roll call. You're a 23-year-old police officer who completed the 720 required hours of police training including 180 hours of field training and the ride alongs with your TO (training officer) over the past four weeks in another district. You're excited about your first assignment. You've waited for this day most of your life and envisioned the day when you stood next to your desk in second grade at Marriot Elementary and told the class, "When I grow up I wanna be a police officer like my Uncle Charley." Throughout your college days, you remember the instructors who encouraged your dream and the few who thought you should go on to law school. Your mind was made up, though: Complete your internship at the police department, finish your criminal justice degree, take the civil service exam, and become a police officer so you could help people and put away the bad guys.

As the sergeant assigned officers to specific duties, the officers left ranks and exited to the police vehicles. After a while it's just you and Serg, "What about me, Serg?"

"Oh yeah," he says and rubes his narrow chin before he speaks. "You're a week early. Sit in the squad room till I figure something out." You feel embarrassed but ask, "Ah, where's the squad room?"

"Rookies," sang Serg, "you're in it." And repeats, "You're in the squad room, kid." An hour or so later when the Serg returns, he says, "We've got a youthful offender in holding tank. Take her to county and process her into the system." Racing through your mind are all the techniques you learned at the academy about prisoner transportation, but you never thought that would be your first duty. Last night when you couldn't sleep, you imagined that your first assignment would be to a young family who would seek your advice about their wayward child. Or maybe you would confront a shoplifter at Walgreens or help a mother find her lost daughter. You had to be prepared for anything—even a car crash and helping with emergency care. The state provided you with the power of arrest and use of force to ensure safety. You found the balance in your heart long ago when you accepted Uncle Charley's ideals of "always doing the right thing." You were always amazed how people respected him at church, in the mall, and at family gatherings. And now finding the balance between protector and enforcer should be easier.

You stop in your tracks, "Hey Serg, I ain't no kid," you say with a snicker, "but where's the holding tank." The Serg laughs and points to the second floor.

"Sign here"—the jailer pushes a form at you upon your arrival. "And remember, the knife in this plastic bag must never leave your sight. Have county sign off on it when you get there."

As you stare at the jailer, he explains, "Chain of custody, officer. If the evidence is tampered with, the DA can't use it."

You nod and the jailer flags you to the cell.

At the end of the shift, you're thinking about the events of your first shift. You think about the mix-up at county when you delivered the prisoner: You weren't even sure which door to use to escort her into the facility. She didn't look like she'd own a knife let alone use it. You smiled recalling that the jailer called you, "officer": that made your day!

1. Which events and circumstance in a child's life might lead him or her toward a police officer's career?
2. After all the training and supervision with a TO, why is there so much more to learn about policing?
3. In what way would balancing the powers of an arrest and use of force provide safety?

Introduction

In recent years, the United States has been caught up in events and circumstance that are as strange as they are confusing.[2] Events and circumstance have altered the way Americans live, shop, and pray. Police practice and policy have changed, too. Today's officer can be described as a *protector* and an *intruder* in the daily lives of the people they encounter, which is the focus of this book.[3] The words of the Secretary General of the United Nations imply that above all others, police officers must work to create a culture of respect for the **rule of law** at home and in the communities where they are empowered.

Traditionally, generously funded academic researchers were often in a position to get the wrong impression about policing. Those misconceptions arose because police officers often cannot provide a full account—even to themselves—of what they see. As a consequence, in the past researchers wrote mostly for each other and themselves.[4] For instance, researchers pointed to many police programs as being ineffective, such as drug awareness (DARE) programs. In reality, the inefficiencies identified in these studies were more often closely linked to the researcher's naiveté than to any flaw of the police programs, at least according to some of the most notable criminologists and practitioners.[5]

This book is written for students who want to learn about police and for those who wish to understand policing—one of the most important endeavors in the United States. The authors who have contributed to this work rise above flashing light icons to present policing for what it really is, rather than as what others think it should be. For that reason, this chapter offers the essentials of policing—the basic principles everyone needs to know. Before discussing officers, however, let's talk about you.

Assumptions about You

More than likely, you are a criminal justice major or in a closely related academic discipline, or your present (or future) occupation is related to the justice profession. You probably completed an introductory course in criminal justice, or you may possess some practical experience in the criminal justice profession. However, even if you do not have those experiences, the stories offered in this book are presented in an uncomplicated fashion, without frills or complexity. It is assumed that many readers currently hold or will hold influential positions in the justice community and will work to further the goals of democracy. This final thought has driven the completion of this book, which assumes that dedicated professionals want to continue to enhance policing and ultimately ensure public safety, but in an informed way.

Flow of This Book

This work is divided into four parts, which collectively go beyond the standard ways of transferring knowledge to bring the realities of the world of policing to life through theoretical links to real-world experiences, issues, careers, systems, and procedures.

- Part 1 introduces the fundamentals of policing, which include police history, accounts from ancient civilizations to early American police activities, traditional strategies modeled after the English strategy, a metaphor that powers community relations strategies, and reasons and evidence-based findings suggesting that police organizations should change to meet the challenges of this new century, particularly as they relate to making "quality-of-life arrests" in public spaces.

- Part 2 focuses on police organizations, describing their management, leadership, and police officer hiring and training.
- The centerpiece of Part 3 is day-to-day police work, including patrol, investigations, strategies, and technological advancements.
- Part 4 explores the challenges of the twenty-first century, such as **police subculture**, discretion, accountability, and officer stress.

Goal of This Book

In recent years, the challenges posed by terrorism, immigration, and a civil rights resurgence have changed the responsibility, character, and accountability of policing from that seen in previous decades. This timely work provides contemporary explanations and logical evaluations of modern-day policing. Its goal is to help you think for yourself, to understand, to imagine, to reflect, and to exercise good judgment associated with the principal issues related to **American policing**. That is, the book is intended to provide relevant theoretical scholarly perspectives and practical police experiences to guide you toward making informed decisions about policing in this new millennium.

One concern that is evident in this book is its focus on "raising the bar" by offering fresh and creative ways to think about policing rather than resorting to the traditional reports and antiquated issues found elsewhere. For instance, the literature generally describes the public as "citizens," which implies a status of citizenship. The relationship between local police and the public—regardless of an individual's actual citizenship status—is protected through the Constitution. Therefore, it might be more appropriate to refer to individuals as constituents.

That said, why become a police officer?

■ Why People Become Police Officers

Many men and women who are attracted to policing are individuals who have wanted to be a "cop" since they were young.[6] For instance, Sergeant Stephen Dickenson of the Palm Beach Florida Sheriff's Department says, "This is always what I had in mind. [I wanted to be] a police officer since I was five." Surveys conducted among police officers consistently echo this view. In a study involving 558 police officers employed by various departments from Boston to Palm Beach, Florida, for example, 71 percent (398) of the participants reported that the reason they became police officers was that they wanted to do things for people—to contribute to the public good, to help. By contrast, only 22 percent (121) reported becoming officers so that they could "put away the bad guy."[7]

Some observers suggest that individuals want to be officers because of the authority, violence, and autonomy that go hand-in-hand with this job. While those desires could certainly be factors in the decision to become an officer, it is equally likely that some people might choose policing because it is comfortable (and sexy) to wear a uniform, officers are not confined to a desk or a cubical, and they meet different types of people on crowded roads, in fancy houses, and little itty-bitty businesses tucked off the road. Demonstrating the unlikelihood of thoughts of enforcing laws or using violence against others being a motivating factor, data show that authority and power are ranked ninth out of eleven variables among officers tested.[8]

According to studies, people chose policing as a career for the following reasons, given here in order from most to least popular:[9]

1. Help people
2. Job security
3. Fight crime
4. Excitement of the job
5. Prestige of the job

A deciding factor for some people who pursue police work is tradition. For instance, if a parent or member of an extended family was a police officer, then it might follow that their children would have an interest in this field, too. Joining the police might even provide a little edge in the family pecking order. Attraction to policing in comparison to non-police parent occupations also pays better, offers more and better benefits, and has more mobility.[10]

The National Center for Women and Policing reports that a police officer's job may be described as follows:[11]

- *Challenging.* When police officers go to work, they never know what they will be doing that day. They could be taking reports of crimes, counseling a runaway girl, arresting a wanted person, helping an elderly or lost person, or any one of a thousand other things. There is certainly no routine in this job.
- *Rewarding.* If you like to help people, policing could be the job for you. Every day, police officers are called upon to assist people in a time of crisis. People often turn to police for the help and advice they need, and police can make a tremendous impact on their lives. You might help a young child avoid becoming a victim (or an offender). You might help people make their neighborhood a safer place. Every day, officers go home knowing that they have made a difference in someone's life.
- *Secure.* Law enforcement jobs provide a great deal of security after police officers complete their probationary period. In addition to job security, the pay is good, the benefits are usually very good, and there is an excellent pension and well-established career ladder.
- *Prestigious.* Most people trust and respect the police. As an officer, you will earn respect as a person who enforces the law while protecting the freedoms guaranteed by the U.S. Constitution.

Some individuals "drift" into police work because of a job opening or an announcement that a civil service examination will be held (discussed in Chapter 8).[12] The civil service exam might also qualify individuals for one of many other municipal jobs, including probation officer, correctional officer, fire fighter, and emergency medical technician (ETM) jobs. (Note that for police officers, job security, pay, and benefits are determined by civil service rules and police unions.)

■ What Officers Say about Their Jobs

According to *Officer Julia Rico, Portland (Oregon) Police Bureau*: "When I was a small child, there was a plane crash. I remember all the crying and confusion; I remember seeing the firemen and the police officers coming to the scene and assisting with the tragic event that had happened. Ever since, I wanted to be a police officer. . . . Every day there are people who I serve. If I can make somebody's day a better day, then I feel good about what I've done, especially when I'm helping families and children. . . . [I] had made a commitment to myself that I wanted to work in a diverse community where I can help the most."[13]

Officer Dana Lewis, Portland (Oregon) Police Bureau, adds: "I actually got into police work by accident. I was going to school full-time. I was an accounting major, and wanted to go to law school. I sought employment and a police officer's position came open. I took the test and have loved it ever since."[14]

Officer Marcus Neider, Wiscasset (Maine) Police Department, used to watch his friend's father, a police officer, get ready for work. What little boy isn't fascinated by all that police and fire fighters represent in our daily lives—courage, the shiny badge, sirens.[15] As a youth, Neider met a Maine State Trooper when Neider became interested in helping out the fire department. The trooper was a member of the volunteer department. "I used to ride with him when I was about 16," Neider says. "I saw what it was all about—how he related to people, how he knew what to do." Those childhood experiences led him to pursue a career in policing.

Sergeant Robert Banks, Stanislaus (California) Sheriff Department, offers this anecdote about one of his first service calls.[16] During one week, Banks received five calls from a resident who was convinced that there were invisible aliens in his attic. Each time Banks arrived at the scene, he looked for signs of drug or alcohol abuse but found none. And each time he assured the resident that there were no aliens in his attic. The sixth time the man called and Banks arrived, he took his speed radar detector into the attic, explaining that it was actually an alien eliminator of sorts and that he had been instructed by high command to seek and destroy aliens that were in the man's home. Carefully, Banks took aim in the attic and moved his weapon from one end to the other. "After about five minutes the special anti-alien radiation-emitting weapons had done its work and the aliens, I'm happy to say, had been destroyed. Weeping with gratitude, the caller thanked me for saving him" and all of mankind, says Banks.[17] Banks never heard from the caller again.

Notes *Officer Jennifer Taylor, Lafayette (Louisiana) Police Department:* "Some things in life bring you pleasure, some things bring you pain. Being an officer brings you both, but in a way that is fulfilling and life-changing. I had no idea when I started working for the LPD that my life would be so affected by the job. During field training, my first training officer told me that to be an officer you had to "have heart." I didn't understand what he meant exactly, but I would soon find out. Someone would ask, "What exactly do you do?" I would answer, "I take care of people; I take care of the good ones and I take care of the bad ones." So, if you want a job that takes you not only physically but mentally to places you never knew, be an officer. I found out what it meant to "have heart." It means to love the job, do the job right, and take care of those who need your help."[18]

Peter Kraska at Eastern Kentucky University says that the chance to strap on a vest, lock-and-load a semi-automatic weapon, and prepare for combat is for some people an exciting reason to join a police department—even if policing as a profession actually involves tactical practices most of the time.[19] When an officer adopts this view, the discrepancy between what the police actually do and what they value can create conflict.[20] In the final analysis, crime prevention is a by-product of effective control of crime.[21] This thought is consistent with the views of Chief Bill Finney (St. Paul, Minnesota), who says that "We also recognize that sometimes, certain people must be policed."[22] This is hardly a new thought linked to policing, of course. What is new are the faces of police department personnel.

■ The Changing American Police Officer

Watching police television dramas from the 1980s and 1990s, it's easy to identify the officers because of their hairdos and tendency to be cigarette smokers. If a female officer appears in the scene in one of these dramas, she is likely to appear in the background. By contrast, in recent police television dramas, male officers have shaved heads, only actors who portray deviant persons smoke, and females and persons of color play leading roles in the action.

Sixty percent of departments hired new officers during the 12-month period ending June 30, 2003. Overall, approximately 35,000 officers were hired, including 29,000 entry-level hires and 5,000 lateral transfers/hires.[23] One estimate is that 10 percent of those candidates failed in their first attempt at classroom training but ultimately passed this training, and 12 percent of the new hires in large agencies and 36 percent in small agencies are officers with previous experience.[24]

In 2003, there were approximately 452,000 local officers and 174,000 sheriff's deputies, or 626,000 total officers. Today's officers are better educated, better trained, and more likely to represent ethnic or racial minorities than their counterparts from earlier eras. As shown in Figure 1–1, racial minorities accounted for 24 percent of local officers in 2003 (up from 8 percent in 1987) and 19 percent of sheriffs in the same year; female officers accounted for 11 percent of local police officers in 2003, up from 8 percent in 1987.[25,26] Since 2000, 13 percent of new hires have been Hispanics (the largest minority in the United States).[27]

To better understand how changes in police hires are changing the profile of the American police officer, consider the example set by the New York Police Department. Over a ten-year period (1986–1996), this department's white hires decreased by 13 percent, while black hires increased by 29 percent, Hispanic hires by 74 percent, Asian hires by 128 percent, and women hires (regardless of race) by 65 percent.[28]

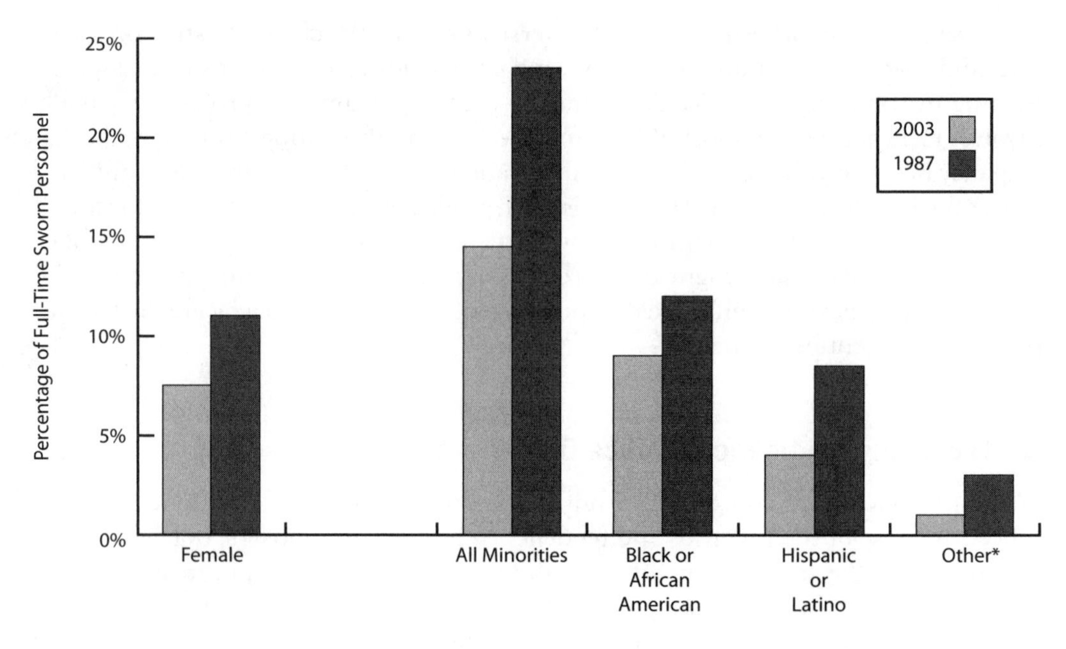

* Includes Asians, Pacific Islanders, American Indians, and Alaska Natives.

Figure 1–1 Female and minority local police officers, 1987 and 2003. *Source:* Bureau of Justice Statistics, *Local Police Departments, 2003*, NCJ 210118 (Washington, DC: Department of Justice, 2006), http://www.ojp.usdoj.gov/bjs/pub/pdf/lpd03.pdf.

■ Key Definitions: Foundations of Police Work

Although many of the following terms and issues will be described in depth in future chapters, building a foundation of key definitions related to policing at this point will serve you well. For instance, how would you describe **American policing**?

> **American policing** is an organization empowered to maintain the rule of law whose members are provided with special legal powers to maintain public order and to solve and prevent crimes.[29]

Special legal powers of the police include the right to perform an arrest, search, seizure, and interrogation; and the right to exercise authority when necessary, including the use of lethal force, whether an officer is on or off duty. The word "police" has its origins in ancient Greek philosophy referring to government or administration, from the Greek *polis*, meaning "city."[30] The word "police" was later coined in France in the eighteenth century.

In a democratic nation such as the United States, the **rule of law** and the **law of criminal procedure** have been developed to regulate **police discretion** so that officers do not exercise their vast powers, which includes police discretion, arbitrarily or in a manner that represents an unjust intrusion into the private affairs of Americans.

Rule of law is the principle that no one is above the law and that government authority is legitimately exercised in accordance with written, publicly disclosed, adopted, and enforced laws in accordance with an established procedure that can be described as due process (discussed in Chapter 10).

The purpose of the rule of law is to safeguard individual privacy against arbitrary government intrusion. The significance of this statement was recognized by the framers of the U.S. Constitution and the Bill of Rights and modern-day proponents acknowledge that the rule of law is intended to limit and guide government intervention.[31] An orderly democratic society can only be governed by established principles and known regulations that apply uniformly and fairly to every member of its population. The public legally, morally, and financially provides the vehicle through which government—including the criminal justice system—serves and protects constituents. As a consequence, the means used to provide governmental services must be held to a higher standard and comply with the rule of law.[32] In a democracy, the welfare of the public takes precedence over government services and objectives. Finally, the rule of law is a core value of the federal judiciary: "[O]ur nation accepts as its ideal that we are governed by the rule of law, which stands in opposition to the personal rule of one individual or body of persons."[33]

Law of criminal procedure is composed of rules governing the series of proceedings through which the substantive[34] (defines how the facts in the case will be handled as well as how the crime is to be charged) criminal law is enforced.[35] State constitutions may change, but they cannot take away from federally mandated protections. Criminal procedure must balance the defendant's rights and the state's interests, for instance, by providing for a speedy and efficient trial characterized by a desire for justice. The rules of criminal procedure are designed to ensure that a defendant's rights are protected.

New recruits learn the norms and expectations of their peers through the **police subculture** beginning from the first day of police training (see Chapter 13 for details).

Police subculture comprises the learned objectives, shared job activities, similar use of nonmaterial and material items, and veteran officers who act as gatekeepers (as opposed to supervisors) and who transmit job obligations, responsibilities, and expectations of the job.

Shared material items include uniforms, equipment, technology, and vehicles. When officers make decisions during the course of their jobs, such as whether to make a "stop" or an arrest, and how to behave, those decisions are most often influenced by their peers—that is, the police subculture. Officers band together to make sense of their experiences and to protect themselves from the uncertainties of the job, which includes the organizational structure and their supervisors. A police subculture enables a wide variety of police activities both on the job and off the job. It links together the experiences of officers for the following purposes:

- To help officers make sense of those experiences
- To provide justification to continue police work
- To help officers balance the inconsistencies of the organization and the job[36]

Some refer to police subculture as a support group for its members. The police organization has little, if any, control over this subculture—a critical issue given that police subculture can be the most influential factor when deciding which action to take or how much police discretion to use.

Police discretion implies that officers make decisions about their conduct when they intrude in the privacy of others (see Chapter 13 for more details).

> **Police discretion** can be characterized as the freedom to act on one's own choices. Although the law legitimizes the police officer's power to intervene, it does not—and cannot—dictate the officer's response in every given situation.[37]

Discretion does not mean simply acting as an officer pleases: Police performance is always bound by procedural guidelines, norms, ethics (professional norms and ethics, community norms, legal norms, and moral norms), and situational and contextual factors. Of course, the department cannot possibly develop a procedure for every scenario confronted by an officer, so the officer must make a reasonable and legal choice about how to respond. Discretion has also been described as the "hole in the doughnut" (doughnut theory of discretion) and "where the law runs out" (natural law theory). According to this perspective, discretion is envisioned as the empty area in the middle of a ring consisting of policies and procedures.[38]

Officers can exercise discretion in choosing the most appropriate disposition in a given situation. For instance, officers who encounter an irrational person creating a disturbance have five choices:

1. Transport the individual to a hospital
2. Make an arrest
3. Resolve the matter informally
4. Call for mental health assistance
5. Walk away

A **mandatory arrest** or *pro-arrest policy* may limit police actions.

> **Mandatory arrest** implies that an arrest is mandated and, therefore, must be made under certain circumstances, such as when officers intervene in domestic violence. Such a policy eliminates police discretion and requires an arrest in all cases where officers have probable cause to believe that an act of domestic violence has occurred.[39]

Nationally, a movement has been established favoring mandatory arrest or proactive arrest and (often linked to a no-drop policy, which means that the victim cannot drop the case, the case proceeds even if the victim does not wish to testify, or automatic prosecution or counseling) practices linked to domestic violence intervention (more on this topic in Chapter 9); such policies are typically intended to counteract the arrest avoidance practiced by the police in the past.[40] In a sense, a mandatory arrest policy limits officer discretion—but it also serves to protect the rights of those victimized, and has the potential to enhance their rights and, thereby, the freedom of those who are battered.

Clearance rates refer to the frequency crimes are "cleared" by police making an arrest. For instance, the Bureau of Justice Statistics reports that the average U.S. homicide rate for the years 1976–2004 was 7.9 victims per 100,000 population, giving a clearance rate of 64 percent.[41] The Uniform Crime Reports (UCR), which tabulates data on reported crimes, is compiled annually by the FBI and is used to determine clearance rates.

Clearance rates, as defined by the UCR program, are identified as situations in which a law enforcement agency reports that an offense is cleared by arrest, or solved for crime reporting purposes, when at least one person is

- Arrested,
- Charged with the commission of an offense, and
- Turned over to the court for prosecution (although such a clearance does not necessarily mean a conviction for the person charged with the crime).[42]

To qualify as a clearance, *all* of the conditions listed above must have been met. In its calculations, the UCR program counts the number of offenses that are cleared, not the number of arrestees. Therefore, the arrest of one person may clear several crimes, and the arrest of many persons may clear only one offense.

Crimes that are reported tend to be discovered at different rates. For example, the Federal Bureau of Investigation (FBI) provides a graph to help better identify "known" crimes that end in an arrest (Figure 1–2).[43] In 2006, 44.3 percent of all violent crimes and 15.8 percent of all property crimes in the United States were cleared by arrest or exceptional means. Of the violent crimes of murder and non-negligent manslaughter, forcible rape, robbery, and aggravated assault, murder had the highest clearance rate—60.7 percent of offenses cleared. Most of these crimes had been investigated by an investigative unit. Unfortunately, clearance rates now appear to be declining. It is reported that crimes are growing faster than arrests for specific crimes, such as sexual assault. However, the attitudes of the people in a jurisdiction can impact reported crimes and arrest rates.[44]

Street justice can be described as unlawful police behavior such as the use of unnecessary force when officers deem the response of the authorities such as the courts to be insufficient. Street justice can refer to officers ignoring due process by administering punishment which can be characterized as "vigilante justice" or retaliation. Alternatively, it can be seen as a way in which an officer restores the peace by choosing tactics that the officer, rather than the rule of law, deems appropriate.

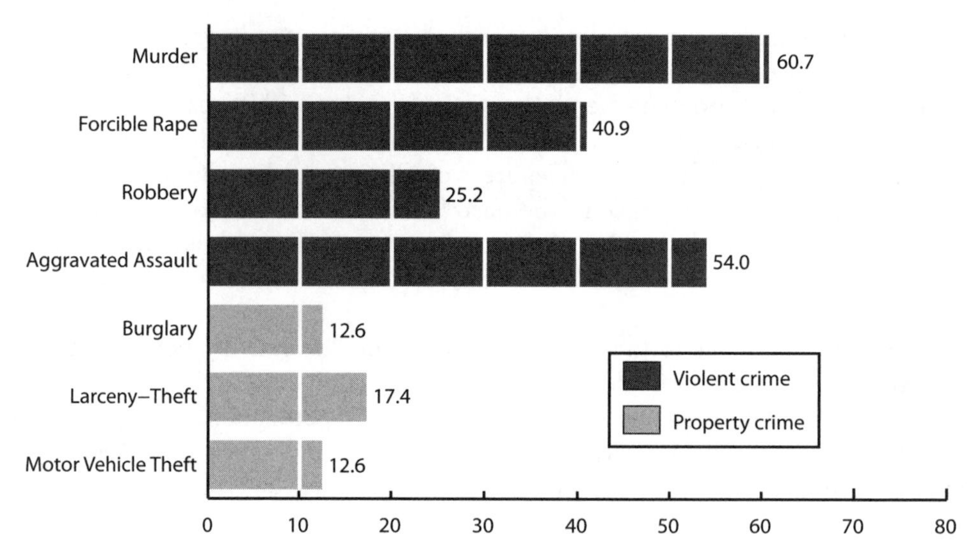

Figure 1–2 Percentage of crimes cleared by arrest or exceptional means, 2006. *Source:* Federal Bureau of Investigation, *Crime in the United States, 2006* (Washington, DC: Department of Justice, 2008), http://www.fbi.gov/ucr/cius2006/offenses/clearances/index.html.

> **Street justice** is police response to a community mandate that something be done about situations in which formal institutions cannot or will not respond for a variety of reasons.[45]

As discussed earlier, the existence of police discretion reinforces the notion that decisions of officers "to intervene, arrest, use force, or issue traffic citations are primarily a function of situational and organizational factors that reflect interpretations of community needs and expectations by the police."[46] As part of this role, officers provided informal peacekeeping services, compensating for the general failure of local governments to provide quality economic programs brought about by the urbanization in a market-based society. One way to think about street justice is to think about reform and quality-of-life issues, which are an important mission associated with the **order maintenance** paradigm.

■ Police Initiatives: A Prisoner in One's Own City

Police policy in New York City and every other jurisdiction is influenced by the **broken windows perspective** (explained here and in Chapter 3), as advanced by **George L. Kelling** and **James Q. Wilson.**[47] Kelling and Wilson argued that New Yorkers had lost control of the city's public spaces: "When you couldn't make a wrong turn without your car getting shot up, when you couldn't go by a park and give a drug dealer a side glance without being intimidated and threatened, things were out of control."[48]

Kelling and Wilson described their observations by using a broken windows metaphor, implying that criminals would rather break into a home (or a neighborhood) that was poorly maintained than attempt to burglarize a well-maintained home (or neighborhood).[49] According to this theory, the guy who successfully jumps the subway gate (turnstile) and gets a free ride assumes that no one cares about his behavior and, therefore, decides that he can commit bigger crimes without fear of apprehension, too.

So, the thinking goes, arrest the thug and crime rates will fall. Take back public spaces with an aggressive arrest policy and quality of life—in areas both big and small—will improve.

The **broken windows perspective** implies that order-maintenance problems such as graffiti, cruising, fare jumping (not paying) transit fares, public drunkenness, and loitering, among other mundane forms of public disorder, are perceived by criminals as an invitation to commit other crimes, including serious crimes; the likelihood of being apprehended or even stopped by the police or community members from committing serious crimes in this environment is perceived as nonexistent.[50]

The framework for their initiative was conceptualized as the idea of rebounding from the "downward spiral of urban decay," a theory advanced by **Wesley G. Skogan**.[51] One idea that was extracted from the broken windows perspective was the notion that aggressive policing in the form of proactive strategies such as **zero tolerance**, **Compstat**, and decentralized command (see Chapter 4) would improve quality of life and lower crime rates. In New York City, a new tax surcharge enabled the training and deployment of some 5,000 new police officers, police decision making was devolved to the precinct level, and a backlog of 50,000 unserved warrants was cleared.

As part of this aggressive and proactive attempt at order maintenance, the New York City Police Department (NYPD; under the leadership of Mayor **Rudolph Giuliani**) also implemented Compstat. The objective of this computer database–driven approach to policing is to control crime and to make command accountable for their response to it.

Compstat ("computer comparison statistics") enables police to identify trouble spots and target the appropriate resources to fight crime strategically. Compstat includes information on victims, times of crime occurrence, and other crime-specific details that enable officials to spot emerging crime patterns. The result is a computer-generated, citywide map illustrating where and when crime is occurring. With this high-technology "pin-mapping" approach, police can quickly identify crime-ridden areas and fashion a comprehensive response.[52]

After NYPD implemented this program, rates of both petty and serious crime in the city fell significantly. It could be argued that the city's proactive aggressive police zero-tolerance rollout was part of an interlocking set of wider reforms, crucial parts of which had been under way for many years. Overall, the quality-of-life initiative produced between 40,000 and 85,000 additional adult misdemeanor arrests per year in New York during the period 1994–1998, and an even greater number of stops and frisks during a period of sharply declining crime rates.[53] Mayor Michael Bloomberg, in the first decade of the twenty-first century, has picked up where Giuliani left off in the twentieth century and continues to support this aggressive policing approach.

Record-setting drops in homicide rates were experienced in New York from 1990 to 1998, and the number of homicides fell dramatically over the same span—from 2,245 (the most ever recorded) to 633.[54] Despite these gains, numerous lawsuits were filed against the NYPD during this period, and Giuliani eventually fired Chief **William J. Bratton**. It was learned that thousands of NYPD's finest habitually avoided radio calls, leaving a small cadre of no more than 20 percent of the working officers to respond to emergency calls.[55] Giuliani has said that the best way to win the war on crime is to take the offensive, rather than defensive, and adds that police departments should collect data and institute zero-tolerance strategies citywide.[56] However, New York's police action—especially zero tolerance—has been criticized because of its punitive approach to maintaining law and order, which makes little or no reference to negotiating acceptable public behavior in the neighborhood, ignores the rights of constituents, and limits police discretion.

> **Zero tolerance** includes strategies that allocate additional resources so as to combat identified crimes or indicators of crime in a particular geographical location such as the subways in Manhattan or the highways in Iowa. When this strategy is employed, officers must follow a specific protocol such as conducting an arrest for petty violations which usually are determined by police discretion. Zero tolerance is designed to prevent antisocial elements from developing to serious crimes.[57]

In July 1994, New York City implemented order-maintenance strategies that emphasized proactive and aggressive enforcement of misdemeanor laws against quality-of-life offenses such as graffiti writing, loitering, public urination, public drinking, aggressive panhandling, subway fare evasion, prostitution, and harassment by "squeegee men."[58] According to Giuliani, the idea behind this program was simple: "Don't let the small guy get away with his wrong-doing; it only encourages him, and others, to do worse."[59] His crime prevention hypothesis (embodied by the zero-tolerance policing strategy) is that the more arrests police make for every petty disorder, the less serious crime there will be.

Order maintenance (also called proactive general police duties) comprises strategies that contribute to social control through the rule of law for the purpose of enhancing quality-of-life experiences of the constituents whom the police are sworn to serve and protect. In large part, order maintenance is a broadened response to routine events, crime prevention techniques, and protective methods of policing when circumstances merit those measures, such as after waves of traffic accidents, increases in criminal victimization, and catastrophic (terrorism and hurricanes) events. This topic will be covered in depth in Chapters 3 and 9. For now, simply recognize that order-maintenance initiatives—like most police strategies—have been in a state of flux and have evolved in many ways from the 1990s to the present.

> **Order maintenance** policing refers to intensified policing of disorder or incivilities; it includes policies such as zero tolerance (automatic arrest) and domestic violence arrest mandates. Arrests under such programs are referred to as quality-of-life arrests because of their association with the broken windows perspective.[60]

Police intrusion at mundane levels is linked to the frequency of public nuisances in public spaces, it is believed, and it can bring policing closer to its mission of ensuring public safety and protecting the freedoms of those preyed upon or victimized, as specified by regulatory mandates.[61] Although police must perform law enforcement duties, they must also "help constituents develop their claim over the local territory, back them up, and assist them in dealing with neighborhood problems" as part of their secondary task of improving quality of life.[62] The potential for private power and for victimization of others can dramatically affect quality-of-life standards; as such, quality-of-life arrests are one way to begin "taking back the streets."

The Newark (Delaware) Police Department (NPD) made a commitment to maintain quality-of-life standards through a strict arrest policy that was designed to counteract any deterioration of those standards.[63] Its order maintenance policy regarded "cruising" as a general nuisance, for example, because it threatens quality-of-life standards for community residents. Cruising is defined as repetitive, unnecessary driving of motor vehicles in the downtown area. Newark's code states that it is unlawful for the person having care, custody, or control of a motor vehicle to drive it past the same point three or more times within a two-hour period between the hours of 8:00 P.M. and 4:00 A.M.; the minimum fine for cruising is $25.00 for the first offense. Most often, order-maintenance initiatives include measures designed to appease "cuff 'm and stack 'em" (i.e., arrest and jail) advocates, and these strategies can often include zero-tolerance strategies.

Uniformed police officers, in particular, perform order-maintenance duties designed to enhance quality-of-life standards—for example, regular patrols, direct traffic, conduct stops, and emergency services on roadways and in homes. They also respond to emergency 911 calls for services such as assistance in cases involving domestic violence.[64] In fact, approximately two-thirds of all calls for police service involve order-maintenance activities (e.g., arguments, fights, medical emergencies, or public nuisances).[65]

Ultimately, police are tasked with two types of duties: (1) preserve the "general good" and (2) protect the public from victimization, which includes protection from others and themselves, events, and an arbitrary intrusive government. In recent years, however, order-maintenance practices have changed from the models followed in the latter part of the twentieth century. Nevertheless, the idea of aggressive, yet informed police practice to enhance quality-of-life standards remains the core of modern-day policing strategy (see Chapter 4).

What Is a Police Officer?

Every state and most municipalities or jurisdictions provide a legal definition of a police officer in their official state code. For instance, a representative description is provided by the state of Pennsylvania: A police officer is any person who is by law given the power to arrest when acting within the scope of the person's employment.[66] The code of the city of Davis, California, explains it this way: A police officer is any officer of the police department of the city, or any officer authorized to direct or regulate traffic or to make arrests for violations of traffic regulations.[67]

A key component of any such description is the idea that police officers are peace officers who are fully empowered by the state in which they are employed. That is, police officers are sworn to uphold the laws of the state and the jurisdiction by which they are employed, and their powers include the ability to conduct an arrest (take an individual into custody), the legal right to use appropriate force, and within limits, the ability to detain suspects. Specifically, police officers are provided with the moral and legal justification to use *police power*, which encompasses detainment and use of force practices.[68]

Most recruits participate in approximately 720 hours of basic training before they engage in unsupervised interactions with the public.[69] (Chapter 8 covers training requirements in depth.) There are approximately 626 state and local police academies across the United States. Once on the job, police officers must participate in in-service training on a regular basis, such as yearly weapons handling and target certification.

Police Responsibilities

People depend on police officers and detectives to protect both lives and property.[70] In addition to order maintenance and law enforcement responsibilities, and as a consequence of the September 11, 2001, police have taken on major homeland security responsibilities as well.[71] These duties include conducting stepped-up surveillance of airports, government buildings, mass-transit systems, and other potential terrorist targets. Police also have new intelligence duties designed to foil terrorist attacks before they occur. Finally, agencies must be ready to respond to possible attacks utilizing chemical, biological, and other unconventional weapons.

According to Police Commissioner Charles Ramsey (Philadelphia Police Department), the police stand between anarchy and lawlessness.[72] Others have observed that countries outside the United States might have less anarchy and government corruption if the human rights of their populations were safeguarded through professional police initiatives. Serving as brakes on their power, which might otherwise run unchecked, U.S. police are accountable to the communities they serve, the taxpayers who pay their bills, and the legal orders governing their authority.[73]

Despite the clearly huge responsibilities taken on by the police, many people do not have a true picture of what modern policing is all about. Indeed, many myths persist about police among both the general population and officers themselves.

Stereotypes about the Police

There are many stereotypes about the police held by the public and the police themselves.

Public Stereotypes

The public's perceptions of police work, as compared to the actual experiences of police officers, are incredibly exaggerated as a result of the distortions and reinforcement offered by the media (e.g., television dramas, news reports, and advertisements). Although surveys show that the public tends to highly rate the honesty and ethics of the police,[74] the consensus is that the public wants the police to change, but they cannot agree about precisely what should be changed.

For example, when 2,010 community members residing in eight jurisdictions across the United States (from Alexandria, Virginia, to Sacramento, California) were surveyed, almost one in five participants characterized police performance as intimidating and frightening.[75] Most of the survey respondents felt unsafe in their neighborhoods and only a few, depending on their race, felt safer than they had a year previously. This finding is consistent with studies that show the fear of crime is greater than the actual crime rate. Thus it is safe to conclude that one of the major distorted

views held by the public is a mythical fear of becoming a victim of crime.[76] Furthermore, if they are victimized, they are not confident in how the police will deal with their victimization.

That said, the gross lack of knowledge by the public about police work implies that the public tends to report fewer crimes and provide less aid to police in solving those crimes that are reported. That perception is consistent with crime data collected by the FBI and published in the UCR.[77]

Public opinion levels related to the police have remained somewhat stable over time, other than experiencing occasional fluctuations produced by local and international events, such as the terrorist attacks of September 11. Yet, significant differences are reported among different socioeconomic, racial, and ethic groups. For instance, lower-socioeconomic-class black communities are less satisfied with police services than are middle-socioeconomic-class black and white communities, and immigrants have less confidence in police services (probably because immigrants are uncertain as to the services provided) than do individuals born in the United States (Table 1–1).[78]

TABLE 1–1 Reported Confidence in the Police

How much confidence do you have in the police—a great deal/quite a lot, some, or very little?

	Great Deal/Quite a Lot	Some	Very Little	None*
National	64%	26%	10%	†
Sex				
Male	64	25	10	1
Female	64	26	10	†
Race				
White	70	22	8	†
Nonwhite	43	39	17	1
Black	41	46	13	0
Age				
18–29 years	61	19	19	1
30–49 years	62	27	10	1
50–64 years	68	30	5	0
65 years or older	71	24	5	0

(Continues)

TABLE 1-1 Reported Confidence in the Police (Continued)

How much confidence do you have in the police—a great deal/quite a lot, some, or very little?

	Great Deal/Quite a Lot	Some	Very Little	None*
Education				
College postgraduate	66	30	4	†
College graduate	72	22	5	0
Some college	61	28	11	†
High school graduate or less	64	23	12	1
Income				
$75,000 or more	69	25	5	1
$50,000 to $74,999	70	21	9	0
$30,000 to $49,999	60	30	9	1
$20,000 to $29,999	57	31	12	0
Less than $20,000	60	19	21	0
Community				
Urban area	60	28	11	1
Suburban area	64	25	10	1
Rural area	69	24	7	0
Region				
East	62	24	14	†
Midwest	68	25	6	1
South	63	28	9	0
West	65	25	10	†

* Response volunteered.

† Less than 0.05 percent.

Source: Table 2.12 in Bureau of Justice Statistics, *Sourcebook of Criminal Justice Statistics 2003* (Washington, DC: U.S. Department of Justice, 2005), 113; http://www.albany.edu/sourcebook/pdf/t212.pdf.

Police Officers' Views of Themselves

The self-image of an officer usually falls somewhere in between how police are portrayed by the media and how the public views those portrayals. During encounters with the public, officers may run head-on into the public's media-influenced stereotypes of police and reinforce those perceptions, with the end result challenging both the self-image an officer holds and the legitimacy of policing as an institution. According to **Sam Walker** and Charles Katz, the "sharp contrast between the crime-fighting imagery of the police and the peacekeeping reality of police activities was one of the first and most important findings of the flood of police research that began in the 1960s."[79] Furthermore, even though the police perform an increasingly wide range of functions (as explored in Chapter 3), it is crime control that remains uppermost in the minds of both the police and the public when they think about the role of the police. In reality, despite the widely held perception that the police spend the majority of their time detecting and apprehending offenders, crime control activities generally occupy less than 25 percent of police officers' time.

Nonetheless, some officers eventually become cynics and adopt an "us versus them" attitude. Cynicism does not necessarily translate into inappropriate behavior, but it can be one sign of a burned-out officer. Officers who concentrate on cynicism, at an extreme, might withdraw from their job and family members.[80] One source of cynicism is forged through the perceptions of disrespect that result from interacting with many violators. Feelings of disrespect have consequences, as described later in this chapter.

Each officer develops his or her own style of performance, influenced by both public perceptions of officers as peacekeepers and police subculture (as explained in Chapter 13). Fact is, the expectations held by the public form the core of police legitimacy—and that legitimacy provides the police with the power (and a budget) needed to exercise their order-maintenance strategies.[81] The public and government agencies, including the judiciary, frequently challenge the legitimacy of police authority, which has emerged as a major problem in the twenty-first century.[82]

How well does an officer's self-image hold up to these legitimacy discrepancies, and in what way do these discrepancies affect the performance of the officer? One answer could be that police officers redefine a rationale for their experiences that is closer to what is suggested by media accounts than to what their job actually is. What the "job" produces for most police officers—other than the satisfaction of helping others and putting away the bad guy—is stress. Indeed, stress is a major occupational hazard for many officers. "The effects of stress—low morale, high turnover, absenteeism, and early death . . . exact a high cost from officers, their families, and their agencies."[83] Stress and its effects on police officers are discussed in depth in Chapter 15.

Some officers see themselves as a person who has been transformed from the individual he or she once was to a completely different person after years spent in police work—specifically, a mechanical bureaucrat.[84] How many officers might describe themselves as a "government drone"? One survey revealed that 20 percent of U.S. officers see themselves as "just going through the motions."[85] Too many officers go through the work week with the goal of just getting to retirement, while doing little to further the police mission of "serving and protecting." Perhaps not surprisingly, then, survey results reveal that officers who don't take the job seriously enough are the number one pet peeve of most officers.[86]

One study found that six of every ten members of the U.S. public rate the honesty and ethics of the typical police officer highly.[87] In other words, most officers respect others and the rule of law. Nevertheless, some members of the public hold a different view of the respect that should be accorded police, and part of that attitude is forged through misperceptions of police characteristics and roles.

Consequences of Inappropriate Stereotypes of the Police

Some officers feel that the public holds a negative view of the police.[88] Public respect relates to public trust:[89] If the public does not trust the police, then one way to demonstrate those feeling is by refusing to report crimes, providing less funding for police operations including equipment and training, and curtailing police power by subjecting police to more civil liability (law suits, discussed in Chapter 14) actions.[90]

For their part, members of police organizations may hold thoughts and advocate policies related to policing that actually escalate public disobedience. Ideally, of course, police will follow the lead of departments that have attempted to do the right thing toward meeting their mission. For instance, the Boston Police Department (BPD) set the standard for public order maintenance at political protests and large demonstrations.[91] The traditional approach to managing large demonstrations was for the police to arrive in full riot gear. The BPD learned that this approach could be a self-fulfilling prophecy; that is, instead of defusing a situation, riot-geared officers may escalate the danger. Recognizing this fact, the BPD developed a three-tiered approach that called for escalating riot and special event deployments as circumstances warranted. Under this approach, fully outfitted Public Order Platoons were used only as a last resort and staged out of view of the public until needed. This strategy was employed successfully at the National Democratic Convention in Boston in 2004 and other events throughout the city.

Stereotypes and Realities of Policing

Stereotypes linger as to why an individual wants to become a police officer despite an applicant pool shortage (see Chapter 8 for more detail). One stereotype is promoted through popular television dramas depicting a James Bond wanna-be driving casually through an emerald city on a summer day in a yellow muscle-car. His smartphone is connected to a strange-haired lab freak, who tracked a stuttering suspect who had purchased a Snickers bar at a dilapidated pharmacy across town. Lab girl, with a single swipe, pushes her hair out of her eyes and informs Bond, "After retrieving a curious candy wrapper wedged in an abandoned vehicle miles from the shining domed metropolis, I have an address for you."

Fade to reality: Officers work (see Chapter 9) rotating shifts. Their time is spent on patrol, and they provide an array of police services at private and public places. Sometimes they're bored. At other times they ask why they became officers after seeing the horrific consequences of people's reckless behavior. They work in cars, on foot, while peddling bikes, mounted on horses and motorcycles, standing in watercrafts, or seated in an aircraft of some type (see Chapter 9) often making places safe while putting themselves at risk.

Legitimacy of the Police

As mentioned earlier, community respect toward the rule of law and community perception of police legitimacy provide the power and budget that enable a police agency to deliver quality police services, assuming the agency has the capability and motivation to do so. For instance, as noted at the beginning of this chapter, the Secretary General of the United Nations recently reported that all countries should work to develop more respect for the rule of law; doing so, he said, would both protect and empower constituents. Put another way, "For policing in a democratic society to be effective, it must be considered legitimate by the vast majority of the people."[92] Police must be seen as having the "right" to exercise force. If a police agency does not have legitimacy, then it must operate by the use of brute force.[93] The public will perceive police as professionals and comply with their directives when the police exercise their legally mandated prerogatives such as discretion and use of force as objectively as circumstances permit.[94] That is, police agencies must be perceived as performing "**judiciously**" during public intrusions.

> **Judiciously** means sensibly and wisely—showing wisdom, good sense, or discretion, often with the underlying aim of avoiding trouble or waste.[95]

When police practice judiciously meets community standards, the community is generally satisfied and most community members attempt to comply with standards and the directives of the police. Simply put, the community will support the police when the community believes that police ideals mirror their own ideals. The greater the legitimacy of a police agency, the fewer the number of violators and the less coercion or force required on the part of officers to maintain social order.[96] *How* the police do their job (i.e., their style) is far more important to their status than *whether* they apprehend the bad guys.

Police professionals do not control police policy, yet are assigned the responsibility of management and setting day-to-day directives to mirror that mandated policy.[97] A compelling argument might be made that police policy, in a democratic society, rightfully belongs outside the control of those who perform the police function. Indeed, policing is being reconstructed worldwide—a trend that has the following distinguishing features:[98]

- The separation of those who authorize policing from those who do it
- The transference of both functions away from government

To understand what is happening to policing, it is essential to distinguish the way in which policing is authorized from the way in which it is provided. Because police have a monopoly on the use of force and detainment power, if they established the laws linked to use of force and detainment, what could happen? Who would guard the guardians?[99]

Compliance with Legal Authority

Constituents comply with police requests in 80 percent of all encounters especially if individuals understand police behavior to be judicious. Other factors affecting compliance rate are the nature of the incident, the behavior of the officer, and the condition of the constituent. However, constituents are *less* likely to comply with officer directives in the following circumstance:

- The officer is too authoritative or disrespectful.
- The encounter occurs in a private place.

Costs of Policing

The costs of policing in the United States were estimated at $40 million per month in 2001, an amount that included funding all federal, state, and local police personnel.[100] Eighty-five percent of that total related to salaries for state and local police personnel, whose number includes 1.3 million employees. In 2003, local police departments spent about $93,300 per sworn officer or $200 per resident to operate for the year.[101] Sheriffs' offices spent about $124,400 per officer or $82 per resident for the year.

On average, larger municipal police departments employ 22 full-time sworn personnel per 10,000 residents.[102] County police departments and sheriffs' offices employed an average of 11 and 10 officers per 10,000 residents, respectively. State law enforcement agencies employed an average of 2 officers per 10,000 residents.

U.S. Attitudes Toward Police Services

What Americans Want from the Police

The best description of Americans' expectations for police in the twenty-first century is probably "ensuring public safety." Public safety can include the professional behavior of the police during the following tasks:[103]

- *Enforcing laws*—the role of officers as crime fighters, even though officers spend a lot of time at court, preparing paperwork, and responding to mundane service calls such as dealing with noise complaints and false alarms.
- *Providing services*—directing traffic, administering emergency medical services, and conducting search and rescue operations at disaster scenes.[104]
- *Preventing crime*—initiatives such as community relations and problem-solving strategies.
- *Preserving the peace*—defusing (as opposed to escalating) danger and improving quality-of-life services associated with controlling public spaces by effectively neutralizing public disorder that place the community at risk (e.g., public drunkenness, urination, and loitering).

At the extreme, the role of police in a democracy should highlight their stature as public servants.[105] That is, we the people, through the Constitution, represent the permanent rulers, and police are agents of the state who are paid to do the job the public wants them to perform.

What Americans Do Not Want from the Police

Americans do not want police services provided at the expense of personal liberty. That preference was confirmed by researchers who interviewed a cross section of community activists and conducted a survey of 400 residents as part of a lawsuit. The lawsuit, which resulted in a federal decree, had alleged that "there is a pattern or practice of conduct by law enforcement officers of the Pittsburgh Bureau of Police that deprived persons of rights, privileges, and immunities secured and protected by the Constitution and the laws of the United States."[106] As part of the settlement, the Pittsburgh police were instructed to make comprehensive changes in oversight, training, and supervision of officers. Among the key elements of the decree was a requirement that the Bureau develop a computerized early-warning system to track individual officers' behavior; document uses of force, traffic stops, and searches; and provide annual training in cultural diversity, integrity, and ethics.

RIPPED from the Headlines

Officers Who Became Off-Duty Victims of Violence

Those who knew Clifton Rife II had nothing but the utmost respect for his abilities and the way he conducted himself. A 13-year veteran of the District of Columbia Metropolitan Police Department, Sergeant Rife, 34, was a hard-working officer who had no fear of the streets. Assigned to the prostitution unit, he was "head and shoulders above any other candidate who applied" for the position, said Commander Hilton Burton.

Sergeant Rife was off-duty when he was confronted by a 16-year-old would-be robber. The two exchanged gunfire. The young assailant, who had run away from a group home earlier in the year after being charged with possession of heroin, died at the scene. Sergeant Rife managed to make it to a friend's apartment before he collapsed; he was flown to a nearby hospital, where he died a short time later.

This incident recalled another case involving a District of Columbia officer that took place seven years earlier. During the early-morning hours, Officer Oliver Wendell Smith, Jr., was on his way home from work when he was robbed at gunpoint as he got out of his car. The thieves were about to flee the scene when one of the suspects found Officer Smith's badge and realized he was a police officer. That's when they stuck a gun to his head and shot him execution-style.

The Washington, D.C., Metropolitan Police Department has had seven officers killed in off-duty incidents. Only two municipal agencies have had more—New York City with 18 deaths and Cleveland with 9 deaths. Another off-duty death occurred on a cold February evening, when a D.C. police officer named James McGee was off-duty and in street clothes when he came upon two men robbing a cab driver at gunpoint. He was trying to make an arrest when two of his fellow officers drove up and failed to recognize Officer McGee as one of their own. When they ordered him to drop his gun, Officer McGee turned toward the sound of their voices with his gun in his hand. One of the officers fired twice. It was a horrible case of mistaken identity. Officer McGee died at the hospital 25 minutes after the shooting.

A look at the records kept by the National Law Enforcement Officers Memorial Fund reveals that more than 300 federal, state, and local officers have lost their lives in off-duty law enforcement incidents. The first was Robert M. Rigdon, a Baltimore City (Maryland) police officer, who was assassinated at his home in an act of retaliation for court testimony the 37-year-old officer had given earlier in the day. This cold-blooded killing took place on November 8, 1858.

Source: "Officers Who Became Off-Duty Victims of Violence," *Policeone.com* (June 1, 2005). http://www.policeone.com/writers/columnists/CraigFloyd/articles/97441/ (accessed June 29, 2007).

Summary

The chapter describes the essential concepts of modern American policing. Most people become police officers because they wanted to help people, job security, and fight crime. In recent years, however, the face of America has changed: New hires have transformed the profile of the police force from all white, working-class males to include educated females and members of many different ethnic and racial groups.

As an organization, American policing is charged with the task of maintaining the rule of law. To do so, its members are provided with special legal powers that enable them to maintain public order and to solve and prevent crimes. The rule of law is the principle that government authority is legitimately exercised in accordance with written, public disclosed laws that have been adopted and should be enforced in accordance with established procedures which includes due process. By contrast, the law of criminal procedure encompasses the rules governing the series of proceedings through which the substantive criminal law is enforced.

Police subculture consists of the learned objectives, shared job activities, and similar use of nonmaterial and material items; it is a major factor in determining precisely how police enforce laws. Police discretion gives officers the freedom to act or not act in certain circumstances, though mandatory arrest policies sometimes set limits on that discretion. In situations where formal institutions fail to meet the community's mandate to rectify problems, street justice may be the police response, as officers solve problems and mete out punishment on their own.

According to Kelling and Wilson, serious crime is a product of willful ignorance of minor violations, with inattention to small crimes leading inexorably to larger crimes. Thus, by pursuing small-time violators, police may be able to better control crime—a policy based on the so-called broken windows theory. When following this approach, police may, for example, take back public spaces by enforcing an aggressive arrest policy (zero tolerance) or use computerized tools such as Compstat to target certain areas for high levels of enforcement.

Order-maintenance initiatives focus on proactive quality-of-life arrests of suspects who commit minor violations linked to public nuisances in public spaces. These efforts, which provide for intensified policing of disorder, are intended to enhance the quality-of-life experiences of residents. Today's order-maintenance initiatives, which represent a large part of current police activity, have been redefined from their original form to better conform with Constitutional guidelines.

A police officer is a peace officer who is empowered to make arrests, use force, and detain individuals, but is also obligated to care and protect for constituents. In sum, officers protect both lives and property and defend the Constitutional rights of the public. Unfortunately, their ability to perform these policing duties is sometimes hampered by biased media and news reports, myths held by the public, the police's view of themselves, and the consequences of inappropriate views of the police, which may include disrespect of officers.

While the police are empowered through legislation, their constituents' willingness to comply with police orders depends on the notion of police legitimacy. Americans want the police to enforce laws, provide services, prevent crime, and preserve the peace, but they are hesitant to give up their liberty, even in the face of threats such as terrorism.

Key Terms and Key People

American policing
Broken windows perspective
Clearance rate
Compstat
Judiciously
Law of criminal procedure
Mandatory arrest
Order maintenance
Police discretion
Police subculture
Rule of law
Street justice
Zero tolerance
William J. Bratton
Rudolph Giuliani
George L. Kelling
Wesley G. Skogan
Sam Walker
James Q. Wilson

Discussion Questions

1. Describe how American police officers have changed from their counterparts in earlier periods of history. In what way is this new profile a positive change? A negative change?
2. Characterize the rule of law, and explain its importance to policing.
3. Describe the law of criminal procedure and its process. How might the law of criminal procedure help or hinder police officers?
4. Explain what is meant by a police subculture. How might peer pressure influence the decision-making process followed by an officer?
5. Explain police discretion and its limits. In what ways do you agree and disagree with those limits?
6. Describe mandatory arrest policies and explain their purpose.
7. Characterize clearance rates. What are their implications for policing?
8. Describe the rationale of street justice. In what way is street justice appropriate or inappropriate for officers?
9. Characterize the broken windows perspective and its objectives. Identify the strengths and weaknesses of this perspective.
10. Describe Compstat and zero-tolerance objectives, including their objectives and their weaknesses.
11. Define order maintenance as a police initiative, and explain its practical use.
12. Describe the stereotypes about policing held by the U.S. public and by police themselves.
13. Characterize the sources of legitimacy.
14. Identify what Americans want and don't want from the police. In what way are these thoughts close to your own thoughts?

Notes

1. "Secretary General Calls for Culture of Respect" (November 16, 2004), http://www.unis.unvienna.org/unis/pressrels/2004/sgsm9592.html (accessed July 4, 2006).
2. Dennis J. Stevens, preface to *Police Officer Stress: Sources and Solutions* (Upper Saddle River, NJ: Prentice Hall, 2008).
3. This thought is consistent with those of former U.S. Attorney General Janet Reno: "A police officer is charged with ensuring public safety, but she or he is also empowered to use force and, if necessary, to take a life to protect others from death or great bodily harm. The police are there to protect us from crime, but they must protect our rights at the same time" (http://www.usdoj.gov/archive/ag/speeches/1999/npc.htm, accessed July 15, 2007).
4. David Weisburd, "Hot Spots Policing Experiments and Criminal Justice Research: Lessons from the Field" (paper presented for the Campbell Collaboration/Rockefeller Foundation's meeting at Bellagio, Italy, November 10–11, 2002).
5. Weisburd, "Hot Spots." Some of those criminologists include David Farrington, James Q. Wilson, and Lloyd Ohlin.
6. Randy Sutton, *True Blue* (New York: St. Martin's Press, 2006), xiii.
7. Dennis J. Stevens, appendix to *Police Stress: Sources and Solutions* (Upper Saddle River, NJ: Prentice Hall, 2008).
8. Virginia B. Ermer, "Recruitment of Female Police Officers in New York City," *Journal of Criminal Justice* 6 (1978): 233–246.

9. M. Steven Meagher and Nancy Yentes, "Choosing a Career in Policing and Female Perspectives," *Journal of Police Science and Administration* 14, no. 4 (1986): 320–327.

10. Ermer, "Recruitment of Female Police Officers."

11. Robert M. Shusta, Deena R. Levine, Herbert Z. Wong, Aaron T. Olson, and Philip R. Harris, *Multicultural Law Enforcement: Strategies for Peacekeeping in a Diverse Society*, 4th ed. (Upper Saddle River, NJ: Prentice Hall, 2008), 52–57. Also see David H. Bayley and Harold Mendelsohn, *Minorities in the Police* (New York: Free Press, 1969), 6.

12. National Center for Women and Policing (2005), http://www.womenandpolicing.org/becominganofficer.asp (accessed June 23, 2007).

13. Bayley and Mendelsohn, *Minorities in the Police*, 30.

14. Officer Julia Rico, Portland, Oregon, Police Bureau, http://www.portlandonline.com/joinportlandpolice/index.cfm?c=eciac (accessed June 23, 2007).

15. Officer Dana Lewis, Portland, Oregon, Police Bureau, http://www.portlandonline.com/joinportlandpolice/index.cfm?c=38409 (accessed June 23, 2007).

16. Paula Gibbs, "DARE Officer Talks about Why He Became a Cop," *Wiscasset News*, July 20, 2006.

17. Robert Banks, "Staging Point for the Alien Invasion," in Sutton, *True Blue*, 19–20.

18. Banks, "Staging Point," 20.

19. Officer Jennifer Taylor is a Ph.D. candidate at the University of Southern Mississippi and is writing her dissertation on women and policing. She contributed to this book—especially Chapters 8 and 9.

20. Matthew Davis, "Death Raises Concerns at Police Techniques," BBC (2006, March 26), http://news.bbc.co.uk/2/hi/americas/4803570.stm (accessed June 24, 2006).

21. Sam Walker and Charles E. Katz, *The Police in America: An Introduction* (Boston: McGraw Hill, 2005), 233.

22. Richard Lumb and Yumin Wang, "The Theories and Practice of Community Problem Oriented Policing: A Case Study," *Police Journal* 79, no. 2 (2006), 177–193.

23. All numbers and percentages have been rounded.

24. Chief Bill Finney, St. Paul, Minnesota, Police Department, http://www.finneyforsheriff.com/ (accessed July 6, 2007).

25. Christopher S. Koper, *Hiring and Keeping Police Officers* (NCJ 202289). (Washington, DC: U.S. Department of Justice, July 2004), http://www.ncjrs.gov/pdffiles1/nij/202289.pdf (accessed June 22, 2007).

26. Of interest, many covers of criminal justice textbooks display female officers. See, for example, William Bennett and Karen Hess, *Management and Supervision in Law Enforcement*, 6th ed. (Belmont, CA: Wadsworth, 2008); its cover features three women, one black male, and one white male. It is unlikely that any pre-2005 criminal justice textbook would showcase women officers.

27. Bureau of Justice Statistics, *Local Police Department, 2003* (NCJ 210118) (Washington, DC: U.S. Department of Justice, 2006), http://www.ojp.usdoj.gov/bjs/pub/pdf/lpd03.pdf (accessed June 20, 2007).

28. U.S. Census (2007), http://www.census.gov/ (accessed June 20, 2007).

29. James J. Fyfe and Robert Kane, "Bad Cops: A Study of Career-Ending Misconduct Among New York City Police Officers" (September 2006), http://www.ncjrs.gov/pdffiles1/nij/grants/215795.pdf (accessed June 6, 2007).

30. MSN Encarta, http://encarta.msn.com/ (accessed June 7, 2006).

31. Wikipedia, the Free Encyclopedia, http://en.wikipedia.org/ (accessed June 7, 2006).

32. John Locke, *Two Treatises of Government* (Published anonymously in 1689). Locke argues that we have a right to the means to survive. When Locke explains how government came into being, he uses the idea that people agree that their condition in the state of nature is unsatisfactory, and so agree to transfer some of their rights to a central government, while retaining others. This idea is known as the theory of the social contract.

33. Hermann Goldstein, *Policing in a Free Society* (Cambridge, MA: Ballinger, 1977).

34. U.S. Courts, "Core Values" (2006), 7, http://www.uscourts.gov/lrp/CH02.PDF (accessed August 3, 2006).

35. The rules of criminal procedure are different from those of civil procedure because the two areas (criminal and civil) have different objectives and results. In criminal cases, the state brings the suit and must show guilt beyond a reasonable doubt. In civil cases, the plaintiff brings the suit and must show only that the defendant is liable by a preponderance of the evidence. Source: Cornell Law School, Legal Information Institute, http://www.law.cornell.edu/wex/index.php/Criminal_procedure (accessed June 25, 2006).

36. Cornell Law School, Legal Information Institute.

37. John P. Crank, *Understanding Police Culture*, 2nd ed. (Cincinnati: Anderson, 2004), 3. Also see Bennett and Hess, *Management and Supervision in Law Enforcement*, 263–264.

38. Linda A. Teplin, "Keeping the Peace: Police Discretion and Mentally Ill Persons," *National Institute of Justice Journal* (July 2000), http://www.ncjrs.gov/pdffiles1/jr000244c.pdf (accessed June 25, 2006).

39. Teplin, "Keeping the Peace."

40. Vito Nicholas Ciraco, "Fighting Domestic Violence with Mandatory Arrest, Are We Winning?: An Analysis in New Jersey," *Women's Rights Legislation Report*, 20 (2001): 169–170.

41. Charlotte A. Watson, *Family Protection and Domestic Violence Intervention Act of 1994: Evaluation of the Mandatory Arrest Provisions* (Albany, NY: New York State Office for the Prevention of Domestic Violence, January 2001), http://www.opdv.state.ny.us/criminal_justice/police/finalreport/index.html (accessed June 28, 2006).

42. Bureau of Justice Statistics, *Homicide Trends in the U.S.* (Washington, DC: U.S. Department of Justice, 2006), http://www.ojp.usdoj.gov/bjs/homicide/hmrt.htm#longterm (accessed July 12, 2006).

43. Federal Bureau of Investigation (2007), http//www.fbi.gov (accessed November 11, 2007).

44. Federal Bureau of Investigation (2007), http://www.fbi.gov/ucr/cius2006/offenses/clearances/index.html (accessed November 26, 2007).

45. Comparing the past ten-year clearance rates at the FBI website shows that more violent crime cases are unsolved than ever before. Also, an estimated 10,300 unidentified human remains were reported to the National Death Index from 1980 to 2004. See Bureau of Justice Statistics, NCJ 219533 (2007).

46. Gary W. Sykes, "Street Justice: A Moral Defense of Order Maintenance Policing," in *The Police and Society: Touchstone Readings*, ed. Victor E. Kappeler (Prospect Heights, IL: Waveland, 1999), 134–149.

47. Sykes, "Street Justice," 135.

48. George L. Kelling and William J. Bratton, "Taking Back the Streets," *City Journal* (Summer 1994), http://www.city-journal.org/article01.php?aid=1428 (accessed June 28, 2006).

49. Jim Waley, "Zero Tolerance New York Style," *NineMSN News* (2006), http://sunday.ninemsn.com.au/sunday/cover_stories/transcript_309.asp (accessed June 27, 2006).

50. James Q. Wilson and George L. Kelling, "The Police and Neighborhood Safety," *Atlantic Monthly* 249, no. 2 (1982): 29–38.

51. Wilson and Kelling, "The Police and Neighborhood Safety," 30.

52. Bernard E. Harcourt, "Illusion of Order: The False Promise of Broken Windows Policing" (2001), http://www.hup.harvard.edu/pdf/HARILL_excerpt.pdf (accessed June 24, 2006). Also see Wesley Skogan, *Disorder and Decline: Crime and the Spiral of Decay in American Neighborhoods* (Berkeley, CA: University of California Press, 2005).

53. New York City Police Department (2006), http://www.nyc.gov/html/nypd/html/chfdept/compstat.html (accessed June 25, 2006).

54. New York State, Division of Criminal Justice Services (2000), http://criminaljustice.state.ny.us/ (accessed June 24, 2006).

55. Patrick A. Langan and Matthew R. Durose (Bureau of Justice Statistics), *The Remarkable Drop in Crime in New York City* (Washington, DC: U.S. Department of Justice, October 21, 2004), http://samoa.istat.it/Eventi/sicurezza/relazioni/Langan_rel.pdf (accessed July 2, 2006).

56. Gareth Griffith, "Zero Tolerance" (1999), http://www.parliament.nsw.gov.au/ (accessed June 27, 2006).

57. "Mayor Rudolph Giuliani," *Toronto Sun News* (November 11, 2005).

58. Norman Dennis, ed., *Zero Tolerance: Policing a Free Society* (London: IEA Health, 1998), http://www.civitas.org.uk/pdf/cw35.pdf (accessed June 28, 2006).

59. After Rudy Giuliani became mayor of New York City in 1993, the New York City police were given orders to take the "quality of life" violations more seriously. The best-known "quality of life" offenders were the people who washed car windshields at the points of entrance into the city, expecting payment. The "squeegee men," as they were known, became an issue in the 1993 mayoral election when Giuliani, the challenger, made them a symbol of everything that was wrong with the city. Newspaper columnists jumped to the "squeegee men's" defense, arguing that they were true entrepreneurs and suggesting that Giuliani was a bully for picking on them in particular. The late William Kunstler, the famous radical lawyer, offered to represent the "squeegee men" whose rights he said were being threatened. The reality was that these "entrepreneurs" were mainly crack addicts, and their windshields cleaners were potential weapons. Most drivers felt intimidated by them; some were terrified.

60. Peter McDermott, "Zero Tolerance New York Style," *ICCL* (May 1997), http://ireland.iol.ie/~iccl/may297.htm (accessed June 28, 2006),

61. Crime Reduction Tool Kits, "Order Maintenance Policing" (2006), http://www.crimereduction.gov.uk/toolkits/pt0304.htm (accessed June 23, 2006).

62. Sykes, "Street Justice," 137.

63. George L. Kelling and Catherine M. Coles, *Fixing Broken Windows* (New York: Free Press, 1996), 103.

64. Newark Police Department (2006), http://newark.de.us/docs/departments/order_maint.html (accessed June 24, 2006).

65. Bureau of Labor Statistics, *Police and Detectives* (Washington, DC: U.S. Department of Labor, 2006), http://www.bls.gov/ (accessed June 7, 2006).

66. For instance, the Fremont, California, Police Department says that, as for most police agencies across the country, 99 percent of the 911 calls it receives are false reports or nonemergency calls. When the department conducted an outreach program to the public concerning these findings, the department reported a 20 to 30 percent reduction in the number of false alarms. In 2004, however, the department still responded to 7,000 alarms, of which more than 6,900 were false alarms (Fremont California Police Department, http://www.fremontpolice.org/alarm/chiefs_ltr.html, accessed June 26, 2006).

Other departments, such as the Baltimore, Maryland, Police Department, have installed 311 nonemergency systems. For the Baltimore police, of the 1.7 million 911 calls received, only 40 percent have been police emergency calls. See National Institute of Justice, *Research for Practice: Managing Calls to the Police 911/311 Systems* (Washington,

DC: U.S. Department of Justice, February 2005), http://www.ncjrs.gov/pdffiles1/nij/206256.pdf (accessed June 26, 2006).

67. Official State of Pennsylvania Code, Rule 10, http://www.pacode.com/secure/data/234/chapter1/s103.html (accessed June 10, 2006).

68. City of Davis California Municipal Code, http://www.city.davis.ca.us/cmo/citycode/detail.cfm?p=22&q=760 (accessed June 10, 2006),

69. For a closer look at this perspective, see Peter K. Manning, *Police Work: The Social Organization of Policing* (Prospect Heights, IL: Waveland Press, 1997), 97–101. For instance, Manning argues that it is a contradiction that "the police are, in symbolic terms, the most visible representation of the presence of the state in every day life and the potential of the state to enforce its will upon citizens" (97).

70. Bureau of Justice Statistics, *State and Local Law Enforcement Training Academies, 2002* (NCJ 204030) (Washington, DC: U.S. Department of Justice, January 2005), http://www.ojp.usdoj.gov/bjs/pub/pdf/slleta02.pdf (accessed June 10, 2006).

71. Bureau of Labor Statistics, *Police and Detectives* (Washington, DC: U.S. Department of Labor, 2006), http://www.bls.gov/oco/ocos160.htm (accessed June 10, 2006).

72. Jerry Wilson, "Police Personnel Crisis Needs Federal Leadership," *Washington Post* (May 23, 2006), http://www.washingtonpost.com/wp-dyn/content/article/2006/05/22/AR2006052200886.html (accessed June 10, 2006).

73. Dennis J. Stevens, *Applied Community Policing in the 21st Century* (Boston: Allyn and Bacon, 2003), 20–21.

74. Jerome H. Skolnick, "Ideas on American Policing: On Democratic Policing," Police Foundation (August 1999), 8; http://www.policefoundation.org/ (accessed May 13, 2004).

75. Table 2.17 in *Sourcebook of Criminal Justice Statistics 2004* (Albany, NY: U.S. Department of Justice, 2005), 119; http://www.albany.edu/sourcebook/pdf/t217.pdf (accessed June 17, 2006).

76. Dennis J. Stevens, *Case Studies in Applied Community Policing* (Boston: Allyn and Bacon, 2003), 17.

77. John H. Schweitzer, June Woo Kim, and Juliette R. Macklin, "The Impact of the Built Environment on Crime and Fear of Crime in Urban Neighborhoods," *Journal of Urban Technology* 6, no. 3 (1999): 59–73. Stephen Farrall and David Gadd, "Evaluating Crime Fears: A Research Note on a Pilot Study to Improve the Measurement of the Fear of Crime as Performance Indicator," *Evaluation* 10, no. 4 (2004): 493–502.

78. Crime in the United States 2004, *Uniform Crime Reports and Clearance Rates* (Washington, DC: U.S. Department of Justice, 2006), http://www.fbi.gov/ucr/cius_04/ (accessed June 17, 2006).

79. Phoenix Police Department, "Hablenos con Confianza" (2001), 5; http://www.prensahispanaaz.com/index.asp?id=7947 (accessed June 17, 2006).

80. Sam Walker and Charles M. Katz, *The Police in America: An Introduction* (Boston: McGraw Hill, 2005), 1. J. P. Brodeur, *How to Recognize Good Policing: Problems and Issues* (Thousand Oaks, CA: Sage, 1998).

81. Stevens, *Police Stress*, x.

82. For instance, while the author was writing this paragraph on April 23, 2005, a Salem police officer was on an extra detail controlling traffic while a roadway was being repaired. A motorist refused to obey the officer's directions and decided to drive through the uniformed officer, sending him to the hospital in serious condition. It is unlikely the officer will be able to report to duty for a year or more. Unusual? In Chapter 5, the number of officers assaulted and killed will confirm the idea that many officers are disrespected daily.

83. National Institute of Justice, *National Institute of Justice 2001 Annual Report* (NIJ 195076) (Washington, DC: U.S. Department of Justice, Office of Justice Programs, February 2004), http://www.ncjrs.org/txtfiles1/nij/195076.txt (accessed May 20, 2005).

84. C. Maslach, "Burned Out," *Human Behavior* 1 (1976): 16–22. I. L. Densten (2001).

85. Personal communication with a Boston police officer who wishes to remain anonymous.

86. "Web Poll," *Police: The Law Enforcement Magazine* 30, no. 6 (2006): 20. Other responses to the question "What is your biggest pet peeve in your day-to-day job?" included "constituents with a lousy attitude toward law enforcement" (32 percent), "having to kowtow to administrators" (23 percent), "officers who take the job too seriously" (8 percent), and "dealing with slimy suspects" (2 percent).

87. *Sourcebook of Criminal Justice Statistics 2004*, Table 2.20, http://www.albany.edu/sourcebook/pdf/t2202005.pdf (accessed June 19, 2006).

88. Vivian B. Lord, "The Stress of Change: The Impact of Changing a Traditional Police Department to a Community Oriented Problem Solving Department," in *Policing and Stress*, ed. Heith Copes (Upper Saddle River, NJ: Prentice Hall, 2005), 59.

89. W. C. Terry, "Police Stress: The Empirical Evidence," *Journal of Police Science and Administration* 9, no. 1 (1981): 61–75. Also see H. Madanba, "The Relationship Between Stress and Marital Relationships of Police Officers," in *Psychological Services for Law Enforcement*, ed. J. Reese and H. Goldstein (Washington, DC: American Psychological Association, 1986).

90. Federal Bureau of Investigation, *Reported Crime, 2003* (Washington, DC: U.S. Department of Justice), http://www.fbi.gov/ucr/cius_03/pdf/03sec2.pdf (accessed June 7, 2005).

91. Boston Police Department, "Annual Report" (2006), http://www.cityofboston.gov/police/pdfs/2004annual_report.pdf (accessed July 28, 2006).

92. David E. Barlow and Melissa Hickman Barlow (2004), 78.

93. Stevens, *Applied Community Policing*, 20–21. Also see Robert C. Trojanowicz and S. L. Dixon, *Criminal Justice and the Community* (Englewood Cliffs, NJ: Prentice Hall, 1974), 147–148.

94. Trojanowicz and Dixon, *Criminal Justice and the Community*, 147.

95. Dictionary.com, http://dictionary.reference.com/search?q=Judiciously (accessed July 26, 2005).

96. Stevens, *Applied Community Policing*, 21.

97. Andrew Cohen, "Justices Get Tough on Crime," CBSNews.com: Court Watch (March 3, 2003), http://www.cbsnews.com/stories/2003/03/05/news/opinion/courtwatch/main542892.shtml (accessed March 15, 2005). Christopher S. Koper, "Federal Legislation and Gun Markets: How Much Have Recent Reforms of the Federal Firearms Licensing System Reduced Criminal Gun Suppliers," *Criminology and Public Opinion* 1, no. 2 (2002): 151–178. "Get Tough on Crime, Public Demands," *Miami Herald* (April 3, 2004), http://www.latinamericanstudies.org/argentina/tough.htm (accessed March 15, 2005). Tomislav V. Kovandzic, John J. Sloan III, and Lynne M. Vieraitis, "Unintended Consequences of Politically Popular Sentencing Policy: The Homicide Promoting Effects of 'Three Strikes' in U.S. Cities (1980–1998)," *Criminology and Public Opinion* 1, no. 3 (2002): 478–498.

98. David H. Bayley and Clifford D. Shearing, *The New Structure of the Police: Description, Conceptualization, and Research Agenda* (NCJ 187083) (Washington, DC: U.S. Department of Justice, Office of Justice Programs, July 2001), http://www.ncjrs.org/pdffiles1/nij/187083.pdf (accessed May 25, 2005).

99. Ronald D. Hunter, "Who Guards the Guardians? Managerial Misconduct in Policing," in *Police Deviance*, 3rd ed., ed. Thomas Barker and David L. Carter (Cincinnati: Anderson, 1994), 169–84.
100. Bureau of Justice Statistics, *Sourcebook of Criminal Justice Statistics 2003* (Albany, NY: U.S. Department of Justice, 2005), Table 1.17, http://www.albany.edu/sourcebook/pdf/t117.pdf (accessed July 2, 2006).
101. Bureau of Justice Statistics, "Operating Expenses: Law Enforcement" (2006), http://www.ojp.usdoj.gov/bjs/sandlle.htm (accessed July 2, 2006).
102. Brian A. Reaves and Matthew J. Hickman (2004, April).
103. Egon Bittner, *The Factors of Police in a Modern Society*, Public Health Service publication 2059 (Chevy Chase, MD: National Institute of Mental Health, 1970), 127.
104. Larry K. Gaines and Roger LeRoy Miller, *Criminal Justice in Action*, 2nd ed. (Belmont, CA: Wadsworth, 2003), 157.
105. Anthony A. Braga and David L. Weisburd, *Police Innovation and Crime Prevention: Lessons Learned from Police Research over the Past 20 Years* (NCJ 218585) (Washington, DC: U.S. Department of Justice, 2006).
106. Robert C. Davis, Christopher W. Ortiz, Nicolel J. Henderson, Joel Miller, and Michelle K. Massie, *Turning Necessity into Virtue: Pittsburgh's Experience with a Federal Consent Decree* (NCJ 200251) (Washington, DC: U.S. Department of Justice, Office of Community Oriented Policing Services, 2002).

Historical Accounts of American Police

"History is always written wrong, and so always needs to be rewritten."

—George Santayana

▸ ▸ LEARNING OBJECTIVES

When you finish reading this chapter, you will be better prepared to

- Characterize the emergence of law
- Describe what is meant by "methods of social control"
- Characterize police performance in ancient Mesopotamia and Egypt
- Describe police activities in ancient Athens and Rome
- Identify methods of social control used by English kings
- Characterize the components of Peel's police principles
- Describe policing in colonial America
- Compare and contrast policing in early urban America and the South
- Describe policing along the American frontier, including the Texas Rangers
- Explain the three eras of American policing
- Characterize the critical reviews of these eras

▸ ▸ CASE STUDY: YOU ARE THE POLICE OFFICER

You're an urban officer who joined the department some five years ago. You've worked through a hurricane's destruction that continues to hamper the progress of a department whose roots were deep in slave patrol activities of earlier generations. You work in a district characterized by high tourist traffic, where your schedule calls for you to work every third week of the month for six hours straight; then you're off for three days. After that, you're assigned to the business district, and a few times you work uptown when that district is short of officers for celebrations.

The most demanding time of the year consists of a two-month period of celebrations; the endless parties are attended by some aggressive participants, which is okay because it's their vacation. It's true: One of your city's biggest enterprises is tourism—with activities ranging from the riverboats to the gambling casinos to the parades.

Sometimes tourists leave their manners at home when they visit the city of your youth (95 percent of the police personnel there are natives of the city). For instance, you tell a group of young adults to back up or stop running up to the parade floats because it is dangerous, but they won't listen. Sometimes these aggressive tourists ignore police directives, thinking they can become invisible in the crowds. You usually ignore agitators because oddly enough they come out of the woodwork later in the evening at any one of the many dance clubs. Unfortunately, a few are disrespectful of other visitors, and sometimes they party until they are totally dysfunctional; that's usually when your uniform becomes stained. (Luckily, your department provides a uniform allotment.)

Sometimes you wonder why you're on the job in a city that ranks high in violent crime and has more than its share of problems. Often visitors taunt you with those stereotypes when you're on the job, especially those who make reference to your race. Thankfully, many more visitors seem pleased with the safety and activities provided by the city's finest. You've learned to overlook the taunts and respond professionally, which includes never taking anything personally.

1. Why are you an officer in this famous tourist city?
2. As an officer, in what way is it different to engage visitors who lack good manners?
3. As an officer, why is the best professional tactic never to take anything personally?

Introduction

This chapter provides a historical account of policing and acknowledges the idea that "history is written from records left by the privileged."[1] Historical accounts are inevitably biased, because they are typically written with the approval of those in power. This thought suggests that the history of the American police—and the history of policing throughout the ages, for that matter—could have been written to compliment who ever or what ever controlled government. In past generations, those in control decided what history would report. There is little question that writers have great difficulty in offering neutral or critical perspectives on their own era, a difficulty that arises because historians write on a moving train, explains Randall G. Shelden.[2] As George Santayana advises, "history is always written wrong."

For all these reasons, you should use this chapter simply as a guide to further your thoughts about policing, rather than assuming it presents the only possible interpretation of events. The chapter provides a description of the police who were employed through ancient civilizations to modern times and emphasizes their function, including the workings of early American agencies in cities and frontiers. The political, professional, and community relations eras of policing are discussed, along with the assessments of critics regarding those eras.

Take everything you read here with "a grain of salt." The task of an author is often difficult, as he or she "peels the onion" of truth in an attempt to make the obvious, obvious.[3]

■ Laws and Their Enforcement

Has the enforcement of the law always been the typical response of most leaders? When social disorder visited bands of nomads 10,000 years ago, what kind of response was expected? There are two schools of thought about these issues.

The first view states that laws and their strict enforcement are the path to social order. Without strict law enforcement, the reasoning goes, civilizations would be characterized by violence, brutality, and chaos because humans, when left on their own, commit criminal acts.[4]

The opposing view says that limiting enforcement power is the right thing to do because too much government power is a destructive path—a view adhered to by the framers of the U.S. Constitution. History is replete with evidence revealing that the state can be an instrument of repression and that individuals in the seat of power have used authority to victimize others.[5] When the framers of the U.S. Constitution asked the question, Restrain people or restrain government?, they ultimately chose to restrain government. Of course, those young Americans were rebelling against English rule, and those actions undoubtedly influenced the form of government they developed. The public's suspicious attitude about police power in early England and Puritan Boston, in fact, influenced the final form of the new U.S. government:

> The U.S. Constitution's primary task is to limit the power of government by restraining arbitrary government intrusion into the private lives of Americans.

That said, many people are now rethinking government's limitations, especially as it relates to policing in the United States of the twenty-first century. Certainly, limiting the enforcement powers of government can lead to an increase in the intensity and frequency of violent crime.[6] Others add that crime control issues should be secondary to the relationship between people and government.

> Arising from the inconsistencies related to public safety through "legalized force" and given the rule of law in the United States, which includes the Constitution and the Bill of Rights, the relationship between government and its constituents takes priority over crime control in the United States.[7]

Recall what was said in Chapter 1 about what America wants from its police: public safety but at an extreme, the police act as public servants who provide that public safety. However, another view was conveyed by **Supreme Court Chief Justice Warren Berger**, who argued:

> When our distant ancestors came out of the caves and rude tree dwellings thousands of years ago to form bands and tribes and later towns, villages, and cities, they did so to satisfy certain fundamental human needs: mutual protection, human companionship, and later for trade and commerce.[8]

In the social sphere, changes can occur because of increases in the population, cultural mixing within a population through immigration and other means, technological changes, and emergence of class divisions, to name just a few driving forces. Public safety just doesn't appear one day out of the blue; instead, laws *need* to emerge. Indeed, laws have traditionally emerged for numerous reasons, including to ensure order and public safety. In the United States, which is a democracy, police are expected to balance the delivery of services with the continuation of freedom. History has played a significant role in shaping these expectations and the methods to deliver police services.

The Significance of History

Police history is far too serious a matter to be left solely to historians. All too often, historians fill in the blanks with the way they *think* police looked rather than with the way early police *really* were. At the same time, be cautioned that those who analyze the present can be prone to mistakes, too.

Nonetheless, the concept of policing was absent among many village- and band-level cultures of the past.[9] Many cultures operated without a king, queen, dictator, president, governor, or mayor—and without police officers, National Guard soldiers, sailors, or marines; CIA, FBI, Treasury agents, or federal marshals. They had no written laws and no formal courts, no lawyers, and certainly no patrol cars, crime scene investigators, or jails or penitentiaries.

Every society defines crime for itself; each society has its own way of establishing order, chasing criminals, and keeping (or at least attempting to keep) youth under control.[10] What is consistent about most societies is that they consistently change and "no society can survive in chaos, and those torn by schisms [divisions] soon disappeared." [11] In summary, law and order are not natural gifts bestowed upon any group of people regardless of the time when that civilization appeared on this planet.

> **Real natural law** is a law of the strongest; it dominates the history of humankind despite the realization that early cultures knew little of law.

Does a society require those in charge to maintain order? In a word—no. Many societies of the past had few expectations of leaders, if any at all.

Acephelous Societies

The term *acephelous* means "without a head." **Acephelous societies** are those that lack any identifiable rules with the power to direct or command other members of society. This type of leadership represents the earliest types of human communities and dates back some 30,000 to 40,000 years since the evolution of modern humans (*Homo sapiens*).[12] Acephelous societies are small, economically cooperative, egalitarian societies with simple technology and divisions of labor. Everyone has largely an equal role in such a society. While these societies lack rulers or governments with power to command and correct behavior, they are nevertheless characterized by more cooperation than chaos.

For our purposes here, the relevance of acephelous societies lies not in whether these societies were free of crime, but rather in their proof that not all societies see law enforcement as a necessary element to further social order. People managed without means of law enforcement for tens of thousands of years. Why, then, are contemporary state-level societies so dependent upon them?

The Emergence of Law

In previous civilizations, small bands of people lived together and behaved pretty much as they were expected to, without any type of direct intervention of "any centralized political authority." [13] In those small groups (and particularly in primitive societies), formal laws were not necessary because, to a large extent, behavior was regulated though informal methods of control such as gossip, criticism, fear of supernatural forces, loss of respect and status, and physical or social isolation.

This means of controlling behavior is effective even by today's standards. Consider what happens in a small town or community where everyone knows everyone else (including everyone else's business): When a scandal involving morality occurs, townspeople stop one another on the streets and in the food stores to talk about it. "A common indignation is expressed." [14] Sentiment or public agreement can affect a criminal justice issue, Durkheim (1933–1984) might argue. The fear of townspeople's sanction can, under certain circumstances, command others to follow their lead, their lifestyles, and their religious ideals—or else. Or else what? Ask Mary Barrett Dyer, whose young body dangled from a huge old elm on Boston Commons in the heat of a summer afternoon in 1660 as her husband and other respected community members watched.[15] A Quaker, she had opposed Boston's Puritan sentiment and became the first female executed in America.

From infancy to childhood to adulthood, we are consistently taught behavior through modeling by our families and other social institutions that influence us; those lessons, in turn, have a lot to do with decisions we make. An anthropologist might say that we are subject to cultural controls—that is, control through values deeply internalized in the minds of individuals who belong to a certain culture.

As populations increase and individuals become less influenced by family or cultural controls, however, face-to-face methods of controls might not deter inappropriate behavior as often as expected. In such circumstances, laws need to be established and written down (so that people can become familiar with them); extending the same thought process, **methods of social control** should be devised to enforce those laws. Thus, as conflicts arise among various members and between various institutions, "a more formal, rationally thought-out method of social control is necessary." [16]

Although there may be some disagreement about the need for and form of laws, at some point, whatever form a society's political organization takes—from chiefdom to states—and whatever else it might do, maintaining social stability emerges as one of a political organization's primary missions. Formal controls intended to resolve conflicts that produce social disorder are called *laws.*

The Code of Hammurabi

One of the earliest systems of law was developed by King Hammurabi (1728–1686 B.C.E.) in Mesopotamia almost 3,000 years ago (1780 B.C.E.). The **Code of Hammurabi** was developed to regulate behavior through 282 codes (collectively called Babylonian

Law) that dictate rules and punishments. The codes focus on theft, agriculture (shepherding), property damage, women's and children's rights, slaves' rights, murder, death, marriage, and injury. Punishment varied depending on the status of the offender and the victim (i.e., free versus slave). Of interest, Babylonian Law made no excuses or explanations for mistakes or fault. The Code of Hammurabi was openly displayed for all to see, although few people of that era could read.

The maintenance of social order through laws (such as the Code of Hammurabi) requires some kind of an institution that can enforce the legal rules created by the state or a method of social control.[17]

Methods of Social Control

Externalized social controls are designed to encourage conformity or compliance to laws.[18] To be effective, any kind of regulations or laws must be enforced, and it is safe to say that early societies with laws probably had enforcers, too. Laws can be considered a social control agent, and enforcing those laws requires a method of social control to make people accountable.

> **Methods of social controls** are systematic initiatives to moderate or suppress conflict in a society or community.[19]

As a way to suppress unwanted conflict and behavior, primitive civilizations relied on gossip, criticism, fear of the supernatural, loss of respect, loss of status, physical isolation, and family intervention. Advanced civilizations are more sophisticated and incorporate informal and impersonal relationships centered in privacy and individualism; written laws and remedies are employed as guides to appropriate behavior and consequences.

Of course, a utopian society might potentially exist that does not require a process of social control, but that is highly unlikely because an analysis of methods of social control must begin with a description of an activity that has been experienced by every community on the earth—crime. The definition of "crime" depends on the civilization and varies from group to group, from nation to nation, and from generation to generation. As a consequence, criminal activity is described in different ways at different times in different groups. Are there universal wrongs—things that are considered criminal in nature for every society, no matter when the society existed? Probably not, but what you need to know right now about crime is that it is a natural kind of society activity and "an integral part of all healthy societies," according to Emile Durkheim.[20]

> All societies have crime, and most societies attempt to do something about it.

At the core of all civilized societies are laws that attempt to regulate behavior including the behavior of the government itself.

Controlling Crime

History informs us that crime has never been stopped. Thus one recommendation for dealing with crime is to control it:[21]

> It stands to reason that if pathological (criminal) behavior in every society varied then the likelihood is that laws and enforcement techniques varied, too. Crime and enforcement techniques or what can be referred to as the police are understood within the historical content of the civilization.[22]

■ Police in Ancient Mesopotamia and Egypt

Institutionalized keepers of the peace were well known in ancient civilizations, particularly in Western and Middle Eastern cultures. For instance, institutions and persons devoted to the security and preservation of their ruling regimes were active 6,000 years ago in Mesopotamia.[23]

Clandestine and covert operations may be among the most intriguing aspects of crime, but the history of espionage is perhaps better described in terms of the evolution of the more mundane components of the trade. Throughout history, *intelligence* has been defined as the collection, selection, analysis, and distribution of strategic information. Its existence implies that keepers of the peace might have duties other than ensuring public safety and social order. For instance, early Egyptian pharaohs employed agents of espionage to ferret out disloyal subjects and to locate tribes outside the "empire" that could be easily conquered and enslaved.

The police force in ancient Egypt (approximately 2000 B.C.E.) were not an extension of the army, but rather was established to enforce the orders of the "gods."[24] Ancient Egyptian police were also tasked with the protection of the weak and control of the criminals through apprehension and administration of justice. They were not looked upon as a hostile body, but rather as guardians and protectors of law-abiding communities. In rural areas, police banished troublemakers and persuaded (physical force was appropriate) the population to pay tribute (taxes). They administered corporal punishment as needed. In addition, some police patrolled desert frontiers with trained dogs to track problematic nomads and escaped slaves.

Nomadic herdsmen called *Medjay* (or *Medjai*) from Nubia became part of the Egyptian police force in the Eighteenth Dynasty (1550–1292 B.C.E., a period during which Tutankhamen ruled, one of Egypt's most famous pharaohs).[25] The Medjay served in Egyptian armies and were respected as archers. Those who entered police service protected towns in western Thebes, officials, and the royal tombs (including those under construction). Their tasks included protecting the safety and ensuring good behavior of the workers. These police were also charged with protecting workers from invaders, acting as messengers, interrogating thieves, being witnesses for administrative functions, and inflicting punishment.

■ Police in Ancient Athens and Rome

The term "police" was derived from the Greek *politiea,* relating to the concept of government citizenship, or *polis,* meaning "city." Both Greek words applied historically to the exercise of civic or collective authority (hence the origins of the word "politics"). In ancient Athens, a body of public slaves known as Scythian archers (from the Caucasus Mountain regions) lived in barracks and were supervised by *eirenarchs* ("magistrates of the peace"). The magistrates' duties covered only certain aspects of criminal activity; however, if their duties mirrored those of their predecessors in early civilizations, one of the magistrates' most important roles was to spy on groups that opposed established power.

Ancient Rome, which had an estimated population of 1 million people, was the largest city in the ancient world. In spite of the glories of the Roman legions and the affluence of the empire's wealthy social class, most Romans were poor and lived in crowded conditions that in recent centuries have engendered well-documented crime and violence.[26] Most often, public slaves performed tasks generally associated with enforcement officers and toured unruly parts of the city. When they came upon trouble, they also administered punishment. Nevertheless, these slaves had neither the authority nor the resources to act as a true police force. To legitimize their authority, the magistrates who controlled the enforcement slaves depended on the acceptance by the citizens of their political institutions and the men who operated them.[27] Disturbances and disputes between citizens were left to them, their families, and their friends. Small-scale disturbances were resolved locally by neighbors and passers-by, who were expected to take sides and usually did so.

In the late Roman republic, emperors bestowed more authority on enforcers.[28] These latter-day forces consisted of urban cohorts, which were commanded by a high-ranking city official, and the *vigils urbani,* a corps of 7,000 freed slaves under an equestrian prefect, whose principal tasks were to act as a fire brigade and to suppress public disturbances. Many Roman cities had fire brigades, but when they were seen as potentially subversive, their power and equipment were limited.

None of the police groups described thus far was involved in routine enforcement responsibilities. Instead, routine enforcement of laws and customs were the province of private citizens and their families.

■ Police in Early Western Europe

For centuries, few social organizations included an enforcement agency. If and when institutionalized forces were ultimately established in early Western Europe, their purpose was typically to protect the church from heretics (as opposed to buildings). Other than raising armies against invaders, from medieval (300–1000 A.D.) to nineteenth-century Europe, inside the boundaries of the community citizens fended for themselves in a manner similar to that followed by their counterparts from the Roman Empire. Families and larger kinship groups assisted and protected their members. As a result, violent crimes went virtually unreported. Even so, it is estimated that violent crime rates were very high in preindustrial Europe.[29]

Institutionalized police systems for individual protection would have to wait until the eighteenth century, when England and France would lead the way in establishing police forces. With the formation of true police systems, violence decreased, but theft and property crimes increased. With the rise of state government, crime control and punishment were taken out of the hands of citizens and placed among the affairs of the state. If the general population held the right to punish those who harmed them, then what role would government play in their lives?

Police and Centralization in England

With the emergence of centralized governments, recognition of the public's responsibility for societal welfare and methods of social control grew, too. For instance, Great Britain's Alfred the Great (870–901 A.D.) developed a system that required citizens to be responsible for their own conduct and the conduct of others.[30] Peasants were financially obligated to catch criminals.[31] The "mutual pledge system" was established. Under this system, upon receiving a report of a crime, everyone in the community had to raise a "hue and cry." If the Crown's subjects failed to catch an offender, the Crown's collectors would punish the incompetents with a royal fine. Clearly, the intent was to force the entire realm to play the role of the police.

American policing is rooted in English tradition. For instance, freedom was the revolutionary spirit that inspired the Magna Carta, which was signed by King John in 1215 as a means to counteract his fear of an uprising among barons and the people.[32] The Magna Carta was a charter of ancient liberties guaranteed by a king to his subjects; the Constitution of the United States is the establishment of a government by and for "We the People" guaranteeing liberty too which guides American policing in the twenty-first century. Also, Edward I (1272–1307) created the first official police force in large English cities to protect residents against fires, particularly those occurring at night. His successor, Edward II, established the Justice of the Peace Act of 1327 to assist the sheriff in his enforcement activities; the enforcers created by this Act served warrants and detained criminals until the arrival of a judge. This system worked well until the growth of large cities, civil disorder, and increased crime led to changes in the system. Unfortunately, changes in law enforcement responses were often swift and brutal.

Policing in Fifteenth-Century England

The use of **frankpledges** compelled all males 12 years of age and older to serve in a quasi-police role.[33] Each person was sworn to protect other citizens and, as part of his duties, held official authority to hold an offender in custody while the accused person was awaiting trial.[34] These individuals formed a group of nine called a *tything*; ten tythings were grouped into a hundred and took direction from a constable appointed by the local nobleman.[35] The constable is considered the first official with true law enforcement responsibilities in England. If any tything member failed to perform, severe fines were levied on all the tything members.

Thus citizen groups were obliged to control any group member who committed an unlawful act. This style of controlling criminal activities emerged in England after 1066 partly as a result of the Norman Conquest and persisted for several hundred years.[36] One way to describe the period's policing is to say it was a collective responsibility, but this would change leading to feelings of suspicion among the population.

Policing in Eighteenth-Century England

Accounts in England (and France) report that the law enforcement system developed after the Norman Conquest was viewed by the public as a sinister organization devoted to enslaving the masses.[37] Citizens looked upon the police with suspicion, and many people believed that the police were secretly surveying and controlling citizens. Some influential British writers suggested that the police should be restrained and their focus should be on civil preventive actions, with justice and humanity being the key underpinnings of the police system, and the general security of the state and its individuals being its ultimate objective.

At the same time, and of equal concern, England (and all of Europe for that matter) was a very violent place to live. For instance, research suggests that there were 20 homicides for every 100,000 inhabitants from the thirteenth to the nineteenth centuries in England—an astounding rate. Eventually, that number declined, reaching approximately 5 homicides per 100,000 population in the twentieth century. Ironically, capital punishment in England declined in tandem with capital crimes, which implies that the death sentence can enhance crimes of violence among certain populations.[38] (Today, the United Kingdom reports 19 homicide victims for every 1 million citizens, with a clearance rate of almost 95 percent.[39])

During the eighteenth century, old systems of law enforcement in England collapsed. In general, industrial progress—such as that experienced during the Industrial Revolution—tends to promote crime, poverty, and conflict between social classes composed of what can be referred to as the "haves" and the "have-nots." London had grown into a large industrial city that overwhelmed the capabilities of citizen-led law enforcement. Policing in the English countryside was in shambles. The watchmen were largely ineffective, the constables in the towns and cities found it impossible to maintain order, and decent people employed bodyguards to protect them and their possessions.[40] Highway robbers infested the roads, and corruption was rife among magistrates and judges. In 1780, the Gordon riots—a clash between Irish immigrants and English citizens—provoked a 50-year debate over the need for social control. The English crown also faced another supreme indignity reflecting its poor enforcement measures—its loss of control over the rebellious Americans.

Policing in Nineteenth-Century England: Establishment of the London Metropolitan Police

Despite the public's suspicion about its solution to the issue of public safety versus individual liberty, Parliament approved the London Metropolitan Police Act in 1829. This Act created the **London Metropolitan (Metro) Police** under the guidance of **Sir Robert Peel** (from whom the term "bobby" originated). Peel, who later became prime minister of England, set forth the following principles on which the police of that day were based.

Police Principles as Dictated by Sir Robert Peel

1. The basic mission for which the police exist is to prevent crime and disorder as an alternative to the repression of crime and disorder by military force and severity of legal punishment.
2. The ability of the police to perform their duties is dependent upon public approval of police existence, actions, behavior, and the ability of the police to secure and maintain public respect.
3. The police must secure the willing cooperation of the public in voluntary observance of the law to be able to secure and maintain public respect.
4. The degree of cooperation of the public that can be secured diminishes, proportionately, the necessity for the use of physical force and compulsion in achieving police objectives.

5. The police seek and preserve public favor, not by catering to public opinion, but by constantly demonstrating absolutely impartial service to the law, in complete independence of policy, and without regard to the justice or injustice of the substance of individual laws; by ready offering of individual service and friendship to all members of the society without regard to their race or social standing; by ready exercise of courtesy and friendly good humor; and by ready offering of individual sacrifice in protecting and preserving life.
6. The police should use physical force to the extent necessary to secure observance of the law or to restore order only when the exercise of persuasion, advice, and warning is found to be insufficient to achieve police objectives; and police should use only the minimum degree of physical force that is necessary on any particular occasion for achieving a police objective.
7. The police at all times should maintain a relationship with the public that gives reality to the historic tradition that the police are the public and that the public are the police; the police are the only members of the public who are paid to give full-time attention to duties that are incumbent on every citizen in the interest of the community welfare.
8. The police should always direct their actions toward their functions and never appear to usurp the powers of the judiciary by avenging individuals or the state, or authoritatively judging guilt or punishing the guilty.
9. The test of police efficiency is the absence of crime and disorder, not the visible evidence of police action in dealing with them.[41]

Peel's London Metropolitan Police exemplified three relevant components of policing that ushered in the era of modern policing:

- New mission associated with the police: crime prevention and deterrence
- Strategy: prevention can be achieved through police visibility—patrol
- Organizational structure: military model—uniforms, rank system, and, most important, the authoritarian system of command and discipline

Peel's idea was to forge policing into a system—that is, to institutionalize it. Prior to Peel's participation, policing was unsystematic, police appeared only to resolve conflicts (if they appeared at all), and citizens achieved justice on their own. In fact, the law in England during the time required males to own weapons to protect themselves and their family members.

Peel's colleagues included Henry Fielding, who made an effort to prevent crime by encouraging the public to report crimes and to give descriptions of the criminals.[42] Field's office was on Bow Street, where he trained six officers. These officers arrested many criminals and became known as the Bow Street Runners. They wore no uniforms, but their reputations made them revered as one of the first criminal investigation units. When Henry Fielding retired, his brother Sir John Fielding (known as "The Blind Beak," because he was blind and could identify more than 3,000 criminals by the sound of their voices) formed the Bow Street Horse Patrol. These mounted patrol officers wore uniforms that resemble the ones worn by their present-day descendants.

For most of the past thousand years, Great Britain had rarely had a visible police agency until Peel came along. It was the British model that the Americans tried to imitate, albeit with a twist—limited police power through fragmentation (see Chapter 6).

Police in Colonial America

Many of the American colonists had departed England in part to escape a tyrannical government, and it goes without saying they were apprehensive of any future government intrusion. Individual liberty took precedence in the New World: freedom, discretion, and participation in all governmental decisions.[43] As part of this perspective, they viewed the organized police force in England (and in early America) with great suspicion because of "its potential for despotic control over citizens and subjects."[44]

From the time the first colonies were established in eastern North America (such as Jamestown, Virginia, in 1607) until the United States became a separate nation, numerous changes took place that affected American attitudes toward crime and crime control.[45] For instance, the New England Puritans attacked family violence with law enforcement that incorporated the combined forces of community, church, and state.[46] As early as 1599, an English Puritan minister argued that a wife beater should be whipped and called upon the biblical verse citing "an eye for an eye" as justification for his punishment.

The colony of New Haven enacted the first American law against family violence (incest punished by hanging) in 1639. Despite the establishment of such laws, there were no paid policing positions in early colonial life. Those who enforced the law and who carried out punishment (often the same person) earned their keep through corruption. Sheriffs collected taxes, set elections, and maintained bridges and roads; and watch patrols dealt with fires, crime, and public disorder. The law enforcement service—regardless of who delivered it—was inefficient, quality of life or order maintenance was nonexistent because officers were so ill equipped, and public services were lacking because officers were paid out of the fees they levied—even jailers produced incomes from those they confined. Not surprisingly, this decentralized and demoralized system was characterized by both corruption and political interference.

> Colonists (including Boston's Puritans) saw government as a potential means of oppression. Consequently, the government that they laid out in the U.S. Constitution limits arbitrary governmental intrusion and safeguards privacy.[47]

Old English suspicions resurfaced in America, as colonists took on the liberty-versus-government debate. At the same time, the colonists brought the preexisting system of justice with them as part of their cultural baggage, including English common law, the court system, and ideas about punishment and enforcement of the law.[48] Chafing under the authority of the English government, they eventually rebelled.

The colonies that later became the United States were established during a wave of English exploration and settlement that saw roughly 10 percent of England's population immigrate to the New World in an attempt to avoid government treachery, circumvent religious restrictions, and compensate for limited opportunities at home. Although Spain, Portugal, and France initially dominated expansion into the Americas (producing conflict and warfare along the way), England saw its colonies as a safety valve—that is, as a means to relieve itself of the pressure exerted by a rapidly growing population. Despite the diversity that now characterizes the United States, those individuals of Anglo (individuals whose first language is English) descent designed the U.S. Constitution.

Attempts to Deal with Urban Uprisings

In colonial America, when civic disturbances occurred, they were handled informally. If disturbances continued, unarmed city watchmen were called upon for help. When large riots occurred, the militia was called.

Urban uprisings and a breakdown of the existing law enforcement system prompted a reconsideration of the means used to deliver police services. For example, dramatic economic conditions were the backdrop for early Boston, which was "inclined to riots and tumults" in 1707.[49] As far as the New World goes, Boston is the 300-year-old leader in violence and rioting: It was the scene of 103 major riots from 1701 to 1976, not including the much-studied Revolutionary-era riots. Early riots in Boston (and New York City) tended to involve immigrants who felt ignored by the political and wealthy elites; the poor and powerless wanted to make their voices heard and demanded social or economic gains. The histories of Boston and New York City are full of stories of riots in which racial, socioeconomic, and religious or ethnic groups were pitted against one another in a battle for power, access, and scarce resources. The ethnic identity of rioters over time followed the timeline of each group's arrival in force to those cities—waves of new immigrant groups led to waves of riots involving the new residents as they sought to carve out their own turf. Given the discontent, an organized police force to control the rioters was the order of the day.

Colonial policing systems also included an incompetent night watch system limited to urban communities such as Boston, Philadelphia, and New York City as early as 1684.[50] The night watch usually consisted of night-time surveillance of "fires, suspicious individuals, riots, and other incidents requiring immediate attention."[51] Members of the watch also "raised the hue and cry" and maintained street lamps, much as their counterparts had done in early British history.[52]

Some cities attempted to improve these local police systems. For example, in 1801, New York passed an ordinance that regulated the fees paid to constables and night watchmen. For instance, serving a warrant was worth nineteen cents and serving a summons was worth twelve and a half cents.[53] If any night watchman slept at his station or committed an act of violence except as was strictly necessary in the execution of his duty, the watchman was severely disciplined.

The Emergence of Private Police

While city night watchmen "hued and cried," private enterprise was encouraged by the rise of bounty (well-paid) hunting. Many heard the siren song of this opportunity, and some developed large enforcement organizations—a trend that continues today in the form of such corporations as Blackwater USA and Wackenhut Services (check their websites and Chapter 4 for details on privatization of policing). In the nineteenth century, private agencies such as Wells Fargo and Pinkerton hunted down offenders and guarded private property and railroads, while public police that were employed by government dealt with menial tasks, producing the impression that they were backward characters—more or less simpletons.[54]

Early American Policing Services

In the nineteenth century through part of the early twentieth century, policing was a hierarchical organization, and the uniformed visibility of officers made them civil servants of general resort. For instance, policing services during this era included a wide range of social services—running soup kitchens, inspecting boilers, standardizing weights and measures, recovering lost children, providing jobs for wayward girls as domestics, and providing overnight housing for thousands of homeless people.[55] Eventually, organized charities took over those tasks. Not until the middle part of the twentieth century did policing focus on crime control.

The primary concern of early police services was to protect property and industry of the elites in New England, New York City, Baltimore, and Philadelphia. Northern industrial cities such as Chicago and Detroit put strike breakers in jail, while western cities such as Austin and Dodge paid marshals to "law up" vagrants and drifters.

Eventually, the elite social classes recognized the advantages to be gained by controlling local police efforts. That is, while the police in the East and Northeast snuffed out fires and cooled down riots, Southern plantation owners realized how influencing the police enterprise could benefit their own operations.

Police and Slavery

Slaves were captured and transported by both French and British slave ships in large numbers, and not just to North America:[56]

- French slave voyages: 4,200; British North America/United States slave voyages: 1,500
- Slaves transported by French slave ships: 1,250,000; slaves transported by British North America/United States ships: 300,000
- Slaves delivered to French West Indies: 1,600,000; slaves delivered to British North America/United States: 500,000

In the history of the Atlantic slave trade, French slavers turned four times as many Africans into slaves as Americans did, they used those slaves far more brutally, and they continued the slave trade long after the rest of Europe had given it up.[57] For example, slaves supplied by the French to the French West Indies worked until they died and were immediately replaced.

Slave resistance took many forms in the colonial period, but it did not often result in outright rebellion until later in American history, particularly in Louisiana (and largely French West Indies) and plantations supplied by French slavers. How could a local peace officer in Louisiana or anywhere in the South deal with a slave-related issue, such as rebellion or runaways? One answer might be that peace officers aided those who could provide them with amenities of sorts—namely, the wealthy plantation owners who relied on slaves to prosper. Ultimately, Southern cities such as New Orleans, Charleston, Mobile, and Atlanta developed "slave patrols" to control the movement of slaves, indentured servants, and poor whites.

The first publicly funded municipal police departments in America were established in the South to protect the property of those who owned property and slaves.[58] Nevertheless, slaves were also present in the Northeast in large numbers, with slaves in Boston accounting for 10 percent of all American slaves, those in New London accounting for 9 percent, and those in New York accounting for 7.2 percent.[59] In fact, prior to the Revolutionary War, New York City was the second-largest urban center of slavery, after Charleston, South Carolina.[60]

Policing helped institutionalize slavery. A letter sent in 1720 to London illustrates the fear produced in those who attempted to contain a large number of people with brute force:

> I am now to acquaint you that very lately we have had a very wicked and barbarous plot of the design of the Negroes rising with a design to destroy all the white people in the country and then to take Charles Town in full body but it pleased God it was discovered and many of them taken prisoners and some burnt and some hang'd and some banish'd.[61]

Slavery in South Carolina and the Stono Rebellion

Slaves did arm themselves in South Carolina in 1739. In September of that year, a group of 20 slaves attacked a general store and severed the clerks' heads, leaving them on the front steps of the store.[62] They then marched in the direction of Spanish Florida, where they believed they would be set free. Along the road, they attacked and burned plantations, killing several white men, women, and children. The band also recruited other slaves to join them and, as they moved south, their numbers grew. Finally, a group on horseback attacked the slaves, killing many of them and driving the rest into the hills.

After the **Stono Rebellion** of 1739, the South Carolina legislature passed a series of repressive laws—including those establishing slave patrols—that sought to eliminate the possibility of revolt by slaves in the future. The fears of some South Carolinians

about the threat posed by slaves were heightened by the fact that blacks were the majority population in that state in the early 1700s. Many of these individuals were immune to the many lowland diseases that led to higher mortality and morbidity rates among European settlers. Interestingly, the sickle cell trait (associated with the disease sickle cell anemia) increased Africans' resistance to malaria and proved to be an advantage in colonial America.[63]

In a letter dated October 5, 1739, less than a month after the Stono Rebellion, Lieutenant Governor William Bull reported to Britian's Board of Trade, informing them of the revolt and updating them on the status of the rebels. Bull, who had personally spread the alarm regarding the revolt, also requested that rewards be offered to Indians who helped re-capture the slaves.[63]

Policing a Slave City: New Orleans and Charleston

The most distinctive features of the early southern police departments in New Orleans (as well as the police departments in Charleston, Mobile, and Atlanta) were the officers' military uniforms, formidable weapons, and wages (rather than fees or compulsory unpaid services); around-the-clock patrolling and unification of day and night forces came later.[64] Southern police interaction with the public—which included whites and blacks alike—was conducted in a systematic manner; that is, the police patrolled in squads resembling military sweeps, rather than having a single officer on patrol, long before Boston or New York City institutionalized their police departments. New Orleans officers were often mercenaries drafted from the Spanish, French, and Germany armies, many of whom spoke little English.

The need to control slaves and a fear of a slave rebellion fueled many of the military police initiatives in the South. To control the slave population, the city guard made mounted night patrols in squads to the beat of the drum, stopping any slave without a pass and dispersing "nocturnal dances." [65] They also inspected slave huts on the plantations and confiscated weapons.

New Orleans was not the only southern city to establish a strong police force with a martial configuration. Charleston created a city guard in 1783. Members of the guard marched forth in mounted squads, wore uniforms, carried weapons, and swept the streets at night. When they confronted any individual (white or black), it usually ended as a tragic experience for that person.

James Stuart, a British visitor to the United States in 1830, said that the "appearance of an armed police, Charleston and New Orleans do not resemble the free cities of America." Some described the Southern police as a "guard of soldiers" who paraded the city during the day and at night assumed the appearance of a great military garrison.[66]

In cities such as New Orleans where free blacks and slaves numerically outnumbered other racial groups (especially the French and Anglos), the reliance on a "city guard" furthered the best interests of the business owners and wealthy. Some reports show to confront the perceived domestic danger, the ratio of "officers" per 10,000 urban population was greater than in the North, particularly during the late antebellum period. However, as brutality, corruption, and violent face-to-face altercations were arbitrarily visited on various individuals, reform of these forces was demanded. Police orientation would ultimately change from a system characterized by military authority and perspective to one governed by civil authority. While slave patrols have certainly vanished today, there are nevertheless strong rumblings among police organizations that suggest they might like to revert back to the military style of previous initiatives.[67]

The *Dred Scott* Decision

The U.S. Supreme Court's decision in the *Dred Scott* case in 1857 legally denied sanctuary to runaway slaves to Canada and Mexico. As it made its way to the western frontier, policing systems sprang up to enforce this ruling. In fact, one of the early primary tasks of the Texas Rangers was to retrieve those slaves and return them to the South.

> In the ***Dred Scott* decision** of 1857, the U.S. Supreme Court, led by Chief Justice Roger B. Taney, declared that all blacks—slaves as well as free blacks—were not and could never become citizens of the United States. The court also declared the 1820 Missouri Compromise unconstitutional, thereby permitting slavery in all U.S. territories.[68]

■ Police in Nineteenth-Century America

Policing Immigrants and "Street Arabs"

In 1853, the United States began surveying railroad routes to the Pacific, mapping four different paths to be followed by trains. Poster, flyers, and advertisements went out to Europe and the rest of the world, extolling the virtues of coming to America and getting "free land."[69] Agents of steamship lines and minions of the railroad companies attracted thousands of immigrants to the United States by describing the country with phrases such as "the land of opportunity," "land of a second chance," and "land of milk and honey." Their efforts were certainly successful: The United States welcomed more than 4.3 million newcomers—many of them children—between the years of 1841 to 1860. Ellis Island in New York processed 2,251 immigrants in its first day of business. In 1907, a record number of immigrants were admitted to the United States—nearly 1.3 million people in a single year.[70]

Many of the new immigrants settled in cities, rather than journeying farther into the western frontier. The jostling for space in major cities, as these new Americans sought to find housing and employment, sometimes resulted in riots and violence, as mentioned earlier in this chapter. From 1849 to 1863, for example, a dozen major riots occurred in Detroit, as well as probably a dozen other minor ones.[71] Most of these disorders occurred in or near the immigrant residential neighborhoods that were then forming in a variegated and compact urban environment. People attacked what they saw as serious threats to the neighborhood communities they were trying to establish. In Detroit, for example, houses of prostitution had no place on the German-dominated outer east side in the late 1850s. By the standards of the time, Detroit in 1880 was a model of law and order. No longer did mobs, corner toughs, and lurking garrotters preoccupy the citizens as they had before.[72] If much of this transition had to do with the greater feeling of security and the image of order provided by the professional police force, it also reflected the fact that serious crime and disorder had declined markedly. In the 1850s and 1860s, Detroit's spatial development had contributed to the breakdown in law and order and increased the anxiety of respectable Detroiters. By the 1870s, the city had alleviated neighborhood tensions, created few new environments especially conducive to criminal activity, and helped make it possible for police to serve as an effective deterrent to crime.

In some areas, the rapid urban growth created numerous problems. Immigrants crowded into urban neighborhoods where their children filled public schools and mastered diverse street trades.[70] Living conditions for immigrants were difficult, in part because jobs for immigrants soon became scarce and labor was cheap (the supply of labor far exceeded the demand for workers in major cities).[73] Without extended family members (grandparents, aunts, uncles) to rely upon in times of need, young families fell apart. Children as young as six years old worked to help support their families. Food became scarce. Job safety was not a priority, and many men were killed in accidents at sea and at other places of work; the women and children they left behind then had to make their own way as best they could. Diseases from living in unsanitary quarters led to the early deaths of many overworked mothers.

Institutions and organizations geared up to deal with the requirements of immigrants and their children, and public funds were spent in every area of urban life to help them cope. Orphanages were built to care for as many children as could possibly be taken in. Adults could pay for their care on a weekly or monthly basis but if the payments stopped, the child became a *ward of the court* and was "disposed" of as the social workers saw fit.

Not surprisingly, perhaps, many children slipped through the cracks. In 1854, the estimated number of homeless children in New York City was approximately 34,000. Many of those children had been thrown out of their homes, were runaways, or were simply trying to survive after their parents died. Their conditions were deplorable: Many were near starvation, few had warm clothing, and almost none had any medical care.[74]

> Among the "Street Arabs" of New York there are many distinct characters of people. Embraced under the head are to be included pickpockets, beggars, and prostitutes that *prey* upon the populace, as well as the itinerant sellers, buyers, etc., who profess to make some return for what they receive; but it is only from the latter class that our present illustrations are selected. These are to be seen every day in the streets of the city.[75]

These homeless children, who became known as "street Arabs," were soon added to the stereotypes of "the dangerous class."[76] Because they faced the threat of street violence, many of them formed gangs as protection; soon, the gangs were a growing problem for police. Of course, some children were arrested; even those as young as five years of age were often thrown in jail with adults.

Policing the American Frontier

Although gangs certainly existed in the Wild West, some notorious criminals actually came from the ranks of the police. The Dalton Gang, for example, comprised of lawmen who became train robbers in the Indian Territory.[77] Other criminals known for their daring exploits along the western frontier included the Younger Brothers, who rode with Jesse James, who made himself a spokesman for the renewal of the Confederate cause during the bitter decade that followed Appomattox.[78]

The ultimate responsibility for maintaining order in the Old West belonged to men armed not with guns but with gavels—that is, the judges. Of these powerful judges, Roy Bean of Texas is perhaps the best known.[79] A corpulent, bull-voiced saloonkeeper and gambler, Bean tried cases between deals of poker, and he regularly recessed trials to sell liquor to counsel, jury, and defendant.[80] Even before Bean became an official justice of the peace, the Texas Rangers brought him suspects for his court. Bean knew a smattering of law and ignored any statutes he personally disliked. While he never held a higher position than justice of the peace in a desert hamlet, he grandly titled himself, with some accuracy, as the "law west of the Pecos"—the river that ran 20 miles east of his saloon.[81]

Law enforcement rode into Dodge City with the advent of respectable officers as W. B. "Bat" Masterson, Ed Masterson, Wyatt Earp, Bill Tilghman, H. B. "Ham" Bell, and Charlie Bassett. The city passed an ordinance stating that guns could not be worn or carried north of the "deadline"—that is, the railroad tracks.[82] Dodge City acquired its reputation for being tough on crime through both the actions of its lawmen and the growth of Boot Hill Cemetery, located just outside the city limits.

Marshal Bass Reeves was a heroic black lawman whom thieves and outlaws, including the Dalton Gang in the Indian Territory, could not kill. Over time, however, he was practically eliminated by scholars of frontier history.[83] Nevertheless, Reeves's capture of many outlaws throughout the Oklahoma Territory is well documented, and some suggest that his "law'en up" activity was the basis for the legendary Lone Ranger.

Elfego Baca, another obscure "lawman," was born just before the end of the Civil War. He appointed himself deputy sheriff in Socorro County, New Mexico.[84] Baca's

goal was to be a top-notch peace officer: He wanted, he said, "the outlaws to hear my steps a block away."[85] In one famous encounter, Baca arrested one of a group of cowboys, resulting in a standoff with 80 outlaws. After 36 hours, the attack ended and Baca walked out of the house unharmed. At the age of 29, Baca became a lawyer. Later he also became a Deputy United States Marshal, an assistant district attorney, and the mayor of Socorro; in 1919, he was elected Sheriff of Socorro County.[86]

The Texas Rangers

The Texas Rangers are the oldest law enforcement organization in the United States with statewide jurisdiction. They played an effective role in providing crime control throughout the early troubled years of Texas, and later served as a model for many other statewide enforcement agencies. Over its long history, the Ranger Service has changed its organization and policies to satisfy varying conditions, different demands for service, and changing state administrations.[87]

Stephen F. Austin, known as the "Father of Texas," first brought 300 families to the then-Spanish province of Texas in 1821. Once there, he organized the settlers to protect and keep order in their new communities. After a period of imprisonment in Mexico, Austin returned to Texas in 1835, where he organized a group of lawmen to "range" and guard the frontier. These Rangers were paid $1.25 per day for "pay, rations, clothing, and horse service," and they enlisted for one year at a time.

After the Texas Revolution of 1835–1836 and until 1840, Rangers were used principally for protection against Native Americans, participating in the Council House Fight in San Antonio, the raid on Linnville, and the Battle of Plum Creek, among other skirmishes. Rangers aided their Southern allies in slave patrols; indeed, their active involvement in this initiative provided a source of solidity for the Rangers. The darkest period in the history of the Rangers came during Reconstruction (1865–1873), when they were ordered to enforce the unpopular carpetbagger laws after the state police fell into disrepute among the war-weary citizens of Texas. Eventually, the Rangers became involved in "detective" work, necessitated by a new group of violators known as fence cutters who perpetrated isolated cases of horse and cattle theft.

On August 10, 1935, the Texas Legislature created the Texas Department of Public Safety. Both the Texas Rangers and the Texas Highway Patrol became members of this agency, with statewide law enforcement jurisdiction.

■ Police in Twentieth-Century America

George L. Kelling and Mark H. Moore divided U.S. police experiences in the twentieth century into three eras: the **political era**, the **reform era**, and the **community-based policing era**. (Chapter 3 describes the fourth era of policing, which includes the twenty-first century.)

The Political Era (1840–1930)

When policing was first developed in the United States, police chiefs were politically appointed by local politicians; in return, many of those police chiefs helped the individuals who appointed them. This was referred to as a patronage system. Police chiefs had little authority over officers, so consequently some officers became involved in inefficient and sometimes corrupt conduct. Foot patrol was the most common police initiative, and the legitimate authority of arrest, search, and seizure was vested in officers. All too frequently, officers' poor decisions, their lack of training, and awkward circumstances led to illegitimate use of force, false arrests, unwarranted searches, and unlawful seizure.

This "spoils era" of politicization promoted reciprocal political favoritism more often than it achieved crime control. The circumvention of Prohibition, fueled by organized crime, is one example of how abuses of political authority were produced by the link between politics and police.

In 1931, President Hoover appointed General George Wickersham to head the National Commission on Law Observance and Enforcement, which was charged with looking into the problems associated with politicization of the policing function. The **Wickersham Commission** blamed both police and politicians for the corruption and incompetence, and reform became the slogan of the day.

The Reform Era (1930–1980)

The Wickersham Commission found that the rise of crime occurred largely because police would not or could not provide effective police services owing to the link between officers and local politicians. Any police work that did not relate directly to "crook catchers" was considered suspect.

August Vollmer and **O. W. Wilson** are credited with advancing the reform movement that ended the political era by calling for a drastic change in the way police agencies provided police services. Vollmer, who was the police chief in Berkeley, California, started the first college education and training program for police at the University of California at Berkeley. Known as the "father of American policing," Vollmer called for police to take the following steps:

- Enhance their accountability for their actions and behaviors related to the public
- Ensure impartiality in law enforcement activities by not favoring (or disfavoring) any group regardless of its members' social or political position
- Increase the honesty of all police officers by attempting to rid the police service of corruption and political influence

Wilson, who was Superintendent of the Chicago Police Department, established his own reform efforts, moving his leadership into national prominence. He minimized the political connections of his officers and tried to professionalize the police organization by adopting the scientific (classical) theory of administration advocated by Frederick W. Taylor. Two key assumptions of Taylor's perspective were particularly pertinent to Wilson's reform efforts:

- Workers (officers) are inherently uninterested in work and, if left to their own devices, will avoid it.
- Workers (officers) have little or no interest in the substance of their work, and the sole interest in common between workers and management is found in economic incentives for workers.

These two principles produced a division of labor and unity of control within policing. A bureaucratic centralized hierarchy of command was established, and the relationship between police and the public took on a professional distance: The public was to be the passive recipient of a professional crime fighter role. Patrol cars, radios, sophisticated initiatives, and federal support helped create the "thin blue line" separating the police from the public.

Despite the gains from increasing the professionalism of police, other problems surfaced for police during the reform era, as Kelling and Moore note:

- Police failed community expectations to control crime.
- Fear of crime continued to enhance social disorder.
- Impartial treatment was not experienced by all citizens.
- Civil rights and antiwar movements challenged police practice and policy.
- "Crime fighting" was an activity that was practiced less among police officers than expected.
- Reform ideology occurred largely at police management levels and failed to rally line officers.
- Financial support for police was reduced by a dissatisfied public.
- The rise of private security took some policing functions away from the public police.

American culture also moved faster than police organizations could respond to the changes. Events in the 1960s and 1970s reshaped American expectations about reform whose centerpiece was police professionalism:[88]

- The rule of criminal procedure reduced police power.
- The "counterculture" movement became characterized by recreational drug use and sexual promiscuity.
- Unarmed college students were killed by members of the National Guard at Kent State University.
- Richard M. Nixon resigned as President of the United States.
- Civil rights and Vietnam War protests encouraged riots at cities and college campuses.
- Technological advances changed police initiatives and enhanced criminal activity.
- Media coverage of violent police practices influenced public sentiment and encouraged riots.

Taken separately, each of these events might seem insignificant. Collectively, however, they were seen as part of the same pattern. The sometimes sensational media coverage of these events led to new problems for police in the 1960s and 1970s:

- Public distrust of police
- Lack of financial support for police
- Obsolete police practices
- Inadequate officer hiring, training, and endurance practices

In 1963, two reports were issued by federal government task forces dealing with the policing issues facing the United States: the National Advisory Commission on Civil Disorders and the National Commission on the Causes and Prevention of Violence. Although the inherent focus of each of these two reports was somewhat different, both yielded very similar findings and recommendations with respect to the police. Four years later, in 1967, the **President's Commission on Law Enforcement and Administration of Justice** released a multi-volume report addressing the wide spectrum of issues facing police agencies. One commonality between the three reports was their suggestion that police required greater empathy with the community to solve community problems and bring the police closer to their mission.

The Community Problem-Solving Era (1980–1995)

In the latter part of the twentieth century, police were encouraged to interact more intimately with the community. Foot, bike, and water patrols were popular police initiatives during this era, and citizens voted to increase taxes to pay for these quality-of-life measures, which were intended to reduce the fear of crime. For most constituents, the higher visibility of officers produced a feeling of greater safety. The greater openness between police and public also enhanced trust levels, resulting in higher arrest rates because the public more willingly provided information to investigators.[89] These findings led to new opportunities that eventually shaped the new breed of community police initiatives and problem-solving strategies.

A Critical Review of Kelling and Moore's Model

Among others, Hubert Williams and Patrick Murphy take issue with Kelling and Moore's historical perspective toward American policing.[90]

Missing from Kelling and Moore's Perspective

Williams and Murphy suggest that Kelling and Moore's interpretation is inadequate because it fails to account for how slavery, segregation, discrimination, and racism affected the development of police strategies, and how those strategies affected police response to minority communities (as is consistent with the historical accounts of Southern policing provided earlier in this chapter). They argue that legal order may have actually encouraged slavery, segregation, and discrimination, allowing it to thrive. The task of the police was to keep minorities under control, and officers had little incentive to protect them as individuals from crime within their communities. As a consequence, members of minority groups benefited less than others from advances made by the police.

Furthermore, Williams and Murphy suggest that Kelling and Moore neglected to consider the effects of the Reconstruction period upon minorities as part of their model of the political era. This period was marked by both legal and political powerlessness for minority communities in both the North and South, and the police response did little to empower those communities in improving their public safety. According to Williams and Murphy, for these reasons lower-class communities might also have been over-represented in police response accounts.

Williams and Murphy argue that Kelling and Moore's analysis of policing was largely based on police departments in the northeastern United States. For instance, Southern slave patrols and racially biased laws were never mentioned in the latter's model, nor was there any mention of the racial and minority police practices in the Northern and New England regions of the country. Most of Kelling and Moore's ideas emphasize the importance of early American history in setting the tone for the rest of U.S. history and reflect an Anglo, white, twentieth-century bias toward urban white conditions such as those found in Boston, Chicago, Detroit, and New York City. The many other police departments and enforcement agencies across the nation receive little, if any, mention in these authors' model. Policing was never really isolated, so the

U.S. system could not have developed within a vacuum or in isolation by itself;[91] many agencies undoubtedly picked up ideas from other agencies.

Missing from Williams and Murphy's Criticism

Other critics add that the reform era, per se, was in effect a time when professional policing became a "true policing strategy,"[92] It developed in the 1920s in response to the failures of the ineffective, corrupt, and brutal policing systems during turn-of-the century America, as mentioned earlier. Kelling and Moore claim that this change represented a turn toward professionalism with a focus on adapting organizational styles employed by corporate America and outside the usual practices of policing.[93] It is argued that true "professions" have standards for hiring, training, disciplinary actions, and promotions, and that they have clear goals and objectives. On the path toward professionalism, policing developed such standards and minimized the role of political cronyism.[94] (Chapter 15 explores professionalism in more depth.) Yet, some wonder if policing will ever truly change.

> Suggested consequences of the trend toward professionalism of traditional police departments include being pushed out of the business through police privatization, until street officers become nothing more than ghetto fighters in urban jurisdictions across the United States.

Privatization of policing (detailed in Chapter 4) is another major concern for police policymakers. Likewise, another trend of concern is the practice of police agencies hiring out to other communities, such as has occurred with the Harris (Houston, Texas) County Precinct 4 Constable's Office (see Chapter 5 for details). The merging of police organizations from city and county agencies has also created concern; such mergers have already taken place between the Chicago Police Department and Cook County Sheriff's Office; the Jacksonville Police Department and Jacksonville County Sheriff's Office (see Chapter 5), and the Charlotte Police Department and Mecklenburg County Sheriff's Office, to name a few. How many other agencies will merge in the near future?

Another source of criticism associated with community policing is that it is seen as a justification for *saturation policing*—that is, an expansion of social control into the lives of populations that is deemed problematic.[95] Is it just another form of institutionalized racism? Or is Neil Websdale correct when he argues, "I identify community policing at the heart of postindustrial apartheid. This latest form of policing constitutes just one element of a wider strategy of controlling the poor"?[96] According to this view, community policing serves as the "lead filter" in the "criminal justice juggernaut," which itself serves to control the urban poor and minorities.[97] This thought is consistent with the findings from a study of the Nashville (Tennessee) Police Department (NPD), in which researchers concluded that the NPD's community policing program targeted poor inner-city people and was simply used as another police justification to make arrests.[98]

Despite the skepticism, many excellent results from community-based programs have been documented, including the outcomes of the child development community police program in Nashville, which was conducted in concert with Yale University and the New Haven Connecticut Police Department. This program went to the heart of juvenile crime and dealt specifically with many of the problems experienced by youths, such as poverty and health issues. According to researchers, it succeeded: It lowered truancy rates, raised participants' grades, and improved their quality-of-life experiences.[99]

***RIPPED* from the Headlines**

Law Enforcement Officers Wanted: Good People for a Thankless Job

Police officers work in situations that most people never experience. They provide 24-hour-a-day, 7-day-a-week protection for their communities. They may work all night, then wait in court all day. Or, they may work all night, when most people sleep, then come home to their families getting ready to start their day. Or, they may work all night, trying to stay awake when things are calm, yet be alert to suddenly respond to a robbery or homicide and handle it properly. Or, they may work all night, aware of the resulting fatigue and poor health that comes from unnatural sleep patterns. Or, worst of all, they may work all night knowing that their families never may see them alive again. Oh, yes, many people work a night shift, but do they face the same situations as police officers?

At times, an officer may be physically tired from trying to subdue a person who will not submit to arrest, from chasing a suspect on foot, from swimming in a cold polluted river to rescue citizens from drowning after their car crashed, from leaning over a ledge on a building high above the ground holding onto a person who was trying to jump off, or from any number of other physical situations that might occur and that most people never experience. These represent only some of the situations that police officers find themselves in at any time. There are many others. How about sitting down to eat lunch, but immediately having to leave it to respond to an urgent call? How about working and not knowing what danger may occur on the next call? How about getting shot at, seeing the bright glint of a knife blade in a subject's hand, being attacked by a crazed drug addict, or facing an attacker who is mentally ill? And, what about that "loose nut" behind the steering wheel of a car? Who's going to stop him? If you're a police officer, it's YOU! You who joined the police department because you cared about other people.

Source: Henry P. Henson and Kevin L. Livingston, "Law Enforcement Officers Wanted: Good People for a Thankless Job," *FBI Law Enforcement Bulletin* 72, no. 4 (2003): 17–18.

Summary

Social order requires a method of social control to enforce laws, which emerged as centralized governments' response to the need to stabilize society. As a systematic way to moderate or suppress conflict, primitive societies relied upon face-to-face means of controlling behavior. Later, more advanced societies turned to written methods of control.

Police performance in ancient Mesopotamia and Egypt was designed largely for the protection and security of ruling regimes. Ancient Athens and Rome utilized the skills of public slaves to control conflict, quell public disturbances, and extinguish fires.

Violence was pervasive in medieval Europe and England. To deal with this escalating problem, centralized regimes moved crime and punishment into the hands of community members through such strategies as frankpledges and tything. Eventually, centralized control of policing became a reality in 1829, when the London Metropolitan Police was established.

Colonists departed England to escape a tyrannical government that curtailed individual freedoms. They brought their skepticism about centralized government with them to the New World, eventually producing a constitution that restrained arbitrary government intervention.

Early American police initiatives sought largely to quell urban disturbances and to control slavery. At the frontier, the ultimate responsibility for order lay with judges, many of whom were untrained and provided makeshift justice.

In sum, law enforcement activity in early America can be described as decentralized, geographically dispersed, and idiosyncratic. Calling the police was considered the last resort. When police intervened in such cases, they often had to use force. At the same time, police provided a variety of social services not related to crime control, though they later relinquished this role.

The political, reform, and community problem-solving eras can be distinguished as means to explain U.S. policing experiences from the 1870s to the latter part of the twentieth century. Critics of this model suggest that it is an inadequate explanation for the evolution of American policing because it fails to take into account the lingering effects of slave patrols and policing in other regions of the country.

Key Terms and Key People

Acephelous societies
Code of Hammurabi
Community-based policing era
Dred Scott decision
Frankpledges
London Metropolitan (Metro) Police
Methods of social control
Political era
President's Commission on Law Enforcement and Administration of Justice
Real natural law
Reform era
Stono Rebellion
Supreme Court Chief Justice Warren Berger
Sir Robert Peel
August Vollmer
Wickersham Commission
O.W. Wilson

Discussion Questions

1. Explain the implications of police history being written by the privileged. In what way could the "privileged" alter actual events?
2. Characterize the emergence of law and its objectives.
3. Describe the rationale that justifies methods of social control.
4. Characterize police performance in ancient Mesopotamia and Egypt. In what way might some of their tasks mirror present-day policing initiatives?
5. Describe police activities in ancient Athens and Rome. What similar conclusions were drawn by the ancient Romans and by the nineteenth- and twentieth-century public when they viewed enforcers of the law?
6. Characterize police and centralization strategies in early England.

7. Identify the methods of social control used by English kings. In what way might a return to any of those practices lead to quality protection in the twenty-first century?
8. Characterize the primary components of Peel's police principles. Explain which two principles have contributed the most to the quality of police services in the twenty-first century.
9. Describe policing in colonial America and the public's reaction to those law enforcement efforts.
10. Compare and contrast policing in early urban America and the South. In what way did both styles of policing seek to advance the best interests of a specific group of individuals?
11. Describe policing along the American frontier, including the founding and mission of the Texas Rangers.
12. Explain the three most recent eras of American policing, and identify the importance of each in shaping modern police strategies.
13. Characterize the criticism of Kelling and Moore's model of the three most recent eras of policing. In what way do you agree and disagree with the critics concerning these three eras?

Notes

1. Randall G. Shelden, *Controlling the Dangerous Classes: A Critical Introduction to the History of Criminal Justice*, 2nd ed. (Boston: Allyn and Bacon, 2007), 1. Actually this comment was first made by Howard Zinn, as Shelden points out.
2. Shelden, *Controlling the Dangerous Classes*, 2.
3. Peter L. Berger, *Invitation to Sociology: A Humanistic Perceptive* (New York: Anchor Press, 1963), 14.
4. Raymond J. Michalowski, *Order, Law, and Crime: An Introduction to Criminology* (New York: McGraw Hill, 1985), 45–47.
5. Roger G. Dunham and Geoffrey P. Alpert, eds., *Critical Issues in Policing: Contemporary Readings* (Prospect Heights, IL: Waveland, 2001), 1.
6. Dennis J. Stevens, *Case Studies in Applied Community Policing* (Upper Saddle River, NJ: Prentice Hall, 2003).
7. Gary Marx, "Technology and Social Control: The Search for the Illusive Silver Bullet," *International Encyclopedia of the Social and Behavioral Sciences* (2001), http://web.mit.edu/gtmarx/www/techandsocial.html (accessed July 9, 2006).
8. American Bar Association, "Chief Justice Warren Berger Address to the ABA" (1981), http://www.abanet.org/ (accessed June 30, 2007).
9. J. Michael Olivero, Cyril D. Robinson, and Richard Scaglion, *Police in Contradiction: The Evolution of the Police Function in Society* (New York: Greenwood Press, 1991). Also see Marvin Harris, *Culture, People, Nature: An Introduction to General Anthropology* (New York: Prentice Hall, 1975), 355–356.
10. Lawrence M. Friedman, *The Roots of Justice: Crime and Punishment in Alameda County, California, 1870–1910* (Chapel Hill, NC: University of North Carolina Press, 1981).
11. Israel Drapkin, *Crime and Punishment in the Ancient World* (Lexington, MA: Lexington Books, 1989) 8.
12. David Hunter and Philip Whitten, *The Study of Anthropology* (New York: Harper & Row, 1976), 62.
13. An detailed report can be found in William A. Haviland, *Cultural Anthropology* (New York: Harcourt, 2000), 357–387.

14. Emile Durkheim, *Durkheim: The Division of Labor in Society*, trans W. D. Halls (New York: Free Press, 1933/1984), 58–65.
15. Kathy Warnes, "Quaker Mary Dyer: Twice Under the Hangman's Noose," in *Famous American Crimes and Trials: Volume 1: 1607–1859*, ed. Frankie Y. Bailey and Steven Chermak (Westport, CT: Praeger, 2006), 1–22.
16. Sue Titus Reid, *Criminal Law* (New York: McGraw Hill, 2001), 3.
17. Michalowski, *Order, Law, and Crime*, 32.
18. A few paragraphs can provide a thumbnail view on the emergence of law. For a full account, see Sue Titus Reid, *Criminal Law*, 1–5, and Michalowski, *Order, Law, and Crime*, 22–38.
19. Titus Reid, *Criminal Law*.
20. Emile Durkheim, *The Rules of Sociological Method*, trans. S. A. Solovay and J. H. Mueller (Glencoe, IL: Free Press, 1958), 67. This work was originally completed in 1895.
21. Kai T. Erikson, *Wayward Puritans* (New York: John Wiley, 1966).
22. Gabi Löschper, "Crime and Social Control as Fields of Qualitative Research in the Social Sciences," *Forum Qualitative Sozialforschung* (*Forum: Qualitative Social Research*) 1, no. 1 (2000), http://www.qualitative-research.net/fqs-texte/1-00/1-00loeschper-e.htm (accessed December 14, 2007).
23. Thomson Corporation, "Espionage and Intelligence Foundations" (2006), http://www.espionageinfo.com/Ep-Fo/Espionage-and-Intelligence-Early-Historical-Foundations.html (accessed July 10, 2006).
24. Anita Stratos, "Feature Story: The Evolution of Welfare" (2003), http://www.touregypt.net/featurestories/war2.htm (accessed July 10, 2006).
25. "Nubian Chronology" (2004), http://touregypt.net/historicalessays/nubiac1.htm (accessed July 10, 2006).
26. Tom Watkins, "Policing Rome: In Fact and Fiction" (1999), http://www.stockton.edu/~roman/fiction/eslaw1.htm (accessed July 10, 2006).
27. T. J. Cornell, "Police," in *Oxford Classical Dictionary* (New York: Oxford University Press, 1996).
28. Seba, "Ancient Roman Police" (2004), http://nefer-seba.net/essays/roman-police.php (accessed July 10, 2006).
29. Pieter Spierenburg, *The Broken Spell: A Cultural and Anthropological History of Preindustrial Europe* (New Brunswick, NJ: Rutgers University Press, 1991), 196.
30. Dennis J. Stevens, *Community Policing in the 21st Century* (Boston: Allyn and Bacon, 2003), 14.
31. Ronald Bums, "Amber Plan: Hue and Cry" (September 2001), http://www.community-policing.org/publications/comlinks/cl16/cl16_burns.htm (accessed July 15, 2006).
32. Magna Carta, "Roger of Wendover" (n.d.), http://www.britannia.com/history/docs/runnymede.html (accessed November 10, 2007).
33. Curtis R. Blakely and Vic W. Bumphus, "American Criminal Justice Philosophy: What's Old—What's New?", in *Crime and Justice in America*, ed. Wilson R. Palacios, Paul F. Cromwell, and Roger G. Dunham (Upper Saddle River, NJ: Prentice Hall, 2002), 16–24.
34. Craig D. Uchida, "The Development of the American Police," in *Crime and Justice in America*, ed. Wilson R. Palacios, Paul F. Cromwell, and Roger G. Dunham (Upper Saddle River, NJ: Prentice Hall, 2002), 97–101.
35. "Frankpledge," *Online Encyclopedia* (2004), http://encyclopedia.jrank.org/FRA_GAE/FRANKPLEDGE_Lat_francum_plegium.html (accessed July 15, 2006).
36. Uchida, "The Development of the American Police."
37. Peter Manning, "The Police: Mandate, Strategies, and Appearances," in *The Police and Society*, ed. Victor E. Kappeler (Prospect Heights, IL: Waveland Press, 1995), 97–126.
38. For instance, there were 3,780 executions from 1509 to 1547, or an average of 99 annually. During the 1548–1553 period, there were 3,360 executions, or an average of 672 per year. The execution rate then declined: In 1625–1649, some 2,160 people were put to

death (90 people per year); in 1650–1658, there were 990 executions. More than 200 crimes were subject to punishment by execution in the seventeenth century. See Leon Radzinowicz, *A History of English Criminal Law, Vol. 1* (London: Stevens and Sons, 1948).

39. Home Office, "Violent Crime in England and Wales" (2004), http://www.homeoffice.gov.uk/rds/pdfs04/rdsolr1804.pdf (accessed July 15, 2006).

40. Devon and Cornwall (U.K.) Police, "The Victorian Policemen: Demise of Old Policing System" (2006), http://www.devon-cornwall.police.uk/v3/about/history/vicpolice/demise.htm (accessed July 15, 2006).

41. W. L. Melville Lee, *History of Police in England* (London: Methuen, 1901), Chapter 12.

42. Devon and Cornwall (U.K.) Police, "The Victorian Policemen: Bow Street Runners" (2006), http://www.devon-cornwall.police.uk/v3/about/history/vicpolice/bow.htm (accessed July 15, 2006).

43. For a closer look at these dynamics, see Blakely and Bumphus, "American Criminal Justice Philosophy."

44. Craig D. Uchida, "The Development of the American Police: An Historical Overview," in *Critical Issues in Policing*, ed. R.G. Dunham & G. P. Alpert (Prospect Heights, IL: Waveland Press, 1993).

45. Cynthia Morris and Bryan Vila, *The Role of Police in American Society: A Documentary History* (New York: Greenwood Press, 1999), 1–4.

46. Elizabeth Pleck, "Criminal Approaches to Family Violence," in *Violence and Society: A Reader*, ed. Matthew Silberman (Upper Saddle River, NJ: Prentice Hall, 2003), 162.

47. John Mack Faragher, Mari Jo Buhle, Daniel Czitrom, and Susan H. Armitage, *Out of Many: A History of the American People* (Upper Saddle River, NJ: Prentice Hall, 2005), 361.

48. Stevens, *Community Policing in the 21st Century*, 3–8; Samuel Walker, *Sense and Nonsense about Crime and Drugs*, 4th ed. (Belmont, CA: Wadsworth, 2001), 11–12; Samuel Walker and Charles E. Katz, *The Police in America: An Introduction*, 5th ed. (New York: McGraw Hill, 2008).

49. Jack Tager, *Boston Riots: Three Centuries of Social Violence* (Boston: Northeastern University Press, 2001).

50. Craig D. Uchida, "The Development of the American Police: An Historical Overview," in *Critical Issues in Policing*, 4th ed., ed. R. G. Dunham & G. P. Alpert (Prospect Heights, IL: Waveland Press, 2001).

51. Blakely and Bumphus, "American Criminal Justice Philosophy."

52. Uchida, "The Development of the American Police: An Historical Overview" (2001).

53. "Our Police Protectors: History of New York Police" (2006), Chapter 3, Part 2, http://www.usgennet.org/usa/ny/state/police/ch3pt2.html (accessed July 22, 2006).

54. David L. Carter and Louis A. Radelet, *The Police and the Community*, 7th ed. (Upper Saddle River, NJ: Prentice Hall, 2001).

55. Eric H. Monkkonen, "History of Urban Police," *Crime and Justice* 15 (1992), 547–580.

56. Douglas Harper, "French Slavery," *The Sciolist* (May 19, 2005), http://etymonline.com/columns/frenchslavery.htm (accessed July 23, 2006).

57. Harper, "French Slavery."

58. David E. Barlow and Melissa Hickman Barlow, *Police in a Multicultural Society: An American Story* (Prospect Heights, IL: Waveland Press, 2001), 19–21.

59. Lee Lawrence, "Chronicling Black Lives in Colonial New England," *Christian Science Monitor* (October 29, 1997), http://www.csmonitor.com/durable/1997/10/29/feat/feat.1.html (accessed July 23, 2006).

60. Stanley L. Engerman, Richard Sutch, and Gavin Wright, "Slavery," in *Historical Statistics of the United States, Millennial Edition*, ed. Susan B. Carter, Scott S. Gartner, Michael Haines, Alan Olmstead, Richard Sutch, and Gavin Wright (New York: Cambridge University Press, 2004).

61. Barlow and Hickman Barlow, *Police in a Multicultural Society*, 20.

62. Barlow and Hickman Barlow, *Police in a Multicultural Society*.

63. Africans shaped South Carolina more than any other group through such things as their knowledge of cattle grazing, rice planting, and clearing.

64. Dennis C. Rousey, *Policing the Southern City: New Orleans, 1805–1889* (Baton Rouge: LSU Press, 1996).

65. Rousey, *Policing the Southern City*, 16.

66. Rousey, *Policing the Southern City*, 135–136.

67. Confidential personal communication between the writer and several officers employed at the New Orleans, Houston, Charleston, and Savannah police departments in spring and summer 2007.

68. The case before the court was that of *Dred Scott v. Sanford*. Dred Scott, a slave who had lived in the free state of Illinois and the free territory of Wisconsin before moving back to the slave state of Missouri, had appealed to the Supreme Court in hopes of being granted his freedom.

> The plaintiff [Dred Scott] . . . was, with his wife and children, held as slaves by the defendant [Sanford], in the State of Missouri; and he brought this action in the Circuit Court of the United States for [Missouri], to assert the title of himself and his family to freedom. The declaration is . . . that he and the defendant are citizens of different States; that . . . he is a citizen of Missouri, and the defendant a citizen of New York.
>
> The question is simply this: Can a negro, whose ancestors were imported into this country, and sold as slaves, become a member of the political community formed and brought into existence by the Constitution of the United States, and as such become entitled to all the rights, and privileges, and immunities, guarantied by that instrument to the citizen? One of which rights is the privilege of suing in a court of the United States in the cases specified in the Constitution. . . .

Taney—a staunch supporter of slavery who was intent on protecting Southerners from Northern aggression—wrote in the Court's majority opinion that, because Scott was black, he was not a citizen and therefore had no right to sue. The framers of the Constitution, he wrote, believed that blacks "had no rights which the white man was bound to respect; and that the negro might justly and lawfully be reduced to slavery for his benefit. He was bought and sold and treated as an ordinary article of merchandise and traffic, whenever profit could be made by it."

Referring to the language in the Declaration of Independence that includes the phrase "all men are created equal," Taney reasoned that "it is too clear for dispute, that the enslaved African race were not intended to be included, and formed no part of the people who framed and adopted this declaration. . ."

Abolitionists were incensed. Although disappointed, Frederick Douglass found a bright side to the decision and announced, "[My] hopes were never brighter than now." For Douglass, the decision would bring slavery to the attention of the nation and represented a step toward slavery's ultimate destruction.

69. What follows is guided by Stevens, *Community Policing in the 21st Century*.

70. David Nasaw, *Children of the City* (New York: Oxford University Press, 1986).

71. John C. Schneider, *Detroit and the Problem of Order, 1830–1880: A Geography of Crime, Riot, and Policing* (Omaha, NE: University of Nebraska Press, 1980), 121.

72. Schneider, *Detroit*.

73. Shelden, *Controlling the Dangerous Classes*.

74. "New York Street Arabs," *Harper's Weekly* (1868), http://immigrants.harpweek.com/ChineseAmericans/Items/Item015L.htm (accessed July 1, 2007).

75. "New York Street Arabs." Also see Shelden, *Controlling the Dangerous Classes.*
76. Shelden, *Controlling the Dangerous Classes.*
77. Robert B. Smith, *Daltons: The Raid on Coffeyville, Kansas* (Oklahoma: University of Oklahoma Press, 1999), 149–152.
78. T. J. Stiles, *Jesse James: Last Rebel of the Civil War* (New York: Knopf, 2002).
79. John R. Wunder, *Inferior Courts, Superior Justice: A History of the Justices of the Peace on the Northwest Frontier, 1853–1889* (Cincinnati: Greenwood, 1979).
80. C. L. Sonnichsen, *Roy Bean: Law West of the Pecos* (New York: Kessinger, 2005).
81. Sonnichsen, *Roy Bean.* Also see Wunder, *Inferior Courts.*
82. Sonnichsen, *Roy Bean.* Also see Wikipedia, the Free Encyclopedia, "The Wild West" (2006), http://en.wikipedia.org/wiki/American_Old_West (accessed July 17, 2006).
83. Art T. Burton, *Black Gun, Silver Star: The Life and Legend of Frontier Marshal Bass Reeves* (Omaha: University of Nebraska Press, 2006).
84. David Santana, Melissa Ann Villela, Rosalynn Torres, and Michael Telles, "Elfefo Baca Lived More Than Nine Lives," El Pasco Community College (2005), http://www.epcc.edu/nwlibrary/borderlands/22_elfego_baca.htm (accessed July 17, 2006).
85. Earl Clark, "Shootout in Burke Canyon," *American Heritage* 22, no. 5 (1971).
86. Elfego Baca Golf Shoot, "Elfego Baca in History" (2006), http://www.hiltonopen.com/Elfego0402.htm (accessed July 17, 2006).
87. Texas Department of Public Safety (2006), http://www.txdps.state.tx.us/director_staff/texas_rangers/ (accessed July 17, 2006).
88. David L. Carter and Louis A. Radelet, *The Police and the Community,* 7th ed. (Upper Saddle River, NJ: Prentice Hall, 2001). Chapter 1 offers a look at some of these thoughts.
89. George L. Kelling and Mark H. Moore, *Evolving Strategy of Policing, No. 4* (Washington, DC: U.S. Department of Justice, 1988). http://www.hks.harvard.edu/criminaljustice/research/community_policing.htm (accessed June 19, 2008).
90. Hubert Williams and Patrick Murphy, *The Evolving Strategy of Police: A Minority View,* Perspective on Policing (Washington, DC: U.S. Department of Justice, 1990). http://www.hks.harvard.edu/criminaljustice/research/community_policing.htm (accessed June 19, 2008).
91. Willard M. Oliver, *Community-Oriented Policing: A Systemic Approach* (Upper Saddle River, NJ: Prentice Hall, 2008), 5.
92. Adam Dobrin, "Professional and Community Oriented Policing: The Mayberry Model," *Journal of Criminal Justice and Popular Culture* 13, no. 1 (2006): 19–28, http://www.albany.edu/scj/jcjpc/vol13is1/Dobrin.pdf (accessed July 22, 2006).
93. O. W. Wilson and R. C. McLaren, *Police Administration,* 4th ed. (New York: McGraw-Hill, 1977).
94. J. J. Fyfe, "Good Policing," in *Critical Issues in Policing,* 4th ed., R. G. Dunham and G. P. Alpert (Prospect Heights, IL: Waveland Press, 2001).
95. Thomas B. Priest and Debra Brown Carter, "Community-Oriented Policing: Assessing a Police Saturation Operation," in *Policing and Community Partnerships,* ed. Dennis J. Stevens (Upper Saddle River, NJ: Prentice Hall, 2005), 111–124.
96. Neil Websdale, *Policing the Poor: From Slave Plantation to Public Housing* (Boston: Northeastern Press, 2001), 6.
97. Craig Hemmens, Book Review: *Policing the Poor,* 11, no. 11 (2001): 503–505, http://www.bsos.umd.edu/gvpt/lpbr/subpages/reviews/websdale.htm (accessed November 11, 2007).
98. Stevens, *Case Studies,* 61–88.
99. Stevens, *Case Studies,* 81–83.

DAVE
MAT A
BABY!

Broken Windows, Fear, and Community Policing

"What we call reality is an agreement that people have arrived at to make life more livable."

—Louise Nevelson

LEARNING OBJECTIVES

Once reading this chapter, you will be able to

- Describe community relations policing
- Characterize the importance of reducing fear of crime
- Identify the historical influences leading to community relations initiatives
- Characterize the thoughts of the New York City subway commuter about crime
- Identify the primary components of the broken windows metaphor
- Describe the criticism of the broken windows perspective
- Characterize the primary components of community policing
- Identify the components of the scanning, analysis, response, and assessment (SARA) model
- Describe the two case studies in community policing
- Identify the neighborhood problems articulated by community members
- Identify the differences between community resolutions and police resolutions to fix neighborhood problems
- Describe the problems faced by community police officers and detectives
- Describe the lessons learned in Madison, Wisconsin, about the drug trade

CASE STUDY: YOU ARE THE POLICE OFFICER

Ever since you graduated from the police academy's aggressive technical training program two years ago, you have wondered whether the rumors were true about police policy moving toward prevention and community relations. You wonder whether police officers assigned to community relations units would be babysitters, because technical initiatives are less of a priority in a **community policing** initiative.

For two weeks, you attended in-service training about community policing strategies and thought about how those strategies could aid the thousands of residents and workers in your district. Now you are scheduled to attend your first neighborhood meeting tonight at six o'clock.

During your training, the police academy instructors didn't encourage typical police intervention strategies; rather, they emphasized a preventive initiative using the SARA model. SARA has four components: scanning (looking and grouping community problems), analysis (the collection of information about offenders and victims), response (deciding which action to take to resolve the problem), and assessment (looking at what we accomplished). It was the "we" that came as the biggest surprise; it wasn't your sergeant and other officers. No, the "we" consisted of the community members who attended community policing meetings and the community officers who locked together on a problem-solving mission to deal with things in the neighborhood that could lead to crime—for example, kids hanging on street corners and abandoned cars on narrow streets.

At six o'clock, the meeting room in the back of a local restaurant is busy with community members anxiously finding a chair after getting a beverage supplied by the restaurant and a little plate of chips. Many of the community members are friendly with the captain, who stands in the back of the room along with the new community police officers, including you. You are a little surprised when a resident casually approaches the podium and speaks into the microphone: "Will everyone please sit down. The captain—you all know Captain Jones—will introduce the officers." It isn't the captain who takes charge, but someone who you recognize as a retail worker down the street. The captain waves at the crowd of maybe 100 people. He introduces the officers, asking them to make a comment about themselves. Finally it is your turn.

"Hi, I'm Officer Towanda Washington, and I'm proud to be here. I'm a graduate of Central Community College and have a criminal justice degree from Central State College. I've been an officer for three years and hope that I can serve our community," you emphasize, "up to and exceeding the expectations of my training and education."

Once the captain exits the room, all eyes shift to the three community officers. "Have a seat," the woman at the microphone said. "Relax, officers, you're among neighbors and friends. The first order of business is . . . ."

1. What evidence from the case study does or does not support the rumors that Officer Washington heard that the officers would become babysitters?
2. In what way could community relations priorities differ from the priorities of typical policing initiatives?
3. In what way could the community officers and members appropriately represent or not represent all their neighbors?

Introduction

Chapter 1 described the essential foundations about policing, and Chapter 2 clarified the historical accounts of policing up to the twentieth century. Chapter 3 explores the broken windows perspective in detail; this theory helps to justify **community relations**

initiatives. In a sense, the history lesson continues in Chapter 4, which describes the modern American police department and the privatization of policing.

In the latter part of the twentieth century, it was argued that policing had entered into a community relations era. Some say that this trend continues in the twenty-first century—but whether community relations, especially its most popular version of community policing, is more alive in rhetoric than reality is another issue. Louise Nevelson's reminder implies that when there is an agreement or a consensus that something exists, it does, regardless of whether that something is actually fact or fiction; individual behavior, in turn, mirrors those beliefs. Thus, if the **fear of crime** is pervasive among constituents, we can expect different behavior from those individuals than from those persons who are not fearful. Have you ever inadvertently driven through a rough part of town? You crank up your windows, double-check your locked doors, and call your mom to tell her you love her. Chapter 3 explores the police strategies that attempt not only to control crime, but also to reduce the fear of crime—a fear that drives community relations initiatives.

■ Overview of Community Relations Initiatives

Your first thought about community relations initiatives might be that community relations is a fashionable police strategy and in the future it will be the initiative of choice. Criminal justice publishers continue to produce community policing textbooks as though this community relations strategy is the benchmark of policing across the United States.[1] Some critics claim that community relations isn't new and revolutionary and complain that, because there are no contemporary "best" models of community relations, a lot is left to the imagination.

This chapter provides a description of community relations and emphasizes one of its most popular initiatives—community policing. Many variations of community relations policing programs exist:

- *Community policing* is ideally a department-wide philosophy characterized by a decentralized agency, empowered committee members, partnerships, problem solving, and proactive attempts toward order.
- In *neighborhood-oriented policing*, officers are assigned to a specific neighborhood and are expected to familiarize themselves with the area, work with area residents, and respond to all calls for service in that area during their shift.[2]
- *Problem-oriented policing* focuses on problem-solving initiatives.
- *Weed and Seed* is a federally initiated and monitored initiative that can assist and oversee local police jurisdictions in their quest to prevent, control, and reduce violent crime, drug abuse, and gang activity.[3] (See Chapter 12.)
- **Order-maintenance policy** provides for police intervention in incidents that do not involve actual criminal activities but often entail interpersonal conflicts or public nuisances (i.e., those occurring in public spaces).
- In *team policing*, officers are assigned responsibility for a certain geographic area; the team of police officers learns about the neighborhood, its people, and its problems, enabling them to provide long-term problem solving rather than simply a rapid response.[4]
- *Quality-of-life initiatives* are designed to improve quality-of-life experiences by enhancing liberty and freedom of choice among constituents in both public and

personal spaces through professional crime control strategies that will also reduce the fear of crime (see Chapter 4).

The common theme among community relations initiatives is the formation of community and police partnerships to achieve the mission of improving community relations. However, four thoughts should be considered before moving on.

First, a fundamental tenet of community relations is that the police acting through the criminal law should be perceived as only one means of policing, rather than the sole goal of policing.[5] That is, in every instance, the police should chose the lawful and ethical means that yield the most efficient and effective achievement of the desired end. Law enforcement is not necessarily the only—or even the principal—technique of policing; thus community relations may be a valuable alternative.

Second, many community members and observers are suspicious of community relations programs, seeing them as devices to control urban populations and, in some cases, to further racial and economic prejudice.[6]

Third, police reluctance to embrace community members as a resource to solve crime originates from universal, unwarranted stereotypes about inner-city residents.[7] Community residents, the thinking goes, are incapable, uneducated, and apathetic.

Fourth, the community relations movement grew out of public dissatisfaction and police corruption, with police departments being perceived as, at an extreme, an "occupying army" rather than as public servants.[8]

Mission of Community Relations Initiatives

The objective of community relations is to identify and resolve issues that directly and indirectly influence pubic safety.[9] Most community relations initiatives include measures to identify and resolve community problems such as street traffic controls, parking, potholes, abandoned building disposition, and gang hangouts.[10] Community relations programs attempt to reduce the fear of crime among community members because, as mentioned earlier in this chapter, fearful people behave differently than those who are not fearful of becoming violent-crime victims.[11]

For example, New York City Metropolitan Transportation Authority (MTA) surveys indicate that although only 11 percent of MTA passengers have had firsthand experiences with crime, 97 percent take some form of defensive initiative before entering the subway to avoid confrontations: 75 percent don't wear expensive jewelry or clothing, 69 percent avoid "certain people," 68 percent avoid certain platform locations, and 61 percent avoid the last car.[12]

Police have often tried to reduce constituent fear of crime by undertaking police community relations efforts and by minimizing the publicity given to reported crime. They have frequently acted on the assumption that concerted efforts to solve crime and arrest criminals would reduce unwarranted fears. Some fear-reduction strategies have been tried over the years, but they were never rigorously tested.[13]

The fear of crime can mean different things to different people but it seems that the consequences are similar: Behavior changes, and often not in positive ways.[14] For instance, in a study among 1,777 public housing residents in Charleston, South Carolina, participants responded to questions about their fear of crime, perceptions of crime, expectations of police service, and participation in crime prevention and education programs.[15] Some 533 respondents (30 percent) were aware of Charleston's community relations efforts and believed it made a difference in crime prevention.[16] Another 800 participants (45 percent) were reasonably optimistic about the program. They ranked important crime threats as child involvement in drugs, random shootings, crimes against children, and robbery. In addition, 1,440 participants (81 percent) reported that they felt safe during the day, and 622 (35 percent) reported that they felt safe at night.

Other respondents to the same survey were less optimistic: 746 participants (42 percent) said the courts were too lenient in sentencing criminals and, consequently, reported that they had not reported a crime because of fear of retaliation. Also, 442 participants (25 percent) said they stayed home at night, 301 (17 percent) had joined a neighborhood watch or installed new locks, 195 (11 percent) requested better lighting from the city, 106 (6 percent) carried mace, and 88 (5 percent) had bought guns.

These finding are consistent with those from another compelling study that examined homicide rates in 63 large U.S. cities.[17] One conclusion from this study was that the public sees homicide rates as being in part due to the inequities of police services and in part due to a lack of understanding by economically poor African Americans, most of whom were law-abiding residents. That is, law-abiding individuals see the use of force (including murder) as an appropriate response in at-risk environments where police strategies are nonexistent or ineffective. In some cases, research on battered women syndrome has produced results consistent with this finding.[18] That is, when an individual feels helpless and lives in constant terror caused by an aggressor who cannot be stopped by the criminal justice community, those institutions open the door to law-abiding people's use of deadly force as a tool toward safety. In light of this and other information, an effort to reduce the fear of crime is probably a worthy enterprise.[19]

The randomness of violence beyond high-crime areas also makes people anxious. For example, after the massacre of 33 students at the Virginia Tech campus in 2007 and the killing of 6 students at Northern Illinois University, many students across the United States felt that they were at risk and took precautions, while others wondered where a safe place existed in their world.[20] After a while, those feelings subsided—but imagine people living in a troubled city environment where crime is a constant. Fear of crime may ultimately result in urban decay and lead to the flight of hard-working people from inner-city areas, as they seek safety elsewhere. In this environment, ideally efforts toward crime prevention through community relations programs would have greater rewards than moving.

Which Community Relations Initiative Is Best?

There are more than 17,000 police departments in the United States, and probably hundreds of varieties of community relations initiatives, too. Yet, only a few philosophies underlie most community relations initiatives, and the most significant of those is associated with the broken windows metaphor (discussed later in this chapter). Before examining this concept and its most popular strategy, community policing, it is helpful to consider the historical influences that inspired community relations outreach programs.

■ Historical Influences on Community Relations Initiatives

The historical influences of the 1960s, 1970s, and 1980s provide an explanation why community relations models were accepted by policymakers, police commanders, and the public.[21] During these decades, many U.S. cities were attempting to recover from destructive riots related to the civil rights movement and the unpopular war in Vietnam. Some members of the public pushed these issues into the streets, demanding changes in foreign and national policy despite the threat of military intrusion and aggressive police efforts to contain them. Parts of New York City, Chicago, and Los Angeles (and many other cities) were battlegrounds punctuated with blazing buildings, shiny bayonets, and crying mothers. Politicians and administrative public officials burned any bridges linked to the disgraced U.S. President Richard Nixon and his administration.[22] The spread of cocaine in urban areas of the United States also caught police off-guard, as their attention was focused on defusing rebellious groups and minimizing collateral damage throughout the decade.[23] Some Americans grew paranoid about the remnants of the 1950s' McCarthyism that persisted in law enforcement—namely, the overbearing tactics, arbitrary arrests, wretched treatment of people of color, and intrusions of local police and federal law enforcement into the simple routines of Americans. Finally, many people had a nagging fear that an armed revolution was imminent, which would dictate change at all levels of U.S. government.[24]

Evidence suggests that right up to the beginning of the twenty-first century, while some American minority groups might have backed off a previously held idea that the cops represented an occupational force, many continued to have a less positive attitude toward police than did whites.[25] One implication of this skepticism—on the part of members of both the majority and minorities—was the notion that the police did not control the streets or consistently provide quality services.[26] Outcry at these perceived failures came from many sources—organizations, special-interest groups, and the public—but few had answers to the dilemma.

Negotiating a Truce

As one possible solution, some parties sought to negotiate a truce that would support both public and police interests.[27] Obviously, the public (or at least those members of the public who mattered) was not happy with traditional policing practices. According to this view, the American public felt, and continues to feel, that there exists a dangerous class of people who represent an immediate threat—a threat that the police cannot control.[28] Ultimately, that threat led to declarations of war, such as "the a war on crime" and "the war on drugs."[29] The fear of crime also prompted a scramble to implement public safety initiatives.

The public had solid reasons to sit down with the police and draw up plans for altering the safety landscape. Likewise, the police came to the realization that they could not control crime by themselves. Other driving forces for negotiation included the following issues:

- The fear of becoming a victim and the fear of crime in general continued to grow in the United States as the homeless, gang, and immigrant populations increased. Those groups had to be dealt with, as opposed to continuing the earlier policy of simply ignoring them.[30]
- The public feared more than falling prey to members of the dangerous class—they feared becoming victims through an evolution of police corruption.[31]
- The public's fears drove a movement toward holding police accountable through community relations programs, which required public oversight of police actions.[32]

A plausible justification for adopting community relations initiatives among those who controlled the public purse strings and consequently police policy, and one that was supported by local police command and concerned community and civic leaders, was that community relations strategies would head off the impending doom faced by their communities.[33] What policymakers wanted was a plan that included community commitment, because such a program promised to relieve some of the political pressure leveled by folks who wanted to be sure their homes and businesses were protected. At the same time, policymakers knew that initiatives employed in community relations, such as use of Compstat data (see Chapter 1), would make police managers more accountable for their actions.

For the police, community relations offered another way to monitor the public without raising concerns over constitutional issues.[34] Conversely, the public felt that their having greater influence over police decision-making processes would curb police corruption—and perhaps get them out of a speeding ticket or two.[35]

A community relations initiative is one way to negotiate a truce, and it is often considered an acceptable remedy by policymakers, the public, and the police. On paper, it looks reasonable, and it can be rationalized through **social control theory**.

Community Relations and Social Control Theory

Community relations programs have their roots in social control theory, which was first developed by Travis Hirschi.[36] Social control theory can be applied or operationalized through police–community partnerships that seek to address the root causes of crime and social order.

In essence, social control theory explains why most people do *not* commit crimes. Hirschi's assumption is that if we were left to ourselves, we would commit crimes; however, because of our individual attachments, commitments, involvement, and beliefs, we don't commit crime.[37] That is, the following qualities largely shape lawful behavior:

- Attachments, such as to family members, friends, and neighbors
- Commitments to education, careers, and our future
- Involvement in church, sports, and the community
- Belief in honesty, morally, and caring for others

When these four factors are in place, the chance that a person will commit a crime (or engage in deviant behavior) becomes less likely.

Although social control is a theory about crime causation, the same concept can be utilized within community relations strategies. For instance, connecting residents to community members would enhance attachment; training and shared information would strengthen commitments; partnerships with police and other agencies would enhance involvement; and helping and sharing with others could foster a belief in the community and in the law. Integrating community, police, and other social agencies into a problem-solving group would rebuild the community both socially and environmentally (if necessary) by controlling crime, reducing the fear of crime, and enhancing quality-of-life experiences.

The Broken Windows Perspective and the Community Relations Movement

One explanation of why rebuilding or at least conceptually rebuilding a community is important is linked to the broken windows metaphor.

Roots of the Broken Windows Perspective and a New York City Subway Commuter

If you watch old movies of New York City set in the 1970s or 1980s, you probably would see a subway system that was too filthy and too frightening to travel on.[38] George L. Kelling routinely rode the New York City subway to and from work during those years,[39] and he collaborated with John Q. Wilson to develop the broken windows perspective based on his experiences.

Kelling, a criminologist, argues that public places and spaces such as subways, parks, and sidewalks of New York City were overrun with thugs, homeless people, and criminals at large during that era. He reports having the following feelings about the quality of life in New York during the 1970s and 1980s, when fear of crime was at an all-time high:[40]

- He felt like a prisoner in his own city.
- He couldn't go to Central Park without fear of being mugged.
- He couldn't drive his car without being carjacked.
- Police and the public lost control of public places,

One conclusion Kelling offers is that because the police lost control of public spaces, the fear of crime was at an all time high. Can the public deal with the "grit, grime, and minor unpleasantness of life demonstrated by thugs and vagrants in public places," Kelling asks?[41] His thoughts are consistent with those of others on this issue. For example, Wesley G. Skogan suggests that disorder, while unpleasant in itself, breeds crime, and he posits that the downward spiral of decay may be almost impossible to turn around unless it's caught in time.[42]

The Broken Windows Metaphor

The idea of "broken windows" is simple, logical, and compelling. Although it is just one of many perspectives that have influenced police policy, the timing of community relations programs was impeccable in terms of coinciding with the rise of this theory. A broken window, an alley teeming with debris, and graffiti enhancing a building do little harm to a neighborhood if these blemishes are addressed promptly.[43] Left untended, however, the signal "whispers" that no one cares about this neighborhood or the folks who live and work there. It suggests that the neighborhood is a safe place to break things, to litter, and to vandalize.[44] Once the miscreants who commit these crimes become established in the neighborhoods a downward spiral occurs in which it becomes acceptable to openly drink, beg, rob, and fight.

People are frightened by serious crime, but they are also scared by the behavior of unpredictable and obstreperous people—youths, drunks, the mentally ill, hustlers, prostitutes, and panhandlers. The average person's fear of disorder is not irrational; indeed, the link between disorder, fear, and serious crime is very strong. This relationship has come to be known as the broken windows perspective.[45] James Q. Wilson and George L. Kelling explain it this way:[46]

> A stable neighborhood of families who care for their homes, mind each other's children, and confidently frown on unwanted intruders can, in a few years or even a few months, [turn into] an inhospitable and frightening jungle. A piece of property is abandoned, weeds grow up, a window is smashed. Adults stop scolding rowdy children; the children, emboldened, become more rowdy. Families move out; unattached adults move in. Teenagers gather in front of the corner store. The merchant asks them to move; they refuse. Fights occur. Litter accumulates. People start drinking in front of the grocery; in time, an inebriate slumps to the sidewalk and is allowed to sleep it off. Pedestrians are approached by panhandlers.

An assumption of the broken windows perspective is that individuals who commit nonserious crimes will commit serious crimes or aid those who want to commit serious crimes.

Experiments Supporting the Broken Windows Theory

The broken windows perspective is consistent with the results of an experiment in which researchers left two identical, vulnerable cars in the street in different neighborhoods—one in the Bronx, New York, and the other in Palo Alto, California.[47] The goal was to see how long it would take for the cars to be vandalized. The car in the Bronx was stripped bare in a day. After the car in Palo Alto sat unmolested for almost a week, the researcher put a hammer through one of its windows; as though this act and its impunity were the starting gun they were waiting for, the Californians then rallied round to destroy the vehicle.

The thought underlying the **broken windows theory** is that, just as a vehicle with a broken window will eventually be destroyed, so too will a community. Urban deterioration promotes urban destruction, and community police initiatives are the perceptions of a proactive remedy to revise deterioration. In a sense, these remedies based on the broken windows metaphor rely on a political process to restore order—for example, through aggressive, proactive arrest policies. The programs make the following assumptions:

- Signals or opportunities are observable to individuals who victimize things, others, and themselves; they can feel safe in those neighborhoods.
- Opportunities are present for individuals to commit crime.
- The central goal is to control public spaces.
- Rebound from the downward spiral of urban decay is possible through aggressive arrest policies.

Criticism of the Broken Windows Perspective

Much of the criticism directed at broken windows programs relates to their mission. Early broken windows strategies were designed to regain political legitimacy and win public respect through team policing initiatives.[48]

Litigation Concerns

Police organizations have been involved in a continual stream of litigation over broken windows strategies.[49] Advocates of aggressive arrest policies were at first pleased with the arrest rates—until they themselves were intruded upon or they became aware of the millions of dollars awarded to individuals who mounted successful legal challenges to policies relying on even temporary detainment (see Chapter 6 for details).[50]

For example, the Legal Action Center for the Homeless won a class-action lawsuit against New York City subway officials and overturned the MTA's rule against subway panhandling. In January 1990, Federal District Court Judge Leonard Sand ruled that begging in the subway is a First Amendment right.[51] That decision quickly touched off a round of debates that continued beyond New York City.

Consider what happened in Baltimore when police there implemented policies based on the broken windows theory. In that city, an era of pervasive violent crime, which included the effects of a crack cocaine epidemic beginning in the 1980s, occurred when local government turned a blind eye to the problem.[52] Baltimore's violent crime rate jumped 53 percent between 1987 and 1994. Rowhouses and abandoned structures throughout Baltimore became fire hazards, and offenders recognized the opportunity to commit other crimes. For example, "Metal Men" removed all the aluminum and other valuable metals from churches, homes, and commercial buildings. One magazine "likened these scavengers to ants, literally eating neighborhoods from the inside out, causing millions of dollars of damage and accelerating the process of urban decline."[53]

In 1994, Baltimore's city planning committee identified 266 distinct neighborhoods in the city, represented by 400 community associations.[54] With the assistance of community associations, neighborhood residents (who conducted numerous vigils and street demonstrations), and vigorous political leadership from the mayor's office, an attempt was made to take back public spaces. This policy, which included measures to curb panhandling and harass homeless people, authorized the police and public safety guides (who served as the eyes and ears of police) to restore order. Perhaps not surprisingly, crime rates dropped after this aggressive policing program began.

Patton v. Baltimore City 1994 was a court action brought by the American Civil Liberties Union (ACLU) on behalf of three homeless individuals. Officers and public guards had asked homeless persons and panhandlers to "move along" and "harassed them when they engaged in activities such as sleeping, eating, or tending to personal needs."[55] Although begging—like crime—has been known as long as humans have been on this planet[56] the Baltimore police's policy toward homeless people, the ACLU argued, constituted cruel and unusual punishment, infringed on their First Amendment right of freedom of association, and attacked their "status" as opposed to their "acts."

Prosecutors in Baltimore neglected to inform the Baltimore Police Department that the city lacked official ordinances to support vagrancy arrests among the homeless and therefore those arrests were illegal.

The city repealed its vagrancy and homeless ordinances and amended its aggressive panhandling ordinance "to redress the constitutional issues" addressed by the court; the city also settled with the plaintiffs over the illegal arrest activity of the police (the details of the settlement were kept confidential).

Criticism of Watchman-Style Policing

Team policing initiatives and neighborhood-oriented partnerships as powered by the broken windows theory represent an attempt to reestablish a "watchman" style of policing. According to proponents of these programs, this style of policing will regain some lost police respect (see Chapter 6). However, the literature does not definitively confirm those assumptions, and it offers little support for the idea that the "watchman" style of policing is efficient and immune to corruption.[57] Indeed, even as historical accounts of "watchmen" became more institutionalized, Americans remained skeptical and suspicious of modern-day police (see Chapter 2).[58]

Police Officer Criticism

Some officers have protested that crackdowns, roundups, and mandatory arrests limit or eliminate their discretion.[59] In addition, some believe that programs based on the broken windows theory do not address the root causes of crime—namely, drunkenness, addiction, poverty, unemployment, hopelessness, mental illness.

Inherent in the broken windows metaphor is the notion that social disorder is a sequenced progression toward serious crime. In reality, only communities with observable indicators of disorder or "hot spots" might be on their way to becoming crime infested.[60] The broken windows perspective makes little mention of trends related to white-collar or corporate crime, such as swept through Adelphia, Enron, and other corporations. Also, disorder and crime alike have been found to stem from certain community social structural characteristics, notably concentrated poverty.[61] For instance, homicide—which is probably one of the best measures of violence—is among the offenses for which researchers have found no direct relationship with disorder.[62]

Criminals' View

The broken windows line of reasoning links conspicuous observation with community destruction. While the need to repair broken windows makes sense in that it provides an appearance of less vulnerability, one reality is that drunks (and criminals) already know where they can get drunk and where they can't, but often they don't care.[63] People who demonstrate public drunkenness may have been drunk for a good part of their lives, so they often know how to rationalize their problems to both themselves and passersby. One reality of drunks, addicts, and bad guys is that they usually engage in their life's work where they live.[64] Equally important, when the police cannot or will not protect vulnerable residents whose communities lack the signs of urban decry, crime will surely manifest itself there as swiftly as in poorer communities.

Organizational Resistance to Change

Implementing new police initiatives without changing the police organizational structure can bring about more inefficiencies than it can solve.[65] Police organizations may well endorse community relations initiatives, yet fail to empower community officers and community members to resolve community problems. In this way, the traditional chain of command remains intact.

Whether the broken windows metaphor or community relations initiatives occupy the predominant spot in policymakers' minds is unclear. What is clear is that community relations initiatives powered from the broken windows perspective provide the decisive support to take action. Even so, most community relations programs, including those involving community policing, have been retooled and now take the form of quality-of-life police initiatives (see Chapter 4).

Community Policing

Community policing is one answer to the question of how to maintain social order.[66] The following facts related to these programs were current as of May 2003:[67]

- Fourteen percent of local police departments, employing 44 percent of all officers, maintained or created a written community policing plan during the 12-month period ending May 30, 2003.
- Nearly half of departments, employing 73 percent of all officers, had a mission statement that included some aspect of community policing.
- Fifty-eight percent of all departments, employing 8 of every 10 officers, used full-time community policing officers (approximately 55,000 officers).
- Thirty-one percent of departments, employing two-thirds of all officers, trained all new recruits in community policing.
- Sixty percent of all departments had problem-solving partnerships or written agreements with community groups, local agencies, or others.

Community policing can be described as a **collaboration** between the police and the community that seeks to identify and solve community problems. This collaboration is not necessarily an equal relationship, however. Rather, the police often play the role of facilitator in a two-way relationship with the public and other agencies such as mental health or social services in an attempt to address public problems. The mission of recent community policing or neighborhood projects (it has as many names as it has models) is to enhance the quality of police service; in contrast, older community policing projects used community policing as a means of giving the community more direct control over police operations.[68]

Community policing was thought to be a direct by-product of the broken windows perspective, though both community policing and order-maintenance initiatives eventually evolved into quality-of-life initiatives (see Chapter 4). For now, simply note that although community policing took different forms depending on who or which jurisdiction designed it, some individuals believe it continues to dominate police strategies.

The Office of Community-Oriented Policing Services (COPS) defines community policing as "a policing philosophy that promotes and supports organizational strategies to address the causes and reduce the fear of crime and social disorder through problem-solving tactics and police–community partnerships."[69] COPS also says that community policing focuses on crime and social disorder through the delivery of police services that include aspects of traditional law enforcement, as well as prevention, problem solving, community engagement, and partnerships. The community policing model balances reactive responses to calls for service with proactive problem solving centered on the causes of crime and disorder. It requires police and citizens to join together as partners in the course of identifying and effectively addressing these issues.

COPS was created as a result of the Violent Crime Control and Law Enforcement Act of 1994. As a component of the U.S. Justice Department, its mission is to advance community policing in jurisdictions of all sizes across the United States. Between 1995 and December 2007, COPS has spent more than $12.4 billion to aid community police efforts.[70] You may find it worthwhile to visit the COPS website at http://www.cops.usdoj.gov/.

Primary Components of Community Policing Initiatives

Community policing programs typically include the following characteristics:[71]

- An organization- and jurisdiction-wide philosophical approach
- **Decentralization** of authority and responsibility
- An empowered partnership consisting of community, government, and police
- A problem-solving orientation
- A proactive stance

Organization- and Jurisdiction-wide Philosophical Approach

Adoption of an organization- and jurisdiction-wide philosophical approach implies that the same policy is followed across the department and across the jurisdiction. It suggests that a specific "hot spot" of criminal activity is not targeted. Instead, all officers (rank-and-file officers, detectives, and commanders) participate in community policing policies, and the entire jurisdiction (all parts of town) is affected by the community policing policy.

Decentralized Authority and Responsibility

Decentralized authority and responsibility refers to a fundamental shift in authority, obligations, and decision-making responsibilities and expectations from the traditional chain-of-command (top-to-bottom) organizational control to a contemporary method of control to fit a community policing philosophy. In other words, the police adopt a bottom-up approach, turning the organization upside down (at an extreme) and producing a leveling of decisions.[72]

Empowered Partnerships

In empowered partnerships, authority rests with collaborative groups consisting of community members, community police officers, and other participants such as representatives from social agencies such as streets and sanitation, education, municipal housing, and so on. Predefined and limited authority is mutually vested in each committee member.[73]

Problem-Solving Orientation

Problem-solving activities pursued by the empowered partnership focus on resolving community problems that could lead to crime, such as deteriorating neighborhood conditions (burned-out buildings and street lights), dangerous roadway conditions (potholes, traffic control signals, and parking problems), and worrisome people situations (gang intimation, drug dealing, and homelessness). Problem solving includes discovering the problem, fixing the problem, and evaluating the solution after it is implemented. Some refer to this type of discovery as **problem-oriented policing**

(POP). That is, a problem-solving strategy can stand alone without a community police umbrella, but community policing cannot be practiced without a problem-solving element.

According to the Police Executive Research Forum (PERF), some of the first tests of POP took place in the Newport News, Virginia, Police Department in the mid-1980s, during PERF research that was conceptualized as a crime analysis study.[74] That early work on POP yielded valuable insights:

- Police deal with a range of community problems, many of which are not strictly criminal in nature.
- Arrest and prosecution alone—the traditional functions of the criminal justice system—are not always sufficient for effectively resolving problems.
- Officers have great insight into the problems plaguing a community, and giving them the discretion to create solutions is extremely valuable to the problem-solving process.
- A wide variety of methods can be used by police to redress recurrent problems.
- The community values police involvement in noncriminal problems and recognizes the contribution police can make to solving them.

Over time, POP initiatives evolved into community police initiatives. For example, the Minneapolis Repeat Call Address (RECAP) program is designed to fight crime through community problem-solving initiatives.[75] As part of this program, a five-member RECAP team analyzes computer-generated data to diagnose the underlying problems in a specific area that has an unusually high number of calls for police help: Sixty-four percent of the calls to the Minneapolis police come from 5 percent of the city's addresses. The team devises an action plan in consultation with the people making the complaints. Weekly follow-up analysis allows each officer to evaluate the plan and modify it if necessary. RECAP enables the police department to better focus its limited resources and to improve the quality of life for neighborhoods served by the program.

The conceptual model of problem solving known as **SARA** (an acronym standing for "**scanning, analysis, response, and assessment**") grew out of the problem-oriented policing project in Newport News. This model has since become the basis for many police agencies' training curricula and problem-solving efforts. The steps in the SARA process are outlined below.[76]

Scanning

1. Identify recurring problems of concern to the public and the police.
2. Prioritize problems.
3. Develop broad goals.
4. Confirm that the problems exist.
5. Select one problem for examination.
6. Identify data collected.

Analysis

1. Try to identify and understand the events and conditions that precede and accompany the problem.
2. Identify the consequences of the problem for the community.

3. Determine how frequently the problem occurs and how long it has been occurring.
4. Identify the conditions that give rise to the problem.
5. Narrow the scope of the problem as specifically as possible.
6. Be creative in identifying resources that may be of assistance in developing a deeper understanding of the problem.

Response

1. Search for what others with similar problems have done.
2. Brainstorm interventions.
3. Choose among the alternative solutions.
4. Outline the response plan and identify responsible parties.
5. State the specific goals for the response plan.
6. Identify relevant data to be collected during the response for evaluation purposes.
7. Carry out the planned activities.

Assessment

1. Determine whether the plan was implemented.
2. Determine whether the goals were attained, and collect qualitative and quantitative data on the outcomes.
3. Identify any new strategies needed to augment the original plan.
4. Conduct ongoing assessment to ensure continued effectiveness.

Many different POP initiatives have applied this SARA model to some degree or another, while other programs have developed their own version of SARA.

- *Alexandria, Virginia, Police Department's (APD) SARA.* The APD community outreach program adds an assessment feature (SARAA) so that the entire process can be assessed to determine the progress made.
- *San Antonio, Texas, Police Department's (SAPD) SARA.*[77] SAPD tweaks POP to produce long-term solutions to problems of crime or decay in communities. That is, SAPD identifies problem solving as being successful only if it produces a long-term solution. For the community outreach committee, SARA is utilized but SARAM is preferred. The "M" added to the acronym stands for "maintenance," reflecting the ultimate objective of developing a long-term solution.

Proactive Stance

Community policing incorporates a proactive approach, meaning that the organization moves from a reactive (after a crime is observed and reported), incident-driven police strategy to a proactive initiative requiring a partnership with the community and other relevant public agencies.[78] Reflecting this orientation, many experts have tried to call community policing "democracy policing." As a tool for addressing the conditions that threaten the welfare of the community, community policing programs may include policies dealing with priority calls and deployment procedures, methods of arrests and limits of force, procedures for stops, hiring, training, discipline, and the promotion process.[79]

Two Case Studies in Community Policing

Two case studies provide a better picture of community policing across the United States. The first study, which took place in 1999, focused on police organizations that were engaged in a transitional move from reactive to proactive initiatives, where each agency had its own ideas about "broken windows" and community policing.[80] Community police strategies were evaluated in nine police agencies:

- Broken Arrow, Oklahoma, Police Department
- Camden, New Jersey, Police Department
- Columbus, Ohio, Police Department
- Fayetteville, North Carolina, Police Department
- Harris County Precinct Four Constable's Office, Spring, Texas
- Lansing, Michigan, Police Department
- Nashville, Tennessee, Police Department
- Sacramento, California, Police Department
- St. Petersburg, Florida, Police Department

The second study was conducted two years later in 2001. It included 76 interviews and 2,010 surveys among community members who described their perceptions of the success of eight police organizations that utilized the community policing model:[81]

- Alexandria, Virginia
- Boston, Massachusetts
- Columbia, South Carolina
- Columbus, Ohio
- Miami–Dade County, Florida
- Midland, Texas
- Palm Beach County, Florida
- Sacramento, California

Researchers had theorized that community policing strategies led to better crime control, reduction in the fear of crime, and enhancements in quality-of-life experiences among constituents.[82] It was suggested that the more extensively community members—especially culturally diverse members—influenced police decisions, the greater the likelihood that public safety and lifestyle experiences would be enhanced.

Identification of Neighborhood Problems

Community members identified their most pressing community problems as street drug activity, home invasion, conditions of their neighborhood, and fear or lack of trust of police. The police identified a neighborhood's serious problems as carjacking, home invasions, and unsupervised youths.

Community-Suggested Solutions for Neighborhood Problems

Suggested solutions to these problems included quality police services and quality municipal/state services. The community defined quality police service as visible patrols, increased arrests of outsiders (selective enforcement), a decreased response time to 911 calls, professional behavior from officers, and severe sanctions against offenders from outside the community.

Quality municipal services, which were deemed to be equally important and in some cases more important than police services, included assistance in the purchase or renovation of residences and enticing large retail businesses to the area; they also included harsh zoning controls over poor local businesses. Additionally, community members wanted community amenities (such as parkways and lighting) enhanced, bus stops moved, and better control over Section 8 (low-income) housing and welfare housing. The consensus was that police have little business in interfering with neighborhood problems: The best way to stabilize the neighborhood, according to community members, is through the people who live there and through municipal public services.

Community members also wanted to shape police decision-making processes associated with routine auto, bike, and boat patrol deployment; placement of mini-stations; building owner notification linked to zoning; use of police force; priorities of calls for service; and police officer disciplinary actions, training, and promotions.

Police-Suggested Solutions for Neighborhood Problems

Solutions for neighborhood problems identified by the police included a greater reliance on Compstat and zero-tolerance initiatives, step-up stops among youths, and greater communication from community members to the police about criminal activity.

Police command saw the development of community policing as a sounding board through which to reach out to community members. Community policing, according to police authorities, was not a partnership in any of the jurisdictions studied at the time. Although sometimes agency websites had the appearance of community policing initiatives, in reality they lacked any of the components of a true problem-solving strategy that partnered residents and officers in a joint venture to identify and solve neighborhood issues. For instance, the Sheriff of Kent County, Michigan, filed the following report concerning community policing is his jurisdiction:[83]

> I would like to discuss the Neighborhood Watch Program. You may have heard of it, or maybe you are already involved with this program. This program has been in place here at the Sheriff's Department and in the community for some time.
>
> Neighborhood Watch is a group of concerned citizens who work together to drive crime from their neighborhood. The people involved in this program get to know their neighbors and work with them in spotting and reporting crime to the Sheriff's Department. The neighborhood and the Sheriff's Department should work together in a partnership for the purpose of eliminating crime.
>
> The Neighborhood Watch program establishes a communication link between citizens in the community and the Sheriff's Department. This system is [intended] to help concerned citizens report emergencies and other problems to

> the Kent County Sheriff's Department. We also discuss tips for crime prevention during the Neighborhood Watch meetings.
>
> When organizing a Neighborhood Watch, it is important to involve as many people as possible from your neighborhood. Once the program is started, a goal of at least 50 percent participation of the neighborhood should be set. The more involvement, the more effective the program becomes. We at the Sheriff's Department also recommend that your neighborhood have an annual meeting where everyone in your community can come together and get to know as many people from the neighborhood. It is important to know your neighbors: The better you know your neighbors, the better the program works.

In both studies, both large and small jurisdictions—from Boston to Miami to Chicago to Sacramento—were found to be similar in their approaches to community policing. Specifically, official websites talked about community partnerships and the use of proactive problem-solving techniques, but the unofficial police response was highly consistent: increase the budget and police authority, beef up patrols, make more arrests. In short, community policing meant "establish a presence."

Study Implications

Official rhetoric says that community policing exists. Nevertheless, in these studies municipal services appear to have more to do with ensuring public safety than with enforcing the law, from the perspective of the residents. It was not policing that was the issue in these jurisdictions, but rather how police service is delivered that counts (confirming the notion of legitimacy, as discussed in Chapter 1). Additionally, it is curious that how well a police agency is doing, from the perspective of the community, depends on how efficiently other public agencies (from garbage pickup to water control to road maintenance) provide services. The police are the windows of government for community members—a thought consistent with the literature indicating that police officers are seen as "snappy bureaucrats." [84]

What was learned from these studies? Community policing is difficult to evaluate because culturally diverse members of the community are rarely part of the community policing committees' membership. Those members who were members of minority groups did not influence police decision-making processes.

This finding is consistent with other research related to community policing. For instance, a study of the Chicago Alternative Policing (CAP) strategy also revealed that many of Chicago's culturally diverse residents never attended a CAP meeting, for various reasons.[85] Therefore, it could be said that culturally diverse members who had participated in community police efforts influence police decision-making processes only to a small extent, if at all.

Major findings from both studies (17 cities) were as follows:

- None of the agencies was engaged in an organization- and department-wide initiative centered on a proactive response system.
- Success and neighborhood problems were measured differently in each jurisdiction.
- Most street and community police officers (CPOs) assigned to community policing committees embraced the tenets of community policing but not its practice.
- Community members traded on personal standards and expectations as opposed to objective standards and expectations.

- Gang members, individuals who didn't speak English, mentally and physically challenged individuals, and persons who had been previously detained or arrested by the police did not attend community policing meetings.
- Persuading leaders from churches, businesses, and community watches to recruit members in a block-by-block effort and to survey neighborhoods in an effort to identify problems enhanced the personal power of those leaders as opposed to enhancing their communities.
- When it came time to take responsibility and participate in what community members wanted, most community members were not available.
- Major issues (e.g., a highway billboard projecting from the community to an above-ground interstate) took center stage at community policing meetings. Once the problem or objective was resolved, community members who focused on it failed to attend later meetings.
- Most community members rejected responsibility for solving problems, but rather expected CPOs to resolve problems, regardless of the magnitude or redundancy of the proposal. For example, CPOs were expected to board up Section 8 housing to keep gang members from using burned-out buildings to trade in drugs, prostitution, and extortion.

These findings are consistent with evaluations revealing that staged dialogues between police and community groups can produce complicated situations of conflict and tension. While the police may work to present a friendly, coherent, and professional demeanor to the community, community members may use the police's ideological language of order to offer a critique of the police themselves and of the sweeping neoliberal economic restructuring of the city around them.[86] All too often, the end result is greater suspicion on both sides. Chapter 4 discusses failures of community relations initiatives in more detail.

■ Community Police Officers: Selection, Responsibilities, and Perspectives

Uniformed patrol officers rarely participate in community police strategies unless they are individually assigned to a community police unit and would be considered community police officers. Largely, CPOs are selected for their agreeable and flexible personalities and their idealized perspectives. Therefore, CPOs are generally younger and less experienced, but better educated (at the time of their hire), than the officers assigned to patrol and service responding units.[87] However, CPO are selected from the patrol ranks, where their training officer and the officers they admired most work.

A Palm Beach community officer explains, "This is the greatest job I've ever had. I get to deal with people and some of their problems, but some of their happy times, too. It's an uplifting experience for me. I never thought I could help so many people at the same time before joining the sheriff's department. After college I had my choice of other jobs, but this one has really kept my interest."[88]

Experiences of Community Police Officers

Some CPOs report that committee members try to intimidate them, police commanders want everything accomplished at the committee meeting "approved by their office," and their police colleagues who try to guide them are generally senior officers who are separated from the CPOs by a gap in education, experience, and priorities.

Think what might happen at the community meetings. A high-powered or civic leader participant might say that he or she would be interested in doing something great for the community, such as providing all the building materials for a new community center. In return, the high-powered community person would like to see this or that done by the police department. The CPO would most likely agree or be intimidated into agreeing with the high-powered community member. Yet the "this or that" fulfillment could easily fall to the CPO.

As a CPO from Chicago notes, "At some of these meetings, a few guys pull out a bottle and take a few snorts. If I arrest 'em, my value as a community advocate hits the toilet, and the boss will be all over me. He'll say, if you can't control these [people] in a meeting, than how could you control 'em on the streets,"[89]

Furthermore, the problems articulated by residents at community meetings from the shores of the Atlantic to the Pacific are often personal problems encountered by individual community members as opposed to community at-large problems that affect crime. Many participants have an unwarranted fear of crime, and their ideas have little to do with enhancing the quality of living experiences of all residents—just their own experiences. In fact, once a personal problem of a committee member is solved, it is unlikely that he or she will return to the meetings.

The Community's Perceptions of Community Policing

Officers applauded by the residents are those officers who most often give the residents what they want as opposed to what the CPO thinks residents should receive. Generally, police command resists providing CPOs with authority, even though commanders might inform the community that they have done so.[90] Once the community committee or community members learn that the CPO has only limited power, bad things can happen—and intimidating a CPO might be easier than it sounds. After all, evaluations of CPOs depend on the approval ratings doled out by community members. Some CPOs think they are engaged in a lose-lose arrangement, especially if community committee members were selected by police command. Experienced street officers rarely provide community members with everything they want and, as might be expected, those officers may never be invited back to community meetings.

If an officer wants to make a career of community meetings, she or he has to make concessions not necessarily in keeping with legal perspectives or police concepts. For instance, what would an officer do if a community committee member arrived at the meeting intoxicated? Make an arrest? Warn the person? If a CPO followed regulations to the letter, how long might the officer keep his or her position as a community policing advocate? In a western jurisdiction, one young offender made an arrest of an intoxicated Mexican American man at a community meeting. By the time the CPO had escorted the suspect to the station and processed him, a delegate of community members was already demanding his release. Disciplinary action was brought against the officer for arresting one of the community's "pillars," and the officer received a five-day suspension without pay for his actions.

Officers' Perceptions of Community Policing

One reality of community policing is that street officers tend to see the centerpiece of community policing as being associated with liberal solutions to social order, and police command perceives such programs as a vehicle to obtain resources—that is, federal and local grants—that the department may then use to further the control of street officers by police commanders.[91] Officers tend to hold conservative (and punitive) thoughts about violators, and their training and directives are consistent with those thoughts.[92] "Go after bad guys but ask if it's okay to control them with handcuffs," responds a Charlotte street officer.[93] "I ain't a social worker. If that's my profession, I would have worked for DSS," adds her partner.

Often the problems presented by community members are seen as distracting police because many community members focus on small, insignificant matters. As a result, officers may be assigned as peace keepers during a neighborhood association meeting.[94] Said one officer of an attendee at one such meeting, "That woman goes on and on about some Asian who parks in front of her house. She called 911 a dozen times. She put chairs and [stuff] in the area where the guy parked during the day, and we had to tell her that she couldn't do that. She threw eggs and hot cooking oil on the guy's car another time, and we were called to the scene. Then when I spotted her at this meeting, I couldn't believe she's still screaming about this incident."[95]

Like most other professionals, CPOs can find it hard to change, especially if they are evaluated using traditional or older systems. Most officers, including CPOs, are evaluated based on response time, arrest rates, and disciplinary actions. Those methods of evaluation would hardly report on the skills employed by many CPOs or officers in general, as you can well imagine.

Street Officers' Perceptions of Community Policing

Street officers are called upon to operationalize **problem-solving strategies**, and they try to support the younger CPOs who turn to them for advice, especially if they were the CPO's former training officer. For instance, in South Boston, through a community relations program of sorts, it was agreed that customers in the business district could double-park on the streets because business leaders had complained about a parking shortage. Driving through South Boston (better known as "Southie"), double-parking is everywhere and has become an institution (i.e., a patterned, expected behavior). In fact, if you don't double-park in Southie, people know you're an outsider. Police vehicles are driven around parked autos, and road chases or traffic control strategies are impossible. "What a mess," one Boston officer said, "but that's what community leaders hammered out. Maybe if they wanted to sell drugs on the street corners, we'd go for that, too."[96]

Two explanations provide a potential pathway to understanding why street officers feel that the problems of community members are isolated and that community policing focuses on specific situations as opposed to at-large problems in the community:

- Officers are correct in their observations that community problems are individualized.
- Officers tend to be more conservative (see Chapter 8 for more information on officer personality characteristics) in their thinking and focus on legal action. They believe that individual accountability can solve most of the problems experienced by community members.

Detectives' Perceptions of Community Policing

Recall that ideally community policing is a department-wide philosophy and, as such, would include detectives. Studies have found that detectives' views of community policing initiatives are consistent with the CPOs' perspectives on these programs.[97] Put simply, they see these models as creating confusion.

For instance, detectives within a community policing context often become uncertain as to their role within the department. As investigators strive to become problem oriented in their work, cases that do not achieve "problem" status fall between the cracks. Detectives wonder if they should be proactive or reactive; they know that paying attention to prevention strategies at the expense of working cases results in lower clearance rates (see Chapter 11). "We're overworked as it is. Can't get to all my caseloads and I gotta give some of the harder cases up because I just don't have time," says a sexual crime's unit detective in Boston.[98]

Thus the question is this for detectives: Do the cases come to us, or do we become proactive and preventive? If the latter is true, exactly how should we do that?

Case Study: Community Support as a Means to Enhance Crime Control

Like most other police officers, CPOs are rarely trained to resolve social troubles. Instead, they tend to focus on individual accountability. Of course, there is nothing wrong with that perspective—indeed, most police officers perform in a specified and expected manner when confronted with individuals they believe to be violators even if those officers are assigned to a community committee (also see Chapter 8).[99] However, under the community policing model, giving in to community pressure can have widespread repercussions that cannot always be anticipated.

Can community policing partnerships sometimes be a mistake? In Madison, Wisconsin, in the early 1980s, the police department bowed to pressure from city hall and established a series of community partnerships. Some members of the city administration had participated in "anti-establishment" riots against the Vietnam War in earlier decades, and some of the protesters were subsequently elected to political office and took control of city hall. As the police department tells it, officers had a long history of adapting and being responsive to changing community needs.[100] The police department, at the behest of community committees, assigned college students to serve as community activists involved in community policing efforts. These college students went from door to door with surveys, armed with high hopes and great intentions of solving the world's problems.

According to Captain Michael Masterson of the Madison Police, in the early 1990s "Madison's high-crime-rate neighborhoods were suddenly deluged with the arrival of massive shipments of crack cocaine."[101] A task force was formed, Weed and Seed money from the federal government was awarded, and many strategies were put into effect to deal with the problem.

A variety of measures were used to assess and evaluate the effectiveness of these efforts. Hundreds of drug charges and convictions, thousands of dollars in recovered drugs and drug money, seizure of drug houses and assets, and the removal of numerous guns for neighborhood drug entrepreneurs were indicators that a traditional task force was doing well. Did the community agree? Police officers felt that much of the initial problem stemmed from the practice of putting community matters into the hands of individuals who do not necessarily see law and order as an important part of enhancing life experiences, controlling crime, or reducing the fear of crime. Had the individuals in the field in the latter part of the 1980s been officers who were trained to spot drug dealers as opposed to anti-establishment college students (who might miss nuances of the drug trade), some of the many problems facing Madison could have been avoided. How many other cities have been overrun by drug traders who exploited the good—or not so good—intentions of community members?

Final Analysis of Community Policing Initiatives

In the final analysis, to restore order, should violators always be arrested? This question arose in Chapter 2 when observers suggested that community policing was a tool to better control or police the poor. Perhaps Neil Websdale's book title explains it even better: *Policing the Poor: From Slave Plantation to Public Housing.*[102] That is, some hold the view that community policing is a tool to better control the lower class rather than help with social issues.

Nevertheless, there is an abundance of literature dealing with community relations policing and its impact on communities, but few studies have explored one of the basic functions of policing—that is, conducting a probable cause arrest—and its impact on the community. One of those few studies asked this question: Does increasing arrests or promoting incarceration in a neighborhood increase participation in the community, and specifically participation in informal social control? The researchers concluded that the answer to this question is both yes and no. "Arrest rates in neighborhoods have positive effects on participation in informal social control by residents. Increases in arrest rates are associated with higher levels of participation in informal social control. The evidence for a positive effect of incarceration on community organization and informal social control is more equivocal."[103]

From the perspective of police officers, a typical response implies that it is illogical to promote community policing and enter into community partnerships with the goal of improving social order if a person engaged in criminal activity is not arrested. "Any police program, regardless of its elegance, must be supported by a specific arrest policy; otherwise, it will fail," reports a Boston police commander.[104]

Another observer points out that a central tenet of community programs has been the **empowerment** of local community leaders to design and implement crime prevention strategies: "This philosophy may amount to throwing people overboard and then letting them design their own life preserver."[105] Policies and market forces causing criminogenic community structures and cultures may, in reality, be out of the control of neighborhood residents, and "empowerment" may not truly include the power to change policies. It is one thing for Section 8 housing tenants to help manage the security guards in their housing project; it is another thing to allow those same tenants to design a new public housing policy and "determine where in a metropolitan area households with public housing support will live."

According to this view, community policing has pitfalls, many of which mask real problems within a community. Equally important, those pitfalls tend to reduce police effectiveness in actual practice. Community policing has never really come to the forefront of police initiatives, but rather has evolved into a number of police strategies such as quality-of-life initiatives. Today, these initiatives appear at the heart of most U.S. policing strategies; they are also the focus of Chapter 4.

RIPPED from the Headlines

Community Policing to Prevent Crime: An Old Idea with a New Name

Police cannot control crime and maintain order by themselves. The shear number ratio of citizens to police officers makes this statement credible. The only way to combat and control crime is through a cooperative effort between the police and the community. While some may say it has been a long time coming, reality is that over the span of less than 15 years (which is a microsecond historically speaking), the United States has gone from seeing community policing as an obscure approach to having the President and Congress allocate billions of dollars to support its development. It is still a relatively new concept to this generation, [though] its roots reach far back into [human history,] when communities were small, held similar beliefs and understandings, and were practicing the concept of community policing, even if they didn't call it that. In our more complex society where there are ethnic differences, religious differences, racial differences, and lifestyle differences all within a few blocks of each other (or even in the same apartment building), community policing is much more challenging, as conflict is more likely.

What is community policing? Wherever you find definitions for this term, there are some differences in the verbiage, but the underlying theme is the same—collaboration between the police and the community to identify and solve crime-related community problems.

In 1979, Herman Goldstein developed and advanced the idea of problem-oriented policing (POP). It encouraged police to think differently about their work and see their role as problem resolver. Goldstein advocated that police identify root causes of problems with an eye toward eliminating or minimizing them, rather than just addressing the outcome. Much of the research in the 1970s was focused on patrol issues; however, Rand Corporation examined the role of detectives [during this era]. It was determined that only a small percentage of crimes analyzed were solved by detectives. Most of the solved cases hinged on information obtained by patrol officers. With appropriate training, patrol officers would be able to perform early investigating and take some of the load off of detectives, who could then focus on more complex investigations.

So with all this research that shows things could be better and offers alternatives, why isn't community policing everywhere? The key word in the answer would be "change." Not everyone is comfortable with change.

There seems also to be a lack of acceptance lower down in the police ranks. In [an] interview with officer X, [the officer] was asked if he thought community policing had an impact on crime. His answer was no. It seems there may also be issues of mistrust between the community policing officer and fellow officers. [The CPO] may appear as the "good guy" within the community, while his or her fellow officers are greeted with the usual less-than-receptive treatment.

One of the biggest stumbling blocks to the successful implementation of community policing is people.

There is an existing 20/60/20 rule that corporations use to define change capability. On average, within any organization, 20 percent of the people are ready, willing, and able to change. Another 20 percent of the people wouldn't change if their life depended on it. Then you have 60 percent of the people waiting to see "which

> way the wind will blow." If an organization focuses attention on the 20 percent change-friendly [people] who make themselves part of the solution, the 60 percent will follow. If they focus attention on the 20 percent who have dug in their heels and won't be budged, 80 percent of the organization will be fighting the change.
>
> *Source:* Morgan Summerfield, "Community Policing to Prevent Crime: An Old Idea with a New Name," *Associated Press* (November 22, 2005), http://www.associatedcontent.com/ (accessed July 4, 2007).

Summary

Some of the historical developments leading to the community relations era were powered by the broken windows metaphor. This perspective says that blight (such as broken windows and graffiti) signals a neighborhood's vulnerability, thereby encouraging individuals who victimize things, others, and themselves. One assumption of this theory is that wrong-doers always have the intent to commit crime and seek opportunities to turn that intention into reality. Once the miscreants are established in vulnerable neighborhoods, it becomes acceptable to be openly drunk, to beg, rob, or steal. According to the broken windows perspective, even the smallest symptoms of antisocial behavior will, if left to fester, breed greater and greater crimes.

Many criminals who commit small crimes are the same ones who commit serious crimes, and many of these small crimes tend to be committed in a specific section of a jurisdiction. As a result, policing initiatives may focus on these "hot spots." One criticism to this approach is that "hot spots" and violators who commit less serious crimes are not always on an inevitable pathway to more serious crimes.

Community policing is defined as an organization- and jurisdiction-wide philosophical approach that promotes a decentralization of authority and responsibility; develops empowered partnerships consisting of community, government, and police members; and has as its centerpiece a proactive approach using problem-solving techniques. For example, it may incorporate a scanning, analysis, response, and assessment (SARA) process to ensure that the initiative's mission and objectives are met. Community policing represents an official move from reactive and incident-driven strategies to proactive initiatives requiring community participation. Community partnerships are viewed as a tool for addressing the conditions (e.g., "broken windows") that threaten the well-being of a community.

Studies have revealed that the more community members (especially culturally diverse ones) influence police decisions, the greater the likelihood that public safety and lifestyle experiences will be enhanced. Nevertheless, most street and community officers embrace the tenets of community policing but not its practice.

Community members often identify their most pressing problems as being street drug activity, home invasion, environmental conditions, and fear or lack of trust of police. By contrast, police tend to identify a neighborhood's most urgent problems as being carjacking, home invasions, and unsupervised youths. Reinforcing the gap between community and police perceptions of crime, community members often suggest the best way to resolve their problems is through quality police and municipal services (with the priority on municipal services), whereas police are typically anxious to increase stops and arrests.

Community police officers tend to be selected for their agreeable and flexible personalities and their idealized perspectives. They are often younger and less experienced, but better educated, than the officers assigned to patrol units. Unfortunately, they are also easily intimidated by committee members and police commanders. Community policing tends to mask real problems and can promote behavior among officers that many commanders and policymakers would rather not know about.

CPO are hired under one set of assumptions, trained accordingly, and are evaluated and move through their careers along what appears to be an organized and clear path. Nevertheless, little of that process relates directly to community relations, leading to uncertainty when they enter into the community. Likewise, detectives who work within a community police context can become uncertain as to their role within the department.

Community relations models can also trigger violations and inappropriate behavior among CPOs. Priorities may shift to less serious crimes (if there is such a thing), and more police involvement in otherwise lower-priority conduct might lead to an escalation of that conduct. Finally, giving into community pressure sometimes has widespread—and unintended—repercussions, as happened in Madison, Wisconsin.

In the final analysis, community policing exists more in rhetoric than in practice and is pursued for various reasons. Over time, many of these strategies have evolved into quality-of-life initiatives, which now represent the principal strategies used by U.S. police organizations.

Key Terms

Broken windows theory
Collaboration
Community policing
Community relations initiatives
Decentralization
Empowerment
Fear of crime
Order-maintenance policing
Problem-oriented policing (POP)
Problem-solving strategies
Scanning, analysis, response, and assessment (SARA) model
Social control theory

Discussion Questions

1. Define community policing, neighborhood-oriented policing, problem-oriented policing, Weed and Seed programs, order-maintenance programs, team policing, and quality-of-life policing initiatives.
2. Characterize the importance of reducing fear of victimization among constituents. In what way do you agree with the idea that fearful people act differently than people who are not fearful?
3. Identify the historical influences that inspired community relations initiatives.
4. Characterize the thoughts of the York City subway commuter about crime. Thinking of your own community, in what way do you agree with this commuter?
5. Characterize the primary components of the broken windows metaphor. In what way does it make sense to you?
6. Describe the criticism of the broken windows perspective. Of the reasons mentioned in the text, which criticism do you think has the most impact, and why?
7. Characterize the primary components of community policing.
8. Explain the components of the scanning, analysis, response, and assessment (SARA) model, and describe some of the variations of SARA used by different police departments.

9. Briefly summarize the results of the two studies of community policing provided in this chapter.
10. Identify the neighborhood problems as articulated by community members in these studies. In your community, what problems do you see related to crime?
11. Identify the differences between community solutions and police solutions to fix neighborhood problems. Why would community members hold different views about problems than police officers? Who is right? Why?
12. Describe the problems faced by community police officers and detectives under the community policing model.
13. Describe the lessons learned in Madison, Wisconsin, after community policing was implemented. In what way do the thoughts of the captain in Madison make sense to you?
14. Summarize the effectiveness of community policing programs. If they are not effective, why do community relations initiatives receive so much attention from communities and funding from the federal government?
15. Explain the 20/60/20 rule as described in the "Ripped from the Headlines" article.

Notes

1. To name a few of the dozen of so new books: Willard M. Oliver, *Community-Oriented Policing: A Systemic Approach to Policing* (Upper Saddle River, NJ: Prentice Hall, 2008). The author states in the preface's first line: "Community-oriented policing is truly an idea whose time has come." See also Linda S. Miller and Karen M. Hess, *Community Policing: Partnerships for Problem Solving* (Belmont, CA: Wadsworth, 2007); Ronald W. Glensor, *Community Policing and Problem-Solving*, 5th ed. (Upper Saddle River, NJ: Prentice Hall, 2007).
2. Joliet (Illinois) Police Department, http://www.jolietpolice.org/JPDNOPT%20Alternative.htm (accessed July 4, 2007).
3. *Weed and Seed Guide to Federal Resources* (NCJ 204808), http://www.ncjrs.gov/App/Publications/abstract.aspx?ID=204808 (accessed April 22, 2007).
4. Jeffrey Patterson, "Community Policing: Learning the Lessons of History," *Lectric Law Library's Stacks* (2004), http://www.lectlaw.com/files/cjs07.htm (accessed July 4, 2007).
5. Gary Cordner and Elizabeth Perkins Biebel, "Problem-Oriented Policing in Practice," *Criminology and Public Policy* 4, no. 2 (2005): 155–180.
6. Neil Websdale, *Policing the Poor: From Slave Plantation to Public Housing* (Boston: Northeastern Press, 2001), 6.
7. Recheal Stewart-Brown, "Community Mobilization," *Federal Law Enforcement Bulletin* 70, no. 6 (2001): 9–17.
8. Anthony A. Braga, *Problem-Oriented Policing and Crime Prevention* (New York: Criminal Justice Press, 2002), 1–5.
9. Community Policing Consortium, "What Is Community Policing" (2005), http://www.communitypolicing.org/about2.html (accessed June 13, 2005).
10. Dennis J. Stevens, *Case Studies in Applied Community Policing* (Boston: Allyn Bacon, 2003), 11–13.
11. Wesley G. Skogan, *Disorder and Decline: Crime and the Spiral of Decay in American Cities* (New York: Free Press, 1990).
12. Ellyn Shannon, "New York City Transit Rider Council" (2004), http://pcac.org/reports/pdf/2004%20Station%20Cond%20Report.pdf (accessed November 11, 2007).

13. Police Foundation, "Research Brief: Police Strategies to Reduce Citizen Fear of Crime" (n.d.), http://www.policefoundation.org/docs/citizenfear.html (accessed July 6, 2004).
14. Katherine Beckett and Theodore Sasson, *The Politics of Injustice: Crime and Punishment in America* (Thousand Oaks, CA: Pine Forge, 2000), 47.
15. Girmay Berhie and Alem Hailu, "A Study of Knowledge and Attitudes of Public Housing Residents: Toward Community Policing in the City of Charleston, South Carolina" (NCJ 182434), *NIJ Research Review* (2002), http://www.ncjrs.org/rr/vol1_3/27.html (accessed May 30, 2004).
16. The percentages given here have been rounded.
17. Dennis J. Stevens, "Urban Communities and Homicide: Why Blacks Resort to Murder," *Policing and Society* 8 (1998): 253–267.
18. Stevens, "Urban Communities and Homicide"; Dennis J. Stevens, "Interviews with Women Convicted of Murder: Battered Women Syndrome Revisited," *International Review of Victimology* 6, no. 2 (1999): 117–136.
19. "Neighborhood Watch: Tracking the Fear of Crime" (September 2004), http://www.crimereduction.gov.uk/neighbourhoodwatch/nwatch03.htm (accessed June 25, 2005).
20. Robert Blendon; see Dennis J. Stevens, *Applied Community Policing in the 21st Century* (Boston: Allyn Bacon, 2002), 39.
21. Police Executive Research Forum, *Themes and Variations in Community Policing: Case Studies in Community Policing* (Washington, DC: Police Foundation, 1996); Dennis J. Stevens, *Case Studies in Community Policing* (Upper Saddle River, NJ: Prentice Hall, 2001).
22. Carl Bernstein and Bob Woodward, *All the President's Men* (New York: Simon & Schuster Adult Publishing Group, 1987), 147–151.
23. Michael F. Masterson and Dennis J. Stevens, "The Value of Measuring Community Policing Performance in Madison, Wisconsin," in *Policing and Community Partnerships*, ed. Dennis J. Stevens (Upper Saddle River, NJ: Prentice Hall, 2002), 77–92.
24. Bernstein and Woodward, *All the President's Men.*
25. Thomas B. Priest and Deborah Brown Carter, "Evaluations of Police Performance in the African American Sample," *Journal of Criminal Justice* 27, no. 5 (1999): 457–465.
26. Dennis J. Stevens, "Stress and the American Police Officer," *Police Journal* LXXII, no. 3 (1999): 247–259.
27. William V. Pelfrey, Jr., "The Inchoate Nature of Community Policing: Differences Between Community and Traditional Police Officers," *Justice Quarterly* 21, no. 3 (2004): 579–602.
28. Randall G. Shelden, *Controlling the Dangerous Classes: A Critical Introduction to the History of Criminal Justice* (Boston: Allyn and Bacon, 2001), 17, 97–98.
29. Samuel Walker, *Sense and Nonsense about Crime and Drugs*, 4th ed. (Belmont, CA: Wadsworth, 2001), 11–12.
30. Shelden, *Controlling the Dangerous Classes.*
31. Frank Anechiarico and James B. Jacobs, *The Pursuit of Absolute Integrity* (Chicago: University of Chicago Press, 1996), 5.
32. Samuel Walker, *Police Accountability* (Belmont, CA: Wadsworth, 2001), 5.
33. Bernstein and Woodward, *All the President's Men,* 147–151; Shelden, *Controlling the Dangerous Classes.*
34. Bernard Harcourt, *Illusion of Order: The False Promise of Broken Windows Policing* (Cambridge, MA: Harvard University Press, 2002).
35. Quint C. Thurman, "Community Policing in a Community Era," in *Crime and Justice in America: Present Realities and Future Prospects*, ed. Wilson R. Palacios, Paul F. Cromwell, and Roger G. Dunham (Upper Saddle River, NJ: Prentice Hall, 2002), 120.

36. Travis Hirschi, *Causes of Delinquency* (Los Angeles, CA: University of California Press, 1969).

37. Oliver, *Community-Oriented Policing*, 95.

38. One of the most memorable movies featuring a dilapidated and crime-infested New York subway is *The Warriors* (1979). The movie is about a Coney Island gang returning home from a gang rally in Upper Manhattan after being accused of shooting a rival gang's leader. They are running from other gangs using the subway. It's a pretty insipid plot but there are some great subway shots: "M"s at Wyckoff, "J"s (mismarked) along the Broadway Brooklyn line, several different incarnations of Hoyt-Schermerhorn. R27s and R30s are the starring cars, among others. Most of the subway cars and stations are filled with graffiti, filth, and decay.

39. George L. Kelling, "Taking Back the Subways," *City Journal* (1991), http://www.city-journal.org/article01.php?aid=1614 (accessed August 26, 2006).

40. George L. Kelling and William J. Bratton, "Taking Back the Streets," *City Journal* (Summer 1994), http://www.city-journal.org/article01.php?aid=1428 (accessed June 28, 2006).

41. Kelling, "Taking Back the Subways."

42. Skogan, *Disorder and Decline.*

43. Robin Skyler, "Fixing Broken Windows" (2003), http://www.ambiguous.org/robin/word/brokenwindows.html (accessed June 14, 2005).

44. James Q. Wilson and George L. Kelling, "Broken Windows: The Police and Neighborhood Safety," *Atlantic Monthly* 127 (March 1982): 29–38.

45. Wilson and Kelling, "Broken Windows."

46. Wilson and Kelling, "Broken Windows," 469.

47. Phillip G. Zimbardo, *The Cognitive Control of Motivation* (Glenview, IL: Scott Foresman, 1969).

48. Samuel Walker, "'Broken Windows' and Fractured History," in *The Police and Society: Touchstone Readings*, 2nd ed., ed. Victor E. Kappeler (Prospect Heights, IL: Waveland Press, 1999), 51–65.

49. Victor E. Kappeler, *Critical Issues in Police Liability* (Prospect Heights, IL: Waveland Press, 2001).

50. George L. Kelling, *Broken Windows and Police Discretion* (NCJ 78259). (Washington, DC: U.S. Department of Justice, October 1999), http://www.ncjrs.gov/pdffiles1/nij/178259.pdf#search=%22broken%20windows%20police%20action%20legal%20challenges%22 (accessed August 27, 2006).

51. Roben Farzad, "Jail Terms for 2 at Top of Adelphia," *New York Times*, June 21, 2005, http://www.nytimes.com/2005/06/21/business/21rigas.html?_r=1&scp=1&sq=Jail+Terms+for+2+at+Top+of+Adelphia&st=nyt&oref=slogin (accessed March 3, 2008).

52. Masterson and Stevens, "The Value of Measuring Community Policing Performance in Madison, Wisconsin."

53. George L. Kelling and Kathleen Cole, *Fixing Broken Windows* (New York: Free Press, 1996), 196.

54. Kelling and Cole, *Fixing Broken Windows.*

55. Kelling and Cole, *Fixing Broken Windows*, 201.

56. Patricia K. Smith, "The Economics of Anti-begging Regulations," *American Journal of Economics and Sociology* (2005), http://www.findarticles.com/p/articles/mi_m0254/is_2_64/ai_n13729927 (accessed July 30, 2006).

57. Walker, "'Broken Windows' and Fractured History," 63.

58. Samuel Walker and Charles M. Katz, *The Police in America: An Introduction*, 5th ed. (New York: McGraw Hill, 2008), 26–31.

59. John P. Crank, *Understanding Police Culture*, 2nd ed. (Cincinnati, OH: Anderson Publishing, 2004).

60. Janis L. Schubert, "Identifying Crime Hot Spots in Space and Time: An Analysis of Dallas Police Department of Criminal Activity," GISC 6387, GIS Workshop, Summer 2004.

61. Robert J. Sampson and Stephen W. Raudenbush, "Disorder in Urban Neighborhoods: Does It Lead to Crime?" (NCJ 186049), National Institute of Justice (2001), http://www.ojp.usdoj.gov/nij (accessed November 11, 2007).

62. Sampson and Raudenbush, "Disorder in Urban Neighborhoods."

63. Michael R. Gottfredson and Travis Hirschi, *General Theory of Crime* (Stanford, CA: Stanford University Press, 1990), 18.

64. Gottfredson and Hirschi, *General Theory of Crime*.

65. Walker, "'Broken Windows' and Fractured History," 63.

66. David L. Carter, "Foreword," in *Case Studies in Community Policing*, ed. Dennis J. Stevens (Upper Saddle River, NJ: Prentice Hall, 2001), xiii.

67. Bureau of Justice Statistics, *Local Police Departments 2003* (NCJ 210118) (Washington, DC: U.S. Department of Justice, May 2006), http://www.ojp.usdoj.gov/bjs/pub/pdf/lpd03.pdf (accessed July 2, 2007).

68. Stevens, *Applied Community Policing in the 21st Century*, 17.

69. Office of Community-Oriented Policing Services, Community Policing Topics (Washington DC: U.S. Department of Justice), http://www.cops.usdoj.gov/ (accessed March 3, 2008).

70. Office of Community-Oriented Policing Services, News and Events (Washington DC: U.S. Department of Justice), http://www.cops.usdoj.gov/Default.asp?Item=2025 (accessed March 3, 2008).

71. Wayne W. Bennett and Karen M. Hess, *Management and Supervision in Law Enforcement*, 5th ed. (Belmont, CA: Wadsworth, 2007), 22. Also see Community Policing Consortium, "What Is Community Policing."

72. Stevens, *Applied Community Policing in the 21st Century*, 17.

73. Community Policing Consortium, "What Is Community Policing."

74. *Excellence in Problem-Oriented Policing: The 1999 Herman Goldstein Award Winners* (Washington, DC: U.S. Department of Justice, August 2000), http://www.ojp.usdoj.gov/nij/pubs-sum/182731.htm (accessed July 3, 2007).

75. Ford Foundation (1988), http://www.fordfound.org/ (accessed July 4, 2007).

76. *Excellence in Problem-Oriented Policing*.

77. San Antonio Texas Police Department, http://www.sanantonio.gov/SAPD/?res=800&ver=true (accessed March 3, 2008).

78. Bennett and Hess, *Management and Supervision in Law Enforcement*, 450–451.

79. A. P. Cardarelli, J. McDevitt, and K. Baum, "The Rhetoric and Reality of Community Policing in Small and Medium Sized Cities and Towns," *International Journal of Police Strategies & Management* 21, no. 3 (1998): 397–415.

80. Stevens, *Case Studies in Community Policing*.

81. Stevens, *Applied Community Policing in the 21st Century*; Stevens, *Case Studies in Community Policing*.

82. Stevens, *Case Studies in Community Policing*.

83. Kent County Sheriff's Department, Michigan, "Community Police Report" (February 2004), http://www.accesskent.com/CourtsAndLawEnforcement/SheriffsDepartment/sheriff_index.htm (accessed June 22, 2005).

84. Egon Bittner, *The Functions of Police in Modern Society* (New York: Aronson, 1975), 53.

85. Jennifer T. Comey, Jill Dubois, Susan M. Hartnett, Marianne Kaiser, Justine H. Lovig, and Wesley G. Skogan, *On the Beat: Police and Community Problem Solving* (Nashville, TN: Westview Publishing, 1999).

86. Benjamin Chesluk, "Visible Signs of a City Out of Control: Community Policing in New York City," *Cultural Anthropology* 19, no. 2 (2004), 250–275.
87. Stevens, *Applied Community Policing in the 21st Century*, 13.
88. Jo Ann Riley, Deputy Sheriff with the Palm Beach Sheriff's Office, Palm Beach, Florida, personal communication with author, June 12, 2002.
89. Officer Gary Palmer, Chicago Police Department, personal communication, July 2006.
90. John P. Crank, *Understanding Police Culture* (Cincinnati: Anderson, 2004), 316.
91. Kent R. Kerley, "Perceptions of Community Policing across Community Sectors: Results from a Regional Survey," in *Policing and Community Partnerships*, ed. Dennis J. Stevens (Upper Saddle River, NJ: Prentice Hall, 2002), 93–110; Walker, *Sense and Nonsense about Crime and Drugs*, 22.
92. Walker and Katz, *The Police in America*, 138–139, 159.
93. Charlotte police officer and former student who wishes to remain anonymous, personal communication with author, February 2004.
94. Boston police officer at a neighborhood association meeting who prefers to remain anonymous, personal communication with author, April 2002.
95. Boston Police Officer Brian Thomas at a community policing meeting, personal communication with author, May 7, 2005.
96. Boston Police Officer Bill Sullivan, personal communication with author, May 4, 2004.
97. Mary Ann Wycoff, "Investigations in the Community Policing Context" (August 6, 2001), http://www.ncjrs.org/pdffiles1/nij/189568.pdf (accessed June 26, 2004).
98. Detective employed by Boston Sexual Assault Unit detective who wishes to remain anonymous, personal communication with author, April 2003.
99. Victor E. Kappeler, Michael Blumberg, and Gary W. Potter, *The Mythology of Crime and Criminal Justice* (Prospect Heights, IL: Waveland Press, 1998).
100. For the full account, see Masterson and Stevens, "The Value of Measuring Community Policing Performance in Madison, Wisconsin."
101. Masterson and Stevens, "The Value of Measuring Community Policing Performance in Madison, Wisconsin." Also see Michael F. Masterson and Dennis J. Stevens, "From Polarization to Partnerships: Realigning the Investigative Function to Serve Neighborhood Needs Rather Than the Bureaucracy's Behest: The Change Experience of the Madison, Wisconsin, Police Department," in *Policing and Community Partnerships*, ed. Dennis J. Stevens (Upper Saddle River, NJ: Prentice Hall, 2002), 137–162.
102. Websdale, *Policing the Poor*.
103. James P. Lynch, William J. Sabol, Michael Planty, and Mary Shelly, "Crime, Coercion and Community: The Effects of Arrest and Incarceration Policies on Informal Social Control in Neighborhoods" (NCJ 195170), (2002), http://www.ncjrs.gov/pdffiles1/nij/grants/195170.pdf (accessed November 11, 2007).
104. Boston police commander who wishes to remain anonymous, personal communication, October 20, 2006. The police commander requested anonymity primarily because there is a directive that police managers are prohibited from providing their professional opinion about police initiatives to the public established by the former Boston Police Commissioner Kathleen O'Toole.
105. Lawrence W. Sherman, "Community and Crime Prevention," in *Preventing Crime: What Works, What Doesn't, What's Promising* (1993), Chapter 3, http://www.ncjrs.gov/works/chapter3.htm (accessed November 11, 2007).

Modern Policing, Quality-of-Life Initiatives, and Privatization

Change has a considerable psychological impact on the human mind. To the fearful it is threatening because it means that things may get worse. To the hopeful it is encouraging because things may get better. To the confident it is inspiring because the challenge exists to make things better.

—King Whitney, Jr.

▸ ▸ LEARNING OBJECTIVES

When you finish reading this chapter, you will be better prepared to

- Characterize compelling reasons why police organizations must change
- Describe the lessons learned from preventive or proactive police strategies
- Characterize the relevance of evidence-based policing and its primary component
- Compare and contrast the four stages of policing in the twentieth century
- Identify studies that influenced the development of modern police practices
- Characterize the centerpiece and components of quality-of-life (QOL) policing initiatives
- Describe the QOL initiatives used by the police in Chicago and New York City
- Describe traffic QOL initiatives
- Characterize the relationship between ordinances and QOL initiatives
- Characterize government incentives to guide privatization

▸ ▸ CASE STUDY: YOU ARE THE POLICE OFFICER

You've been an officer in an urban American city for the past six years since leaving college. You love your job. Every day on the job is exciting, but every day you can see the changes in the area where you patrol. Over the course of the last few years, the population has changed from predominately African Americans to Latino, and there are pockets of residents from Southeast Asia as well.

Policing equipment and techniques have changed, too. The new computers and video equipment in your police vehicle requires more training. Protocols and regulations changed, too, so that you had to return to the academy to learn the new "use of force" protocols. When you were at the academy the last time, you also learned about evidence-based studies that supported police performance.

In addition, you have noticed that policy changes associated with everyday practices—that is, the things you can do and the things you can't do—require a great deal of attention. What you and other officers were told to do last week is not what you were told to do today.

Early roll calls in your career were built around aggressive police interventions, which were geared toward diffusing potentially violent situations and preventing incidents from escalating into dangerous situations. Then a new chief directed that new strategies be laid out at roll call. Aggressive police actions were replaced with "swipes" into gang areas. Now, before you can make an arrest for a street violation such as vagrancy, if the applicable regulation wasn't in your roll call handouts, you have to communicate with your supervisor to see if an ordinance exists. Fleeing-felon chases also have to be approved by your supervisor, which seems to reduce police reaction time. Today the buzzwords relate to quality-of-life arrests and protocols covering everything from stop and search to conducting an arrest.

Looks like you'll be a regular at the police academy if you want to keep up with all these changes.

1. In what way do population changes lead to changes in police practices?
2. In what way do the changes referred to in this case represent a positive step?
3. What would make quality-of-life arrests different than other arrests?

Introduction

Chapter 3 provided a look into the broken windows metaphor, a perspective that helps shape community relations strategies. With this understanding, the stage is set to consider why police organizations must change. That is one focus of this chapter, which will characterize the way policing initiatives changed from the short-lived community relations era to the quality-of-life era, whose centerpiece is prevention and initiatives targeting public spaces. Unlike community policing efforts, quality-of-life (QOL) policing does not include an empowered community partnership associated with the strategy—that is, it does not emphasize collaboration between the community and police strategy. Instead, QOL policing favors the legislation of ordinances that aid aggressive police strategies to manage public spaces and intrude upon violators, leading to lawful arrests of these offenders.

Organizational Resistance to Change

One product of an enhanced police organization is the notion that **probable cause arrests**[1] can translate into higher clearance rates and, ultimately, higher conviction rates.

Probable cause can be defined as a reasonable belief, based on facts and circumstances that a person is committing, has committed, or is about to commit a crime. Also see Chapter 10 for more details.

This thought is consistent with the viewed espoused by researchers who define "reinventing government" as "the fundamental transformation of public systems and organizations to create dramatic increases in their effectiveness, efficiency, adaptability, and capacity to innovate."[1]

How innovative are police agencies? Ten years ago or even two years ago, there was a lot of discussion about myspace.com and its potential to further crime, especially by pedophiles. By 2008, however, only a few police departments had profiles on myspace.com, including police departments in Chicago, Illinois; Newton, New Hampshire; Smithville, Ohio; Haverhill, Massachusetts; Newburyport, Massachusetts; and Miami–Dade County, Florida.[2] Most say they started the myspace pages to provide crime tips. The Miami–Dade County Sheriff's website urges parents to have their teens select the department as an electronic "friend" to scare off sexual predators. If the police department is chosen as one of the top friends, its logo appears when the page opens. To date, more than 1,000 people have asked the department to be their friend. "It acts as a deterrent for an online predator," says Sergeant Erik Palmer, who works in Miami's Internet crimes unit and created the page. "We're just a click away."

Haverhill, Massachusetts, police launched a myspace.com profile in the fall of 2007 to help investigate crimes. Crime analyst Kristina Perocchi says she gets two or three tips each week from the site, some of which have led to drug-related arrests.

Myspace.com and other avenues of change have been around for a while, so why haven't more police organizations sought to use them to enhance the delivery of police services and bring police closer to their mission? Many police departments resist change and are traditional in their approach.

> Despite the challenges, demands, and mandates faced by twentieth-century police organizations, most continue to demonstrate organizational inflexibility, follow an unprofessional or personal approach toward management, and are resistant to external influences of change.[3]

Compelling Reasons for Police Organizations to Change

Police organizational change would allow departments to keep pace with the changes occurring in American life. The thought that thousands of police agencies across the country operate as though they were little kingdoms[4] may have a certain appeal—but the reality is that, in the twenty-first century, there is little room for individuality among agencies. Let's consider some compelling reasons recommending changes in how U.S. police departments operate.[5]

Terrorism Changes

The advancement of both domestic and foreign terrorism argues for a coordinated police approach. Willard Oliver suggests that the demands placed on police recently have changed the way they can deliver police services and bring a new urgency to update training and jurisdictions in delivery of routine performance.[6] Consider, for example, the standard operating procedure of addressing a suspect and telling him or her to drop the weapon and put his or her hands up or freeze. That procedure will not work with a suicide bomber; in fact, officers who use this tactic will essentially be signing their own death warrant. Improvements in "intervention" techniques geared toward stopping terrorists before they can execute a threat could mean the difference between successfully taking a suspect into custody and facing serious collateral damage.

Criminal Changes

The sophistication and tactical advancement of offenders are often beyond the reach of most highly trained police organizations, as evidenced by the declining trend in major case clearance rates (see Chapter 1 for more details or the FBI-UCR website) and the effectiveness of interdiction and immigration strategies currently utilized by many police agencies, including those that are part of the federal government.[7] In 2004, violent crime clearance rates averaged 46 percent, based on 1.1 million crimes; by contrast, in 1994, the index was 45 percent—but was based on 1.7 million crimes.[8] In theory, the commission of fewer crimes should result in higher clearance rates.

Demographic Changes

Needs of communities have changed in recent years, as increasing racial, ethnic, and sexual diversity have made community orientation more complex and led to different policing priorities. Population shifts and new arrivals complicate the ability of departments to provide police services in an appropriate manner. For instance, the Boston police reported that they received more than 600,000 911 calls in 43 languages in 2004.[9] Furthermore, individuals not represented in the police organizational chain of command want to be part of this hierarchy—and should be. Thus police departments face new challenges in dealing with evolving communities and in their hiring and disciplinary processes.[10]

Technological Changes

Changes in technology are coming at a fast and furious pace. Computers, Compstat (discussed in Chapter 1), geographic information services (GIS) crime mapping, cell phones, and forensic science techniques such as DNA analysis and automated fingerprint identification systems (AFIS) are all part of the modern-day police arsenal; residents have real-time information relating to crime; and some departments provide officers with their own computers.[11] **COPLINK** displays data rapidly in a computer web-based format that has revolutionized police communications across the country (as discussed in detail in Chapter 10).[12]

Of course, technological change is a double-edged sword, in that criminals may also use new technologies to commit crimes. This fact may motivate police departments to rethink the deployment of their personnel. For instance, the U.S. government spends approximately $4.5 billion per year on defense against cyberterrorism.[13] Symantec, the world's largest provider of virus protection software, reports that "software holes" leading to vulnerabilities in programs and systems increased by 80 percent in 2002. Local police departments realize they must aid the federal government in combating cybercrime because the victims of cybercrimes are local residents.

Economic Changes

The United States has enjoyed an era of unparalleled growth in recent decades, and many local governments—especially those relying upon property and sales taxes for their financial support—have enhanced their tax base. Concurrently, revenues available for law enforcement agencies, including federal funding, have dramatically increased but only in certain areas such as military technology. At the same time, with unemployment rates at one of the lowest levels in history, law enforcement finds itself competing with the higher pay and better benefits offered by the private sector—and by other police departments—in the quest to hire the best and brightest young people beginning their professional careers.[14]

Environmental Changes

Environmental changes now pose a major concern to policing. In such states as California, Texas, Florida, and Arizona, the infrastructure sometimes cannot handle the explosion in the local population. Urban sprawl, traffic congestion, and water restrictions have all become police matters. Disasters, from hurricanes to tornadoes to fires, increasingly occupy the attention of police agencies and personnel.[15]

Political Changes

Political change has a tremendous effect on police agencies and their personnel. For instance, in the 1990s and early 2000s, federal funding was made available to local agencies in unprecedented amounts.[16] (See Chapter 3 for more details.) Since the turn of the century, however, federal spending and funding accountability have changed dramatically.[17] For example, in 2006, the George W. Bush administration sought to cut the Edward Byrne Memorial Justice Assistance Grant program, which doles funding to local agencies; and in 2007 the program was "not able to demonstrate an impact on reducing crime."

At the neighborhood level, police have increasingly encountered an increased focus on communitarianism and the emergence of strong grassroots involvement in recent years. Put bluntly, citizens want to be involved in the community governance. As a result, community-linked criminal justice (i.e., policing, victim services, corrections, and prosecution) is increasingly the norm, and criminal justice agencies continue to remold their philosophy, structure, and tactics to meet community expectations and needs.[18] Additionally, political change may be driven by politically aggressive organizations such as the American Civil Liberties Union (ACLU).[19]

Institutional Changes

Most American social institutions are changing, and those changes often directly affect police policy and practice in ways that make it more difficult for local agencies to meet their mission of ensuring public safety. For instance, a typical strategy that is gaining acceptance among local and state police agencies is to partner with other state agencies, such as community corrections and private health providers, to support discretionary programs among constituents. Some police agencies are also collaborating with religious leaders and their organizations to combat street gangs and to further community relations programs in places such as Chicago and New York.[20] Local police agencies and federal enforcement agencies may work together in "sting" and undercover enforcement initiatives that provide local agencies with access to federal agencies, training, and funding (see Chapter 11 for more details). Linking homeland security and emergency management is also seen as a logical step in formulating an effective local policy, which includes emergency relief linked to national disasters such as hurricanes, forest fires, and terrorist destruction.[21]

Many of these institutional changes are directly related to the political changes mentioned earlier, especially changes related to funding for police organizations. In particular, partnerships between police and either private providers or religious institutions were rare in the past.

Legal Changes

The legal changes that have taken place in the last few decades and have been driven by private agencies such as the ACLU which affects the way police provide services. Equally important, they reflect the higher level of sophistication of the general public, whose members often resort to litigious methods to safeguard human rights.[22] (See Chapter 14 for more detail on police accountability.) The legal dos and don'ts of street officer performance are different than they were 10 or 20 years ago, and officers are more likely to be defendants in civil rights or police misconduct cases than at any other time in police history.[23]

Professional Changes

Changes within the police profession are observed and monitored by personnel, other institutions, family members, researchers, and collective bargaining agents such as unions and fraternal organizations such as Fraternal Order of Police (see Chapter 7). Also, numerous changes have altered the law associated with personnel, their performance, and the human rights of criminals and suspects. Police organizations must address those changes or face litigation, state or federal intervention, or some other form of intervention from outside organizations.[24]

Preventive or Proactive Strategies

Funds to aid in the development of proactive police agencies were provided by the 1994 Violent Crime Control and Law Enforcement. In recent years, the funds made available for these endeavors have been declining.

> When police agencies develop and practice preventive or proactive strategies, crime has been more efficiently deterred, the professional nature of officers has been enhanced, and compliance by the public to police directives has been more likely.[24]

Scientific Police Management

The trend of scientific police management (SPM) began with the passage of the Omnibus Crime Control and Safe Streets Act in 1969 at the behest of the Law Enforcement Assistance Administration (LEAA). The Exemplary Projects Program created by this Act led to numerous studies of policing (see details later in this chapter and in

Chapter 9). The use of the scientific method in these investigations has produced an extensive body of professional research, which can in turn guide policymakers through an informed decision-making process. Because they lacked such hard evidence, policymakers in the past often proposed subjective policies that unwittingly reduced the quality of police services.

> **Scientific police management (SPM)** is the application of social science techniques to the study of police administration for the purpose of increasing effectiveness, reducing the frequency of citizen complaints, and enhancing the efficient use of available resources.

■ Evidence-Based Policing

Evidence-based policing practices utilize the results of scientific studies as advanced by researchers such as Lawrence W. Sherman and others.[26] They also grew out of the application of the scientific method of discovery, which is used in many academic disciplines—for example, in business management and the social sciences (e.g., psychology, sociology). Today, evidence-based policing is fast becoming a standard in most academic criminal justice departments.

> **Evidence-based policing** is the use of the best available research on the outcomes of policing to implement policy, regulations, and training among police agencies. The process employs the scientific method of discovery, which is best suited to support an informed decision-making process.

The Kansas City Experiment

The most famous application of evidence-based principles was the **Kansas City Preventive Patrol Experiment** (see Table 4–1).[27]

Until the results of this experiment became known, police patrol strategies had centered on two assumptions:

- Visible police presence prevents crime by deterring potential offenders.
- The public's fear of crime is diminished by such police presence.

The Kansas City experiment revealed that traditional routine preventive patrol has no significant impact either on the level of crime or on the public's feeling of security. It was also learned that random patrols are not as relevant as directed patrols (see Chapter 9).

Domestic Violence

The experiment in Minneapolis related to domestic violence in 1984 was one of the first scientifically engineered social experiments to test the impact of arrest (versus forms of disposition) on domestic abuse (Table 4–2).[28] Domestic violence was a "hands off" approach up to this time. Arrest was a last resort. Domestic violence was

TABLE 4–1 The Kansas City Experiment

The experiment began in October 1972 and continued through 1973; it was administered by the Kansas City Police Department and evaluated by the Police Foundation.

Patrols were varied within 15 police beats. Routine preventive patrol was eliminated in five beats, labeled "reactive" beats (meaning officers entered these areas only in response to calls from residents). Normal, routine patrol was maintained in five "control" beats. In five "proactive" beats, patrol was intensified by two to three times the norm.

The experiment asked the following questions:

- Would citizens notice changes in the level of police patrol?
- Would different levels of visible police patrol affect recorded crime or the outcome of victim surveys?
- Would citizen fear of crime and attendant behavior change as a result of differing patrol levels?
- Would their degree of satisfaction with police change?

Information was gathered from victimization surveys, reported crime rates, arrest data, a survey of local businesses, attitudinal surveys, and trained observers who monitored police-citizen interaction. "Ride-alongs" by observers during the experiment also revealed that 60 percent of the time spent by a Kansas City patrol officer typically was noncommitted. In other words, officers spent a considerable amount of time waiting to respond to calls for service. And they spent about as much time on non-police related activities as they did on police-related mobile patrol.

TABLE 4–2 The Minneapolis Domestic Violence Experiment

The Minneapolis Domestic Violence Experiment looked at the effectiveness of methods used by police to reduce domestic violence, using 25 randomized field experiments. Cases used in the study were misdemeanor assault calls, which make up the bulk of domestic violence calls. Both the victim and the offender needed to be present when the police arrived, in order to be included in the study. Fifty-one patrol officers in the Minneapolis Police Department participated in the study. Each was asked to use one of three approaches for handling domestic violence calls, in cases where officers had probable cause to believe an assault had occurred:

1. Send the abuser away for eight hours.
2. Advice and mediation of disputes.
3. Make an arrest.

The study lasted approximately 17 months and included 330 cases.

considered a private family matter, one that did not require police intervention. Findings showed that arrests deterred domestic violence, which in turn supported the adoption of mandatory arrest policies that kick in when police are called to domestic violence scenes (see Chapter 9 for more details).

Cars and Drugs

Problem-oriented policing in Newport News, Virginia, was a scientific test comparing traditional incident-driven policing to problem-oriented policing, which mobilizes police agencies to respond to citizen complaints and offenses.[29] In this case, constituents had complained about a shipyard where more than 36,000 cars were parked while awaiting delivery. Numerous strangers were seen in the area, and many of them broke into those cars. Police subsequently discovered that one purpose of those break-ins was to retrieve drug deliveries from foreign ports, proving that police solutions to community problems can, indeed, aid crime control.

Community Revitalization

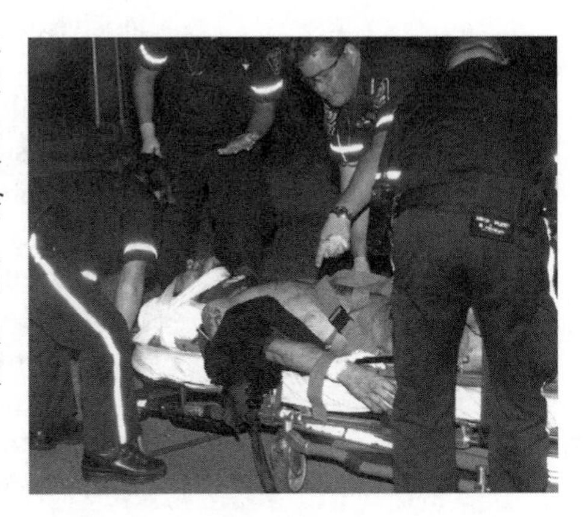

Weed and Seed studies are community anticrime approaches that link intensified geographically targeted enforcement efforts involving police and prosecutors with local neighborhood improvement initiatives and human services programs.[30] For example, in Wilmington, Delaware, community revitalization has been accomplished by implementing a broad array of social, economic, and recreational programs in targeted neighborhoods, once violent felons and drug traffickers have been removed from the area.[31] Achieving community revitalization requires substantial cooperation between federal, state, and local agencies, along with local business and community organizations. This type of initiative focuses on four key areas:

- Law enforcement
- Community policing
- Prevention, through intervention and treatment
- Neighborhood restoration

The U.S. Department of Justice collaborated with the Ministry of Caring, Inc., to initiate the "seeding" efforts of the Weed and Seed strategy in Wilmington. The Ministry of Caring is a community-based nonprofit organization that has been serving the poor and the homeless throughout Wilmington for 25 years.

Guns and Homicide

Efforts to reduce gun violence in Indianapolis were based on a study that found targeted patrols effectively decreased gun-related crime to a substantially greater extent than did routine patrols, a finding consistent with the goals of **directed patrol** initiatives.[1,32]

> **Directed patrol** refers to high police visibility and concentrates on police visibility where there is crime. The theory behind directed patrol is that the more police presence in hot spots the less likely crime will occur because it increases the threat of detection and punishment for criminal activity.

The Indianapolis results were congruent with the results of studies of targeted patrols' effects on homicide rates in California.[33]

Two variations of this strategy exist: general deterrence and specific deterrence. General deterrence involves saturating a geographic area with police presence, including stops of as many persons as possible for all (primarily traffic) offenses.

Specific deterrence focuses police patrols on certain behavior, individuals, and/or places. With both strategies, perceptions of the certainty of punishment for crime tend to increase, thereby deterring individuals (at least temporarily) from illegally carrying guns and reducing the likelihood of lethal violence in the streets.[34]

Suicide Bombers

Analysis of terrorist-related studies has demonstrated a need for a centralized, comprehensive database that will enhance police strategies for dealing with both foreign threats and local criminal activities.[35] For instance, understanding that certain personality traits are typically encountered among suicide attackers is relevant to the identification and apprehension of suicide terrorists (and predators). According to evidence-based studies, suicide-intent terrorists possess weak personalities; are socially marginalized; are subject to rigid, concrete thinking; and demonstrate low self-esteem. Motivating factors often cited by suicide attackers include national humiliation, religion ("to do God's will"), personal revenge, and admittance to paradise in the afterlife.

The influence of the terrorist group over particular individuals who are planning suicide attacks is also instrumental in success. Following their recruitment into a terrorist organization, individuals make a commitment to the group in the form of a contract, which leads to a personal commitment to the mission.

Computer and Digital Crime

Studies on digital evidence have shown the way for investigators to leap ahead in the war on cybercrime and develop procedures for identifying and processing electronic evidence.[36] Police departments must educate officers on the myriad ways in which computers can be used to facilitate criminal acts, resulting in stronger cases and leading to more convictions. First on the agenda is an assessment of the tools that officers require to catch cybercriminals and, at the same time, allow them to stay ahead of the electronic technology curve. The point here is that police must look beyond the immediate horizon, and research can aid them in accomplishing that task and thereby providing quality police services.

Police Integrity

Studies on police integrity conducted in numerous jurisdictions in 2005 can provide police agencies with the tools and the techniques they need to discover the level of understanding that their officers have about agency rules *before* officers engage in wrongful acts (see Chapter 14).[37] Research in this area has revealed that an agency's culture of integrity, as defined by clearly understood and implemented policies and rules, is more important in shaping the ethics of police officers than is the hiring process.

Unconstitutional Searches

In one study, researchers who were assessing police behavior during searches asked three questions:

- How frequently do patrol officers engage in searches?
- How often do their searches meet constitutional standards?[38]
- What explains the proclivity to search unconstitutionally?

The results of this research paint a disquieting picture, with nearly one third of the searches unconstitutionally performed and almost none were visible to the courts. There are substantial costs when police search unconstitutionally—namely, diminishing of the rights of individuals and the legitimacy of policing. What was learned from this and other studies is that police need to learn how to reduce constituent unconstitutional practices, which will in turn reduce civil litigation against both officers and departments.

Methamphetamine

Everything about "meth"—from its composition to its manufacturing and distribution to the physical effects upon its users—is unique.[39] These distinctions require that police officers adopt specialized approaches to criminal investigations and arrests involving methamphetamine. Unlike heroin or cocaine, methamphetamine is easy to manufacture domestically using information widely available on the Internet or in underground publications. In fact, anyone with high school chemistry experience can "cook" methamphetamine, and the chemicals used in its production are household or farm products that are not feasible to regulate. Without research into the simple chemistry underlying this drug epidemic, policing might still be in the dark ages when combating the meth problem.

Sexual Assault

A study that examined the attitudes of first responders toward 911 calls led to guidelines intended to improve conviction rates for sexual assaults.[40] Because sexual assault often goes unreported or is under-reported, first responders may not initially recognize the extent of the person's victimization. When sex offenders are apprehended, many cases are dismissed because of a lack of evidence or a credible witness. The findings of the previously mentioned study revealed that training and supervision styles affect sex offender detection and conviction rates, and probably most criminal convictions. It is recommended that Total Quality Management (TQM) styles replace the hierarchy of command in dealing with these types of offenses (also see Chapter 7).

> **Total quality management** is a personnel and constituent focused management system that aims at continual increase in satisfaction through quality circles which translates to continually lower real costs and greater compliance.

Foot Patrol

A study in New Jersey found that foot patrols had little effect on violent crime rates but did reduce the fear of crime among community members.[41] In this study, foot patrol officers immersed themselves in the lives of their neighborhoods and often redefined the conditions and manners under which prohibited activities could be conducted. For example, drinking in public was okay if the bottle was hidden in a brown paper bag.

The use of the scientific method to examine policing strategies has already provided much valuable information that has then been used to guide police policymakers through an informed decision-making process. In particular, evidence-based policing has modernized police agencies by pointing out the benefits of quality-of-life initiatives.

■ The Evolution of Community Relations Initiatives

Community relations strategies such as community policing initiatives play a role in police outreach. Nevertheless, this approach has often failed to reach expectations because community members avoid meetings and responsibility and because police officials are reluctant to empower officers and community members to carry out problem-solving strategies.[42] Reasons for community member avoidance can include language barriers, cultural barriers, and personal privacy issues.[43] Community Alternative Police Strategies (CAPS) in Chicago, for instance, discovered after lengthy community problem-solving sessions were conducted throughout the city (at the cost of millions of dollars) that large populations within the city were rarely represented at meetings, despite extensive efforts by Chicago police and community members to enlist their attendance.

One problem facing police in implementing community relations strategies was that empowerment "contradicts many of the tenets that dominated policing thinking for a generation."[44] This thought is consistent with other findings that reveal commanders routinely resist delegating authority and responsibility to community policing officers.[45] Some critics have suggested that many departments accept federal funding to implement a community policing program, only to find that differing priorities and lack of funding prevent adequate fulfillment of community policing goals.[46] Others add that because federal funding efforts have little empirical evidence behind them to guide their implementation process, it is difficult to tell how well or how poorly federal funding affects crime rates.[47] The federal government is attempting to remedy flaws in its evaluation policy linked to federal funding among police agencies, but it may take some time for those studies to catch up with current trends.

Lawrence W. Sherman shares his thoughts about community relation programs, which sums up matters quite well:

> Ironically, a central tenet of community prevention programs has been the empowerment of local community leaders to design and implement their own crime prevention strategies. This philosophy may amount to throwing people overboard and then letting them design their own life preserver. The scientific literature shows that the policies and market forces causing criminogenic community structures and cultures are beyond the control of neighborhood residents, and that "empowerment" does not include the power to change those policies. It is one thing, for example, for tenants to manage the security guards in a public housing project. It is another thing entirely to let tenants design a new public housing policy and determine where in a metropolitan area households with public housing support will live.[48]

Official Rhetoric versus Actual Practice

The community problem-solving movement, which had begun in the 1980s, took a different path around 1990.[49] It has been suggested that the community policing era of the 1980s was dominated by police rhetoric, rather than by police practice accompanied by facilitative police leadership and community partnerships, contrary to the strategy suggested by the best information available about the uniqueness and dominance of community policing.[50]

One way to think about the difference between official rhetoric and actual practice is to think of the data produced by evidence-based police studies versus "window dressing" claims. Few studies support the contention that community policing is particularly effective.[51] Even the studies mentioned earlier in this chapter support the effectiveness of only some components of community policing, as opposed to community policing directly. It is unlikely that community policing in its purest form exists along a continuum.[52] Instead, "community policing" is a buzzword that has the power to attract federal grants and garner some community cooperation for a short period of time. Indeed, some experts argue that community policing is simply used as window dressing to obtain federal funds.[53]

In the latter part of the 1990s, this situation changed somewhat. Order-maintenance policies in various forms dominated police practices during that time period. In addition, community relations was retooled into a workable police initiative: quality-of-life policing, which continues to dominate police strategies in most jurisdictions today.

The Quality-of-Life Era

The quality-of-life (QOL) era includes the period from 1990 to present (see Table 4–3). In this approach to policing, leadership tends to develop a decisive, mandated policy about accountability; this policy typically emphasizes detection and arrest of suspects who commit less serious crimes in an effort to prevent them from graduating to more serious crimes. Rather than centralizing command and mandates in one location, QOL initiatives spread command responsibilities out, resulting in a flattening of the hierarchy of command and an increase in the **span of control** (the number of officers reporting to a single supervisor). For example, officers may participate in broadly based task forces with a single objective (e.g., tri-county drug stings).

QOL policing encourages cooperation and communications with other public police agencies, including federal agencies.[54] It also encourages private providers to train police officers in communications and field training. Given the civil liability issues associated with local training, private concerns and corporate training centers may well provide superior training for officers and minimize departments' exposure to public lawsuits (see Chapter 14 for more details on civil liabilities suits).[55]

In QOL programs, top command typically orders officers to target minor infractions that might potentially lead to more serious crimes, such as those committed by gun-toting thugs who intend to use those weapons to fight rival gangs. The tactics pursued as part of QOL initiatives can include zero tolerance (although swooping down on any and all in-transit vehicles is a thing of the past). At an extreme, officers may pursue a policy of arresting all persons for any infraction of the law, with no warnings being given. Other tactics also include aggressive action against loitering by gang members, camera surveillance, empowering officers to implement the pro-arrest policy, tactical units, superior equipment, use of computers, and improvements in communication. Expectations include order maintenance, resident satisfaction, officer satisfaction, and stabilized social order. Outcomes have been seen in reduced crime rates, enhanced quality of life, and development of professionally trained managers and officers.

TABLE 4–3 Four Stages of American Policing

	Political Era (1840s–1930s)	Reform Era (1930s–1980s)	Community Era (1980s–1990s)	Quality-of-Life Era (1990s–Present)
Leadership	Primarily political	Law and professionalism	Community support (political), law, professionalism	Decisive command Managerial accountability
Objectives	Crime control* Quality of life Broad social services	Crime control Isolation from politicians Expected community passiveness	Crime control Crime prevention Problem solving	Detect and proactive arrests of less serious offenders to control serious crimes Public space control
Organizational design	Decentralized Geographical	Centralized Classical	Decentralized Task force matrices	Reduced hierarchy of command Increased span of control
Relationship to community	Close and personal Face-to-face intimacy	Professionally remote	Consultative Police defend values of law and professionalism, listen to community concerns	Minimal community input into strategies Cooperation and communication with police agencies Partnerships with federal law enforcement agencies Private communications and training
Demand	Managed through links between politicians and precinct captains	Channeled through dispatching activities	Channeled through analysis of underlying problems	Determine minor offenses leading to serious crimes
Tactics	Incident driven Foot beat patrol Rudimentary investigation	Preventive patrol Rapid response	Problem solving Community relations partnerships	Swipes and gang loitering initiatives, camera surveillance Empower officers toward pro-arrest policy Tactical units, equipment, computers, and communication
Expectations	Maintaining citizen and political satisfaction Neighborhood norms	Number of arrests, response time, number of passings* Crime rate	Quality of life Citizen satisfaction	Lower crime rates Higher arrest rates Aggressive police action Resident satisfaction Officer satisfaction Order in public places
Outcome	Political satisfaction Discrimination Police corruption Lack of agency control	Crime rose Fear of crime rose Unfair treatment Less financial support	Control of public spaces	Reduced crime rates Professionally trained managers and officers Less community participation More police power

*Ranked as a priority.

Source: The first three stages were adpted from George L. Kelling and William H. Moore, "The Evolving Strategy of Policing," *Perspectives on Policing Sreies* 4 (1988). The last stage was developed by the author based on the best available information.

Rudolph W. Giuliani, the former mayor of New York City, defined his police department's (NYPD) QOL initiative as follows:

> [Quality-of-life policing is] about focusing on the things that make a difference in the everyday life of all New Yorkers in order to restore this spirit of optimism—a spirit which hardly existed five years ago in New York City.[56]

The major difference between QOL policing and its predecessor, community relations, is the fact that QOL programs incorporate little, if any, community participation. In community relations programs, police relied on the good-faith cooperation of the communities they served to control crime. When this important link of trust is broken—as it already has been in many American minority communities—and those programs exist in rhetoric only, the impact on the effectiveness of the police can only be negative, meaning that police need to find new methods to serve and protect.[57] This final thought may be alarming to some who are concerned with policing policies; a lack of public participation in police policy could ultimately open the door to organized police power and less police accountability.[58]

Even though not all police initiatives in recent years have been formally labeled QOL strategies, the same characteristics appear in most aggressive police strategies of the present day, even when police ostensibly take different approaches. For instance, in the eyes of many experts, the tactics of the New York City, San Diego, and Boston police departments represent a new national model of policing, with reduced crime rates attesting to their effectiveness.[59] New York City has stressed tough enforcement, San Diego pioneered community and problem-solving policing, and Boston developed an approach combining evidence-based policing with community participation and local ministers while taking selective actions against the worst criminals. In all three cities (plus Baltimore, as discussed in Chapter 2), no matter what label was attached to the strategy—community policing, problem solving, community outreach, or get-tough approach—police have pursued a policy of proactive arrests. These arrests are considered justified (from the perspective of police officials such as Los Angeles Police Chief William Bratton[60]) because they "break the window" of opportunity for offenders.

> The primary goal of **quality-of-life policing** is to maintain order by having police take an active role in controlling public spaces—for example, by arresting suspects who commit nonserious and serious crimes under the aegis of proactive ordinances that prohibit behavior that influences crime and the fear of crime.

Does it matter what a police policy is called? Today, television advertisements routinely announce QOL programs with slogans such as "You buzz—you lose," or "Click it (seatbelts) or ticket." These campaigns reveal which specific behavior is being targeted and justify the police intrusion into the private lives of the public.

One underlying concept of QOL initiatives is the notion that social order cannot be maintained if the conduct of individuals leads to antisocial behavior or circumstances. Nonetheless, policing is being reconstructed across the United States—and globally for that matter—to reflect the QOL perspective. These police restructuring efforts all share two distinguishing features:[61]

- Separation of those who authorize policing from those who do it
- Transference of both functions away from government

Restructuring of the police is changing the way the police deliver services and the way they implement policing strategies.

Components of Quality-of-Life Policing

At the underlining belief of the QOL strategy is that crime is a sequenced order of events that starts in public places. Given this underpinning, QOL strategies are primarily designed to achieve the following goals:[62]

- Deter crime by reducing minor public disorder (on the premise that it sets off a chain of events), thereby leading to a reduction in serious crime and less crime overall
- Increase surveillance of targeted public places that are "hot spots" of criminal activity
- Conduct legal arrests through official ordinances that allow prosecutors to charge offenders with less serious crimes, and courts to render guilty verdicts in these cases when applicable
- Conduct pre-emptive intervention in public spaces
- Gather intelligence about those who use or frequent public spaces

Note that these strategies do not eliminate police's concerns with and focus on private places of crime.

Examples of Quality-of-Life Strategies

The Chicago QOL Initiative

Chicago's Anti–Gang-Loitering Order-Maintenance Policy, 1992

In 1992, the Chicago Police Department (CPD) began a QOL program that was intended to provide order maintenance by targeting gang members who congregated in public. Failure to obey a first-time order of a police officer to disperse was punishable by a fine of up to $500, by incarceration of up to six months, by community service of up to 120 hours, or by any combination of the three.[63] The city council held hearings prior to enacting the ordinance that mandated this policy and incorporated its findings in a preamble that set forth the underlying logic of the ordinance: "In many neighborhoods throughout the city, the burgeoning presence of street gang members in public places has intimidated many law-abiding citizens." The council explained, "Members of criminal street gangs avoid arrest by committing no offense punishable under existing laws when they know the police are present, while maintaining control over identifiable areas by continued loitering."[64] As a result, the council found, "aggressive action is necessary to preserve the city's streets and other public places so that the public may use such places without fear."[65]

Chicago Police Department

The CPD vigorously enforced the anti–gang-loitering ordinance, resulting in the issuance of more than 89,000 orders to disperse and the arrest of more than 42,000 people during the period 1993–1995. Nonetheless, the ordinance was challenged by the ACLU and was subsequently ruled unconstitutional in *City of Chicago v. Jesus Morales*.[66] Police officials then had to figure out a new way to resolve the problem and thereby improve QOL for community members.

Effectiveness of the Chicago QOL Program

Evidence-based research conducted by Dennis P. Rosenbaum and Cody Stephens looked closely at the issues of violence and crime control in Chicago and the effectiveness of the city's QOL initiatives.[67] In particular, homicide rates began to rise in 2002. To combat this trend, Chicago officials chose to focus on the role of gangs, guns, and drugs as contributing to violent street crime, especially homicide: More than half of all homicides in Chicago are gang related, and gangs control the multi-million-dollar illegal drug markets. Additionally, more than 70 percent of Chicago's homicides are committed with a firearm. The CPD's primary objective in 2004 was to crack down on these criminogenic activities by introducing a wide range of aggressive and visible enforcement programs in problematic locations.

Criminogenic activities are activities that are directly related to causation of crime. Gang association and drunkenness, for instance, are criminogenic activities that often lead to crime. Criminogenic risk factors are directly related to the probability of offending or reoffending. Risk factors are used in offender management to predict future criminal behavior and to assign levels and types of treatment services.

The CPD initiatives included the following pro-arrest and prevention strategies (for which the CPD's rating of the strategy's effectiveness toward meeting its mission is also given):

- Dispersals in hot spots (a tactic that dates back to the 1992 strategy, which was subsequently contested in court): 88 percent very effective or somewhat effective
- Curfew enforcement: 94 percent effective
- Gang cards: 91 percent effective
- Investigative alerts: 94 percent effective
- *Terry* stops and frisks:[68] 95 percent effective
- Vehicle impoundments: 98 percent effective
- Directed patrols: 98 percent effective
- Street-corner conspirators: 99 percent effective

All of these measures are discussed in depth later in this book.

Also as part of the CPD's effort to take back public spaces through QOL initiatives, the city employed civil remedies to clean up physical dilapidation, drug trafficking, and social disorder on private property, in a manner consistent with the broken windows theory. Such code enforcement has substantial benefits, in that it reduces the window of opportunity for crime. Indeed, one study found that narcotics-related crimes declined in targeted areas when individual drug houses were cleaned up.

Overall, Chicago's homicide rates dropped from more than 600 homicides per year in the five years before 2004 to 447 homicides in 2004; most other types of violence declined as well. Rosenbaum and Stephens report that although the CPD's "hot spots policing, targeted deployment, and enhanced activities were likely key contributors to the overall reduction of public violence and homicide,"[69] more data would help establish the fact that pro-arrest initiatives were responsible for the reduction in violent crime in Chicago (and probably elsewhere). The CPD generated exponentially higher levels of gun recoveries, hot spots dispersals, and contact card activities, to name a few. Nevertheless, Rosenbaum and Stephens remain reluctant to say that QOL is an effective strategy.

Today, QOL initiatives are broadly defined to include 90 percent of the work police officers perform, including the gang-control initiatives once outlawed by the court. For example, during the summer of 2006, 400 Chicago police officers and officers from the U.S. Drug Enforcement Agency made a "swipe" at one of Chicago's projects (Dearborn Homes) and made arrests of 47 Mickey Cobra gang members who distributed heroin and held the community in a "stranglehold."[70]

The New York City QOL Initiative

Once New York City enacted its proactive arrest policies in the 1990s, serious crime declined in the city.[71] But was the drop in crime attributable to the QOL policy or to some other cause? Some argue that crime dropped broadly across the United States as a result of socioeconomic factors, such as the improving economy, falling numbers of teenaged males, and declining use of crack cocaine. To determine whether crime reductions were truly related to QOL programs, researchers recommended that scholars look into police interventions—particularly the enforcement of laws against minor crimes, known as "broken windows" policing—as playing a major role in serious crime reductions.

One study reached the following conclusions:[72]

- More than 60,000 violent crimes were prevented from 1989 to 1998 because of the "broken windows" strategies employed in New York City.
- Other changes in police tactics and strategy could also be responsible for some of the decline in crime. For example, studies conducted in six New York police precincts in 2000 revealed that precinct commanders often used Compstat as a policing tool.
- As implemented by the New York Police Department (NYPD), "broken windows" policing is not the rote and mindless "zero tolerance" approach that critics often say it is. Case studies show that police vary their approach to QOL crimes, with their actions ranging from citation and arrest at one extreme to warnings and reminders at the other extreme, depending on the circumstances of the offense.

Nonetheless, QOL strategies removed common thugs from public spaces in New York by forcing them to be accountable: Offenders either stopped committing criminal activities on a small scale or faced arrest for doing so. Today, zero-tolerance policies generally translate into "swipes" (as practiced by the CPD, for example) for the purpose of making an arrest with less probable cause than expected, rather than consisting of the wide-scale intrusions into everyone's privacy in a specific area, such as occurred in the 1990s.

Traffic Enforcement Programs

Although stepped-up traffic enforcement is the most common type of QOL policing activity, so far research has failed to detect a relationship between intensive traffic enforcement and lowered crime rates.[73] One suggestion is that traffic enforcement may actually enhance QOL standards and, therefore, is most often associated with keeping the peace rather than with crime fighting. Think of it this way: Most adults drive cars, but only an estimated 9 percent of all licensed drivers (representing nearly 17 million drivers age 16 or older out of the 193 million drivers in the United States) were stopped by police in 2002.[74] One half of those people were stopped for speeding, and 84 percent of those drivers felt the officers were justified in stopping them.[75]

This snapshot undoubtedly presents an oversimplified version of police activities, yet current information indicates that most officers do, indeed, spend less time on crime-related matters than might be expected. Instead of dealing with crime, police officers often devote their time to providing first aid and ensuring the safety of accident victims, preparing paperwork documenting their activities, and attending court hearings. In large police departments, most officers are usually assigned to a specific type of traffic duty or other non-crime-related assignment.

■ Quality-of-Life Policing Models

Police QOL initiatives can be classified as following one of four order-maintenance models.[76]

In the *crime prophylactic model*, police intervention is intended to diffuse potentially violent situations and prevent them from escalating into criminal violence. An example relates to the previously mentioned loitering ordinance targeted at gangs by the city of Chicago. The law's purpose was prophylactic: to prevent crime before it occurred by banishing certain people from public places. It gave police wide powers to disperse any group of two or more persons who were in a public place "with no apparent purpose" if one of the persons was "reasonably suspected" of being a gang member. Failure to disperse resulted in arrest.

In the *police knowledge model*, non-crime-related calls give officers broader exposure to the community that results in the police having more knowledge that may help them solve crimes.

In the *social work model*, the covert coercive power of the police is used to help steer potential lawbreakers toward law-abiding behavior.

In the *community cooperation model*, effective responses to non-crime-related calls help the police build greater credibility and access to and with the public.

■ Use of Ordinances to Implement Quality-of-Life Policing

The police are generally clear about which behaviors to target (regardless of the public's position on the matter), but an ordinance or statute prohibiting problematic behavior is not always on the books.[77] An essential part of QOL police strategy is linked to the existence of an applicable statute, which can be created either by adapting a statute to fit the need and by passing new ordinances.[78]

For instance, Seattle—like many urban centers—perceives homelessness as being linked to a variety of problems for tourists, businesses, and residents (especially the elderly). The undesirable behaviors of homeless people, such as begging, lying on sidewalks, sleeping in parks, and urinating in public places, are part of the problem identified by the broken windows perspective. In the 1990s, Seattle police arrested many homeless people under a pedestrian interference ordinance to satisfy parents who were forced into the streets to avoid running over the homeless with strollers filled with children and elderly persons who used walking aids. When it came time to prosecute the individuals arrested, however, prosecutors had to show intent to commit a prohibited act because many of those arrested "for obstructing were mentally ill, drunk, or drugged so as to be indifferent to the effects of their behavior."[79] The city attorney had to develop a narrow ordinance aimed specially at the undesirable behaviors for police to make legitimate arrests. Of course, there was an outcry from advocates concerning the new ordinance. Nonetheless, when the police followed through on QOL arrests, they were performing within the legal limits of the law.

When the goal is to prosecute QOL violations, five concerns arise:

- Inapplicable statutes
- Need for new ordinances
- Making an arrest
- Charging the suspect
- Finding the defendant guilty of the prohibited behavior

Courts' acceptance of cities' QOL ordinances has varied widely. In the 1990s, for example, the NYPD successfully expanded its focus on reducing behaviors that detract from the overall QOL through aggressive enforcement of a variety of ordinances.[80] By contrast, in Los Angeles, a federal court deemed the city's "sleeping on sidewalks" ordinance to be cruel punishment when officers arrested the homeless, even through Los Angeles had patterned its ordinance on Seattle's rules. There were specific differences in the two ordinances, however: In Seattle, it is not illegal to sleep on the sidewalks.[81] In an ironic twist, the same district court judges who ruled in favor of Seattle's ordinance ultimately struck down the Los Angeles ordinance.

Implications of the Quality-of-Life Policing Movement

The adoption of QOL and order-maintenance activities helped ignite a virtual "revolution in American policing," also known as the "Blue Revolution."[82] The popularity of this approach in the United States is apparently matched by its appeal abroad. In 1998 alone, for instance, representatives of more than 150 police departments from foreign countries visited the NYPD for briefings and instruction in QOL or order-maintenance policing and its justification (i.e., the broken windows perspective).[83] In the first ten months of 2000, another 235 police departments—85 percent of them from outside the United States—sent delegations to One Police Plaza. QOL arrests are endorsed by police chiefs across the United States, and proponents of the QOL policing movement are sought out for lectures and consulting around the world.[84]

To examine the outcomes associated with this approach more closely, researchers conducted a study of 539 New York City arrestees in 1999. Almost all of the participants reported that they were aware that the police were targeting various disorderly behaviors, resulting in their individual decision to stop or cut back on their illegal behaviors. According to the arrestees, a police presence (i.e., QOL policing) was the most important factor behind their behavioral changes. One conclusion that can be drawn from this study is that QOL policing has a deterrent effect on criminal behavior.

Another study on the impact of QOL policing on crime and disorder was conducted by the Chandler, Arizona, Police Department.[85] It used data on calls for police service related to 10 categories of offenses. The findings in the Chandler study suggest that the QOL initiative exerted the strongest effect on two categories of crime and disorder: public morals and physical disorder.

Lessons Learned about Quality-of-Life Policing

George Kelling and Catherine Coles provide a lesson plan for the implementation of QOL policing initiatives. They suggest that although QOL policing might well be a path to restoring order, four elements in particular will enhance the success of efforts undertaken by city officials and police departments to develop legislation and implement QOL programs to retake public places:

- Claim the high moral ground.
- Learn to problem solve.
- Prepare to win in court.
- Involve the community.

Claim the High Moral Ground

The purpose of QOL policing is to prevent fear, crime, and urban decay. Often, the strategies employed to achieve this goal put police in contact with homeless persons who are emotionally disturbed, youths, and substance abusers. QOL initiatives are not designed to deal with the root causes of the behavior that is prohibited. Even so, policymakers must recognize this inevitable outcome, and must develop a policy that restores order and can benefit everyone in the community, including both those who are needy and "troublemakers." One goal is to ensure the fundamental interests and liberties of all members of the community. To realize this goal, officials must frame ordinances in proper terms from the beginning rather than creating a political trap, like that created by New York City's original subway ordinances or Chicago's anti-gang ordinances. As Kelling and Coles report, to retake public places and restore order, police in urban areas cannot ignore circumstances such as homelessness. Instead, these issues must be dealt with in a legal and appropriate manner the first time around.

Learn to Problem Solve

To resolve problems, policymakers must identify those problems correctly at the outset. This need suggests that they follow the SARA model (discussed in Chapter 3).

Prepare to Win in Court

Ordinances must be able to survive legal challenges if cities are to reach the goal of improved QOL. To ensure that ordinances can withstand these sorts of challenges, policymakers, community members, police managers, and other concerned social agencies must work together in a coordinated effort to frame ordinances that are moral, utilitarian, and legal. This perspective includes being prepared to convince the court with factual data that the ordinance is crucial for restoring order. Also, advocates must be able to convince the court that the legislation has been narrowly and precisely drafted so as to avoid vagueness challenges. For instance, disorderly behavior is not protected by the First Amendment, whereas begging might be. If an ordinance is passed that prohibits begging, then it is unlikely to prevail if someone mounts a challenge to it in court. (More on this point can be found at findlaw.com.)

Keep the Community Involved

According to Kelling and Coles, "no efforts at restoring order in the community will be successful in the long run without the development of a full partnership between citizens in the community and the criminal justice institutions that affect conditions in their neighborhoods."[86] Nevertheless, QOL seems to rely less on community participation for its success, which has stirred up a lot of criticism.

Criticism of Quality-of-Life Policing

The restructuring undertaken within most police agencies in recent years emphasizes, to some degree or another, QOL components or policies. For instance, some observers conceptualize QOL "as targeted at specific minor behaviors in public locations. QOL policing is an explicit strategy of enforcing minor laws and ordinances that are thought to contribute to incivility and disorder in communities."[87] However, the issue is not how frequently QOL standards are employed, but in what way QOL furthers the mission of public safety. In particular, "[the QOL policing] agenda is becoming increasingly that of government rather than individuals; it is specializing in criminal investigation and undercover surveillance; its operations are undertaken in groups; and it is increasingly militarized in equipment and outlook."[88]

Consistent with this rhetoric is the idea that QOL contributes to an erosion or diminishment in the QOL experiences among the targeted community membership. For instance, the components of QOL initiatives do not necessarily further the democratic process but rather increase the power of the state, especially police power.[89] Of course, the literature does support the idea that there is a strong link between arresting suspects for minor offenses and reductions in rates of serious crime.[90] But are the persons arrested and the persons who commit serious crimes necessarily the same offenders?

In response, advocates of QOL policing say that this movement's purpose is

> to teach the citizenry to follow the various laws and regulations governing the use of public space. In policing circles, quality-of-life enforcement has a broad intention of "sending a message" to violators that a specific behavior is wrong, and arrest processing is intended to punish those who "push the envelope" by engaging in minor disorders. Social scientists refer to the rules and regulations as civil codes (or civic norms) that police are expected to enforce—especially upon those who apparently violate these codes on a regular basis.[91]

Consistent with this thought, another group of researchers confirm the thoughts that street people are aware that police were targeting various disorderly behaviors:

> Among those that engaged in disorderly behaviors, about half reported that they had stopped or cut back in the past six months. They reported a police presence was the most important factor behind their behavioral changes. These findings support the idea that QOL policing has a deterrent effect.[92]

Although the validity of these findings is not suspect, evidence-based studies among violators also show that they are often successful at manipulating their environments for their own gain—that is, they lie.[93] Furthermore, Robert Sampson and Stephen Raudenbush argue that petty arrests do not necessarily reduce the rate of serious crime.[94] Their study was based on careful data collection using trained observers.[95] The researchers selected 15,141 streets of Chicago at random for analysis. The researchers found that disorder and predatory crime were moderately correlated, but that when antecedent neighborhood characteristics (e.g., neighborhood trust and poverty) were taken into account, the connection between disorder and crime "vanished in 4 out of 5 tests—including homicide, arguably our best measure of violence."[96] Sampson and Raudenbush conclude that attacking public order through tough police tactics may be politically popular but perhaps it is—on an analytical basis—a weak strategy for reducing crime.

Similar doubts have been voiced by other researchers about adopting the New York style of policing, thus attacking both the broken windows perspective and the QOL arrest strategy. After making a cross-city comparison of policing strategies and homicide rates, one observer thought the attention has not been positive. Anne Joanes reveals that many New York City residents and observers have blamed QOL policing for a rise in police brutality and racial tensions and a loss of trust and respect for the police.[97] Joanes goes on to say that New York has not achieved a greater crime reduction than all other cities across the United States. In fact, three cyclical measures indicate that New York City's decline in crime was either equal to or less than the decline in crime over the same period experienced by several other large cities, including San Francisco, San Jose, Cleveland, San Diego, Washington, St. Louis, and Houston.

Root Cause Argument Critics also say that QOL initiatives have failed to address the root causes of violence and that they have not neutralized the illegal markets that fuel violence.[98] The demand for and availability of illegal drugs and guns remain high in cities across the United States, regardless of those jurisdictions' petty-crime arrest patterns. The number of high-risk individuals willing to provide these criminal services and the number of youth gangs have also remained relatively stable.

More importantly, neighborhood structural characteristics, life as part of a street family and discrimination are significant predictors of whether a person will adopt the "street code." Studies show that violent delinquency is influenced by all of these factors: Neighborhood effects have a 20 percent effect, racial discrimination has a 20 percent effect, and family characteristics have a 4 percent effect on whether a youth engages in violent delinquency.[99] One implication is that the "street code" has more influence over small-time thugs than does the threat of arrest. Arrests for petty crimes can actually have an unintended effect on street codes, in that the modeling of disrespect toward the police may enhance the fear of crime among community members.

To guarantee continued and substantial reductions in violence, argue Sampson and Raudenbush, the United States should look beyond police resources and seek to build a comprehensive strategy that calls upon the community, social service agencies, business, and government to play more substantial roles in crime prevention strategies.[100] Yet, despite these criticisms of QOL programs, in the final analysis, crime is down in the United States, including drug addiction. For instance, in New York City, homicides reached a high of 2,605 in 1990 but fell to 596 in 2006.[101]

Privatization of Policing

It can be argued with some confidence that some police departments refuse to adopt appropriate police initiatives, that police misconduct and police brutality continue, and that the public and corporate America often turn to other providers of protection such as private security firms. If public agencies cannot provide the desired police protection, taxpayers and business inevitably find—and pay—the organizations that will.

Is private policing a service that will become a reality only in the distant future? Not quite. In 1971, there were 5,000 federal police providing security at U.S. government buildings. Today there are only 409 of these positions, with private-contract guards making up the difference. Private officers have served at U.S. Department of Energy sites across the United States and overseas for the past 30 years.[102] Often, these officers hold similar powers as sworn officers. At the Federal Law Enforcement Training Center (FLETC) at Glynco, Georgia, the police department is a privately owned organization whose officers have mandates for detainment and use of excessive force similar to those that govern the actions of public officers (see Chapter 8 for specifics about FLETC and Glynco).[103] FLETC instructs some 30,000 federal and local enforcement officers each year. Likewise, the Office of Personnel Management (OPM) contracts with a private organization to perform background investigations into applicants for federal jobs.

Rent-a-cop models have been around for some time. This global phenomenon was first widely noted by a 1972 Rand Corporation study commissioned by the National Institute of Justice,[104] and the trend toward privatization of police services is growing worldwide.[105] For example, this "quiet revolution" toward private policing is apparent in Canada as well as in both Western and Eastern Europe.[106] In fact, private security companies are used so extensively in the United Kingdom to perform the duties of sworn officers that they outnumber their public-sector counterparts.[107] An update of the original Rand assessment in 1985 concluded that private security outspent public police by 73 percent and employed two and a half times as many people.[108]

Recall from Chapter 2 that Pinkerton agents were hired to provide protection to early U.S. railroads and that Pinkerton helped develop the U.S. Secret Service.[109] Of particular concern to modern society is the fact that these earlier private concerns were unable to suppress the riots, demonstrations, and strikes that became frighteningly frequent during the 1830s without the assistance of the militia.[110] The Coal Police, Railroad Police, and company town police agencies hired by industrial concerns were used against Irish and German immigrants in the urban seacoast slums to break strikes and suppress hunger riots. More often, military personnel were deputized for guard duty or for the duration of a strike.

Even today, private guards (and military personnel) pop up everywhere—patrolling shopping malls, workplaces, apartment buildings, and neighborhoods. The phenomenal growth of massive private shopping malls, the never-ending expansion of gated communities, and the steady loss of public police patrols could eventually mean that members of the public are more likely to encounter private security forces than public police on a daily basis.[111] The business and industrial communities already pay for security in malls, stores, offices, banks, manufacturing plants, and highly congested public places such as New York City's Grand Central Station. With the threat of terrorist activity at airports (e.g., the attacks in London and Glasgow in 2007), railroad properties (where highly toxic chemicals are stored),[112] and nuclear power plants, private professional police are in great demand.

As rent-a-cops supplant public police in a variety of functions, the private security industry is creating a separate and unequal system under which the rich protect their privileges and guard their wealth from "barbarians at the gate." These heavily armed private guards are accountable not to the public, but rather to the well-manicured hands that feed them.[113]

Privatization Accountability

Public accountability of private enterprise is the primary argument against privatization, along with the need to comply with federal and local laws and regulations. Private security firms, some say, are inherently a law unto themselves, accountable only to the corporate bottom line.[114]

When researchers at Vera Institute of Justice (VIJ) examined police privatization in New York, Johannesburg, and Mexico City, they explored the nature of internal monitoring and disciplinary structures, media coverage of the security service, civil and criminal actions filed against private police officers, the role of the public police in monitoring misconduct among private police, and the role of trade associations in setting and overseeing performance standards.[115] Their results show that private police can be made accountable for their actions. While governmental controls may play a limited role in shaping these individuals' actions, they can still be powerful incentives for private security firms to develop internal controls. The firms examined in this study cited strong incentives to avoid embarrassment to their clients, and they displayed a clear aversion to civil lawsuits, especially when the suits were filed against their officers. Government oversight that threatened embarrassment to their clients when private security firms misbehaved seemed to be a potentially effective means of control, as were laws that held clients and security firms liable in civil courts for the misconduct of their officers.

External accountability may have symbolic importance in its own right, but its functional importance appears to lie in its ability to provoke effective management and internal controls. This may turn out to be true of external accountability for public police as well.[116]

Privatization Assessment

More than a decade ago, an assessment of private policing concluded that it might prove more benign than some experts had feared. The trend toward privatization was seen as irreversible,[117] but the researchers predicted that privatization of some services would allow public police to concentrate on activities requiring more education and training. Five years later, however, another report revealed that private police had begun to perform all activities once thought to be the exclusive domain of the public police.[118]

Even though public and private providers may perform similar tasks, they employ distinctively different practices when doing so. Specifically, governmental providers tend to prevent crime by administering punishment; nongovernmental providers do so through exclusion and the regulation of access.[119] Meanwhile, it is left to public police forces to maintain coercive order within deteriorating inner cities.

If the present trends continue, public officers could eventually be limited to functioning as "ghetto fighters," with private police covering all other jurisdictions.[120] The other option is for public police agencies to upgrade services and organizational structures in keeping with democratic guidelines, educate the public, and become the leaders, mentors, and managers of the public safety enterprise.

Government Incentives to Guide Privatization Accountability

The government can align incentives to private police agencies through the following policies:

- Require periodic review by an oversight agency that aggressively seeks public comment on the performance of the private security firm
- Subject private security firms to civil liability in the courts for the misconduct of their personnel; requir private security firms to provide training to personnel appropriate to the roles they will play
- Require private security firms to document their annual review of individual personnel against written standards
- Require private security firms to maintain records of their personnel's movement and activities for occasional external review
- Encourage private security firms to participate in professional associations that certify firms' compliance with industry standards

These safeguards may misconduct on the part of private police, but the strongest incentive remains the accountability of the organization to its clients. The question remains, however: Can a private police agency, with jurisdiction over a large land mass and population within a municipality, operate independently without public oversight or review?[121]

One answer to this question came in a case involving Mercer University in Georgia, where a legal complaint sought a temporary restraining order against Mercer and its private police department. The plaintiff asked that the court declare immediately that the Mercer University Police Department is, under Georgia's legal definition, a public agency and therefore subject to the Open Records Act, or that it force the Mercer police to stop acting as a public agency. Similar lawsuits are pending: Harvard University, Cornell University, and Taylor University (Indiana) claim, as Mercer does, that they are private institutions and that their campus police records are not subject to their respective states' open records laws.

The private security industry's rapid growth presents new challenges for those seeking progressive solutions to problems of crime and violence. Calling for more authority for public police is not an appealing remedy in communities where police-inflicted beatings (like that of Rodney King in Los Angeles) are the rule rather than the exception. Community organizations recognize the dangers of placing too much trust in either public or private police, while acknowledging the need for action to combat crime, which strikes disproportionately at low-income neighborhoods. In one attempt to bridge this gap, the Oakland-based Center for Third World Organizing has helped to bring some sponsors of locally focused initiatives together to share strategies and resources.

The nationwide Campaign for Community Safety and Police Accountability (CCSPA) is addressing the need to hold security forces accountable to the public while implementing programs designed to reduce crime by meeting social needs. Global organizations involved in the military affairs of the United States (such as Blackwater, which has played a role in policing Iraq) have come under fire from various groups. Ultimately, it may take decades to sort out the relationship between private and public police groups.

One conclusion that can be offered based on the evidence in the "Ripped from the Headlines" feature is that the role of the police is actually guided by law and how well the police respond to the law. The public has legal remedies that define the role of police—and rightfully so in a democratic nation—and might more often turn to private police.

RIPPED from the Headlines

One Very Bad Cop

More than any other policeman in Vermont, Paul D. Lawrence, 30, had a reputation in drug circles as one tough officer. In his six years on the drug beat, first as a state trooper and then on various local forces, Lawrence racked up some 600 drug convictions. He never failed to turn up incriminating evidence. Police and counterculturists agreed that his record was almost too good—or too bad—to be true. As it turned out, "bad" was the right word.[122]

The Governor was contemplating an extraordinary letter from Francis Murray, the Chittenden County (Burlington) prosecutor, asking him to pardon all 600 of those convicted on Lawrence's testimony. Lawrence, the prosecutor pointed out, faced up to 16 years in prison after being found guilty of turning in false arrest affidavits and giving false information to a police officer. The big drug buster apparently arrested anyone he was suspicious of, often supplying the narcotics evidence himself and claiming he had made a buy from the alleged pusher. Judges, and in a few cases even juries, simply took Lawrence's word.

As his story unraveled in court, it became clear that Lawrence had rarely been more than one quick step ahead of discovery. Somehow, he became an officer, despite a youthful arrest for illegal possession of liquor and an Army discharge for "behavioral disorders" after three AWOL [absent without leave] incidents in seven months of service. He was with the Vermont state police for four years and quit shortly after his squad-car windshield was apparently shot out from the inside when he was alone on patrol. His record also included the beating of a man he had arrested. After that, a brief stint as a tobacco salesman came and went amidst claims by his employers that a cache of cigarettes had mysteriously disappeared. Lawrence then managed to get a job as chief of the four-man police force in the small town of Vergennes, Vermont; he left a year later, this time just before being fired for questionable drug arrests and hyperactive enforcement of speed limits.

He next became an officer assigned to the drug beat in St. Albans, Vermont. There, defense attorneys soon noted that an unusually high percentage of their clients claimed that Lawrence had framed them. The lawyers persuaded the state defender general to hire a private detective, who filed a 30-page report that was highly critical of Lawrence's activities. The state attorney general, who was busy running for reelection, shelved the charges as unsubstantiated. By then, the St. Albans police had lent Lawrence to Burlington to work on undercover drug enforcement there. That was the end of the line.

At about the same time that a fellow policeman became suspicious of Lawrence's too-easy arrests, a reporter approached the local prosecutor's office with stories of Lawrence's past. A special undercover policeman brought in from Brooklyn (for his expertise and to assure that he would not be recognized) was pointed out to Lawrence as a suspected drug dealer. Then, while police watched from hiding, Lawrence approached the planted "pusher," but exchanged neither words nor money with him. Nonetheless, Lawrence later filled out an arrest affidavit and claimed he had bought a nickel ($5) bag of heroin. At Lawrence's trial, one girl he had arrested testified to another Lawrence technique. She was hitchhiking, she said, and he picked her up, offered her dinner and a few sniffs of cocaine, then asked her to spend the night with him. She refused; two weeks later he busted her for selling him drugs upheld by the court.

In his ruling, U.S. District Judge James Judge Whittemore wrote, "A generalized fear of terrorism should not diminish the fundamental Fourth Amendment protections envisioned by our Founding Fathers. Our Constitution requires more."

Source: "One Very Bad Cop." (1975, May 10). *Time,* CNN. http://www.time.com/time/printout/0,8816,917201,00.html (accessed June 27, 2007).

Summary

The issue of police change has arisen because of the need to deliver quality police services and to increase clearance rates. Most police agencies in the twentieth century were inflexible, followed unprofessional or personal approaches toward management, and were resistant to external influences of change. Today, however, police agencies are more willing to change for reasons ranging from the need to respond to technological changes to changes in the legal environment.

Evidence-based policing makes use of the best available research on the outcomes of policing and can be used to implement policies, regulations, and training among police agencies. This process utilizes the scientific method of discovery and provides objective information that can be used effectively in decision-making processes. The goals when using the scientific method include increasing police effectiveness, reducing the frequency of citizen complaints, and enhancing the efficient use of available resources.

Community relations strategies continue to play a role in today's police outreach efforts, albeit only a limited one. They have largely fallen out of favor owing to the widespread adoption of quality-of-life (QOL) policing initiatives. The latter approaches assume that control of serious crimes can be achieved through aggressive actions against suspects who commit relatively minor offenses. QOL levels authority and increases the span of control; it also focuses on outreach to other agencies and private providers, including religious leaders.

The centerpiece of QOL policing is a strategy of maintaining order through active control of public spaces by arresting suspects who commit any offense through proactive ordinances. Its primary mission is to reduce minor disorder. Critics complain that QOL initiatives enhance police power at the expense of community participation, suggesting that this strategy may not be in accord with democratic principles.

Privatization of policing is a trend that is growing worldwide; its expansion serves as a warning that public agencies must change or consider the consequences. One argument against private policing is that it avoids public accountability, though several government incentives that might guide privatization accountability are under consideration.

Key Terms

COPLINK

Criminogenic activities

Directed patrol

Evidence-based policing

Kansas City Preventive Patrol Experiment

Minneapolis Domestic Violence Experiment

Probable cause arrest

Quality-of-life policing

Scientific police management (SPM)

Span of control

Total Quality management

Discussion Questions

1. Characterize the compelling reasons why police organizations must change. Which two of these are the most important reasons for police to change? Why?
2. Describe the lessons learned from preventive or proactive police strategies. In what way do you agree with the lessons learned?
3. Characterize the relevance of evidence-based policing and its primary component.
4. Compare and contrast the four stages of twentieth-century policing.
5. Summarize several studies that influenced the development of modern police practices. Which two of these studies most strongly influenced the evolution of QOL policing? Why?
6. Characterize the main thrust and components of QOL policing initiatives.
7. Describe the QOL initiatives used by the police in Chicago and New York City.
8. Describe traffic-related QOL initiatives.
9. Characterize the relationship between public ordinances and QOL initiatives. In what way do you agree with the idea that ordinances must be in place to bring a department closer to its mission?
10. Describe the concerns related to privatization of police. In what way do you see privatization as a tool to enhance public policing?

Notes

1. David Osborne and Peter Plastrik, *Banishing Bureaucracy. The Five Strategies for Reinventing Government* (Reading, MA: Addison Wesley, 1997), 4.
2. Policeone.com. http://www.officer.com/article/article.jsp?id=35290&siteSection=1 (accessed May 18, 2007).
3. David L. Carter and Louis A. Radelet, *Police and the Community*, 7th ed. (Upper Saddle River, NJ: Prentice Hall, 2001), 16.
4. Peter Manning, *Police Work: The Social Organization of Policing* (Prospect Heights, IL: Waveland, 1997), 97–100.

5. James D. Sewell, "Managing the Stress of Organizational Change," *FBI Law Bulletin* 71, no. 3 (2002): 15–17.
6. Willard M. Oliver, *Homeland Security for Policing* (Upper Saddle River, NJ: Prentice Hall, 2007).
7. Federal Bureau of Investigation, "Clearance Rates: Table 25–28" (2006), http://www.fbi.gov/ucr/cius_04/offenses_cleared/table_28.html (accessed July 24, 2006).
8. Bureau of Justice Statistics, *Sourcebook of Criminal Justice Statistics 2004* (Albany, NY: U.S. Department of Justice, 2004), Table 4.20, http://www.albany.edu/sourcebook/pdf/t4202004.pdf (accessed July 24, 2006).
9. Boston Police Department, "Annual Report: 2004" (2005), http://www.cityofboston.gov/police/pdfs/2004annual_report.pdf (accessed July 28, 2006).
10. U.S. Census Bureau (2006), http://www.census.gov/ (accessed July 19, 2006).
11. Raymond E. Foster, *Police Technology* (Upper Saddle River, NJ: Prentice Hall, 2005).
12. COPLINK (2006), http://www.COPlink.com/ (accessed July 28, 2006).
13. Tom Ladsford, Robert J. Pauly, Jr., and Jack Covarrubias, *To Protect and Defend: US Homeland Security Policy* (Burlington, VT: Ashgate, 2006), 155.
14. Bureau of Labor Statistics (2006), http://www.bls.gov/ (accessed July 19, 2006).
15. David Brinkley, *The Great Deluge: Hurricane Katrina, New Orleans, and the Mississippi Gulf Coast* (New York: Time Life Books, 2006).
16. COPS, "Police and Communities Partner to Fight Meth" (2006), http://www.officers.usdoj.gov/print.asp?Item=1307 (accessed August 18, 2006).
17. Jack Thompson, "Drug Grant in Bush's Cross Hairs: Rift Over Results Has Neb. Officers Defending Program," *World-Herald Bureau* (July 10, 2006), http://www.policeone.com/Grants/articles/386862/ (accessed August 18, 2006)
18. Wayne W. Bennett and Karen M. Hess, *Management and Supervision in Law Enforcement* (Belmont, CA: Thomson Wadsworth, 2007), 71.
19. For instance, the ACLU filed a federal discrimination lawsuit on behalf of the surviving family members of a man who died of a heart attack after the police chief physically prevented his friend from performing cardiopulmonary resuscitation (CPR). The police chief blocked the CPR because he falsely assumed that the man, who was gay, was HIV positive and therefore a health risk. ACLU, "ACLU Sues West Virginia Police Chief Who Blocked Life-Saving Measures for Gay Heart Attack Victim Assumed to Be HIV Positive" (March 2, 2006), http://www.aclu.org/hiv/gen/index.html (accessed July 28, 2006).
20. Officer One (2007), Policeone.com (accessed July 5, 2007).
21. Ladsford, Pauly, and Covarrubias, *To Protect and Defend*, 159.
22. Rolando V. del Carman, *Civil Liabilities in American Policing* (Upper Saddle River, NJ: Prentice Hall, 1991); Roland V. del Carmen, Charles Williamson, William C. Bloss, and J. Coons, *Civil Liabilities and Rights of Police Officers and Supervisors in Texas* (Huntsville, TX: Law Enforcement Management Institute of Texas and Sam Houston State University Press, 2003).
23. Dennis J. Stevens, "Civil Liabilities and Arrest Decisions," in *Policing and the Law*, ed. Jeffrey T. Walker (Upper Saddle River, NJ: Prentice Hall, 2002), 53–70. In particular, 42 USC Section 1983 is a vehicle for suing defendants acting under the color of state law for violating another person's federal rights. By its very terms, Section 1983 applies only to persons acting under the color of law. It is applicable only to state and local law enforcement officers who exert authority derived from state law.
24. Bennett and Hess, *Management and Supervision in Law Enforcement*, 20.
25. Vivian Lord, "An Impact of Community Policing: Reported Stressors, Social Support, and Strain Among Police Officers in a Changing Police Department," *Journal of Crimi-*

nal Justice 24 (1996): 503–22; Kenneth J. Peak, *Policing in America* (Upper Saddle River, NJ: Prentice Hall, 2006), 88–97; Central Florida Police, "Stress Unit" (2006), http://policestress.org/index.htm (accessed April 7, 2006).

26. Lawrence W. Sherman, *Evidence-Based Policing* (Washington, DC: Police Foundation, 1998), 3.

27. The Kansas City Preventive Patrol Experiment (Washington, DC: The Police Foundation, 2003), http://www.policefoundation.org/docs/kansas.html (accessed March 15, 2008).

28. Lawrence W. Sherman and Richard A. Beck, *Minneapolis Domestic Violence Experiment* (Police Foundation Report 1) (Washington, DC: Police Foundation, 1984).

29. National Institute of Justice, *Newport News Tests Problem-Oriented Policing* (Washington, DC: Author, January 1987).

30. Terrance Dunworth and Gregory Mills, *National Evaluation of Weed and Seed* (Washington, DC: National Institute of Justice, 1999).

31. State of Delaware, Community Capacity Development Office (2006), http://www.state.de.us/cjc/weedseed.shtml (accessed October 1, 2006).

32. Edmund F. McGarrell, Steven Chermak, and Alexander Weiss, *Reducing Gun Violence: Evaluation of the Indianapolis Police Department's Directed Patrol Project* (Washington, DC: National Institute of Justice, 2002).

33. Jeremy M. Wilson, John M. Macdonald, Clifford Grammich, and K. Jack Riley. *Reducing Violence in Hayward, California: Learning from Homicides* (Washington, DC: U.S. Department of Justice and Rand Corporation, 2004), http://www.rand.org/ (accessed October 1, 2006).

34. An analysis of homicides in Hayward suggests that violence is predominately a problem among young adults and involves the use of firearms and killings in public places. Also, homicides are more common during summer months, in the early evening hours, and through the Saturday–Monday period of the week. This pattern is important to note because it suggests how law enforcement resources could be targeted. In addition, a larger proportion of homicides in Hayward than elsewhere involve a victim and an offender of different ethnic backgrounds. This finding is important because it suggests how changing demographics may affect future violence in the city.

35. Michael S. Hronick, *Analyzing Terror: Researchers Study the Perpetrators and the Effects of Suicide Terrorism* (NCJ 214113) (Washington, DC: U.S. Department of Justice, 2006).

36. Nancy Ritter, *Digital Evidence: How Law Enforcement Can Level the Playing Field with Criminals* (NCJ 214116) (Washington, DC: U.S. Department of Justice, 2006).

37. *Enhancing Police Integrity: Research for Practice* (NIJ 209269) (Washington, DC: U.S. Department of Justice, Office of Justice Programs, December 2006), http://www.ncjrs.gov/pdffiles1/nij/209269.pdf (accessed October 1, 2006).

38. Jon B. Gould and Stephen D. Mastrofski, "Suspect Searches: Assessing Police Behavior under the U.S. Constitution," *Criminology & Public Policy* 3, no. 3 (2004): 315–325.

39. Dana E. Hunt, "Methamphetamine Abuse: Challenges for Law Enforcement and Communities," *NIJ Journal* 254 (2006), http://www.ojp.usdoj.gov/nij/journals/254/methamphetamine_abuse.html (accessed October 1, 2006).

40. Dennis J. Stevens, "Police Training and Management Impact Sexual Assault Conviction Rates in Boston," *Police Journal* 79, no. 2 (2006): 145–155.

41. George L. Kelling and Catherine M. Coles, *Fixing Broken Windows* (New York: Free Press, 1996), 17–18.

42. Wesley G. Skogan, "Evaluating Community Policing in Chicago," in *Policing and Program Evaluation*, ed. Kent R. Kerley (Upper Saddle River, NJ: Prentice Hall, 2005).

43. Wesley G. Skogan and Susan M. Hartnett, *Community Police, Chicago Style* (New York: Oxford University Press, 1997).

44. George L. Kelling and Mark H. Moore, *Evolving Strategy of Police, No. 4.* (Washington, DC: National Institute of Justice, 1988), http://www.ksg.harvard.edu/criminaljustice/publications/pop4.pdf (accessed November 21, 2007).

45. Kent Kerley, "Perceptions of Community Policing Across Community Sectors: Results from a Regional Survey," in *Policing and Community Partnerships*, ed. Dennis J. Stevens (Upper Saddle River, NJ: Prentice Hall, 2002), 111–124.

46. Kerley, "Perceptions of Community Policing."

47. Fox Butterfield, "Efforts to Stop Crime Fall Far Short, Study Finds," *New York Times* (April 16, 1997), http://www.ncpa.org/pi/crime/april97b.html (accessed July 22, 2006). Also see Lorie Fridell and Mary Ann Wycoff (eds.), *Community Policing: Past, Present, and Future* (Washington, DC: Annie E. Casey Foundation and Police Executive Research Forum, 2004); Wesley G. Skogan, "Prospects and Problems in an Era of Police Innovation: Contrasting Perspectives," in *Police Innovation: Contrasting Perspectives,* ed. David Weisburd and Anthony A. Braga, (Cambridge, UK: Cambridge University Press, 2005), 39–78, http://www.northwestern.edu/ipr/publications/policing_papers/caps30.pdf (accessed March 14, 2008).

48. Lawrence W. Sherman, "Communities and Crime Prevention" (1998), http://www.ncjrs.gov/works/chapter3.htm (accessed November 21, 2007).

49. George E. Capowich, "A Case Study of Community Policing Implementation: Contrasting Success and Failure," in *Policing and Program Evaluation*, ed. Kent R. Kerley (Upper Saddle River, NJ: Prentice Hall, 2005).

50. Dennis J. Stevens, *Applied Community Policing in the 21st Century* (Boston: Allyn and Bacon, 2003), 12, 20–21. Also see Dennis J. Stevens, *Case Studies in Applied Community Policing in the 21st Century* (Boston: Allyn and Bacon, 2003).

51. George L. Kelling and William J. Bratton, "Implementing Community Policing: The Administrative Problem" (1993), http://www.ksg.harvard.edu/criminaljustice/publications/pop17.pdf (accessed November 21, 2007).

52. Mark L. Dantzker, "An Important Ingredient in Police Program Evaluation: Making Sure There Is Something to Evaluate," in *Policing and Program Evaluation*, ed. Kent R. Kerley (Upper Saddle River, NJ: Prentice Hall, 2005).

53. Kent R. Kerley (ed.), *Policing and Program Evaluation* (Upper Saddle River, NJ: Prentice Hall, 2005).

54. Samuel Walker and Charles M. Katz, *The Police in America: An Introduction*, 6th ed. (Boston: McGraw Hill, 2008).

55. Willard M. Oliver, *Community Oriented Policing: A Systematic Approach to Policing* (Upper Saddle River, NJ: Prentice Hall, 2008), 269.

56. Rudolph W. Giuliani, "Giuliani Archives: The Next Phase of Quality of Life" (1998), http://www.nyc.gov/html/rwg/html/98a/quality.html (accessed November 13, 2007).

57. Eric Manch, "Throwing the Baby Out with the Bathwater: How Continental-Style Police Procedural Reforms Can Combat Racial Profiling and Police Misconduct," *Arizona Journal of International and Comparative Law* 19, no. 3 (2002): 1026–1058.

58. Bernard Harcourt, *Illusion of Order: The False Promise of Broken Windows Policing* (Cambridge, MA: Harvard University Press, 2001). Also see Gary T. Marx, "Police and Democracy," in *Policing, Security, and Democracy: Theory and Practice, Volume 2,* ed. M. Amir and S. Einstein (Hampshire, UK: Ashgate, 2001).

59. Fox Butterfield, "Cities Reduce Crime and Conflict without New York-Style Hardball," *New York Times* (March 4, 2000), http://www.ncpa.org/pi/crime/pd030700e.html (accessed July 22, 2006). Also see Dantzker, "An Important Ingredient"; M. Smith and R. V. Clarke, "Crime and Public Transport," *Crime and Justice: A Review of Research* 27

(2000): 169–233, http://www.crimereduction.gov.uk/toolkits/pt0304.htm (accessed July 29, 2006).

60. Daniel B. Wood, "William Bratton: Lauded Chief of Troubled LAPD," *Christian Science Monitor* (July 6, 2007), http://www.csmonitor.com/2007/0706/p01s05-usju.html?page=1 (accessed November 11, 2007).

61. David H. Bayley and Clifford D. Shearing, "The New Structure of Policing: Description, Conceptualization, and Research Agenda Series: Research Report" (2001), http://www.ncjrs.gov/txtfiles1/nij/187083.txt (accessed November 11, 2007).

62. Kelling and Moore, *Evolving Strategy of Police*; George L. Kelling and William J. Bratton, "Taking Back the Streets," *City Journal* (Summer 1994), http://www.city-journal.org/article01.php?aid=1428 (accessed June 28, 2006). Also, QOL components were reviewed by James McCabe, NYPD Inspector (retired) and currently a professor at Sacred Heart University, Fairfield, Connecticut, who made suggestions, some of which were adapted (November 21, 2007).

63. Chicago Municipal Code §8–4–015 added June 17, 1992.

64. *City of Chicago v. Jesus Morales, et al.* (October 1, 1998), http://www.aclu.org/scotus/1998/22661prs19981001.html (accessed June 24, 2006).

65. *City of Chicago v. Jesus Morales, et al.* (1998).

66. *City of Chicago v. Jesus Morales, et al.*, 1998.

67. Dennis P. Rosenbaum and Cody Stephens, "Reducing Public Violence and Homicide in Chicago: Strategies and Tactics of the Chicago Police Department" (June 15, 2005), http://www.icjia.state.il.us/public/pdf/ResearchReports/ReducingPublicViolenceandHomicideinChicago.pdf (accessed July 22, 2006).

68. "*Terry* stop-and-frisk" refers to the classic *Terry v. Ohio* (1968) case where the Supreme Court established the right of a police office to stop and frisk someone for "reasonable suspicion" in the absence of facts that would establish "probable cause" for making a lawful arrest.

69. Rosenbaum and Stephens, "Reducing Public Violence and Homicide in Chicago," 36.

70. Monique Bond, Director of News Affairs, City of Chicago (June 26, 2006), http://www.chicagopolice.org/MailingList/PressAttachment/release.snakebite.pdf (accessed July 7, 2006).

71. George L. Kelling and William H. Sousa, Jr., *Do Police Matter? An Analysis of the Impact of New York City's Police Reforms* (New York: Manhattan Institute, 2001), http://www.manhattan-institute.org/html/cr_22.htm (accessed August 2, 2006).

72. Kelling and Sousa, *Do Police Matter?*

73. Alexander Weiss and Sally Freels, "The Effects of Aggressive Policing: The Dayton Traffic Enforcement Experiment," *American Journal of Police* 15, no. 3 (1996): 45–64.

74. Bureau of Justice Statistics, *Characteristics of Drivers Stopped by Police* (Washington, DC: U.S. Department of Justice, 2002), http://www.ojp.usdoj.gov/bjs/abstract/cdsp02.htm (accessed June 9, 2006).

75. Chicago, for instance, reported 206,000 traffic stops in 2005 (http://www.chicagopolice.org/).

76. Stephen Mastrofski, "The Police and Non-crime Services," in *Evaluating the Performance of Criminal Justice Agencies*, ed. G. Whitaker and C. Phillips, (Beverly Hills, CA: Sage, 1983), 44–47.

77. Commission on Accreditation for Law Enforcement Agencies (2006), http://www.calea.org/ (accessed August 15, 2006).

78. Kelling and Coles, *Fixing Broken Windows*.

79. Kelling and Coles, *Fixing Broken Windows*, 216.

80. Rebecca Renwick, "Quality of Life" (2006), http://www.utoronto.ca/qol/ (accessed August 14, 2006).

81. Tracy Johnson, "A Tale of Two Cities' Laws on Homeless," *Seattle Post-Intelligencer* (April 18, 2006), http://seattlepi.nwsource.com/local/267057_homeless18.html (accessed August 20, 2006).

82. Bernard Harcourt, "Policing Disorder: Can We Reduce Serious Crime by Punishing Petty Offenses?", *Boston Review* (April/May 2002).

83. Elissa Gootman, "A Police Department's Growing Allure: Crime Fighters from Around World Visit for Tips," *New York Times*, p. 6 (October 24, 2000).

84. "On Crime as Science: A Neighbor at a Time," *New York Times*, p. 31 (January 6, 2004).

85. C. M. Katz, V. J. Webb, and D.R. Schaefer, "An Assessment of the Impact of Quality of Life Policing on Crime and Disorder," *Justice Quarterly* 18, no. 4 (2001): 825–876.

86. Kelling and Coles, *Fixing Broken Windows*, 234.

87. Bruce D. Johnson, Andrew Golub, and James McCabe, "Quality of Life Policing in New York City: International Implications," *Police Quarterly* (2008, in press).

88. Bayley and Shearing, "The New Structure of Policing," 2.

89. Bayley and Shearing, "The New Structure of Policing"; Harcourt, *Illusion of Order*.

90. John L. Worrall, "Does Targeting Minor Offenses Reduce Serious Crime?", *Police Quarterly* 9, no. 1 (2006): 47–72. Also see Samuel Nunn, Kenna Quinet, Kelley Rowe, and Donald Christ, "Interdiction Day: Covert Surveillance Operation, Drugs, and Serious Crime in Inner-City Neighborhoods," *Police Quarterly* 9, no. 1 (2006): 73–99.

91. Johnson, Golub, and McCabe, Quality of Life Policing in New York City."

92. Andrew Golub, Bruce D. Johnson, Angela Taylor, and John Eterno, "Quality-of-Life Policing: Do Offenders Get the Message?", *Policing: An International Journal of Police Strategies & Management* 26, no. 4 (2003): 690–707.

93. Michael R. Gottfredson and Travis Hirschi, *A General Theory of Crime* (Stanford, CA: Stanford University Press, 1990), 14. Also see Stanton E. Samenow, *Inside the Criminal Mind, Revised and Updated* (New York: Crown, 2004); Dennis J. Stevens, "Predatory Rapists and Victim Selection Techniques," *Social Science Journal* 31, no. 4 (1994): 421–433.

94. Robert J. Sampson and Stephen W. Raudenbush, "Systematic Social Observation of Public Places: A New Look at Disorder in Urban Neighborhoods," *American Journal of Sociology* 105 (2000): 637–638.

95. Johnson, Golub, and McCabe, "Quality of Life Policing in New York City."

96. Robert J. Sampson and Stephen W. Raudenbush, *Disorder in Urban Neighborhoods: Does It Lead to Crime* (Washington, DC: U.S. Department of Justice, 2001), http://www.ncjrs.gov/pdffiles1/nij/186049.pdf (accessed December 16, 2007).

97. Ann Joanes, "Does the New York City Police Department Deserve Credit for the Decline in NYC's Homicide Rate? A Cross-City Comparison of Policing Strategies and Homicide Rates," *Columbia Journal of Law and Social Problems* 33 (2000): 256–303.

98. Harcourt, *Illusion of Order*.

99. Eric A. Stewart and Ronald L. Simons, "Structure and Culture in African American Adolescent Violence: A Partial Test of the 'Code of the Street' Thesis," *Justice Quarterly* 23, no. 1 (2006): 1–33.

100. Sampson and Raudenbush, *Disorder in Urban Neighborhoods*, 42.

101. Federal Bureau of Investigation, "Uniform Crime Reports" (2007), http://www.fbi.gov/ucr/cius2006/data/table_08_ny.html (accessed December 19, 2007).

102. For example, Wachenhut Services, Inc., at Oak Ridge and other government facilities; http://www.oakridge.doe.gov/media_releases/1999/r-99-031.htm (accessed June 29, 2004).
103. The author visited FLECT March 9, 2006, to gather information for this and other publications.
104. James Kakalik and Sorrel Wildhorn, *Private Police in the United States* (Santa Monica, CA: Rand Corporation, 1971).
105. Vera Institute of Justice, "The Public Accountability of Private Police: Lessons from New York, Johannesburg, and Mexico City," (August 2000), http://www.vera.org/publication_pdf/privatepolice.pdf (accessed July 28, 2005).
106. Phillip C. Stenning and Clifford D. Shearing, "The Quiet Revolution: The Nature, Development, and General Legal Implications of Private Security in Canada," *Criminal Law Quarterly* 22 (1980): 220–48; N. South, "Privatizing Policing in the European Market: Some Issues for Theory, Policy, and Research," *European Sociological Review* 10 (1994): 219–233.
107. Harry R. Dammer and Erick Fairchild, *Comparative Criminal Justice Systems*, 3rd ed. (Belmont, CA: Wadsworth/Thomson, 2007), 109.
108. Vera Institute of Justice, "The Public Accountability of Private Police."
109. "Detective History," http://www.crimelibrary.com/gangsters2/pinkerton/ (accessed June 29, 2004).
110. David E. Barlow and Melissa Hickman Barlow, "A Political Economy of Community Policing," in *The Police in America*, ed. Steven G. Brandl and David S. Barlow (Belmont, CA: Wadsworth, 2004), 68–75.
111. Mike Zielinski, "Private Police: Armed and Dangerous," *Covert Action Quarterly* (2000), http://mediafilter.org/caq/CAQ54p.police.html (accessed June 29, 2004).
112. PBS, *American Investigative Reports* (2006), http://www.pbs.org/wnet/expose/episode202/index.html (accessed July 1, 2007).
113. Mike Zielinski, "Private Police."
114. Gary Marks, "The Interweaving of Public and Private Police Undercover Work," in *Private Policing*, ed. C. Shearing and P. Stenning (Thousand Oaks, CA: Sage, 1987).
115. Vera Institute of Justice, "The Public Accountability of Private Police."
116. Private security services can be highly accountable, although not necessarily to all of the same governmental agencies that oversee the public police. Sometimes private agencies are accountable to elected officials, but they are also subject to criminal investigation and prosecution, civil liability, and various public reporting requirements. These external controls create incentives for the employers of the private security forces to establish their own strong internal accountability mechanisms.
117. Marcia Chaiken and Jan Chaiken, *Public Policing—Privately Provided* (Washington, DC: National Institute of Justice, 1987).
118. L. Johnston, *The Rebirth of Private Policing* (London: Routledge, 1992).
119. Bayley and Shearing, "The New Structure of Policing."
120. Stevens, *Applied Community Policing in the 21st Century*.
121. "Private Police Powers Need Public Scrutiny," *The Telegraph* (January 25, 2004), http://www.macon.com/mld/telegraph/7783263.htm (accessed July 1, 2004).
122. One point about "One Very Bad Cop": Private police officers have not been caught up in corruption at the same level as Paul D. Lawrence.

POLICE
YORK

Local Police and Federal Enforcement Agencies

If you are just, don't fear the warrior when he walks among you. If you are not righteous or inflict pain and suffering and exploit the weak, then fear the law, and the warrior it brings.

—Anonymous member of the Boston Crime Gang Unit

▸ ▸ LEARNING OBJECTIVES

When you finish reading this chapter, you will be better prepared to

- Compare and contrast local police activities with federal enforcement activities
- Identify the various types of local police agencies
- Describe municipal police organizations and their primary tasks
- Characterize county police agencies and their functions
- Describe state police objectives and jurisdiction
- Distinguish between the roles of the coroner, the medical examiner, and the forensic pathologist
- Describe tribal police agencies, and identify issues among their constituents
- Characterize the largest federal law enforcement agency
- Discuss the primary issues, remedies, and limitations of fragmentation of policing authority
- Characterize trends related to assault and murder of local and federal officers

▸ ▸ CASE STUDY: YOU ARE THE POLICE OFFICER

You have been a campus police officer at a western college for the last three years. When you attended a small college and studied American policing, you were impressed the first time you learned about police careers. A police officer's job had the elements of an excellent career, including little time sitting at a desk or a cubicle during business hours, which suited your preferences: You favored a country setting rather living in an urban environment where most of the jobs were. You wanted a people-contact job helping individuals—and in particular college-aged people—make good choices about their lives. But social work, counseling, and teaching seemed like only part of what you wanted to do with your life, at least for right now.

It was in your last year of college that you saw a recruiting poster for the campus police department, which was interviewing job candidates during the week before Thanksgiving break. You asked your adviser about the job posting, and she recommended that you talk to the campus police recruiters to get interview experience. She also mentioned that your parents might be happy that you actively sought employment after graduation but before you went on to graduate school.

When you and the other students arrived at the appointed time for the interview, the campus police department presented a short media piece about the college, the police department, and the students on the campus. What caught your eye was that the college buildings were situated in rambling hills rather than built on flat ground, the police officers had the same authority and responsibilities as most municipal police organizations, and their challenges were unique given the small, nationally recognized college community they served. Also, the students pictured at the university seemed to represent many different ethnic groups. But what hooked you was the fact that campus police officers were encouraged to partner with faculty, staff, and students on crime prevention and law enforcement issues. You also liked the idea that officers were on bike patrol throughout most of the school semesters. The benefits seemed great, and the free tuition for the policing masters' program only added to the position's attractiveness. Oddly, everything about the job seemed to match your expectations about a career—and it didn't even require you to move to an urban area.

One concern you had was that the local municipal police department in the area where the campus was situated conducted all of the investigations linked to crimes of violence rather than the campus police. The recruiters said that, similar to the practice followed by many police departments that lacked the expertise or budget to conduct investigations, the campus police's policy permitted seeking another police agency's help. When another student asked if the local police department could come on campus and raid a dorm, the answer was that public police department responsibilities were generally fragmented. You remember checking that out after you completed the application for the job.

1. Why would most campus police officers have similar authority as local or municipal officers?
2. In what way might the limitations of campus police employment outweigh the limitations of employment at a local police department for some prospective officers?
3. Explain the advantages of fragmentation among police departments in the United States.

Introduction

Chapter 5 describes the roles and responsibilities of local and state police, campus police, capital police, coroners and medical examiners, forensic pathologists, and tribal police. All of these personnel, to some extent or another, are subject to oversight by local authority, are governed by state regulations, and are responsible for the police services in a specific jurisdiction. This chapter focuses on several specific types of police departments to provide a better understanding of contemporary police organizations. Special units within police departments such as harbor, motorcycle, and mounted police are covered in Chapter 9; investigative and forensic units are discussed in Chapter 11; and undercover police are examined in Chapter 12. Few real-world police organizations or units fit perfectly with the idealized model presented in this chapter for a variety of reasons—for example, differences in funding, jurisdiction, geographic anomalies (location, and land and water ways to service), historical relationships (between constituents and agency), personnel, and policies.

This chapter also profiles federal law enforcement agencies, whose personnel are charged with enforcing federal laws, and describes the fragmentation in policing that occurs because of the overlap of authority in police jurisdictions. Finally, trends related to assault and murder of local officers and federal special agents are explored so as to provide a better understanding of these jobs.

■ Local Police Agencies

In the United States, there are approximately 17,876 police departments (employing 731,903 sworn officers) below the state level (12,766 local departments; 3,067 sheriff's departments [employing 175,018 sworn officers]; 513 constables [employing 2,323 sworn officers—largely in Texas]; by comparison, there are 49 state police agencies and 1,481 special-jurisdiction agencies (employing 49,393 sworn officers) as of September 2004.[1]

In 2004 there were more than 800,000 full-time sworn law enforcement officers in the United States (see Table 5–1).

Public organizations (as opposed to private security agencies) employed more than 732,000 sworn full-time officers (individuals with detainment and use of force authority) and approximately 46,000 sworn part-time officers as of September 2004.[2] In 2004, there were an average of 2.5 **sworn officers** per 1,000 population in the nation's cities,[3] though this ratio varied depending on the agency.[4]

TABLE 5–1 Number of Full-time Sworn Officers

Type of Agency	Number of Agencies	Number of full-time sworn officers
Total		836,787
All State and local	17,878	731,903
Local police	12,766	446,974
Sheriff	3,087	175,018
Primary state	49	58,190
Special jurisdiction	1,481	49,398
Constable/marshall	513	2,323

Source: Bureau of Justice Statistics, Law Enforcement Statistics (Washington, DC: U.S. Department of Justice, 2004).

The Bureau of Justice Statistics provides additional highlights of local police departments for the year ending 2003:[5]

- In 2003, local police departments cost approximately $93,300 per sworn officer and $200 per resident to operate for the year.
- Departments had total operating budgets of $43.3 billion.
- Sheriffs' offices cost about $124,400 per officer and $82 per resident in 2003.
- Starting salaries for local police officers ranged from an average of about $23,400 in the smallest jurisdictions to about $37,700 in the largest jurisdictions. For example, Boston's starting pay for officers was $47,000 in 2006; New York City's starting pay was $55,000; and San Jose's starting pay was $77,000, with the maximum salary capping out at $139,000.
- Racial and ethnic minorities accounted for 24 percent of full-time sworn personnel.
- Approximately 62 percent of all departments had policies related to racial profiling.
- Some 39 percent of departments had a terrorist attack plan.
- The .40-caliber semiautomatic was the most commonly authorized sidearm.
- Approximately 7,500 officers were called up for reserve military service.

Municipal Police

Almost immediately after Hurricane Katrina destroyed much of the Gulf Coast, including New Orleans, on August 29, 2005; after 17,000 acres burned in California wildfires in 2007; and after the destruction of the World Trade Center in New York City, it became obvious of the U.S. public that municipal police and fire personnel are first responders providing emergency care and protection. It is fair to say that police officers throughout the country—regardless of the size of their home departments—would perform just as bravely if they were called upon to deal with such emergencies. **Municipal police** are the providers of general police services to cities and towns and other urban or populated areas. One estimate is that 70 percent of the police agencies in the United States employ more than two thirds of all employed officers in the country.[6]

The U.S. states include dozens of major cities, including 11 of the 55 largest cities in the world and three "alpha" global cities: New York City, Los Angeles, and Chicago. According to Bureau of Justice statistics, these three alpha global cities jointly employed more than 59,000 sworn officers in 2003. The 50 largest U.S. police agencies, each of which employed more than 1,000 officers, employed more than one third of all the employed sworn officers in the country. By comparison, rural departments account for less than half of all municipal departments in the United States; many of these departments employ fewer than 10 officers, including 561 agencies with just a single officer.

The 50 largest departments in the United States can be considered "typical" municipal departments, although each department differs somewhat in terms of its politics, laws, budget, experiences (of officers and populations), and personnel. Nevertheless, population size does not always dictate the number of sworn officers, the ratio of the police to the population, and the number of reported offenses (Table 5–2).

Patrol officers are the backbone of the departments; they will be discussed in depth in Chapter 9. In the typical department, of every 10 officers on the payroll, 6 have regularly assigned patrol duties, which include responding to calls for 911 emergency service. The proportion of officers responding to 911 calls ranges from about 6 in 10

TABLE 5–2 Fifteen Largest City Populations: Officer–Population Ratios and Known Crime Rates, 2003

City	Population within City Limits	Region	Officer Number*	Number of Officers per 10,000 Residents†	Violent Offenses Known to Police per 100,000 Residents‡
1. New York City	8,143,197	Northeast	35,973	44	673
2. Los Angeles, California	3,844,829	West	9,307	24	821
3. Chicago, Illinois	2,842,518	Midwest	13,469	47	NA
4. Houston, Texas	2,016,582	South	5,350	27	1,173
5. Philadelphia, Pennsylvania	1,463,281	Northeast	6,853	46	1,467
6. Phoenix, Arizona	1,461,575	West	2,808	19	729
7. San Antonio, Texas	1,256,509	South	2,056	17	637
8. San Diego, California	1,255,540	West	2,103	17	519
9. Dallas, Texas	1,213,825	Southwest	2,943	24	1,254
10. San Jose, California	953,679	West	1,408	16	384
11. Detroit, Michigan	886,671	Midwest	3,837	42	2,361
12. Las Vegas, Nevada	545,147	West	2,640	17	744
13. Indianapolis, Indiana	784,118	Midwest	1,170	15	993
14. Jacksonville, Florida	782,623	South	1,624	21	830
15. New Orleans, Louisiana	454,863	South	1,622	35	NA

*Bureau of Justice, "Local Police Departments, 2003" (May 2006).

†Bureau of Justice, "Local Police Departments, 2003" (May 2006).

‡"Statistical Abstracts," http://www.census.gov/compendia/statab/law_enforcement_courts_prisons/crimes_and_crime_rates/ (accessed September 30, 2006).

among departments serving 100,000 or more residents, to about 9 in 10 in departments serving fewer than 10,000 residents.

For instance, the New York City Police Department (NYPD) serves the most populous city in the United States (an estimated 8.1 million residents) and employs approximately 36,000 sworn officers (see Table 5–2). Thus the NYPD has 44 officers for every 10,000 residents (i.e., 4 officers for every 1,000 residents). When you add in the vast number of business visitors, workers, and vacationers who travel to the city each day, the actual ratio would quickly change to something like 1 or 2 officers for every 1,000 individuals in New York City.

By comparison, the Los Angeles Police Department (LAPD) serves a population of 3.8 million residents with approximately 9,300 police officers, which averages out to 24 officers for every 10,000 residents. This city experienced a (reported) crime rate of 821 violent crimes per 10,000 residents in 2005.

Clearly, population size, numbers of officers, and number of crimes reported to police differ for different jurisdictions. Compare, for example, Indianapolis, Indiana, with Jacksonville, Florida. While the sizes of these cities' populations are similar, the number of officer–population ratio is higher in Jacksonville (21 officers per 10,000 population versus 15 officers per 10,000 population in Indianapolis), while the (reported) crime rate is lower in Jacksonville.

Municipal police departments in particular face a relatively high serious crime load as compared to other jurisdictions. For example, the 15 largest departments in 2003 served approximately 8 percent of the U.S. population but handled 24 percent of all violent crime in the country.

Indeed, the jurisdictions with the largest populations (e.g., New York City, Los Angeles, and Chicago) do not have the highest crime rates per capita in the United States. For instance, the cities with populations of more than 500,000 that had the most violent crime reported in 2007 were Detroit, Baltimore, and Memphis, according to the FBI Crime Index.[7] By contrast, the safest cities with large populations (more than 500,000) were San Jose, Honolulu, and El Paso.

Municipal departments dominate the public's images of the police. Think about popular television dramas, which are generally filmed in New York City, Los Angeles, or Miami. Similarly, news stations tend to cover events that occur in the dominant cities. A disproportionate amount of research related to these departments also appears in the literature because studies tend to be conducted in New York City, Chicago, Los Angeles, Boston, and Washington, D.C. Given this media and literature domination, metropolitan police departments set the standards for police experiences. Less is known about middle-sized and small agencies—unless you work there, of course.

Despite this pervasive image of large-city departments, the typical municipal police department is actually located in small-town America, and half of those agencies employ fewer than 10 sworn officers. Police intervention in those communities tends to focus on less serious crime than occurs in large cities: Traffic control accounts for 25 percent of most calls, family disturbances for 18 percent, public disturbances for 19 percent, and stray dogs for 11 percent; the remaining 27 percent of calls involve miscellaneous types of incidents.

Profile of a Municipal Police Department: San Jose Police Department

The chief of police, Rob Davis, at the San Jose Police Department (SJPD) is pleased to work in one of safest cities in the United States. Because his department serves a highly diverse population, he provides messages to his constituents in Spanish, English, and Vietnamese. After welcoming constituents in this manner, the chief hopes, the "calls for service" information will provide better services for individual neighborhoods to reduce crime. The chief urges neighborhood members to actively participate in community policing strategies.

Vision

The SJPD is a dynamic, progressive, and professional organization that is dedicated to maintaining community partnerships that promote a high quality of life for the city's diverse population. It is committed to treating all people with dignity, fairness, and respect; protecting their rights; and providing equal protection under the law.

Mission

The SJPD has documented that it has the following mission:

- To promote public safety
- To prevent suppress and investigate crimes
- To provide emergency and non-emergency services
- To create and maintain strong community partnerships
- To adapt a multidisciplinary approach to solving community problems
- To develop and promote a diverse, professional workforce

Minimum Requirements for Employment

All applicants for employment as SJPD officers must meet the following minimum requirements prior to taking the run/physical agility test:

- Applicants must be at least 20 years old by the (civil service) exam date and 21 years old by date of appointment.
- Applicants must possess a high school diploma from an accredited U.S. high school or a general equivalency degree (GED).
- Applicants must have completed 40 semester or 60 quarter units at an accredited college or university.
- Applicants must be citizens of the United States or permanent resident aliens who are eligible for and have applied for U.S. citizenship.
- Applicants must possess a valid driver's license.
- Applicants may not have any felony or domestic violence convictions.
- Applicants must have at least 20/40 vision (uncorrected).

As of October, starting pay for SJPD officers was $70,307.27. Benefits included almost five weeks of paid vacation per year, a uniform allowance, and other fringe benefits (http://www.sjpd.org/). Figure 5–1 shows the organizational chart for the SJPD.

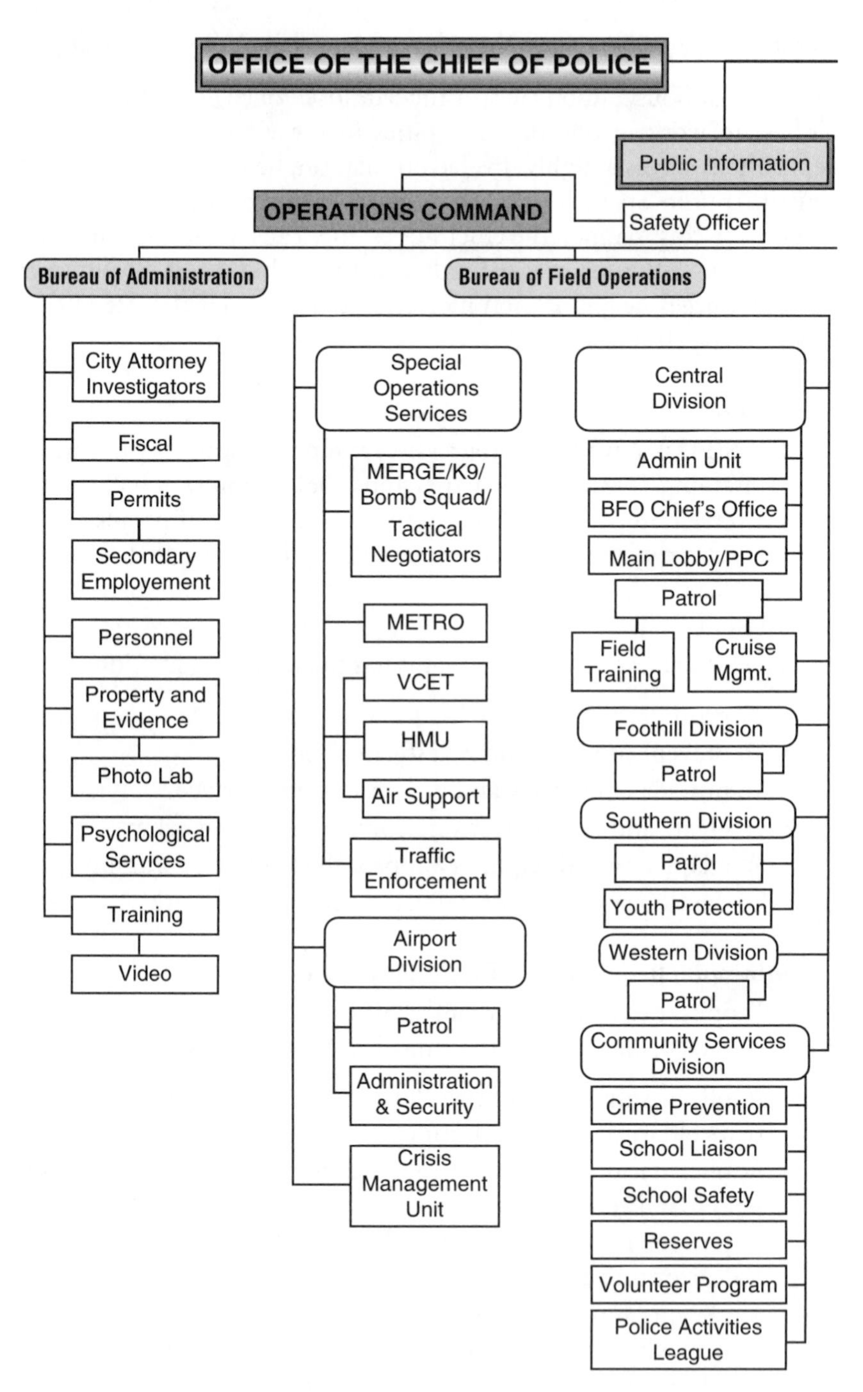

Figure 5–1 Organizational chart for the San Jose Police Department. Courtesy of the San Jose Police Department, Chief Rob Davis.

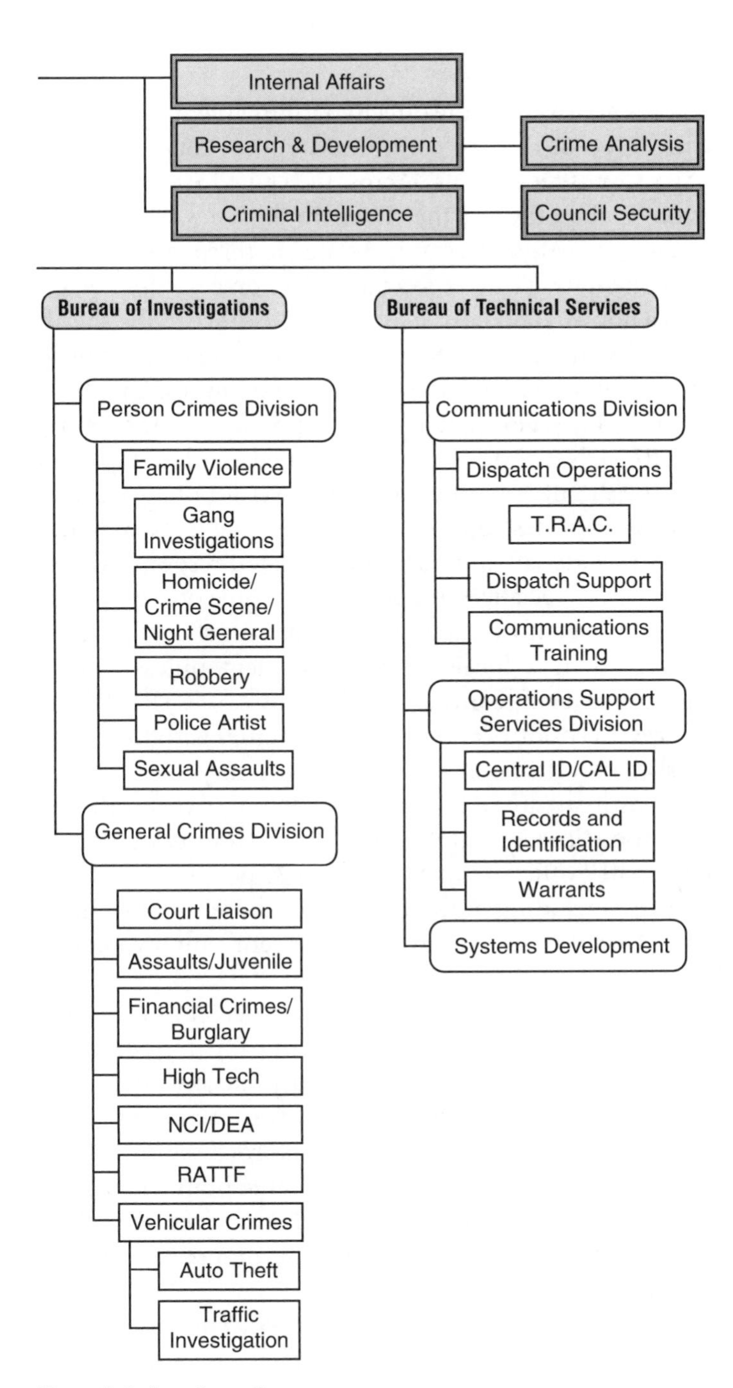

Figure 5–1 (continued)

County Police (Sheriffs)

Sheriffs are generally elected officials and, as such, are directly accountable and responsible to eligible voters; thus sheriffs are not directly controlled by local county boards, commissioners, supervisors, mayors, or other officials.[8] Across the United States, popular election is the means of selection of the office of the sheriff in 46 states. Sheriffs are elected to four-year terms in 41 states, to two-year terms in 3 states, to three-year terms in one state, and to six-year terms in one state. The election takes place on a partisan ballot in 40 states; sheriffs are elected on a nonpartisan basis in 6 states. The office of the sheriff is the "chief law enforcement office" within most U.S. counties. There are no sheriffs in two states: Alaska and Connecticut. In Rhode Island, the governor appoints the sheriff. In two Colorado counties and in Dade County, Florida, sheriffs are appointed by the county executive. In New York, the sheriff of New York City is appointed by the mayor; in Westchester and Nassau counties, the sheriff is appointed by the governor of New York.

Sheriff's offices typically serve counties and municipalities that lack a police agency.[9] They often operate a countywide jail and have court-related responsibilities. Some also operate fire departments and specialized units (see the profile of the Jackson Sheriff's Office later in this chapter). The enforcement jurisdictions for sheriff's offices typically exclude county areas served by a local police department unless a local department or constituency-based authority contracts with or hires the sheriff's or constable's offices for police services. In counties and independent cities with their own separate police department, the sheriffs' office may not have any primary jurisdiction.

Some sheriff's offices employ as many as 4,000 people, including sworn officers, fire fighters, jailers, and other personnel; for example, Palm Beach County in Florida matches this description. Other sheriff's offices employ as few as 20 people.

One fourth of all sheriffs' offices, including more than half of those serving 1 million or more residents, routinely employ foot patrols; 20 percent employ bicycle patrols, mounted units, and marine units on a regular basis. In 2003, 94 percent of sheriffs' offices participated in a 911 emergency system. Thirty-six percent of sheriffs' offices had deputies assigned on a full-time basis to a special unit for drug enforcement, with approximately 4,000 officers being assigned to this duty nationwide. Nearly half of all sheriffs' offices had officers assigned to a multiagency drug task force.

Most often, the sheriffs' offices is responsible for serving civil processes, providing court security, and operating the county jail. Some jails confine large numbers of suspects and guilty defendants; for instance, Cook County Jail in Chicago has more than 10,000 detainees. In contrast, Madison County Jail in London, Ohio, houses 27 inmates (and will be shut down by the time this book is in print).

Each sheriff's office has different responsibilities (see Table 5–3). The Los Angeles Sheriff's Department employs approximately 8,600 sworn personnel, of whom 42 percent are assigned to patrol duty, 6 percent to investigation, 25 percent to jail operations, 15 percent to court security, and 1 percent to process-serving duties. Contrast this with the 61 percent of sworn personnel in Cook County and the 62 percent of sworn deputies in Erie County (Buffalo), New York, who are assigned to jail operations, and the 10 percent in Cook County and 18 percent in Erie County sworn deputies assigned to patrol assignments.

During fiscal year 2003, total operating budgets for U.S. sheriff's offices amounted to $22.3 billion. Expenditures during the same year averaged $124,400 per deputy, with

TABLE 5–3 Twenty-Five Largest Sheriff's Offices, 2003

County or Parish	Full-Time Sworn Officers	Patrol Officers (%)	Investigative Officers (%)	Jail Operations Officers (%)	Court Security Officers (%)
Los Angeles County, California	8,622	42	06	25	15
Cook County, Illinois	5,555	10	01	61	27
Clark County, Nevada	2,640	48	18	25	0
Harris County, Texas	2,517	30	06	27	05
Orange County, California	1,755	29	07	40	12
Jacksonville County, Florida	1,624	61	12	0	05
Broward County, Florida	1,605	84	13	0	02
Riverside County, California	1,542	69	09	13	09
San Bernardino County, California	1,541	32	12	22	11
Sacramento County, California	1,525	36	11	26	06
San Diego County, California	1,320	35	22	04	14
Orange County, Florida	1,294	62	15	0	10
Palm Beach County, Florida	1,177	65	09	0	09
Hillsborough County, Florida	1,126	65	13	0	0
Alameda County, California	974	15	02	48	16
Wayne County, Michigan	893	11	03	45	06
East Baton Rouge Parish, Louisiana	875	14	09	36	03
Pinellas County, Florida	868	48	16	0	16
Oakland County, California	840	41	08	48	03
San Francisco County, California	824	0	01	67	09
Erie County, New York	813	18	04	62	16
Ventura County, California	760	30	12	15	10
Jefferson Parish, Louisiana	747	42	27	25	0
Contra Costa County, California	707	37	06	36	09
King County, Washington	705	58	22	0	04

Note: The missing percentages are in process serving.

Source: Bureau of Justice, *Sheriff's Officers, 2003* (NCJ 211361) (Washington, DC: U.S. Department of Justice, 2006).

starting salaries for entry-level deputies ranging from an average of $23,300 in the smallest jurisdictions to $38,800 in the largest districts.

An issue that has been the subject of considerable debate is whether sheriffs should continue to be elected or should be appointed. One concern is that sheriffs typically are not required to be sworn officers.

Profile of a County Police Department/Sheriff's Office: Jacksonville Florida Sheriff's Office

The merger of the Duval County Sheriff's Office and the Jacksonville Police led the creation of a new jurisdiction—the Jacksonville Sheriff's Office, which has since grown significantly in capability and reputation. (Mergers of county and city police jurisdictions are becoming common across the United States and is one solution to the fragmentation problem discussed later in this chapter.) For a discussion of the new department's patrol division, see Chapter 9.

Department of Corrections

The Jacksonville Sheriff's Office's Department of Corrections is made up of 650 certified corrections officers and civilian personnel. These employees operate three correctional facilities for the secure, humane, and corrective detention of individuals incarcerated in Duval County. The largest of these facilities is the John E. Goode Pretrial Detention Facility (PDF), which is located in downtown Jacksonville, adjacent to the Police Memorial Building and next door to the Duval County Court House, where many of the inmates travel for adjudication of their cases. A state-of-the-art facility when opened in April 1991, the PDF is a 12-story building with an inmate capacity of 2,189.

Department of Police Services

The Department of Police Services is managed by a director who is responsible for the department's overall operation; it is funded by a budget of $203.4 million and federal/state grants totaling almost $21 million. The department manages the critical support functions for all Jacksonville Sheriff's Office personnel and other members of the criminal justice community. The Department of Police Services is composed of two divisions: the Budget and Management Division and the Police Services Division.

Other units at the Jacksonville Sheriff's office include Aviation, Bomb Squad, Canine, Career Criminal, Criminal Apprehension, Dive Team, Homeland Security, Juvenile Intervention, Mounted Patrol, Narcotics, Seaport Security (see Chapter 6), and Vice and Investigation units.

Constables

The office of **constable** arrived in America in the mid-seventeenth century from England, and with it continued the divergence between constable and sheriff. In America, as in England, the main qualification for the person holding the office of sheriff was "that he be of sufficient estate."[10] This limited the choices for sheriff to a relatively small and elite group of planters in each county. Consequently, the constable and the justice of the peace were the only law and order most rural American settlers ever saw. Perhaps not surprisingly, a constable's duties and jurisdiction varied depending on the circumstances and the times.

There is no consistent use of the term "constable" even today, and the use of constables may vary not just between states, but also within the same state. A constable can be an official responsible for process servicing or court-related tasks, which includes summonses and subpoenas in states such as in Mississippi and Louisiana. Most often state law permits counties to establish constables as officers who can be either elected or appointed by the Chief Magistrate Judge of the county's magistrate court and then serve at the pleasure of the Chief Magistrate; as in New York, they serve at the pleasure of local towns and villages. In these cases, the powers and duties of constables may include the following tasks:

- Attending all regular sessions of the court
- Collecting and paying money owed to the court
- Executing and returning warrants, summonses, and other processes directed to them by the magistrate court

In some states, a constable may be appointed by the judge of the court that he or she serves; in other cases, the constable is an elected or appointed position at the village, precinct, or township level of local government. However, unlike police officers and deputies, constables can make arrests only with a warrant or at the direction of a judge—unless those constables are in Texas.

Shortly after Texas became a state, an act passed by the newly formed Texas legislature specified that the constable should be "the conservator of the peace throughout the county," adding that "it shall be his duty to suppress all riots, routs, affrays, fighting, and unlawful assemblies, and he shall keep the peace, and shall cause all offenders to be arrested, and taken before some justice of the peace." Today, Texas constables serve a term of four years and number around 3,500, including clerical and other personnel.[11] Their law enforcement roles vary widely, but in general constables' police powers are no different from those of other peace officers in the state of Texas. Complete records do not exist, but the most recent estimate is that at least 93 Texas constables have died in the line of duty, including 67 in the twentieth century. Often personnel from the constable's office work side by side with the sheriff's department and municipal police agency personnel.

Profile of a Constable's Office: Harris County Constable's Office

Harris County, Texas, has eight Constable Precincts; two sheriff's offices serve the area as well. Precinct 4 of the Harris County Constable's Office is headed by Constable Ron Hickman, who had been the assistant chief under the previous constable. Precinct 4 serves 16 justices of the peace and serves a geographic area of more than 520 square miles that is home to an estimated 750,000 residents. Since Hurricane Katrina, the population has increased dramatically owing to resettlement of hurricane refugees in Harris County, though there are few reliable estimates of the actual number of new residents available. Several other police agencies operate in overlapping, concurrent jurisdictions with Precinct 4, including the Houston Police Department.

The Patrol Division, in addition to being the largest division within Precinct 4, is the most visible part of the Harris County's Constable's Office. There are almost 200 full-time and 100 reserve sworn deputies on patrol;[12] the patrol fleet numbers 146 cars. The duties of the Patrol Division include high-profile marked patrol, emergency and non-emergency response to calls for service, on-scene and follow-up investigations, crime prevention, and traffic enforcement. In addition to monitoring the public roadways, deputies are responsible for policing county parks, the Hardy and Sam Houston toll roads, and the Cy-Fair ISD school facilities.

Many subdivisions opt to participate in the contract deputy program offered by the Harris County Constable's Office as a way to supplement their existing policing services. Precinct 4 has 75 contracts with various subdivision associations within and outside of Harris County, one school district, and a shopping mall.

■ State Police

In the United States, the 49 state police agencies (troopers, highway patrol, department of public safety) employ approximately 58,000 sworn officers with statewide jurisdiction over limited violations such as those related to traffic control on state highways and often incidents occurring at airports.[13] In many states, the state police also have the following responsibilities:

- Operate the state's crime lab
- Perform specific investigations for local departments such as homicide investigations
- Deliver statewide services to local departments such as stress management, tactical units, or special weapons and tactics (SWAT) units
- Participate with other state agencies such as the Department of Corrections whereby appropriately trained troopers are parole officers (e.g., Massachusetts)

All state police agencies are administered by state government, except in the state of Hawaii. In that state, the duties of a highway patrol are carried out by an appointed sheriff (although regulation is provided by state government). State troopers have general law enforcement assignments, including maintaining regular patrols and responding to calls for service, especially those related to traffic control on state highways. Uniformed officers are best known for issuing traffic citations to motorists who violate traffic laws on state highways. At the scene of an accident, state troopers provide first aid, call for emergency equipment including fire equipment, and direct traffic.

Most states also have a state-run bureau of investigation or identification that gathers, analyzes, and publishes statewide crime information. For example, in Massachusetts, state troopers investigate homicides and other felony cases in many jurisdictions; in addition, members of one Massachusetts state trooper unit serve as parole officers. Many state trooper or highway patrol divisions investigate organized crime and missing persons reports, conduct interdiction (drug control) and sting (see Chapter 12) operations, and perform background checks on police, liquor license, and gun permit applicants. Many state trooper agencies operate the state's sex offender registry, issue "Amber alerts," run crime laboratories (including those that perform DNA and trace evidence assessments), and act as the agency that seizes, dismantles, and disposes of chemicals, equipment, and waste from clandestine laboratories such as those for manufacturing methamphetamine (see Chapter 12). Some troopers might operate aviation, port or water authority, and homeland security units.

State police officers generally undergo training in the state's police academy for 12 to 14 weeks, plus a year's additional training. In 2002, median annual earnings for state police officers were $47,090. In 2005, new hires in Louisiana earned about $2,600 per month during training and about $3,000 per month after one year on the job. In Virginia, new hires earned about $32,000 annually while training and $36,000 annually after one year.

Profile of a State Police Department: Illinois State Police

Approximately 1,700 uniformed and plainclothes officers in 21 districts work together to provide comprehensive law enforcement services for the Illinois State Police (ISP).[14] Troopers patrol state highways and tollways, conduct truck weight inspections, and oversee hazardous materials control. ISP operates nine forensic science laboratories statewide and provides investigators with DNA identification experience and the examination of nearly anything collected at a crime scene. It also operates a unit that removes hazardous materials from clandestine laboratories such as those in which methamphetamine is manufactured.

Mission

According to ISP documents, the department's mission is to promote public safety and to improve the quality of life in Illinois. ISP reports that its goals include the following:

- Improve the quality of life for citizens through unimpeachable integrity, public service, training, and education
- Safeguard the public by assisting law enforcement, decreasing traffic fatalities and injuries, and reducing crime and the fear of crime
- Provide leadership through innovation as a dynamic, diverse, learning organization that promotes personal and professional growth

Starting Salaries

While at the state's academy, state police cadets' pay is $2,570 monthly ($30,840 yearly) as of 2006. However, cadets are paid overtime, which brings their salary to approximately $37,000 yearly. Fringe benefits include health benefits and retirement programs.

Once on active duty, the first-year trooper's base salary is $44,876 yearly (for cadets starting in 2007). With pay for extra duty, troopers can earn as much as $70,000.

Training

ISP officers must complete the Basic Training Course for Local Law Enforcement Officers, which includes 480 hours of Illinois law enforcement training, and obtain standards board certification. The basic training course prepares trainees for the performance of their duties as officers and includes the following instructional units:

- Illinois Vehicle Code
- Physical Training
- Criminal Offenses in Illinois
- Civil Rights and Civil Liability
- Firearms
- Law Enforcement Driving
- First Responder

Evaluations of cadets are based on written examinations, performance in field practical exercises (Physical Training, Control and Arrest Tactics, and Firearms), classroom participation, and behavior.

State troopers continue to receive training after they pass the basic training course. For example, the Critical Incident Response Training course provides officers with the skills required to safely respond to and manage the initial stages of all types of critical incidents. A simple "game plan," called the Seven Critical Tasks, allows the first responding supervisor to take control of a rapidly unfolding incident, stabilize the scene, and prepare for the arrival of specialized assets. Topics covered include the role and responsibility of first-responder officers/supervisors, an overview of hazardous materials/weapons of mass destruction issues, and critical incident stress. This class also provides a basic understanding of the Incident Command System (ICS) and the National Incident Management System (NIMS). Each student has the opportunity to try out the role of incident commander in highly realistic exercises on the state's model city simulator.

Campus Police

In 2002, there were approximately 16 million students enrolled in 4,200 colleges and universities across the United States. Between 1995 and 2002, college students ages 18–24 were victims of approximately 479,000 crimes of violence annually, including rape/sexual assault, robbery, aggravated assault, and simple assault.[15]

One study shows that three fourths of all U.S. public college and university campuses with 2,500 or more students and private campuses with 10,000 or more students have sworn officers on duty, while a number of campuses rely on nonsworn security officers.[16] Sworn campus police officers typically have to pass through many more screening measures and undergo three to four times as much training as their nonsworn counterparts.

The same study revealed that per-student employment and expenditures for police officers at private institutions were nearly twice that at public campuses.[17] Nearly all agencies operated general crime prevention programs on campus, and two thirds had programs aimed specifically at rape prevention. (Other topics covered by the report include agency functions, personnel characteristics, equipment, computers and information systems, and policy directives.)

A nationwide campus survey about law enforcement agencies serving four-year universities and colleges with 2,500 or more students reported the following findings:[18]

- More than 9 in 10 public institutions used sworn police officers, compared to less than half of the private institutions.
- Most sworn campus police officers were armed, and overall about two thirds of the campuses had armed officers.
- Police agencies serving private campuses had operating costs of approximately $181 per student during fiscal 1994 compared to $94 per student for public campuses.
- Nearly all of the agencies operated a general crime prevention program, and about two thirds had rape prevention programs. About half operated programs to combat drug and alcohol abuse.
- Larceny and theft were the most frequently reported crimes on campus, and aggravated assault was the most serious violent crime reported.

Promoting Safety on College and University Campuses

College and university campuses traditionally have provided a special environment in which students can explore new ideas and learn about the world. Creating a safe and supportive campus community is both an obligation and a challenge for college and university administrators, faculty and staff, other campus personnel, and students.

At many college campuses, aside from traffic and special-event control, serious issues handled by police include sexual assault, dating and domestic violence, and stalking.[19] College women are at high risk for all forms of violence. For instance, half of all stalking victims are between 18 and 29 years old, and the highest rate of intimate partner violence is found among women ages 16–24.[20] Sexual assault is the second most common violent crime committed on college campuses (after aggravated assault). Most perpetrators of sexual assault are students known by the victim.[21] Half of these sexual assaults occur in the victim's residence, and another one third take place in off-campus student housing such as fraternities. Given the prevalence of these problems (and the fact that alcohol plays an important role in many campus-related assaults), it is clear that campus police officers require specialized training geared toward the unique populations that they serve.[22]

On many campuses, sexual assaults against women are gravely underreported, with some estimates suggesting that as many as 84 percent of assaults go unreported to the police. Unlike their counterparts in the larger community who experience such assaults, female students who are victimized by other students often face challenges specific to life on a "closed" campus environment. Given the unique and progressive nature of many stalking cases, student victims are often unaware or unsure when they are being stalked, or they may have difficulty convincing others that a problem exists. The stalker may have seemingly "legitimate" reasons for remaining in contact with or in proximity to the victim in class, the dining hall, or the library. Victims of sexual assault or dating violence may continue to encounter their assailants in residence halls or at campus events. Even changing one's living arrangements or class schedule may not eliminate the threat or additional trauma caused by ongoing contact.

What can be done about this problem? While sexual assault, dating and domestic violence, and stalking have implications for campus and student life, they often involve criminal acts that necessitate the involvement of the criminal justice system. Campus adjudication procedures are critical for increasing the safety and security of women on campus, but they cannot substitute for criminal investigation or prosecution. Indeed, sexual assault cases—especially among female students—are rarely resolved through strictly university administrative avenues without public law enforcement participation.

Profile of a Campus Police Department: Florida State University Campus Police

The Florida State University Police Department (FSUPD) is dedicated to supporting the educational mission of the Florida State University (with an enrollment of 38,000 students) by striving to create and maintain the safest possible environment for its students, staff, and visitors.[23] This includes taking a proactive approach to law enforcement and using aggressive problem-solving techniques for identified crime trends. The FSUPD received accreditation in 2002 from the Commission on Florida Law Enforcement Accreditation and employs 60 sworn police officers for the main and Panama City campuses.

Qualifications for Employment

Candidates of the FSUPD must meet the following criteria:

- Be a citizen of the United States
- Be at least 19 years of age
- Have a high school diploma or equivalency certificate
- Have a valid Florida driver's license
- Have obtained a Law Enforcement Standards Certificate
- Meet other requirements of Chapter 943, Florida State Statutes

Duties

New officers are assigned to the Patrol Division of the FSUPD. Officers in training receive 15 weeks of in-house orientation and field training. Officers are instructed, evaluated, and observed on a daily basis.

FSUPD officers apprehend criminals, prevent criminal activities through proactive patrol techniques, and enforce federal, state, and local laws. Additionally, Florida State University police officers assist other agencies when necessary and work closely with university officials to enforce university rules and regulations.

In 2005, the starting salary for a FSUPD officer was $35,181 and included a salary incentive program that could produce another $130 per month upon successful completion of advanced education and approved police training programs. Overtime hours are paid at the rate of time and a half. Officers receive uniforms, service weapons, and duty belts furnished by the department, plus a $350 uniform maintenance allowance and $150 shoe allowance.

■ Capital Police

There are 50 state **capital police** divisions in the United States, whose primary responsibility is to preserve order, prevent crime, and protect the public and property of the state at the state capital complex and at most of the state buildings and agencies. In every state, a state code provides the capital police with their authority. For instance,

Alabama's capital state police force was established by state code §41-4-182 as a division of the Finance Department of the state of Alabama; the state Finance Director is the head of the Capitol Police Division. In other states, the individual in charge varies depending on the state policy.

Most often, capital police officers have police powers anywhere within the state. Their organizational structure usually consists of a chief of police (who generally answers to a specific official, which in some states means the lieutenant governor's office), chief operations manager, a few investigators, administrative staff, and three shifts of officers and patrol officers. Each shift consists of a sergeant and an assigned number of patrol officers.

Profile of a Capital Police Department: Division of Capital Police in Delaware

Delaware's Division of Capitol Police is dedicated to the protection of elected and appointed officials, state employees, visitors to state facilities, and state facilities and properties as directed by the secretary of the Department of Safety and Homeland Security, through the delivery of professional law enforcement services, utilizing sworn police officers, security officers, and civilian staff.[24] As part of its goals, the department strives to accomplish the following tasks:

- Provide police services in an efficient and professional manner
- Institute and maintain procedures that allow state agencies the ability to provide services and perform without interruption
- Initiate and maintain procedures that provide for the safety of state employees and visitors while at state facilities and properties
- Provide specialized police services to the General Assembly to ensure the safety of the legislators, staff, constituents, and visitors, while providing members of the General Assembly with the ability to perform their duties without interruption
- Provide specialized police services to the judiciary to ensure the safety of the justices, judges, commissioners, staff, clients, and visitors, while providing the members of the judiciary with the ability to perform their duties without interruption
- Provide specialized police services, in coordination with the Division of State Police, to the governor's office to ensure the safety of the governor, lieutenant governor, staff, and visitors to the governor's and lieutenant governor's offices and the official residence, Woodburn, thereby providing staff of the governor's and lieutenant governor's offices with the ability to perform their duties without interruption
- Initiate and maintain procedures that provide mechanisms to prevent and respond to threats against state facilities, including acts of terrorism, natural disasters, or other situations requiring a coordinated response by law enforcement and non-law enforcement entities
- Interact with the community by involving leaders, organizations, and the public in a proactive law enforcement approach to prevent and deter criminal activity at state facilities

Coroners, Medical Examiners, and Forensic Pathologists

The term **coroner** has been in use in England since approximately the year 900. Its origin lies in the Latin word *corona*, meaning "crown"; thus a coroner is an officer of the crown—he or she works for the king or queen. The office of the coroner (or medical examiner) is considered a law enforcement unit because it has evolved over the centuries to become officially responsible for investigation and certification of cause and manner of cases of sudden and unnatural death, including epidemics related to infectious disease.

In the United States, infectious disease deaths substantially declined during the first eight decades of the twentieth century as a result of public health interventions. The end of the century was marked by an increase in infectious disease deaths, however—primarily owing to acquired immune deficiency syndrome (AIDS), pneumonia, and influenza. In fact, the increasing number of deaths from known infectious diseases, the emergence of new infections, and bioterrorism have made surveillance for infectious diseases a public health concern.[25]

The Centers for Disease Control and Prevention (CDC) reports that an estimated 20 percent of all deaths are investigated by a coroner or medical examiner. In small counties, the elected prosecuting attorney often serves as coroner as well. In medium-sized counties and larger jurisdictions, the coroner is a separate elected office.

In addition to investigating all deaths suspected of resulting from violence, criminal means, or suicide, or any unattended death whatever the cause, the medical examiner/coroner provides identification, orders autopsies or medical examinations, conducts inquests when necessary, and carries out any other requirements related to deaths that fall into the previously mentioned categories. Generally, most medical examiner's or coroner's offices provide the following services:

- Report the cause of death, as determined by the coroner's investigation, autopsy, and toxicology report
- Return personal effects of the deceased to survivors
- Compile statistics on violent deaths in the jurisdiction
- Conduct the coroner's inquest, including setting the time and place of the inquest

In many jurisdictions, such as the state of Washington, there are no occupational requirements for the coroner.

Some states have separate coroners, medical examiners, and forensic pathologists who perform the tasks highlighted above. **Medical examiners** were first used in 1877 in Massachusetts. At the time, the public was dissatisfied with the work of layman coroners, so the Massachusetts system was changed to provide for investigation of suspicious deaths by physicians—that is, medical examiners. One medical doctor was appointed to each district (similar to a county jurisdiction) to be the person or public official responsible for investigation of sudden and unnatural deaths. Medical examinations were a part of the investigation—hence the term "medical examiner." The medical examiner concept is now used in many states. All medical examiners are appointed (not elected) positions, and most medical examiners are physicians, although they are not necessarily trained in forensic pathology.

The modern medical examiner system developed in 1915 in New York City, when a forensic pathologist was appointed to be the city's medical examiner, imbued with statutory authority to investigate deaths, and provided with a facility, support staff, and a toxicology laboratory. Forensic pathology is a branch of medicine that applies the principles and knowledge of the medical and related sciences to problems that concern the general public and issues of the law. A **forensic pathologist** is a physician with specialized medical and forensic science training and knowledge.

What determines whether a jurisdiction has a coroner, a medical examiner, or a forensic pathologist? The answer might depend on the jurisdiction's policy, budget, and available personnel.

Tribal Police

As of June 2000, Native American tribes operated 171 law enforcement agencies employing the equivalent of at least one full-time sworn officer with general arrest powers.[26] In addition, the U.S. Bureau of Indian Affairs (BIA) operated 37 agencies providing law enforcement services in **Indian country**.

> There are 341 federally recognized American Indian tribes in the lower 48 states, reports BJS (2002). Indian country is Indian tribes with tribal land with tribal administered police departments. However, the federal government (Bureau of Indian Affairs and the FBI) investigate crimes and directly enforce the law in the Indian country. But public safety on reservations is the primarily responsibility of local and tribal law enforcement.

Tribally operated agencies employed 3,462 full-time personnel, including 2,303 sworn officers (67 percent) and 1,159 nonsworn officers (33 percent). These agencies also employed 217 part-time personnel, including 88 sworn officers (41 percent) and 129 nonsworn officers (59 percent). Tribally operated agencies had a combined service population of 1,016,188 residents in 1999. Thus the police–population ratio in Native American–controlled areas was approximately 2.3 full-time sworn officers per 1,000 residents, across all agencies.

Tribally operated agencies provided a broad range of public safety services and functions in 2000. Nearly all **tribal police** officers responded to calls for service, and a large majority engaged in crime prevention activities, executed arrest warrants, performed traffic law enforcement, and served court papers (see Figure 5–2). More than half of the agencies provided court security and conducted search and rescue operations. Approximately one fourth of these agencies operated one or more jails.

Criminal jurisdiction is a major difference that distinguishes tribally operated agencies from their state and local counterparts.[27] Jurisdiction over offenses in Indian country may rest with federal, state, or tribal agencies depending on the particular offense, the offender, the victim, and the offense location.

To better understand the challenges of tribal policing, it is necessary to look at Public Law 280 (PL 280).[28] PL 280 transferred federal criminal jurisdiction in Indian country to six states that could not refuse jurisdiction, known as "mandatory" states:

- Alaska
- California

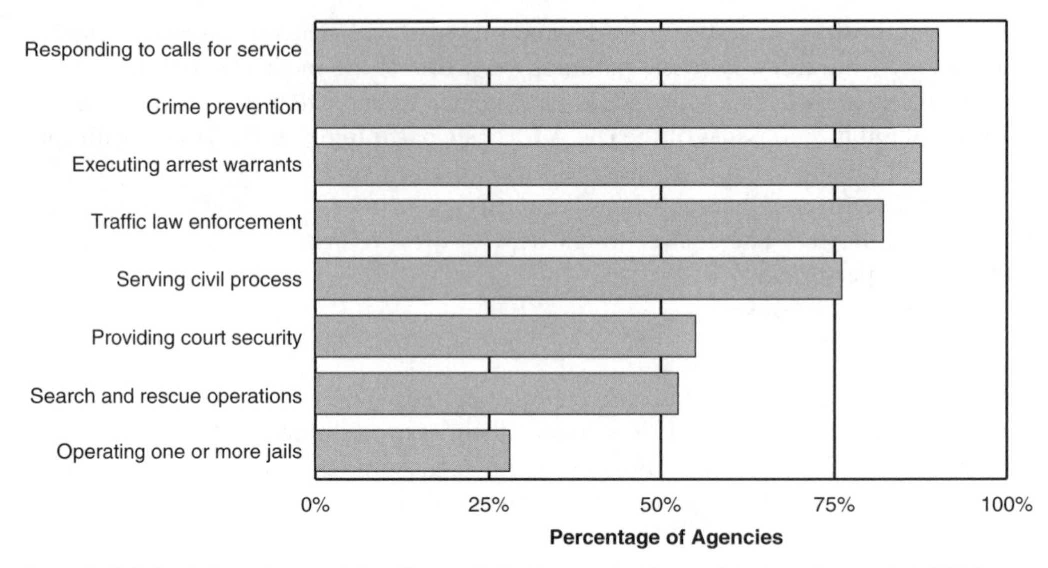

Figure 5–2 Selected services and functions of tribally operated law enforcement agencies, 2000.

Source: Bureau of Justice Statistics, *Tribal Law Enforcement, 2000* (NCJ 197936). (Washington, DC: U.S. Department of Justice, 2003), http://www.ojp.gov/bjs/pub/pdf/tle00.pdf

- Minnesota
- Nebraska
- Oregon
- Wisconsin

The same law permitted other states, at their option and without consulting tribes, to choose to assume complete or partial jurisdiction over crimes committed by or against Indians in Indian country. Ten states chose to do so:

- Arizona
- Florida
- Idaho
- Iowa
- Montana
- Nevada
- North Dakota
- South Dakota
- Utah
- Washington

Of interest, PL 280 did not provide for the consent of affected tribes.[29] Therefore the criminal laws in those states became effective over the Indian populations within as well as outside Indian country. That said, you can imagine the many challenges confronting both tribal police and the people they serve to protect (see Figure 5–3). One huge issue relates to criminal jurisdiction, which could lie with federal, state, or tribal

Aside from jurisdictional issues, policing on Indian reservations faces many difficulties that law enforcement elsewhere generally need not confront, at least to the same extent. Data collected by the Bureau of Justice, for example, suggest that violent victimization among American Indians and Alaska Natives exceeds that of other racial or ethnic subgroups by about 2.5 times the national average.

According to a National Institute of Justice–supported study, a typical police department in Indian country serves a population of 10,000 residing in an area about the size of Delaware patrolled by no more than 3 officers at any one time. Even so, many reservation residents live in areas with characteristics of suburban and urban locales. Researchers found that the overall workload of Indian country police departments has been increasing significantly in intensity and range of problems—driven by rising crime, heightened police involvement in social concerns related to crime, and increased demand for police services.

The study reported that most police departments in Indian country are administered by tribes under contract with the Bureau of Indian Affairs (BIA). The second most common type of department management is direct BIA administration. Under the former arrangement, law enforcement personnel are tribal employees; under the latter, they are federal employees. State and local authorities supply police services to tribes not affected by retrocession in PL 280 states.

Of Indian country police departments surveyed, the researchers found:

Officers who were Native American	66%
Officers who were women	12%
Native American officers who were members of the tribe they serve	56%
Officers who were unable to speak the language native to the community they serve	87%

Figure 5–3 Overview of policing in Indian country
Source: Bureau of Justice Statistics (2005). Public Law 280 and Law Enforcement in Indian Country. Research Priorities, NCJ 209839. Washington, DC: U.S. Department of Justice.

agencies depending on such issues as the identity of the alleged offender and the victim and the nature and location of the offense.

Table 5–4 provides an idea as to the hard work performed by tribally operated police agencies. For instance, the Navajo Nation in Arizona employs the largest number of sworn officers (321) who serve and protect a population of 170,000 individuals living in an area of 22,000 square miles. This staffing provides for a ratio of 2 officers for each 1,000 residents, but only 1 officer for every 100 square miles.

■ Federal Law Enforcement Agencies

As of September 2004, federal agencies employed approximately 105,000 full-time personnel who were authorized to make arrests and carry firearms in the 50 states and the District of Columbia, and another 1,500 officers in U.S. territories, primarily in Puerto Rico.[30] (Chapter 8 provides details on qualifications and training for federal officers.) The largest number of federal officers (40,408 officers) performed criminal investigation and enforcement duties (38 percent) (see Figure 5–4). The next largest category (22,278 officers) was engaged in police response and patrol (21 percent). A total of 17,280 officers performed inspections (16 percent) related to immigration or customs laws, and 16,530 officers performed corrections- or detention-related duties (16 percent). Others had duties related to court operations (5 percent) or security and protection (4 percent).

TABLE 5–4 Twenty Largest Tribally Operated Police Agencies, 2000

Agency Name and State	Full-Time Sworn Officers	Service Population (Rounded)	Square Miles of Reservation (Rounded)	Full-Time Sworn Officers per 1,000 Population	Full-Time Sworn Officers per 100 Square Miles
Navajo Nation, Arizona	321	170,000	22,200	2	1
Tohono O'Odham, Arizona	76	17,000	4,500	4	2
Seminole, Florida	67	2,600	NA	26	NA
Gila River, Arizona	58	15,000	600	4	10
Oglala Sioux, South Dakota	58	41,000	3,200	1	2
Cheyenne River, South Dakota	53	11,000	4,300	5	1
Salt River, Arizona	51	6,700	80	8	63
Choctaw, Mississippi	38	7,000	30	5	152
Saginaw, Michigan	37	1,000	220	36	17
White Mountain Apache, Arizona	36	13,000	2,700	3	1
Rosebud Sioux, South Dakota	35	19,000	1,400	2	3
Oneida Indian Nation, New York	33	2,000	NA	17	NA
Warm Springs, Oregon	33	4,000	1,000	9	3
Colorado River, Arizona	32	2,000	400	16	9
Assiniboine and Sioux, Ft. Peck, Montana	31	7,000	3,300	4	1
Yakima, Washington	31	16,000	2,200	2	1
Cherokee, North Carolina	30	7,500	80	4	36
Miccosukee, Florida	30	600	130	51	23
Turtle Mountain Band of Chippewa, North Dakota	26	11,000	70	2	38
San Carlos, Arizona	25	11,000	3,000	2	1

Source: Bureau of Justice Statistics, 2006.

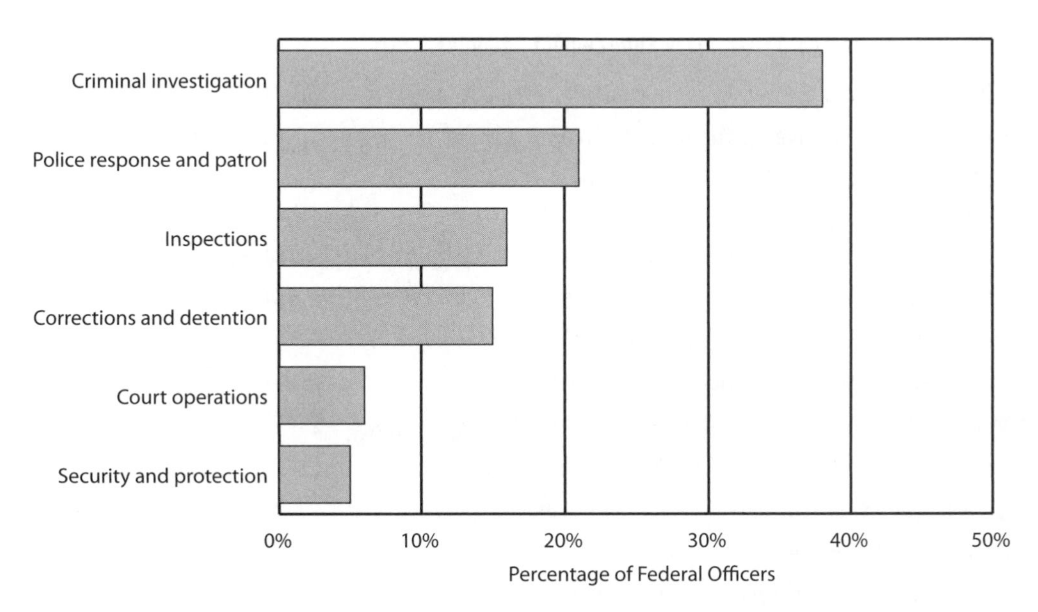

Figure 5–4 Primary function of full-time federal officers with arrest and firearm authority, September 2004.

Source: Bureau of Justice Statistics (2006 July). Federal Law Enforcement Officers, 2004.

Major Employers of Federal Special Agents

Federal agencies with 500 or more special agents who have authority to carry firearms and make arrests employ approximately 80,816 federal enforcement personnel (see Table 5–5). After the terrorist attacks on September 11, 2001, federal law enforcement agencies were reorganized and continually increase in number of personnel. The largest federal enforcement agencies came under the authority of a new organization called the Department of Homeland Security (DHS).

U.S. Customs and Border Protection

The largest agency within the DHS is the U.S. Customs and Border Protection (CBP), which employs an estimated 34,251 enforcement personnel (as of 2007):[31]

- 18,000 officers
- 13,300 border agents
- 1,800 agriculture specialists
- 651 air and marine officers
- 500 pilots

CBP combines old inspectional workforces and broad border authorities of the U.S. Customs Service, U.S. Immigration and Naturalization Service (INS), the U.S. Border Patrol, and the Animal and Plant Health Inspection Service. The new agency is tasked with stopping terrorists, terrorist weapons, illegal drugs, aliens, and materials harmful to agriculture from entering at, or between, ports of entry. To better understand the magnitude of CBP's tasks, consider that its personnel police 7,100 miles on the Mexican and Canadian border, monitor 95,000 miles of shoreline, and manage 325

TABLE 5–5 Largest Federal Law Enforcement Agencies, 2005

Agency	Officers
U.S. Customs and Border Protection (CBP), www.cab.gov	33,105 (July 2006)
Federal Bureau of Investigation (FBI), www.fbi.gov	12,617 (September 2006)
U.S. Immigration and Customs Enforcement (ICE), www.ice.gov	10,399
U.S. Secret Service, www.secretservice.gov	4,769
Drug Enforcement Administration (DEA), www.dea.gov	4,400
U.S. Marshal Service, www.usmarshal.gov	4,161
U.S. Postal Inspection Services, www.usps.com	2,976
Internal Revenue Service (criminal investigation), www.irs.gov	2,777
Veterans Health Administration (VHA), www.va.gov	2,423
Bureau of Alcohol, Tobacco, Firearms, and Explosives, www.atf.gov	2,373
National Park Service, www.nps.gov	2,148
U.S. Capital Police, www.uscapitalpolice.gov	1,535
Bureau of Diplomatic Security, Diplomatic Security Service, www.state.gov/m/ds	825
U.S. Fish and Wildlife Service, Division of Law Enforcement, www.fws.gov	708
U.S. Department of Agriculture, Forest Service, Law Enforcement and Investigations, www.fs.fed.us	600

ports of entry. The agency's website (http://www.cbp.gov/) provides information about CBP's activities, including current enforcement activities (e.g., the seizure of 567 pounds of cocaine valued at $6.5 million in November 2006), announcements (e.g., the newest weapon to secure U.S. borders—an unmanned aircraft capable of providing an "eye in the sky" for border patrol agents on the ground), and hiring requirements.

In 2004, the CBP workforce included 10,895 border patrol officers. These officers perform patrol and response functions along, and in the vicinity of, all U.S. boundaries. In 2004, 89 percent of all border control officers were stationed in the four states bordering Mexico.

CBP employs nearly 17,000 federal officers who were immigration (INS) or U.S. Customs Inspectors before the creation of the DHS. Those CBP officers are now included in the inspections category.

Federal Bureau of Investigation

The Federal Bureau of Investigation (FBI) includes 12,617 special agents who have arrest and firearms authority and are employed by the Department of Justice. Except for 240 FBI police officers, all are special agents, responsible for criminal investigation and enforcement. The FBI's top priorities are as follows:

> To protect and defend the United States against terrorist and foreign intelligence threats, to uphold and enforce the criminal laws of the United States, and to provide leadership and criminal justice services to federal, state, municipal, and

international agencies and partners.[32]

Other priorities include combating cybercrime, white-collar crime, violent crime, organized crime, and public corruption. Protecting civil rights also continues to be a priority.

In fiscal year 2005, the total budget for the FBI was approximately $5.9 billion, including $425 million in new funding intended to enhance counterterrorism, counterintelligence, cybercrime, information technology, security, forensics, training, and criminal programs.

U.S. Immigration and Customs Enforcement

U.S. Immigration and Customs Enforcement (ICE) is another law enforcement component of the DHS. ICE is organized into four major offices: Investigations, Intelligence, Detention and Removal Operations, and Federal Protective Service (FPS, transferred from the General Services Administration in 2003). The primary mission of ICE is to prevent acts of terrorism by targeting the people, money, and materials that support terrorist and criminal activities.

ICE is the largest investigative arm of DHS and is responsible for identifying and eliminating vulnerabilities in the United States' border, economic, transportation, and infrastructure security. In September 2004, this agency employed 10,399 officers with arrest and firearms authority. This number included 8,000 agents involved with criminal investigations and enforcement, 900 officers performing detention and deportation functions, 800 officers providing response and patrol services, and 600 officers providing security and protection services. Included among these ICE officers were 1,119 Federal Protective Service (FPS) officers who perform security, patrol, and investigative duties related to federal buildings, properties, and personnel.

ICE initiates many enforcement strategies. For example, Operation Predator was a comprehensive initiative designed to protect young people from foreign national sex offenders, human traffickers, and other predatory criminals. It marked its one-year anniversary of operation on July 9, 2006, and at that point had been responsible for more than 3,200 arrests.

ICE has the following stated mission:

> Our mission is to protect America and uphold public safety. We fulfill this mission by identifying criminal activities and eliminating vulnerabilities that pose a threat to our nation's borders, as well as enforcing economic, transportation and infrastructure security. By protecting our national and border security, ICE seeks to eliminate the potential threat of terrorist acts against the United States.[33]

U.S. Secret Service

In 2004, the U.S. Secret Service included 4,769 personnel who were authorized to make arrests and carry firearms. Approximately two thirds of Secret Service officers are special agents with investigation and enforcement duties primarily related to counterfeiting, financial crimes, computer fraud, and threats against dignitaries. Most other Secret Service officers are in the Uniformed Division and provide protection for the White House complex and other presidential offices, the Main Treasury Building and

Annex, the President and Vice President and their immediate families, and foreign diplomatic missions.

The U.S. Secret Service was founded in 1865 as a branch of the U.S. Treasury Department. The original mission was to investigate counterfeiting of U.S. currency—at the time, one third to one half of all U.S. currency in circulation was counterfeit. In 1901, following the assassination of President William McKinley, the Secret Service was assigned the responsibility of protecting the President. This agency was moved under the control of the DHS as part of the reorganization of federal law enforcement.

Drug Enforcement Agency

Also within the U.S. Department of Justice is the Drug Enforcement Agency (DEA), which employed an estimated 4,400 personnel with arrest and firearm authority as of September 2004 (10 percent, or 400 employees, more than in 2002). DEA special agents investigate major narcotics violators, enforce regulations governing the manufacture and dispensing of controlled substances, and perform other functions to prevent and control drug trafficking. The mission of the DEA is to enforce the controlled substances laws and regulations of the United States. It is also tasked with bringing to the criminal and civil justice system of the United States (or any other competent jurisdiction) those organizations and principal members of organizations that are involved in growing, manufacturing, or distributing controlled substances appearing in or destined for illicit traffic in the United States. Finally, the DEA recommends and supports non-enforcement programs aimed at reducing the availability of illicit controlled substances on the domestic and international markets.

U.S. Marshals Service

The Department of Justice's U.S. Marshals Service employs 94 marshals (one for every federal court district) and 3,067 deputy marshals and criminal investigators totaling 4,161 agents with arrest powers and firearm authority in 2006. The Marshals Service takes custody of all persons arrested by federal agencies and is also responsible for transportation until sentencing and moving federal inmates between federal prison facilities. They have jurisdiction over federal fugitive matters concerning escaped prisoners, probation and parole violators, persons under DEA warrants, and defendants released on bond. The agency makes more than half of all federal fugitive arrests. Other responsibilities include managing the Federal Witness Security and Federal Asset Seizure and Forfeiture Programs and security for Federal judicial facilities and personnel. One strategy conducted by U.S. Marshals was Operation FALCON III: a massive fugitive operation, covering the eastern half of the United States that took place October 22–28, 2006. As a cooperative effort, Operation FALCON III resulted in the arrest of 10,773 fugitives.

U.S. Postal Inspection Services

In 2004, the U.S. Postal Inspection Services employed 2,976 officers. Approximately two thirds of these officers were postal inspectors, who are responsible for criminal investigations covering more than 200 federal statutes related to the postal system.

The others were postal police officers, who provide security for postal facilities, employees, and assets, and escort high-value mail shipments. An abbreviated list of the postal service's areas of jurisdiction follows:

- Assaults and threats that occur to postal employees while they are performing their official duties or as a result of their employment
- Bombs and bomb threats (a rare crime that is given a high investigative priority)
- Burglary of federal postal equipment and mail
- Child exploitation enforcement efforts to combat the production and distribution of child pornography and other crimes exploiting children through the mail and, when it involves the mail, over the Internet
- Controlled substances, including initiating investigations related to transporting and distributing narcotics through the mail or at postal facilities
- Counterfeit stamps and money orders
- Identity fraud when postal services are used to commit this crime

Internal Revenue Service

In 2006, the Internal Revenue Service (IRS) employed 2,777 special agents with arrest and firearm authority within its Criminal Investigation (CI) Division. This law enforcement arm of the IRS is charged with enforcing U.S. tax laws. IRS investigative agents' jurisdiction includes tax, money laundering, and Bank Secrecy Act laws. While other federal agencies also have investigative jurisdiction for money laundering and some Bank Secrecy Act violations, the IRS is the only federal agency that can investigate criminal violations of the Internal Revenue Code. Compliance with the tax laws in the United States relies heavily on self-assessments of what tax is owed, a system known as voluntary compliance. When individuals and corporations make deliberate decisions not to comply with the law, they face the possibility of a civil audit or criminal investigation, which could result in their prosecution and possibly jail time.

As financial investigators, CI special agents fill a unique niche in the federal law enforcement community. Today's sophisticated schemes to defraud the government demand the analytical ability of financial investigators who can wade through—and make sense of—complex paper and computerized financial records. Owing to the increased use of automation for financial records, CI special agents are trained to recover computer evidence. Along with their financial investigative skills, these agents use specialized forensic technology to recover financial data that may have been encrypted, password protected, or hidden by other electronic means.

Veterans Health Administration

As of September 2004, the Veterans Health Administration (VHA) employed 2,423 officers with arrest and firearms authority. This number represented an increase of nearly 50 percent over the level of employment in 2002, as the VHA completed its program to expand firearms authority to its entire force. VHA officers provide police services for VHA medical centers nationwide.

Bureau of Alcohol, Tobacco, Firearms, and Explosives

In 2004, the Bureau of Alcohol, Tobacco, Firearms, and Explosives (ATF) employed 2,373 full-time officers with arrest and firearms authority. This agency enforces federal laws related to alcohol, tobacco, firearms, explosives, and arson. It was moved from the Department of the Treasury to the Department of Justice in 2003.

The ATF is the principal law enforcement agency within the Department of Justice dedicated to preventing terrorism, reducing violent crime, and protecting the United States. ATF agents have the dual responsibilities of enforcing federal criminal laws and regulating the firearms and explosives industries. ATF works directly and through partnerships to investigate and reduce crime involving firearms and explosives, as well as illegal trafficking of alcohol and tobacco products.

National Parks Service

The National Parks Service employed 2,148 full-time personnel with arrest and firearms authority in 2004. This number included 1,536 park rangers, who must provide police services for the entire national parks system. The National Parks Service total also includes 612 U.S. Park Police officers, most of whom work in the Washington, D.C., area.

U.S. Capitol Police

The U.S. Capitol Police employed 1,535 officers to provide police services for the U.S. Capitol grounds and buildings in 2004. In 1992, Congress granted the Capitol Police full law enforcement authority covering the area surrounding the Capitol complex.

Bureau of Diplomatic Security

In 2004, the Department of State's Bureau of Diplomatic Security employed an estimated 825 officers, whose primary function is to protect visiting dignitaries. The agency's special agents also investigate passport and visa fraud and threats against foreign missions in the United States, foreign dignitaries, or federal employees.

U.S. Fish and Wildlife Service

The Department of the Interior's U.S. Fish and Wildlife Service employed 708 full-time personnel with arrest and firearms authority in 2004. Nearly three out of every four of these personnel were refuge officers, with duties related to patrol and enforcement of federal wildlife conservation and environmental laws in the National Wildlife Refuge system. The others were special agents who investigate violations of federal wildlife protection laws and treaties.

USDA Forest Service

The U.S. Department of Agriculture's (USDA) Forest Service division of law enforcement and investigations employed 600 law enforcement officers in 2004. These officers are charged with protecting the people and natural resources in the National Forest System, whose lands are visited by hundreds of millions of people each year. Forest Service law enforcement officers respond to incidents encompassing a wide range of criminal and noncriminal activity, including search and rescue, drug trafficking, and archaeological resource theft.

Federal Salary Schedule

Federal law provides special salary rates to federal employees who serve in law enforcement. Additionally, federal special agents and inspectors receive law enforcement availability pay (LEAP) equal to 25 percent of the agent's grade and step, awarded because of the large amount of overtime that these agents are expected to work. For example, in 2007, FBI agents who entered federal service at the GS-10 pay grade in Dallas, Texas, started with a base salary of $50,169, yet earned about $62,711 per year with LEAP.

The starting pay depends on the locality pay area. For example, the starting salary for the GS-10 pay grade ranges from a low of $48,159 to a high of $55,723 in the San Francisco, California, area.

Special agents can advance to the GS-13 grade level in field nonsupervisory assignments at a base salary of approximately $75,414, which is worth almost $94,267 with LEAP. Federal agents may also be eligible for a special law enforcement benefits package.

Fragmentation in Policing Across the United States

After reviewing this chapter, it might appear that there are so many agencies that the mission of policing must inevitably be accomplished. Yet, many observers suggest that agencies are **fragmented** both internally (i.e., their internal divisions conflict with other divisions) and externally (i.e., they conflict with other agencies), or that agencies lack unity and cohesion. In fact, in 1967, the President's Crime Commission concluded that there is a sense of "fragmented crime repression efforts resulting from the large number of uncoordinated local government and law enforcement agencies."[34] Critics suggest that there are many problems related to this phenomenon:

- Lack of *cooperation* exists between agencies.
- Lack of *coordination* exists between agencies in the same geographic area.
- Police services and initiatives do not *operate* in the same way in similar jurisdictions.
- There is a lack of *consistent standards* linked to hiring, training, and discipline.

The development of cooperation, coordination, operation, and consistent standards would probably improve police services. The failure to provide coordination leads to duplication in crime labs, investigations, and training academies, and it sometimes produces difficulty with prosecution and court proceedings. Poor operation of police services might mean that one jurisdiction maintains a zero-tolerance policy toward hand guns, while another jurisdiction does not; this discrepancy could produce displacement of criminal activities, such that violators would operate in "safe" jurisdictions. The failure to develop consistent standards means that officers are hired, trained, and disciplined in a different manner, leading to disparities in policing various communities, and fewer opportunities for detection and control of police corruption.

Critics of fragmentation recommend at an extreme that police agencies merge and limit leadership or command to a few parties. In contrast, advocates suggest that fragmentation strengthens the democratic process and experiences of Americans.

As discussed in Chapter 4, the police department in Jacksonville, Florida, and the Duval County Sheriff's Office merged into the Jacksonville Sheriff's Department in an attempt to address some of these concerns. Likewise, fragmentation drove the merger of the Charlotte, North Carolina, and the Mecklenburg County Sheriff's Office. Encouraged by the results of their merger, the Clark County Sheriff's Office and Las

Vegas Police Department merged for efficiency reasons. Of course, not all mergers of police and sheriff's departments lead to effective public service organizations, particularly when personnel fight to keep their autonomy.[35]

Another recommendation intended to decrease fragmentation is that fire departments and police departments merge. This merger is premised on the notion that where there are fires, crowd control is a police matter. According to advocates of this view, such a consolidation would allow for greater flexibility and more effective deployment of personnel.

Another avenue to reduce fragmentation is for small departments to contract with larger departments for specific services or to contract with private industries for private security. The Harris County Constable's Office, Precinct 4, sets an excellent example in actively seeking contracts for private security, as detailed earlier in this chapter.[36]

Of course, the police are not autonomous social actors: They are constrained by the political/social/economic structures of the communities they police.[37] Also, the customary patterns of interaction between the police and the communities they serve shape the possibilities for police reform. Making the police more effective as well as more humane requires reform of more than police agencies.[38] Also, the framers of the U.S. Constitution purposely overlapped authority to provide a system of "checks and balances" and prevent the concentration of power in any one branch of government.

■ Violence Against Police Officers

Although most police officers are rarely attacked during their entire careers, the threat or actual violence (both intended and unintended) can occur "on the job." For that reason, the selection and training of police officers are paramount issues related to the safety of officers and the public they serve (discussed in Chapter 8).

Assaults on Police Officers

Police officers were the targets of more than 58,000 assaults in 2006; of that number, more than 15,000 attacks resulted in serious injury (e.g., stabbing, shooting).[39] The South is not only the deadliest region in the United States for police officers, but also the most violent: Officers in Southern police departments were almost four times more likely than members of Northeast departments to be assaulted.

In general, disturbance incidents are the most dangerous for patrol officers (see Table 5–6). Also, late-night (between 12:01 A.M. and 2 A.M.) hours are when most assaults occur. Prisoner handling and traffic stops are ranked next in terms of being occasions when injuries for an officer are more likely.

In 2006, the rate of assaults was 11.8 events per 100 sworn officers, and 26.8 percent of those officers who were the subject of assaults suffered injuries. Most of these assaults occurred while the officers were assigned to one-officer vehicle patrols or were working alone. Police cleared 90.7 percent of these assaults, however.

Deaths of Police Officers

According to the FBI, the number of police officers killed feloniously is decreasing. In 2006, 48 officers were killed as the result of felonies.[40] Research has not proved definitively why officer deaths have decreased, but we do know that line-of-duty deaths occurred most often in California.

TABLE 5–6 Law Enforcement Officers Assaulted, 2006: Circumstance at Scene of Incident by Type of Assignment

Circumstance	Total	Two-Officer Vehicle	One Officer Alone	One Officer Assisted	Detective Alone	Detective Assisted
Total	58,634	9,820	15,324	21,598	948	1,929
1. Disturbance calls (family quarrels, bar fights, person with gun)	18,135	3,250	4,572	8,200	246	246
2. Burglaries in progress/pursuing burglary suspects	844	150	240	364	8	26
3. Robberies in progress/pursuing robbery suspects	517	120	128	199	10	24
4. Attempting other arrests	9,233	1,478	2,406	3,462	170	553
5. Civil disorders (e.g., mass disobedience, riots)	669	112	138	250	22	40
6. Handling, transporting, custody of prisoners	7,133	709	1,333	1,834	95	170
7. Investigation of suspicious persons/circumstances	5,568	1,280	1,504	1,837	107	260
8. Ambush situations	179	43	49	30	10	8
9. Handling mentally deranged persons	1,058	164	258	498	7	17
10. Traffic pursuits/stops	6,490	1,199	2,202	2,535	60	129
11. All other	8,808	1,307	2,494	2,389	219	456

Source: Adapted from FBI Law Enforcement Officers Killed and Assaulted, Table 67 (2006). Washington, DC: U.S. Department of Justice. Percentages were rounded and "other" circumstances were deleted. Reporting agencies totaled 10,346 agencies, providing services to 74.4 percent of the U.S. population.

According to the FBI, most of the police officers killed fit the following profile: white, male, approximately 38 years of age, with an average of 11 years of police experience. Most were assigned to patrol duties and were killed during arrest situations. Ten officers were fatally assaulted when ambushed. Among the officers murdered, 36 of the murderers used handguns and 2 used vehicles. Of the 26 officers who were killed while wearing body armor, 15 officers suffered fatal gunshot wounds to the head and 7 were mortally wounded after receiving shots to the torso.

In 2006, there were 66 accidental deaths among officers.[41] The victim profile in these cases was similar to that of officers killed feloniously: white, male, approximately 36 years of age, with an average of 10 years of experience. The most common circumstance in these accidental deaths was automobile accidents. Other incidents included deaths that occurred when officers were struck by a vehicle, accidental shootings, and motorcycle accidents.

A review of incidents in which LAPD officers were killed feloniously reveals the following consistent observable characteristics:[42]

- Friendly to everyone; service-oriented
- Tended to use less force than other officers
- Hard-working
- Used force as the last resort
- Did not follow procedures and rules (e.g., for arrest, transport)
- Did not wait for backup

These findings suggest that the characteristics of an officer may have something to do with whether a patrol officer is victimized.

Assaults on and Deaths of Federal Law Enforcement Special Agents

Overall, some 1,273 federal law enforcement officers were assaulted in 2006; 212 of these officers were injured in the attack.[43] By department, federal agents were injured and killed at the following rates in 2005–2006:

- The Department of Homeland Security employed 1,159 of the federal officers who were assaulted. Of these officers, 139 were injured.
- One sworn officer (assigned to ICE) was feloniously killed.
- The Department of the Interior employed 528 of the officers who were assaulted, and 109 of these officers were injured.
- The Department of Justice employed 162 of the officers who were assaulted, and 42 of these officers were injured.
- The U.S. Postal Inspection Service employed 18 of the officers who were assaulted, and 12 of these officers were injured.
- The U.S. Capitol Police employed 12 of the officers who were assaulted, and 7 of these officers were injured.
- The Department of the Treasury employed 3 of the officers who were assaulted, and none of the officers were injured.

***RIPPED* from the Headlines**

Routine Patrol Ended with an Ambush and Two Members of the Patrol Wounded

Unfortunately, this is not a news bulletin from Iraq—it is the headline from a new and disturbing criminal trend that has increasingly become a threat to the Los Angeles Police Department during the past year—unprovoked attacks on police officers.

This past year, officers were repeatedly attacked while on routine patrol. Although this situation is unparalleled in LAPD history, there has been little media attention or community outrage over the issue.

LAPD officers have been fired upon without warning 40 times this year, compared with 29 assaults with firearms on officers reported in 2002.

These are not exchanges of gunfire between criminals and officers—these are cold-blooded attempts to murder unsuspecting officers without any warning.

The latest incident occurred in the early-morning hours of December 14, when suspected gang members fired on two officers patrolling the Watts area. Both ended up in the hospital. One was treated and released; the other stayed overnight. One of the suspects in the incident was seriously wounded. Whatever the motivation for these attacks, their frequency and brazenness should be cause for major concern.

Local hoodlums are becoming so bold as to step out and fire on police officers without the benefit of cover, and in the presence of witnesses.

Even more alarming is that no citizens are stepping forward with information on which parties are responsible for these attempted assassinations. These attacks have evoked nothing but a roar of silence. Not outrage, not activism, not action—nothing.

Will it take the murder of a police officer before the community responds? Would the community even react to that outrage? Will our citizens take an active role in seeking out those who are targeting police officers with their violent intentions?

The LAPD isn't the only law enforcement agency affected. On November 15, Burbank Police Officer Matthew Pavelka was murdered, and the suspect in the killing, David A. Garcia, escaped. Officer Pavelka and Officer Gregory Campbell were greeted by a volley of gunfire as they approached Garcia's vehicle in a hotel parking lot. Local police agencies and the Burbank community banded together to find Garcia and ensure he faced prosecution for this horrific crime.

Officer Pavelka is the only Burbank police officer ever murdered in the department's 82-year history.

The LAPD has not been as fortunate—196 LAPD officers have been killed in the line of duty since the department was formed more than 100 years ago. Chief William Bratton believes there is no other law enforcement agency in the United States whose officers routinely face so much violence.

Source: Bob Baker, "Alarmingly, Unprovoked Attacks on Police Are Rising" (2006), http://www.apbweb.com/articles-z47.htm (accessed June 21, 2006).

Summary

Local police agencies include municipal and county sheriffs and constables. They are complemented by state police; capital police; coroners, medical examiners, and forensic pathologists; tribal agencies; and federal law enforcement agencies.

The authority extended to each local police officer—which includes management, budget, and regulation—is provided by the state and the jurisdiction where the officer is employed. Municipal officers work in large cities and account for two thirds of all sworn officers in the United States; the remainder of local police are employed in rural or small jurisdictions. Sheriffs are (usually) elected county officials whose responsibilities can include processing of warrant services, court security, and delivery of police, fire, and jail services. Constables, at least in Texas, are typically elected and deliver full police services; in other states, constables are employed by court systems.

State police are administered through state government, have statewide jurisdiction, and are largely limited to traffic control duties on state highways. Often, they operate state crime labs; perform investigations for local departments; provide other local services such as stress units and SWAT units; and participate with other state agencies, such as serving the department of corrections as parole officers.

Campus and capital police have full sworn powers of local officers. However, campus police jurisdiction is limited to college and university campuses, and capital police jurisdiction is limited to property of the state such as the state capital.

Tribal police are subject to oversight by state government and, in some jurisdictions, by the Bureau of Indian Affairs. These officers possess the authority of detainment and use of force much like any other sworn officer, but their jurisdiction is limited to Indian country.

Coroners, medical examiners, and forensic pathologists are officially responsible for investigation and certification of cause and manner of death in cases of sudden and unnatural death. They are playing an increasingly important role in ensuring public safety given the recent rise of infectious disease deaths, emergence of new infections, and bioterrorism.

For federal law enforcement agencies, their jurisdiction tends to focus on violations of federal crimes (as opposed to state and local violations).

Given the many different local police agencies and federal law enforcement agencies, a great deal of duplication of police and enforcement services exists—a phenomenon known as fragmentation. Opponents of fragmentation suggest that police departments should merge; advocates argue that fragmentation strengthens the democratic lifestyle by providing "check and balance" practices.

Key Words

Capital police
Constable
Coroner
Forensic pathologist
Fragmentation
Indian country
Medical examiners
Municipal police
Sheriffs
Sworn officers
Tribal police

Discussion Questions

1. Characterize the primary differences between local police agencies and federal law enforcement organizations.
2. Identify the various types of local police agencies.
3. Describe municipal police organizations and their primary responsibilities. In what way do they differ from county agencies?
4. Characterize sheriff's departments and their primary responsibilities.
5. Characterize constable's departments and their primary responsibility.
6. Describe state police objectives and jurisdiction. In what way might the typical routine of a state police officer differ from that of a municipal police officer?
7. Explain capital police objectives and jurisdiction.
8. Distinguish between the primary differences and enforcement responsibilities of coroners, medical examiners, and forensic pathologists.
9. Describe tribal police agencies, and identify issues that might be the source of discontent among those policed. How might those issues be resolved?
10. Identify the largest federal law enforcement agencies and explain their objectives.
11. Define the concept of fragmentation, and explain the primary issues related to it, remedies for fragmentation, and limitations of those remedies.
12. Characterize trends related to the assault and murder of police officers and federal law enforcement officers. How might these numbers be reduced?

Notes

1. Bureau of Justice Statistics, "Census of State and Local Law Enforcement Agencies 2004" (NCJ 212749) (Washington, DC: U.S. Department of Justice, 2007), http://www.ojp.usdoj.gov/bjs/pub/pdf/csllea04.pdf (accessed March 15, 2008).
2. Bureau of Justice Statistics, *Local Police Departments, 2003.*
3. The 2008 Statistical Abstract, "Law Enforcement, Courts, and Prisons. Crime and Crime Rates," http://www.census.gov/compendia/statab/tables/08s0302.pdf (accessed March 16, 2008).
4. Bureau of Justice Statistics, *Law Enforcement Management and Administrative Statistics, 2000: Data for Individual State and Local Agencies with 100 or More Officers* (NCJ 203350) (Washington, DC: U.S. Department of Justice, March 2004), http://www.ojp.gov/bjs/pub/pdf/lemas00.pdf (accessed October 15, 2006).
5. Bureau of Justice Statistics, "State and Local Law Enforcement Statistics" (2006), http://www.ojp.usdoj.gov/bjs/sandlle.htm (accessed October 29, 2006).
6. Bureau of Justice Statistics, *Local Police Departments, 2003.*
7. Federal Bureau of Investigation, "Crime Index" (2008), http://www.fbi.gov/ucr/prelim2007/index.html (accessed January 18, 2008).
8. National Association of Sheriffs, http://www.sheriffs.org/GovtAffairs/OfficeofSheriff/Preserving_the_Office_of_Sheriff_Through_Election.pdf (accessed November 4, 2006).
9. Matthew J. Hickman and Brian A. Reaves, *Sheriff's Offices, 2003* (NCJ 211361) (Washington, DC: U.S. Department of Justice, Bureau of Justice Statistics, May 2006), http://www.ojp.usdoj.gov/bjs/pub/pdf/so03.pdf (accessed October 7, 2006).
10. Allen G. Hatley, "Texas Constables: A Frontier Heritage," in *The Austin Papers* (3 vols., 1924–28), ed. Eugene C. Barker (Lubbock, TX: Tech University Press, 1999).
11. Wikipedia Online, http://en.wikipedia.org/wiki/Constable (accessed November 1, 2006).
12. "Precinct 4" (2006), http://www.cd4.hctx.net/ (accessed November 3, 2006).

13. "State Troopers" (2006), http://www.statetroopersdirectory.com/ (accessed October 8, 2006).
14. "State of Illinois State Troopers" (2006), http://www.isp.state.il.us/academy/localtraining.cfm (accessed October 8, 2006).
15. K. Baum and P. Klaus, *Violent Victimization of College Students, 1995–2002* (NCJ 206836) (Washington, DC: U.S. Department of Justice, Office of Justice Programs, Bureau of Justice Statistics, January 2005).
16. Bureau of Justice Statistics, *Campus Law Enforcement Agencies, 1995* (NCJ 161137) (Washington, DC: U.S. Department of Justice, 1996), http://www.ojp.usdoj.gov/bjs/pub/pdf/clea95.pdf (accessed October 14, 2006).
17. Bureau of Justice Statistics, *Campus Law Enforcement Agencies, 1995.*
18. Bureau of Justice Statistics, *Campus Law Enforcement Agencies, 1995.*
19. Violence Against Women Office, "Promoting Safety and Nonviolence on College and University Campuses Series: Toolkit to End Violence Against Women" (October 2002), http://toolkit.ncjrs.org/files/chapter7.txt (accessed October 14, 2006).
20. Violence Against Women Grants Office, *Stalking and Domestic Violence: The Third Annual Report to Congress Under the Violence Against Women Act* (NCJ 172204) (Washington, DC: U.S. Department of Justice, 1998), 10.
21. Bureau of Justice Statistics, *Violence by Intimates: Analysis of Data on Crimes by Current or Former Spouses, Boyfriends, and Girlfriends* (NCJ 167237) (Washington, DC: U.S. Department of Justice, 1998), 13.
22. C. A. Presley, J. R. Meilman, J. R. Cashin, and L. S. Leichliter, *Alcohol and Drugs on American College Campuses: Issues of Violence: A Report to College Presidents* (Core Institute Monograph) (Carbondale, IL: Southern Illinois University, 1997), 4.
23. Florida State University Campus Police (2006), http://www.police.fsu.edu/employment.cfm (accessed October 8, 2006).
24. State of Delaware (2006), http://www.state.de.us/capitolpd/ (accessed October 15, 2006).
25. Mitchell I. Wolfe, Kurt B. Nolte, and Steven S. Yoon, *Fatal Infectious Disease Surveillance in a Medical Examiner Database* (Atlanta, GA: Centers for Disease Control and Prevention, Emerging Infectious Diseases, 2004), http://www.cdc.gov/ncidod/EID/vol10no1/02-0764.htm (accessed November 1, 2006).
26. Bureau of Justice Statistics, *Tribal Law Enforcement, 2000* (NCJ 197936) (Washington, DC: U.S. Department of Justice, 2003), http://www.ojp.usdoj.gov/bjs/pub/pdf/tle00.pdf (accessed October 29, 2006).
27. U.S. Department of Justice, *Indian Country Law Enforcement Review* (December 1999); see also *Jails in Indian Country, 2001* (NCJ 193400).
28. National Institute of Justice, *Public Law 280 and Law Enforcement in Indian Country—Research Priorities* (NCJ 209839) (Washington, DC: Office of Justice Programs, U.S. Department of Justice, December 2005), http://www.ncjrs.gov/pdffiles1/nij/209839.pdf (accessed October 29, 2006).
29. National Institute of Justice, *Public Law 280.*
30. Bureau of Justice Statistics, *Federal Law Enforcement Officers, 2004* (NCJ 212750) (Washington, DC: U.S. Department of Justice, July 2006).
31. U.S. Customs and Border Protection (CBP), http://www.cbp.gov/ (accessed June 9, 2008).
32. Federal Bureau of Investigation, "About-US-Quick Facts." http://www.fbi.gov/quickfacts.htm (accessed June 10, 2008).
33. U.S. Immigration and Customs Enforcement (Washington, DC: U.S. Department of Homeland Security, 2008), http://www.ice.gov/about/index.htm (accessed March 17, 2008).

34. President's Commission on Law Enforcement and Administration of Justice, *Task Force Report: The Police* (Washington, DC: Government Printing Office, 1967), 68.

35. Samuel Wilson and Charles M. Katz, *Policing in America: An Introduction*, 5th ed. (Boston: McGraw Hill, 2005).

36. Deputy Constable at Precinct 4, personal communication with the author (October 30, 2006).

37. Hung-En Sung, *The Fragmentation of Policing in American Cities: Toward an Ecological Theory of Police–Citizen Relations* (Westport, CT: Greenwood Press, 2001).

38. Albert J. Reiss, Jr., "Police Organization in the Twentieth Century," *Crime and Justice* 15 (1992): 1–47.

39. Federal Bureau of Investigation, "Law Enforcement Officers Killed or Assaulted: 2006" (2007), http://www.fbi.gov/ucr/killed/2006/index.html (accessed December 13, 2007).

40. Federal Bureau of Investigation, "Law Enforcement Officers Killed or Assaulted: 2006."

41. Federal Bureau of Investigation, "Law Enforcement Officers Killed or Assaulted: 2006."

42. Los Angeles Police Department (2007), www.lapdonline.org (accessed December 13, 2007).

43. Federal Bureau of Investigation (2007), http://www.fbi.gov/ (accessed December 23, 2007).

Police Organizations, Management, and Police Officers

You always pass failure on the way to success.

—Mickey Rooney

LEARNING OBJECTIVES

When you finish reading this chapter, you will be better prepared to

- Define an organization
- Describe the evolution of police organizations
- Characterize the primary components of a police organization
- Provide an analysis of the primary components of a police organization
- Articulate the rationale of a police organization
- Characterize the Peter Principle
- Describe the various organizational styles of policing
- Characterize the three major police organizational theories
- Describe the influences that shape police organizations
- Articulate the chief results of the influences that shape police organizations

CASE STUDY: YOU ARE THE POLICE OFFICER

You have been a patrol officer for almost five years. On your twenty-sixth birthday, the chief asked you to take the sergeant's exam, to work under your current sergeant for a year, and, when the sergeant retired, to take his place. Usually young patrol officers don't have these opportunities.

You talked to your pops about it, and he said, "Go for it!" Your mom was more cautious about the move. She said she read an article about rank-and-file personnel who were promoted to a level of management that they really didn't understand, simply because they had never been a manager before. You explained that the organizational structure has specific career paths for patrol officers. According to the chief, the number of patrol officers reporting to you would be minimal at first, but would increase as you learned more. Your mom was also concerned with what the police hierarchy would have to say about a young patrol officer being promoted so fast and reminded you that once you become part of management, you lost your police union standing. Sergeants aren't as secure in their job as patrol officers, she added.

You talked with the chief about the issues raised by your parents, and he agreed that there is a risk to every promotion. Even so, he felt confident that, based on your evaluations, you would master the skills needed for the management position. The chief informed you that one of his future plans for the department was to flatten the bureaucratic structure: His plans were to decentralize the hierarchical flow of "orders" from his office, so that new sergeants like you would become responsible for their patrol officers. As part of the new management vision, the chief suggested that you should think not only about the issues brought up by your parents, but also about the cultural barriers to police performance and corrective values that you might implement.

1. Your mom pointed out that patrol officers have little experience as supervisors and cautioned you that a rapid promotion (Peter Principle) can sometimes lead to failure. Explain in what way her advice might be well founded, and describe the steps the chief plans to take to head off this possibility.
2. Explain why the number of patrol officers (span of control) reporting to a new sergeant could have a great deal to do with his or her success as a supervisor.
3. Explain the merits and consequences of the chief's vision of a decentralized command.
4. Explain the relevance of cultural barriers to police performance and why they would be an issue for a sergeant.

Introduction

The public is informed through popular television dramas and police reports that the old image of policing—that on which the "Keystone Cops" were based—belongs to another age.[1] Rationalized police organizations deploy well-equipped officers throughout their jurisdictions in an organized and quick fashion to calls for service. Nevertheless, putting officers on the streets in a timely and organized fashion and getting them to particular locations rapidly is quite different from shaping police behavior once officers are in the community dealing with the problems, needs, and conflicts of individuals.

This chapter describes and characterizes the typical contemporary police organization and its components, acknowledging that there are few—if any—police organizational models that fit every police organizational structure because of the huge variety of organizational structures that mark the hundreds of police agencies across the United States. Police organizations tend to be unique because they are largely products of local histories, resources, relationships, and contingencies. There is also some ambiguity over what constitutes a police organization.[2] "Police come in a bewildering variety of forms . . . moreover, many agencies that are not thought of as police nonetheless possess 'police' powers."[3]

The focus of this chapter is on public police organizations in the United States whose primary mission is to provide generalized police services, including responses to calls for service and emergency care to an unspecific constituent in a specific jurisdiction. Most police administrators could probably write a textbook on police effectiveness, as they have passed failure several times on their way to success in achieving their stated mission (as Mickey Rooney says). It takes time to institute changes, but the changes affecting police departments (e.g., budget cuts, litigation, and terrorism) happen fast. By the time a police agency reaches its organizational objectives, the changes it made along the way probably aren't enough to ensure that it continues to achieve its mission in the future.

Chief William Bratton of the Los Angeles Police Department (LAPD) explains another wrinkle related to the time factor:

> I can remember during my time in New York City that once we had a plan, we did everything, everywhere, all at once because with 38,000 officers—for the first time in my career—I could do that. According to the experts, this type of approach did not allow for valid experiments or a perfect research setting. Well, I'm sorry, but I'm sure that the thousands of people whose lives were saved are grateful that we didn't wait to experiment here and there. This difference in mindset contributes to what I believe is part of the divide between some researchers and some practitioners.[4]

Bratton's statement highlights the difference between research and practice. He suggests that in the twentieth century, researchers and police tolerated each other and talked "past each other without connecting." That said, transformation can be accomplished by revising organizational systems based on the findings provided through evidence based policing initiatives (see Chapter 4 and Chapter 9 for more details).[5]

Definition of an Organization

The need for organizing is as old as humankind. Humans realized early in their history that a group can usually outperform an individual. Both ancient societies and today's advanced societies acknowledge that organizations are more productive in terms of their ability to protect a population from nature and from those who would harm them or their possessions.

> An **organization** is a legal artificial structure created to coordinate personnel and resources to achieve a mission or a goal.

Organizations develop rules, establish methods to enforce rules, and attempt to provide punishment to those who don't obey those rules. They are the framework for dividing, assigning, and coordinating work. Experts report that the rationale for an organization relates to four primary characteristics:[6]

- Ability to create order
- Accountability
- Accomplishment
- Predictability

Being organized creates an appearance of order (things and people have a place) and puts the focus on personnel performing in a certain way. Managerial positions have an inherent and legal right to give an order, with the expectation that those orders will be followed. Other members of the organization are obligated to perform in keeping with those expectations. If individuals fail to do so, they are accountable to both the official directives and the unofficial directives of the organization.[7] Because performance is expected to move along a certain path toward, for instance, an organizational objective, qualitative measures—such as those relating to services rendered by police officers—can be formulated as a way to assess progress. In this way, an organized group, guided by a similar mission or goal, should be able to accomplish more than an individual.[8]

There is little doubt that modern-day police organizations are a result of the need for "law and order."[9] Police organizations are not merely a product of our times, but rather arose from a historical context (see Chapter 2). There is a rich tradition linked to the evolution of police organizations dating back to eighteenth-century England. During the Industrial Revolution, a large population fled rural areas in England for urban areas, leading to waves of unemployment, poverty, and crime in the larger cities. One result of this trend was the creation of *institutionalized* (established as normal and dependent on routine) policing through the Metropolitan Police of London in 1829 (see Chapter 2).

Components of Organizations

Organizations in general—such as educational and human services systems, corporations and trade associations, and criminal justice systems including police departments—evolve through design and a rational plan that can produce unexpected outcomes such as the emergence of a bureaucracy and personnel stress.[10] Most organizations are characterized by the following ten components:[11]

- A clear purpose
- Precise communicated statements
- Structure
- Different levels
- Division of labor
- Degrees of specialization
- Hierarchy of command
- Bureaucracy
- Management
- Clear career paths

The next sections explore these components as they relate to police organizations.

A Clear Purpose

A clear purpose refers to the organization's mission and goals. Generally, the purpose of an organization is the primary focus of and driving force for that organization.[12] The glue bonding the organization together is its intention to reach specific goals; that is, organizations are goal driven.

For instance, Chief Jacqueline Seabrooks[13] of the Santa Monica, California, Police Department says that the department's goals are to provide the community with the highest-quality police services available and to control any detrimental forces on public safety. These goals are accomplished through proactive and reactive police services. The proactive initiatives are furthered through public visibility, community participation, and community programs. The reactive initiative is to respond to emergency and non-emergency calls from the public, Chief Seabrook advises.

Traditional View

The traditional philosophy among police officers across the United States has been that the purpose of policing relates to officers' control and arrest powers. According to this perspective, the idealism that was once seen as part of policing is now viewed as the "civics-book picture" of justice.[14] The predominant philosophy of policing and, therefore, its clear purpose are powered by what policymakers, interest groups, and the public see as diligent and hard-working police officials enforcing the law as it is written in the statutes. Law breakers should be arrested, according to this view.[15] An assumption linked to this philosophy is that social order can be maintained only through control of the criminal element.

Traditional police efforts include crime-control objectives and assumptions that reflect shared beliefs held by policymakers, interest groups, and the public:[16]

- Police action can prevent crimes from occurring.
- Police intervention during the commission of a crime can influence the outcome.
- Police efforts after the crime has occurred can resolve the situation, ideally by solving the crime, arresting the perpetrator, and, in appropriate cases, restoring stolen property.

Thus traditional police efforts rely on patrol as the first line of offense and defense, special units, high-tech gadgets, sophisticated law analyses (e.g., Compstat, COPLINK), and investigative follow-up to complement crime control efforts on the streets.[17]

Opposing View

On the other side of the coin, policing has been portrayed as a chaotic system where neither the law nor justice prevails.[18] According to this perspective, officers are out of control: They see arresting members of certain racial or socioeconomic groups as always appropriate and employ whatever behavior suits the situation best—including beating, shooting, or killing suspects.[19] The purpose or intent of the police, according to some critics, is to guard the interests of the privileged and to control lower-class members (otherwise known as the dangerous class) through the process of an arrest, even if the police have little or no evidence on which to make a probable cause arrest.[20]

According to this view, nothing succeeds like failure. That is, the police must fail to succeed because the only way to maintain order is through the rule of law.[21] The police "fail in the fight against crime while making it look as if serious crime and thus the real danger of society is the work of the poor," argues Reiman.[22] He calls this notion the "pyrrhic defeat theory"—that is, a military victory purchased at such a cost in troops and treasure that it amounts to a defeat. According to this perspective, officers tend to control the lower classes more often than the affluent classes because one supposed purpose of policing, advocates of this viewpoint claim, is to enforce the law; the reality, says Reiman, is that police enforce only certain laws in certain jurisdictional matters—specifically, they target the behavior of the lower classes.

Precise Communicated Statements

Precise communicated statements are usually written and available for personnel and others to read. These statements lay out the purpose and goals of the organization. They also include mission statements and objectives, work plans (including day-to-day operations), and values (including rules, regulations, and expectations). These statements are interdependent and necessary to carry out the organizational purpose or mission (a mission statement reveals the values of an organization, too).

Contemporary traditional police organizations can be described as utilizing rulification practices concerning precise and communicated statements. Explained another way, **rulification** assigns greater importance to rules and departmental regulations than to common sense.[23] Policy, procedures, and directives are distributed in a systematic process, even if they duplicate previous efforts to disseminate information. Personnel can review these written documents to better understand organizational mandates and then follow specific directives as opposed to arbitrary perspectives.

As an example, think about felony pursuits employing police vehicles. If documents are available for officers to review before patrol, each would understand the standing orders of the department related to such pursuits. Orders that might counter those written directives would also be provided in the documentation, which is often the case: Directives change frequently based on litigation and policy. Consequently, an officer would know what to do in each situation and would feel assured that he or she is in compliance with agency directives or new litigation.

Of course, legal jargon is not always clear, and street officers might require an interpretation or application to police performance if they are to perform as well as expected. Nonetheless, the importance of rulification is linked to police accountability, personnel control, and "doing the right thing" while delivering police services.[24]

Structure

Organizational structure can be described in terms of its levels of organizational authority (policymakers, managers, and rank-and-file officers). Earlier, we suggested that many police departments share certain aspects of organizational structure and administrative style.[25] For instance, the formal structure of most police organizations is based on divisions and lines of authority (see Chapter 7). In such a structure, roles of personnel fall into one of two categories: line personnel (field or supervisory personnel) and staff personnel (support personnel who perform administrative duties).

Of course, although most large police organizations perform a similar function, a tremendous variation exists in terms of how individual organizations are structured. To account for this variation, a new theory was developed that attributes the formal structures of large municipal police agencies to the contexts in which they are embedded. According to Edward R. Maguire, the most relevant features of an organization's context are its size, age, technology, and environment.[26]

While police agencies are clearly different in matters such as size, technology, and environmental attributes, a hierarchal structure with top-down directives tends to dominate police structures. This pattern implies that differences arise only in the sense of the degree of organizational components—the primary elements are always present. For instance, municipal police departments are typically headed by a chief of police; below the chief's position are various other levels of personnel who hold ranks similar to those found in the military (i.e., recruits, sergeants, captains, majors, and so on). In smaller agencies, the rank of major or captain might not exist, but the *authority of the rank* probably does exist. In the typical traditional police structure, levels of authority are crucial to policing methods and operations.

Different Levels

In police organizations, levels are represented by various ranks of authority: commanders, majors, captains, lieutenants, sergeants, and so on. In this type of division of labor, authority is vested in the position or rank as opposed to the specific person occupying the position. Upper-level ranks are associated with more authority than are lower-level ranks. Thus, the higher up the ladder, the more authority and power over subordinates vested in that position. Figure 6–1 shows the various levels found at a typical police department in the United States.

The existence of such levels does not necessarily mean that the person occupying a higher position than another has more experience, knowledge, or expertise. Indeed, promotion is not always based on experience, knowledge, or expertise. Sometimes promotion is based upon the "who you know" principle. Knowing and being liked by the right top commanders plays a role in promotions, yet sometimes knowing the right commanders could easily work against an officer. That is, some watchdog groups

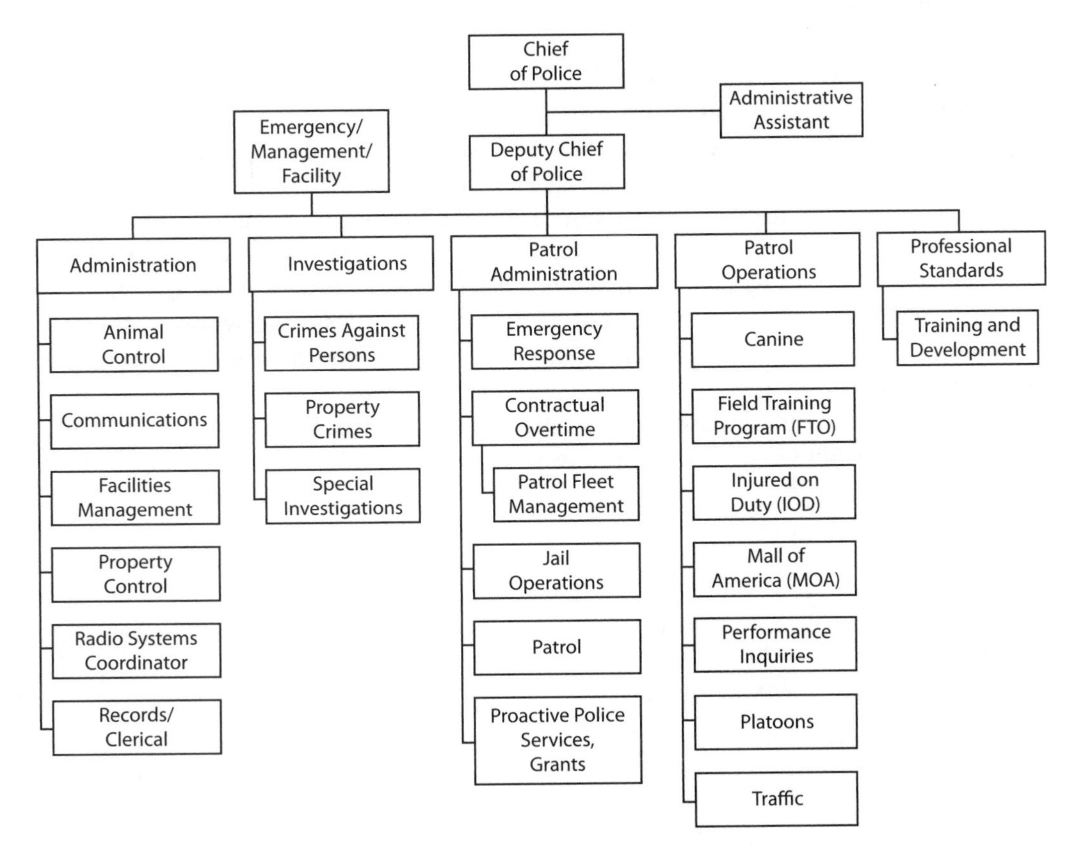

Figure 6–1 Typical Police Department Organization Chart

watch promotion processes closely, and improprieties can make the front page of the local newspaper.

At the same time, candidates for promotion may exemplify the Peter Principle.[27]

> The **Peter Principle** advances the idea that in a bureaucratic organizational hierarchy, personnel can rise to his or her level of incompetence through occupational promotion.

The Peter Principle argues that a hierarchical bureaucratic organization defends its status quo (traditions) long after the time when the status quo has lost its status. This thought is consistent with the idea that super competent performance can lead to dismissal because super performance by any individual employee disrupts the hierarchy, and the hierarchy must be preserved at all costs, explain Peter and Hull. In essence, the Peter Principle implies that great students don't necessary become great teachers and highly competent patrol officers don't necessarily become competent chiefs.

Division of Labor

A division of labor is represented by different occupational positions within the organization situated at various levels within the structure, whereby personnel can develop a **degree of specialization**.

Contemporary traditional law enforcement agencies can be described as utilizing levels for **delegation of authority**, a scheme that seeks to maintain the integrity of a police agency by clearly defining the tasks, responsibilities, and assignments linked to specific positions occupied by personnel. Delegation of authority also provides police personnel with the authority, resources, and skills needed to complete those tasks, responsibilities, and assignments. That is, each officer knows what is expected, and each is given documented power (through written directives or verbal instructions at roll call) and the opportunity to enhance his or her professional working skills or degree of specialization through means such as in-service training at a justice academy or advanced education at a university.

Division of labor is supposed to limit the scope of an officer's duties. Nevertheless, many officers feel under siege when things go wrong. Although the following example presents an extreme case, it doesn't take a level five hurricane to produce a hostile environment in many U.S. cities.

Officer David J. Lapene
New Orleans Police Department

Patrolling the city I grew up in after Hurricane Katrina was a very strange experience for me and my partner Officer Tommie Felix. It was like one of those movies I had seen as a kid: *Land of the Living Dead*. We worked narcotics (Uptown) unit. We had little idea how bad things were the day after the storm because Uptown didn't get hit hard. But flooding started four days later (backup flood control devices failed). My city was underwater. I had lost every one of my possessions, as did my father, who was so proud of his tiny home in the mid-city section. Unlike many of my coworkers, I evacuated my wife and other family members to Memphis, but had no way to communicate with them once they left New Orleans.

On patrol, we came to the intersection where we observed a corner store being looted by a band of thugs. We scared them off, knowing we could take no further action because the Orleans Parish Prison was ten feet under water. Fifteen guards held hundreds of inmates on the Board Street overpass, which was flooded on three sides.

We stopped a vehicle that was stolen and the driver fought my partner, at which point we brought him to compliance and arrested him. Not sure what to do with [the suspect] afterward, we drove around with him in backseat for a while and then released him.

Later we helped get a shooting victim into the back of a five-ton military vehicle; but he died. He was looting a clothing store when another looter shot him for his flashlight.

Officer Kevin Thomas was shot in the head by a looter on the west bank of the city and we could nothing to help him—or any other officer.

We watched looters use garbage cans and inflatable mattresses to float away with food, blue jeans, tennis shoes, and TV sets from opposite overpasses and were helpless to stop them other than to shoot them dead. Outside one food store, thieves commandeered a forklift and used it to push up the storm shutters and break through the glass. The driver of a nursing-home bus surrendered the vehicle to thugs rather than be murdered.

For two weeks, I watched my city burn and go underwater. And what parts were not flooded or burned were torn apart by civil unrest and political uncertainty.

Source: Told to the author, who put Officer Lapene's story into words (June 29, 2007).

Degrees of Specialization

Degrees of specialization incorporate the skills, training, and certifications held among police personnel regardless of their rank or organizational level, such as patrol officers, investigators, tactical specialists, and training officers. Contemporary traditional policing, which employs degrees of specialization within the organization, also relies on a strategy known as a **span of control.**[28]

Span of control refers to a specific number of police personnel who can be adequately supervised by any one individual.

The organization's span of control dictates, for example, the ratio of sergeants to officers. As you expect, the validity of a specific number for any such ratio has been hotly debated among scholars and practitioners alike. For instance, how many patrol officers should a sergeant supervise during the graveyard shift? Would that number be different than the ratio during a Saturday evening shift?

Sergeants often develop close relationships with their officers, which can affect decisions about deployment, discipline, and the grievance process. These relationships can become a focal point in police subcultures.[29] Sergeants control the activities of the station house and often have more unofficial power over subordinates than do top commanders, because sergeants are engaged in every aspect of daily routines among street officers.[30]

Consistent with this thought, in Luther Gulick's classic work on span of control, it was argued that span of control structures relationships between leaders and subordinates in organizations.[31] In addition, Gulick identified three key determinants of span of control:

- Diversification of function
- Time
- Space

Each of these determinants depends on the organization, the resources, and the expertise of management.

Sometimes police management can miscalculate a situation related to span of control. When they do, the results can be detrimental to both constituents and street officers. For example, in the mid-1990s, the Los Angeles County Sheriff's Department experienced a number of officer-involved shooting incidents assigned to the Century Station.[32] An investigation by Special Counsel Merrick Bobb—the department's form of citizen oversight—found that the source of the problem was more than a few bad officers; it derived from bad management, too.[33] At times, each sergeant was supervising 20 to 25 officers, "a ratio that far exceeded the department's own standard of 8 to 1."[34]

Hierarchy of Command

Hierarchy of command—also known as top-down authority and responsibility—refers to the arrangement of personnel so that some can legally and morally issue orders to other personnel, who are legally and morally obligated to follow those orders.[35] In the traditional police hierarchy of command, ultimate responsibility rests with the chief of police and each position farther down the hierarchy has a relatively smaller amount of responsibility. (See Chapter 7 for details on police management and leadership issues.)

Most police organizations can be described as hierarchical, quasi-military organizations. Such a hierarchy of command encourages bureaucracy (discussed later in this chapter). Because of the quasi-military nature of police operations, the primary method of organization is by formation of rank.[36]

Rank Structure

Figure 6–2 depicts the rank structure that is most commonly found in police organizations. At the top are administrators, who include the chief, deputy chiefs or seconds in command, and majors; these individuals might be called *command-level personnel.* Captains and lieutenants are often called *middle-level management*, whereas

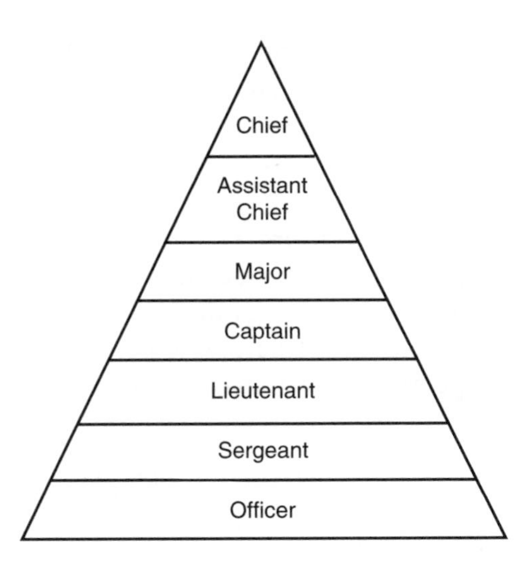

Figure 6–2 Bureaucracy

sergeants and first-line leaders (officers who actually have contact with officers on an ongoing basis) are referred to as *lower-level management.* Officers are referred to as *line personnel.* Of course, the precise rank structure depends on the size of the department; some departments have commissioners who have legal and moral control over a chief.

Lieutenants tend to be morale specialists who do not exercise regular supervision, but rather are called in when a street officer is having a problem. Their talents and leadership come into play when something unexpected happens.

Sergeants are generally first-level supervisors. They are either station house oriented or street oriented, depending on how much they have abandoned the patrol officer's mentality. The sergeant rank is one of the most sought-after positions in policing organization, and perhaps the hardest one to obtain if not gained by seniority alone.

Corporals tend to be field-training officers (FTOs). They are rewarded for their length of service by being tasked with breaking in rookies or given other duties related to employee development.

Officers make up the rest of the department, with patrol officers being the backbone of the department. It is also worth considering the concept of Master Patrol Officer (MPO) in some departments.[37]

Contemporary traditional police agencies have a **defined command**, referring to the assignment of a specific person to be in charge of a specific task or tasks and to be in charge of specific employees. For instance, during a critical incident such as deployment to a hostage situation, the ranking detective on site is in command. In this and other matters, the clarity associated with having a defined command ensures that there is less confusion about whose orders to follow.

Official communication moves vertically within the organization, similar to communication within a military structure. For instance, orders and policy directives emanate from the top and are sent downward through the chain of command. Information and reports on the activities of subordinates or problems are sent upward to the appropriate levels.[38]

Implications of Paramilitary Command Structures for Police Organizations

Both supporters and critics of the paramilitary command structure for police organizations have long debated the intrinsic worth of this kind of top-down hierarchy. Supporters want to reinforce the paramilitary structure by providing greater resources for aggressive patrol strategies,[39] whereas critics see paramilitary structures as inconsistent with twenty-first-century strategies such as decentralization and problem-oriented strategies.[40] Advocates argue that it provides more control, obedience, and loyalty to supervisors and the organization.

Regardless of the outcome of the debate on these issues, the reality is that most police organizations are operated through a hierarchy of command because of two assumptions:

- Individuals in higher positions have more experience than individuals in lower positions.[41]
- Hierarchy of command has greater predictive powers.

Bureaucracy

Bureaucracy refers to the structured layers of the organization as originally defined by Max Weber.[42] Before we explore the implications of this characteristic, here's something to note about the word: *Bureau* (French, borrowed into German) means "desk." Thus *bureaucracy* is rule conducted from a desk or office—that is, by the preparation and dispatch of written documents (and these days, their electronic equivalents).[43] In the office, records of communications are sent and received, files are stored, and archives are consulted in preparing new documents.

Weber argues that bureaucracy is a servant of government and is the means by which government rules. Bureaucracies are objective, are impersonal or neutral servants of society, and are above politics. Weber emphasizes that every bureaucracy has interests of its own, as well as connections with other social strata (various groups) and other bureaucracies. He argues, however, that bureaucracy is an "iron cage" and is an enviable component of government.

The Relationship Between Bureaucracy and Change

Formally and informally, bureaucracy is a means to an end in the traditional police agency. The hierarchical flow of "orders" can give way to an imbalance: The average officer has ideals and knows how best to perform the job, yet organizational dictates must have their way. For example, the New York Police Department (NYPD) crime-control model and Compstat were innovations, and as such represent change in the NYPD and any other agencies that initiate a similar strategy.[44] The effort that must be asserted to accommodate a Compstat initiative includes a significant change in the attitudes of patrol officers and a significant change in the way the department is managed (as discussed in Chapter 1). Communication is important to officers when forging ahead with this kind of initiative; otherwise, problematic situations occur. For instance, an officer for the Sacramento (California) Police Department writes, "There were changes made by the department that none of us knew about until roll call, and suddenly we were faced with a completely new procedure that was explained in less than a minute by the sergeant." All too often, officers first learn of new procedures, policies, and expectations only minutes before engaging the public.

Police managers should embrace change and make it work for their department if the agency is to flourish.[45] Managers should develop new techniques to move street officers closer to the organizational objectives without pressuring those personnel who resist change.[46] However, there is a point when a resister has to come around, like it or not.

In the final analysis, bureaucracies are rigid, inflexible, and unable to adapt to external changes.[47] In our modern era characterized by economic, spiritual, and technological changes (as described in Chapter 1), it is uncertain how long the bureaucratic process will continue to serve U.S. interests effectively. Indeed, some researchers and police practitioners have already demonstrated ways to break down the bureaucratic hierarchical structure of police organizations.

Breaking Down the Bureaucratic Hierarchical Structure

David Osborne and Peter Plastrik identify five key strategies as necessary to transform a public bureaucratic organization into a high-performing, twenty-first-century organization through what they call the five C's: core, consequences, customers, control, and culture.[48] The application of Osborne and Plastrik's strategies is exemplified by the efforts of William Bratton when he was commissioner of the NYPD (1991–1994).

Core Strategy and Setting Goals The first "C" in Osborne and Plastrik's model relates to identification of a core strategy. Bratton found "a fearful, centralized bureaucracy with little focus on goals" when he first arrived at the NYPD. Bratton claimed that the aim of the police in the past had been to avoid criticism from the media, politicians, and the public, and he made it clear that his department would have one overriding purpose (core strategy): to reduce crime. The first-year goal would be a 10 percent reduction in crime rates. Every decision would be judged on how it affected that goal, which included personnel changes, and would be made with that goal in mind. While the core strategy appeared to meet its goal of reducing crime, it also resulted in a decentralization of command by providing the authority for divisional and unit commanders to implement policy and provide personnel supervision without top command oversight.

Consequences The second "C" in Osborne and Plastrik's model focuses on the creation of consequences (incentives) for performance. For instance, Bratton used the "broken windows" rationale (see Chapter 3) when he established consequences for officers and constituents who would not comply with the new rules and regulations governing the NYPD. For instance, the new rules stated that subway fare jumpers (i.e., people who refused to pay to ride the subway) and the "squeegee men" who descended upon drivers at key street corners in Manhattan would be arrested. Next Bratton launched, in rapid succession, new strategies to get guns off the streets, prevent youth violence, and reduce drug use, domestic violence, auto crime, and corruption. He subsequently broadened this strategy, going after other signs of community breakdown, from aggressive panhandling to urinating on the streets. His concern was that the efficiency of police units would be evaluated by measured activities such as arrests, with arrest rates serving as a method to keep score of how well units were progressing toward their goals. As part of his plan to use statistics to prevent crime, Bratton implemented the use of Compstat to map crime hot spots in the city. Because of the way Compstat was implemented (weekly meetings with Chief Bratton), police divisional commanders became accountable for their actions and the actions of their officers. In part, Compstat paved the way for a huge reorganization of the commanders in various divisions and units.

Customers The third "C" in Osborne and Plastrik's model is concerned with making public organizations accountable to their "customers" (as distinguished from constituents). That is, putting customers in the "driver's seat" of the police department through community meetings (see Control below) and public support with the help of the mayor's office. For instance, Mayor Giuliani supported police reform and the media seized on police initiatives as the primary cause of the crime decline which garnered tremendous public approval for police initiatives (despite the idea that much of the credit for the decline could have been a result of the accountability fostered through Compstat meetings, departmental reorganization, and decentralization [see Control below] of the NYPD).[49]

Furthermore it should be mentioned that public support was also gained through decentralization of the department which brings services closer to the constituents and enables a faster reaction to the changes on the ground.[50]

Control The fourth "C" in Osborne and Plastrik's model is control, which involves employee and community empowerment. Bratton felt that the NYPD was very military oriented, with a strict chain of command, such that information did not flow easily from one bureau or unit to another. Each bureau was like a silo: Information entered at the bottom and had to be delivered up the chain of command from one level to another until it reached the chief's office, where it would wait to be dealt with. Even when a memo finally arrived, there was a less-than-acceptable level of cooperation between bureaus. At some points, the system seemed to resemble a bakery, where one bureau would call another and have to take a number before it would get a response.

Bratton recruited 300 officers to work on 12 different reengineering teams, which were assigned the task of designing more productive methods to manage the NYPD. Bratton eventually eliminated an entire layer of the organization—namely, the divisions that sat between precinct commanders and the boroughs. He reduced staff at headquarters added personnel to the precincts, and empowered the precinct commanders. In short, Bratton decentralized the NYPD organization and changed the hierarchy so that it could provide for a quicker managerial response:

> I encouraged the precinct commanders to use their own initiative, and I told them I would judge them on their results. . . . I did not penalize them for taking actions that did not succeed, but I did not look kindly on those who took no action at all.[51]

Because of unreliable information and fear of corruption, precinct commanders had been told to leave all vice, drug dens, houses of prostitution, and automobile chop shops to specialized units. As a result, Bratton reports,

> Precinct commanders went to community meetings and got their heads handed to them about all the crime locations in their precincts, but they didn't have the power to address those issues. . . . We changed that. . . . We freed them from old

restraints, gave them responsibility, held them accountable, and were very pleased with the results.[52]

Culture: Changing Habits, Hearts, and Minds of Personnel The final "C" in Osborne and Plastrik's model relates to culture. One of Bratton's first moves was to hire a consulting firm to perform a "cultural diagnostic" test, to pinpoint the cultural barriers to performance and the corrective values that the new leadership needed to implant. After surveying nearly 8,000 officers, his reengineering teams reported that the culture was rampant with distrust and actively discouraged creativity.

The biggest impact that leaders can have on an employee's culture comes in their first weeks and months on the job. To ensure that the right values were instilled in officers, Bratton's team redesigned the Police Academy, where recruits were trained, raised the minimum age for officers from 20 to 22, required applicants to have two years of college, and mandated that all applicants pass a tough physical exam.

Bratton also went to work on the cynicism that afflicts most officers after a few years in the trenches: "I went after the culture to make the NYPD a more proactive police force and a more respectful one."[53] He started by taking the 75 top leaders on a two-day retreat, at which he outlined his goal of reducing crime by 10 percent in year one and laid out the new strategies and new culture he wanted to create. "This is how we're going to go," he told them. "If you can't deal with it, you're going to have to get out. If you stay, and I find you are still not with the program, then I'm going to have to get rid of you."[54]

Many isolated innovations among organizations might include these five strategies but "a continuously improving, self-renewing system" cannot be created without them.[55] Also, note that the first four strategies will not stick over the long run unless they become an integral part of the organizational culture.

Management

Management can be described as those occupational positions that legally administer the policy of the organization and control its affairs or a particular sector of the organization.[56] Positions such as police commanders, chiefs, and supervisors control the degree to which rules and policies are formalized as well as the organization's degree of centralization—that is, the degree to which the decision-making capacity within an organization is concentrated in a single individual or small select group.[57] In sum, police management includes personnel who have the responsibility for creating, maintaining, protecting, and perpetuating its systems.

On an unofficial level, there is far more to management than just filling a position within the organization. Managers are the role models most often emulated by personnel. The behavior of a police manager should reflect professional behavior,[58] and the manager's day-to-day routines should exemplify the qualities of honesty and integrity.

Public servants typically have authority to make decisions over others and make decisions that influence the public good.[59] "Ethical mentoring and role modeling should be consistent, frequent, and visible" at all times if managers expect ethical behavior from their subordinates.[60] Unfortunately, some personnel fall short in this regard because of moral blindness.[61]

Moral blindness often exists in police organizations whereby fundamentally ethical people commit unethical blunders because they are blind to the implications of their conduct.

Sometimes moral blindness could be the result of not knowing that an ethical dilemma exists; at other times, it results from willfully ignoring police corruption. It could also arise from a variety of mental defense mechanisms. Some see police officers as enjoying a "double standard" rather than adhering to a "higher standard."[62] Some critics note that politicians routinely receive honorariums and "fees" from those parties whom their decisions affect; similarly, some police officers, often following the lead of their supervisors, routinely excuse themselves from some laws even while enforcing the laws.[63]

Because supervisors are the guardians of organizational directives geared toward fulfilling the organizational objectives, most employees—regardless of the type of organization—are not asked to contribute to the development of those directives or objectives.[64] Consequently, personnel hostility toward organizational goals is a common reaction especially when those dictates do not mirror lower-level employees' perspectives.[65] Even so, the actions and values espoused by supervisors often influence the career paths of younger officers.

Clear Career Paths

Clear career paths refers to the existence of a process for examining an officer's progress from recruitment to retirement.[66] "Good hires" are not enough unless the training and other organizational amenities are available.

As discussed earlier, the hierarchical method of command entails the vertical flow of orders down the hierarchy and the vertical rise of reports of patrol officers up the hierarchy. Career paths in policing are similar in that commanders and supervisors usually rise from the lower ranks. However, not all officers are eligible for, nor do they always want, such promotions.

In the traditional police organization, clear career paths are sometimes obscure because bureaucracies are not conducive to career paths. Lateral movement is the exception rather than the rule, which means that those in field positions do not easily obtain desk jobs.[67] Upward mobility is also somewhat difficult to achieve, as it is easy to "lock in" to some "dead-end" job. Obviously, many officers do rise through the ranks, but they often follow different routes to a position of higher command.

Promotion and assignment to special units often is based on a network-based decision. For example, a Boston police officer reports that she applied several times for a special unit in the Boston Police Department, only to be told that she did not have enough field experience. The officer who did ultimately get the promotion to the juvenile unit had less experience, but the officer's father and the commander of the unit were friends.[68]

Organizational Components: A Summary

While numerous sizes and types of organizations are often found in the social services field (and in corporate America), police organizational structure tends to be more consistent in following a paramilitary model. Most police agencies are designed around a similar function—to manage police officers who have the legal and moral authority to use force and detainment. Unlike their military counterparts, however, officers are not sequestered on military bases or in prisons, nor do they tend to operate in groups under close command. Instead, officers are dispersed throughout a jurisdiction. As they patrol the streets unsupervised, whether singly or in pairs, their power can expose them to temptations. As a consequence, some may deliver inadequate police services.[69] Ultimately, the quality of police services often depends on the quality of the police organization.[70]

■ Organization Rationale

There is a consensus among experts that when the police organizational structure encourages officers to focus on the department's crime-reducing mission, and when it provides the appropriate legal ordinances and tools to allow officers to perform their jobs, officers will comply more often with rules, regulations, and the expectations of command.[71] Police command and policymakers, therefore, established a rigidly bureaucratic hierarchical structure to better control officers, to enhance police services, and ultimately to enhance the quality-of-life choices among constituents.[72] Police organizational structures include rules and regulations in every conceivable aspect of organizational life, and the bureaucratic hierarchical structure is the most pervasive police organizational structure regardless of a department's size, resources, or policies.[73]

For example, most police organizations, through techniques such as civil service requirements, attempt to sort through the personalities and private lives of police candidates to find those individuals who are passively inclined to follow organizational dictates and policies.[74] Required basic law enforcement training and in-service training also socialize officers toward that aim. Until recently, departments kept officers in cars to prevent "contamination" by citizen contact, and regulations limited officers from conducting drug (possession) arrests so as to forestall the seduction presented by the mountains of cash involved in drug dealing.[75] In many jurisdictions, a centralized 911 emergency service call center has been set up to screen requests for police service to ensure that individual officers are not asked to do improper things and their behavior can be more closely monitored. The secretive internal affairs bureau penetrates every aspect of police work to guard against corruption and to further enforce organizational control of police personnel. Many police administrators have tried to restrict officers to dealing with only the most serious crimes, as enforcement of laws against minor crimes such as panhandling and disorderly conduct plunges patrol officers into ambiguity and requires them to exercise considerable discretionary judgment.

Some experts suggest that organizational scholarship in policing has not progressed in an orderly or cumulative fashion.[76] Although classic studies of police organizations remain well read, they are rarely replicated or their findings enhanced. There is some evidence that police organizations are beginning to reevaluate their police initiatives to deliver police services because police leaders realize that the styles of delivering police services must chance to meet the challenges of the twenty-first century.[77]

Organizational Styles of Policing

Many styles of policing exist, all of which are influenced by organizational and environmental dynamics.[78] As you can imagine, police style—that is, the way in which police services are delivered—varies from jurisdiction to jurisdiction and between urban and rural areas.[79] These organizational styles can be loosely characterized as watchman, legalistic, service, community relations, and other styles.[80]

Watchman Style of Policing

The **watchman style of policing** is generally found in jurisdictions that emphasize informal police interventions such as persuasion, threats, and roughing up of suspects—rather than arrest—because the priority is maintaining the peace.[81] Watchman policing is characteristic of police initiatives targeting largely lower-class communities, in which police intervention tends to be more informal as it relates to the lives of residents. For instance, rather than arresting a suspect who is engaged in disorderly conduct, an officer might say, "If I have to come back to deal with you, I will arrest you." The watchman style generally relies on police officer discretion to maintain control within the jurisdiction.[82]

Legalistic Style of Policing

The **legalistic style of policing** represents a commitment to enforcing the letter of the law and features frequent use of arrests and other formal police interventions. Unlike their counterparts who follow the watchman style of policing, officers who operate under the legalistic model would make an arrest the first time around with a suspect.

Legalistic initiatives tend to be more militaristic than other styles of policing and routinely deploy critical incident units (i.e., SWAT units) to situations in which other departments might simply deploy patrol officers. This policing style tends to produce an enormous number of arrests by each sworn officer as compared to the watchman, service, and community relations styles of policing. Zero tolerance, roadblocks, and pursuit of a felon even through school zones might be examples of legalistic performance. Obviously, some departments employ a legalistic approach to some criminal activities and pursue a service style in response to other criminal behavior.

The legalistic style of policing attempts to control officer behavior through a rule-bound, by-the-book administrative approach.[83] Some departments manage police officer activities by enforcing quotas for traffic tickets, arrests, or field investigations, despite the fact that such quotas do not necessarily translate into quality police services.

Service Style of Policing

The **service style of policing** focuses on helping the community by working closely with social services agencies and by using referrals (to other agencies such as substance abuse programs) rather than conducting arrests. This style of policing is marked by its concern with helping rather than strict enforcement. Service-oriented police agencies are more likely to take advantage of community resources and make fewer arrests than do other types of police agencies.

Service-leaning police advocate the use of discretion to help solve social problems. In contrast, police who follow a legalistic style oppose discretion because it interferes with their responsibility to enforce the law equitably. Service-oriented discretion reinforces the idea that crime is caused by the individual's disposition or the offender's situation. In contrast, watchman-related discretion is associated with authoritarianism, ethnocentrism, and a belief in individual crime causation. Most often, the service style of policing is found in suburban police departments where crime rates are relatively low.

Community Relations Style of Policing

The **community relations style of policing** (also known as problem-solving policing) entails a collaborative partnership between the police, the community, and other public agencies, which can include schools and substance abuse clinics. Working together, these parties try to identify social problems and social disorder that can lead to crime, search for solutions to those problems, and ferret out the implications of those solutions. (See Chapters 3 and 4.)

Other Styles of Policing

Some criminal justice experts find the study of police organizational style to be a futile attempt to stereotype or pigeonhole departments and officers. Others find it an illuminating and useful area of study. It has been proposed that individual officers have their own individualized style of policing. A typology (classification of types) of those individual styles appears in Chapter 9. In some cases, however, police officers handle situations involving homeless and mentally disturbed individuals, prostitutes, juveniles, and people under the influence of alcohol or drugs in an informal way, by "dumping them."[84]

> **Dumping** refers to the practice of dropping troublesome people off somewhere. Also referred to as police-initiated transjurisdictional transport (PITT), this practice involves transporting a troublesome person out of the officer's jurisdiction and releasing the person into his or her own recognizance somewhere else.

■ Police Organizational Theory

Attempts to explain police organizational behavior can be articulated through many theories, including contingency theory, institutional theory, and resource dependence theory, among others.[85]

Contingency Theory

Contingency theory says that the performance of a police organization depends on how well it fits the context within which it is embedded.[86] It is an inclusive theory of structure, process, and performance, which holds that police organizations will be effective only if they remain dynamic, adapting to changes in technology and environment.

Here the term "technology" is used in the broadest sense, referring to those tools and strategies used by the organization to process raw materials. Contingency theory posits that certain leaders perform better under certain factual circumstances that complement their leadership style (see Chapter 7).[87] Additionally, it includes the material technologies that are having such a profound influence upon policing[88] as well as the social technologies used by the police to process and manage people and communities.[89]

The environments of police organizations are complex, yet contingency theory focuses predominantly on the "task environment." That is, this theory emphasizes those elements of the environment that directly affect the police organization—the people serviced, the justice agencies that interact with the police, patterns of crime and criminality, sources for recruiting and training officers, physical and social attributes of the community, and other external forces that shape the structure and function of police agencies.

For example, a recent study examined 6,000 police agencies operating in jurisdictions with populations less than 50,000.[90] Property offenses, domestic violence, and drugs were the most frequently reported concerns, whereas gangs and violent crime were often ranked as less serious concerns. Rankings of public safety concerns varied across agencies and were affected by population density, violent and property crime rates, type and size of police agency, and region. The study concluded that the context in which a police organization is located plays a role in shaping its public-safety practices and is important in understanding the priorities, goals, and strategies of the police organization. For all these reasons, contingency theory is the dominant theoretical framework for understanding police organizational policy.[91]

Institutional Theory

Institutional theory is an organizational perspective that competes with contingency theory. Its proponents suggest that police organizations are not necessarily rational, dynamic, or adaptive, as contingency theory proclaims.

Institutional theory has its roots in early studies of organizations that described institutionalization as the process by which organizations develop its own character and culture. Advocates of this theory argue that the environment is more than a source of raw materials, clients, technologies, and other technical elements essential to its function; values, norms, and standards are also crucial sources of organizational legitimacy.

Because police organizations require legitimacy to survive and prosper, they are often more responsive to institutional concerns than to technical concerns. For instance, traditional police departments have clarified their primary function as ensuring public safety through the enforcement of specific laws. It is through pursuit of this crime-fighting mission that departments maintain their legitimacy. Yet without computers, communication systems, and other technological enhancements, a police organization might be hard-pressed to maintain its legitimacy, let alone actually achieve its mission.

Resource Dependency Theory

Resource dependency theory is essentially a theory of power and politics, including the methods used by a police organization to secure the flow of resources (e.g., information, budgets, and status). Resource dependency theory holds that the outside environment is a source of information and resources. That is, sectors of the environment contain "pools" of resources and information that are critical to police operations. Resource dependency theory has not yet been applied to policing in a comprehensive way, although studies have described its relevance to police organizations.[92]

Organizations process the information they obtain from the environment in a way that decreases "information uncertainty." For example, one observer has outlined the links between organizations, environments, and information-processing technologies such as computer-aided dispatch systems, centralized call collection (911) mechanisms, "expert" systems, management information systems, and other tools designed to increase the organization's capacity to take in and process information.[93]

These relationships are consistent with the views espoused by researchers who view police organizations as part of a larger network of institutions responsible for identifying, managing, and communicating risks—especially homeland security risks.[94] These researchers argue that policing (at multiple levels) is shaped by external institutions and their need for information about those risks.[95] For instance, when the Department of Justice paid out $8.8 billion to local police agencies for the purpose of community policing, every department across the United States with grant-writing abilities applied for funds, which led to the birth of community policing (at least on paper). However, the incoming funds were also used by departments to fulfill other obligations.

Despite such shortcomings in the application of resource dependency theory, given the emergence in policing of sophisticated technologies for collecting and processing information, this perspective deserves further study.

Influences That Shape Police Organizations: Police Unions

Numerous influences shape police organizations. For instance, oversight committees and civil liabilities are discussed in Chapter 14, police officer stress is the focus of Chapter 15, and community relations is discussed in Chapter 4. Here we explore one major influence that is related to police organizational structure: police unions.

Definition of Police Unions

A **police union** is an organization authorized to bargain collectively or represent police officers in negotiating matters such as wages, fringe benefits, grievance hearings, conciliation proceedings, and other conditions of employment, which includes shift work and safety (especially in regard to police equipment such as police vehicles). A union shop refers to a situation in which candidates for employment must belong to or join the union to be hired.

Like many other employees across the United States, police personnel have the legal right to be represented by a union that will collectively bargain on their behalf. The **National Labor Relations Act of 1935** (Wagner Act) legalized collective bargaining and required employers to bargain with the elected representatives of their employees. The purpose of unions is to improve employment conditions for their members by using the size of the group to exert pressure on the employer, rather than having employees negotiate individual deals in which they would be in a relatively weak position vis-à-vis the employer. An estimated 73 percent of all police departments and 43 percent of all sheriff's departments in the United States work with some type of collective bargaining agent.[96] Many unions represent officers across the nation, including the American Federation of Labor and Congress of Industrial Organizations (AFL-CIO) and the Fraternal Order of Police (FOP).

Historical View of Police Unions

Early in the experiences of police officers, local police agencies were understaffed and underpaid. Economics and society's fear of "police authority" frequently restricted the departments and their budgets. As early as 1889, police officers fought for adequate compensation that would cover their basic needs, but early attempts to organize the police into unions met with defeat. Other groups, such as unskilled laborers, proved more successful at forming unions and using them as leverage to obtain better pay and working conditions.

The environment was decidedly hostile to police offers' initial attempts to unionize. For example, in 1919, Boston police accepted a charter from the American Federation of Labor (AFL) to form a union. Nineteen officers protested their working conditions and salary; all 19 were subsequently fired. When more than 1,000 of their fellow Boston police officers went on strike in their support, they, too, were fired.

Even the International Association of Chiefs of Police at first came out against collective bargaining among police officers, though they did support the idea of membership in approved social, fraternal, or benevolent police associations. By the mid-1900s, the Fraternal Order of Police had formed a few social groups as a precursor to formal unions. In the late 1930s and early 1940s, when police officers again attempted to establish unions, they were more successful. By the early 1960s, police unions were largely local and independent. At that time, approximately 30 to 40 percent of department personnel were represented by a local union, 40 to 45 percent belonged to a regional or state union, and 15 to 20 percent belonged to a national labor group.

In July 2007, more than a century after the first attempts to organize police, the U.S. Congress finally passed a bill allowing all first responders—such as police officers and fire fighters—the right to join a collective bargaining agent such as a union, though it prohibits first responders from going on strike.[97] Two states, Virginia and North Carolina, continue to hold out on this issue, forbidding their first responders from joining such an organization.

Police-Specific Unions

Unlike the case for some other personnel, there is no one union that represents all police officers across the United States. For instance, the United Automobile Workers (UAW) represents most auto workers in the country. By contrast, police unions are fragmented, though there are four primary police unions.

Fraternal Order of Police

Today, the FOP is the world's largest organization of sworn law enforcement officers, with more than 324,000 members in more than 2,100 lodges. The stated mission of this union is to be "the voice of those who dedicate their lives to protecting and serving communities." (See the FOP's web site at http://www.grandlodgefop.org.)

Patrolmen's Benevolent Association

Several labor unions representing police officers carry the name of Patrolmen's Benevolent Association. For example, the Patrolmen's Benevolent Association of the City of New York (NYCPBA) represents 33,000 officers of the New York City Police Department. (See the NYCPBA's web site at http://www.nycpba.org.)

International Union of Police Associations

IUPA is one of 89 chartered unions affiliated with the AFL-CIO. It is estimated that the IUPA represents 35,000 officers. (See the IUPA web site at http://www.iupa.org/members/iupalocals.html.)

International Brotherhood of Teamsters

The IBT is a controversial union because of its early alleged affiliation with organized crime. Like the IUPA, IBT is linked with the AFL-CIO. It represents approximately 15,000 officers in 225 police organizations found in suburban and rural jurisdictions. (See the IBT's web site at http://www.teamsters.org.)

Union Issues

Sources of controversy related to police unions include the rank at which a police officer can join a union. In most cases, management personnel refrain from joining unions. In policing, 85 percent of all officers are represented by unions, including both patrol officers and sergeants. This would not necessarily represent a conflict of interest, except that negotiations sometimes include areas under the control of sergeants and other middle police managers, such as one or two-officer patrol cars, transfers and promotions, and duty assignments. In addition, on-the-job sources of conflict may arise between patrol officers and their chain of command.

Another source of tension is union strikes, which represent a last resort for many unions when negotiating with management. Although many departments and policies call for a no-strike policy, when confronted with unsuccessful union bargaining results, officers have sometimes called in sick in large numbers (i.e., "blue flu"). Other actions short of outright strikes are also possible. For example, in Boston during the 2004 Democratic National Convention, the police union drew local and national attention when its members picketed early-arriving delegates—a dilemma for Democrats, given that the party has traditionally been regarded as an ally of organized labor. Having worked without a contract for two years, the Boston police union finally struck a deal with Boston Mayor Thomas Menino for a new contract, who sought to avoid the embarrassment of pickets occurring during the high-profile event.

Unions also exert considerable influence over day-to-day operations in some departments. Noted one commander from an urban police department, "Unions stop us [officials] from disciplining our officers. Sometimes I can't even fire or suspend an officer unless I have the approval of the union to do it."

A vote of no confidence is a measure police union officials use to strengthen their position at the negotiating table. Police officers typically gain many benefits from collective bargaining, but often in the midst of tough negotiations, union officials and officers might agree on a vote of no confidence concerning police management. For instance, many police chiefs across the nation have been the targets of a no-confidence tactic.[98] In many cases, unions use this tactic in an attempt to discredit or remove a chief. At other times, no-confidence votes are held to enhance the union's collective bargaining position. Whatever the motivation for this step, police management must move swiftly to deal with the issue because a no-confidence vote could easily undermine the quality of police services by affecting police morale negatively.

For instance, Chief Robert T. Murray of the Falls Church, Virginia, Police Department received a vote of no confidence after officers became concerned that the weapons provided by the department were defective.[99] Previously, the union had complained that commanders were requiring officers to meet a high quota for traffic tickets and that failure to meet the quota could result in demotions or pay cuts. Since the no-confidence vote, officials have relaxed their ticket-writing requirements, but officers have not been told exactly how their performances will be evaluated.

■ Organizational Culture

Police officers do not voluntarily accept responsibility for fulfilling police organizational objectives.[100] Instead, police officers are supervised and monitored to determine whether they are performing appropriately. Typical means of monitoring and managing their performance include time clocks, shift rosters, hourly wage rates, and benefits schedules.

In most public organizations, similar to the situation encountered in police organizations, accountability has come to mean compliance with the will of the "boss."[101] Accordingly, rather than being associated with the police organization, officers are the "troops" whose role is "not to reason why." The gap between officers and the police organization is further widened by interference from oversight committees, civil liabilities issues, community relations groups, and unions.

For their part, police officers tend to be unwilling to regulate the behavior of their peers and are resistant to change.[102] Attempts to uncover violations of the police organizational rules are inevitably frustrated by the "code of silence." Discipline is seen as a weapon of police management. Therefore, the organization takes on a culture of its own, which often does not resonate with, or generate the respect of, most patrol officers.

Because officers are salaried workers, many are preoccupied with monetary rewards (because pay scales tend to be low) rather than trying to fulfill the organizational objectives. They work to maximize returns in relation to hours worked. Officers might work holidays for double-time rates and then take two sick days off—a practice that earns them an extra vacation at no monetary loss (the so-called Spanish practice).

In altercations or other situations that could implicate an officer in wrongdoing, most officers redefine the circumstances to support their actions. In this way, officers protect their own.

As a result of all of these forces, the police organization develops its own identity—its own culture. All too often, formation of the organizational culture occurs without police officer participation. In response, officers may develop their own police subculture (Chapter 13) and resists the organization's efforts to hold them accountable for their actions (Chapter 14).

RIPPED from the Headlines

Sometimes police agencies ask for the federal government's help in handling particularly difficult or complex situations. For instance, in New Orleans, federal agents patrol along with New Orleans Police Department patrol officers. Why? Dual jurisdiction. Local officers have limited jurisdiction as compared to federal agents, as a result of the passage of the U.S.A. Patriot Act.

Feds Hit New Orleans Streets

The federal government is in the midst of an unprecedented intervention into the local criminal justice system, adding more than 40 agents and prosecutors to regain some basic daily operations lost to Hurricane Katrina in 2005.

The effort comes as the city's violent crime increased 107 percent the first quarter of 2007 over the same time in 2006, according to police statistics. "The citizens of this city desperately need protection," U.S. Attorney Jim Letten said. "There are elements of the local justice system that have failed. That is flat unacceptable, and it has to change."

Federal and local agents have assembled an unusual watch list of 71 criminal suspects whose links to unsolved murders, robberies, and assaults represent serious threats. The suspects—some linked to multiple killings—have at least one violent-crime conviction and might be eligible for mandatory federal sentences of five to ten years if charged with new gun offenses.

- About 22 FBI agents are assigned to combat violent crime—at least 13 embedded with homicide detectives, Riley said.
- Drug Enforcement Administration agents got a temporary expansion of their authority to deal with gun crimes and other violent offenses, spokesman Michael Sanders said. The DEA also reassigned eight agents to work directly with police to help curb illegal drugs in a city where crack cocaine continues to fuel turf battles among returning dealers.
- Six federal prosecutors were added to the U.S. Attorney's Office to handle the increased focus on both federal gun and drug cases.
- The Bureau of Alcohol, Tobacco, Firearms, and Explosives, which deployed six additional agents, expects to boost federal gun cases 27 percent, said David Harper, the ATF's top agent in New Orleans.
- Mayor Ray Nagin announced that the Louisiana Air National Guard would begin nightly patrols over the city through 2007.

Source: Kevin Johnson, "Feds Hit New Orleans Street," *USA Today* (June 4, 2007), http://www.officer.com/article/article.jsp?siteSection=1&id=36379 (accessed June 4, 2007).

Summary

An organization is as a legal, artificial structure created to coordinate personnel and resources to achieve a mission or a goal. The components of an organization include a clear purpose, precise communicated statements, structure, different levels, division of labor, degrees of specialization, hierarchy of command, bureaucracy, management, and clear career paths.

The rationale behind police organizations is to encourage officers to focus on the department's crime-reducing mission, and to provide them with the appropriate legal ordinances and tools to enable them to perform their jobs. That is, police organizations seek to better control officers, to enhance police services, and ultimately to enhance the quality-of-life choices among constituents. Organizational styles of policing include the watchman, legalistic, service, and community relations styles, among other. Three theories dominate discussion of how organizations function and evolve: the contingency theory, the institutional theory, and the resource dependency theory. A variety of influences shape police organizations, including police unions. Taken collectively, all of these forces result in a unique organizational culture.

Key Words

Bureaucracy
Community relations style of policing
Contingency theory
Defined command
Degree of specialization
Delegation of authority
Dumping
Institutional theory
Legalistic style of policing
Management
Moral blindness
National Labor Relations Act of 1935
Organization
Peter Principle
Police union
Resource dependence theory
Rulification
Service style of policing
Span of control
Watchman style of policing

Discussion Questions

1. Define an organization, and explain the benefits expected from it.
2. Describe the evolution of police organizations.
3. Characterize the primary components of a police organization, and explain your perspective on those primary components.
4. Analyze the primary components of a police organization, and explain in what way you agree and disagree with the analysis provided in this chapter.
5. Characterize the rationale underlying police organizations, and describe in what way you agree and disagree with this rationale.
6. Describe the various organizational styles of policing. Which style in best suited for the community where you permanently reside?
7. Characterize the three major police organizational theories, and identify which theory best describes the police organization where you permanently reside.
8. Characterize the Peter Principle and explain how it can affect police bureaucracy.
9. Discuss the influences that shape police organizations.
10. Describe the issues arising from police unions, and explain your position on those issues.

11. Articulate the chief results of the many influences that shape police organizations. In what way do you agree and disagree with those results?

Notes

1. George L. Kelling, "Broken Windows and Police Discretion" (1999), http://www.ncjrs.gov/txtfiles1/nij/178259.txt (accessed June 4, 2007).
2. Edward R. Maguire, Jeffrey B. Snipes, Craig D. Uchida, and Margaret Townsend, "Counting Officers: Estimating the Number of Police Departments and Police Officers in the USA," *Policing: An International Journal of Police Strategies & Management* 21, no. 1 (1998): 97–120.
3. David H. Bayley, *Patterns of Policing* (New Brunswick, NJ: Rutgers University Press, 1985), 7.
4. William Bratton, "LAPD Chief Bratton Speaks Out: What's Wrong with Criminal Justice Research—How to Make It Right," *NIJ Journal* 257 (2007): 29–32, http://www.ncjrs.gov/pdffiles1/nij/jr000257.pdf (accessed July 8, 2007).
5. John P. Crank, "The Influence of Environmental and Organizational Factors on Police Style in Urban and Rural Environments," *Journal of Research in Crime and Delinquency* 27, no. 2 (1990): 166–189.
6. Stephen P. Robbins and David DeCenzo, *Fundamentals of Management*, 4th ed. (Upper Saddle River, NJ: Pearson Education, 2005). Also see Samuel Walker and Charles Katz, *Police in America. An Introduction*, 6th ed. (Boston: McGraw Hill, 2008), 233.
7. Dennis J. Stevens, *Police Officer Stress: Sources and Solutions* (Upper Saddle River, NJ: Prentice Hall, 2008).
8. Wayne W. Bennett and Karen M. Hess, *Management and Supervision in Law Enforcement*, 5th ed. (Belmont, CA: Thomson/Wadsworth, 2007), 3.
9. Bennett and Hess, *Management and Supervision in Law Enforcement*,
10. Bennett and Hess, *Management and Supervision in Law Enforcement*, 11. Concerning bureaucracy, see Walker and Katz, *Police in America*, 91.
11. Bennett and Hess, *Management and Supervision in Law Enforcement*, 11; David L. Carter and Louis A. Radelet, *The Police and the Community*, 7th ed. (Upper Saddle River, NJ: Prentice Hall, 2001), 50–52; Peter Drucker, *Management: Challenges for the 21st Century* (New York: Harper Collins, 1999); Stan Stojkovic, David Kalinich, and John Klofas, *Criminal Justice Organizations: Administration and Management*, 3rd ed. (Belmont, CA: Wadsworth, 2003), 6–7; Joycelyn M. Pollock, "Ethics and Law Enforcement," in *Critical Issues in Policing: Contemporary Readings*, ed. Roger G. Dunham and Geoffrey P. Alpert (Prospect Heights, IL: Waveland, 2001), 356–73; Robert Trojanowicz, Victor E. Kappeler, Larry K. Gains, and Bonnie Bucqueroux, *Community Policing: A Contemporary Perspective*, 2nd ed. (Cincinnati: Anderson, 1998), 84–93.
12. Bennett and Hess, *Management and Supervision in Law Enforcement*, 11.
13. Santa Monica California Police Department (2006), http://santamonicapd.org/information/mission.htm (accessed December 3, 2006).
14. Samuel Walker, *Sense and Nonsense about Crime and Drugs: A Policy Guide* (Belmont, CA: Thomson/Wadsworth, 2001), 27.
15. For a better read on this perspective, see Dennis J. Stevens, *Applied Community Corrections* (Upper Saddle River, NJ: Prentice Hall, 2006), 21–35. In part, this perspective is linked to a punishment philosophy held by the criminal justice community. Punishment is justified through just desserts or retribution, deterrence, and incapacitation actions conducted by the justice community in the name of social order. Criminals get what they deserve.

16. Trojanowicz, Kappeler, Gains, and Bucqueroux, *Community Policing*, 87.
17. Trojanowicz, Kappeler, Gains, and Bucqueroux, *Community Policing*, 87.
18. Walker, *Sense and Nonsense about Crime and Drugs*, 27.
19. Jeffrey Reiman, *The Rich Get Richer and the Poor Get Prison: Ideology, Crime, and Criminal Justice* (Boston: Allyn and Bacon, 1998).
20. Randall G. Shelden, *Controlling the Dangerous Classes: A Critical Introduction to the History of Criminal Justice* (Boston: Allyn and Bacon, 2001), 3–10.
21. Shelden, *Controlling the Dangerous Classes*, 8. Also see Reiman, *The Rich Get Richer*, 4, 5, 59.
22. Reiman, *The Rich Get Richer*, 5.
23. Max Weber, *From Max Weber: Essays in Sociology*, trans. H. H. Gerth and C. Wright Mills (New York: Oxford University Press, 1946). Also see, for this interpretation, Gary Cordner and Gerald Williams, "Community Policing and Police Agency Accreditation," in *Policing Perspectives*, ed. Larry K. Gaines and Gary W. Cordner (Los Angeles: Roxbury, 1998), 315–345.
24. Roy R. Roberg, Jack Kuykendall, and Kenneth Novak, *Police Management*, 3rd ed. (Los Angeles: Roxbury, 2002), 265–278.
25. Walker and Katz, *Police in America*, 90.
26. Edward R. Maguire, *Organizational Style in American Police Agencies* (Albany, NY: SUNY Press, 2003).
27. Laurence J. Peter and Raymond Hull, *The Peter Principle* (New York: Morrow, 1972), p. 2.
28. Geoffrey P. Alpert and Roger G. Dunham, *Policing in Urban America*, 3rd ed. (Prospect Heights, IL: Waveland, 1997), 79–81; Walker and Katz, *Police in America*, 478.
29. Walker and Katz, *Police in America*, 478.
30. Wesley Skogan, Susan M. Hartnett, Jill DuBois, Jennifer T. Comey, Marianne Kaiser, and Justine H. Lovig, *On the Beat* (New York: Westview, 1999), 191–223.
31. Kenneth J. Meier and John Bohte, "Span of Control and Public Organizations: Implementing Luther Gulick's Research Design," *American Society for Public Administration* (January 2003), http://unpan1.un.org/intradoc/groups/public/documents/ASPA/UNPAN007281.html (accessed May 15, 2005).
32. Walker and Katz, *Police in America*, 478.
33. Walker and Katz, *Police in America*, 478.
34. For a closer look at these findings, see: Merrick Bobb, *Ninth Semiannual Report* (Los Angeles: Police Assessment Resource Center, 1998), 22–23, http://www.parc.info/ (accessed May 15, 2005). Also available at that website are updates as of February 2005.
35. Stojkovic, Kalinich, and Klofas, *Criminal Justice Organizations*, 24–25.
36. Thomas O'Connor, North Carolina Wesleyan College (2005), http://faculty.ncwc.edu/toconnor/205/205lect07.htm (accessed May 15, 2005).
37. For instance, elevation to Master Police Officer in the Hampton, Virginia, Police Department includes a 5 percent increase in pay and is given after at least three years of consecutive service as a Senior Police Officer. MPO elects must have received at least a rating of "surpassed expectations" on their latest annual performance evaluation, and must have participated in at least two community events. They must have a minimum of three specialties such as K-9, field training instructor, firearms instructor, marine patrol, or school recourse officer, and three competencies such as radar certification, dive team, SWAT, or hostage negotiator (http://www.hampton.va.us/police/recruiting/career_paths.html, accessed March 6, 2006).
38. Stojkovic, Kalinich, and Klofas, *Criminal Justice Organizations*, 24.

39. William Bratton, "New York Crime Rate Down Forty-Five Percent," *New York Times* (February 12, 1996). http://topics.nytimes.com/top/reference/timestopics/subjects/c/crime_and_criminals/index.html?offset=60&s=newest&query=BRATTON%2C+WILLIAM+J&field=per&match=exact (accessed March 19, 2008).
40. Herman Goldstein, *Problem-Oriented Policing* (New York: McGraw Hill, 1990). Also see Dennis J. Stevens, *Case Studies in Applied Community Policing in the 21st Century* (Boston: Allyn and Bacon, 2003).
41. Stojkovic, Kalinich, and Klofas, *Criminal Justice Organizations*, 4.
42. H. H. Gerth and C. Wright Mills (trans. and ed.), *From Max Weber* (New York: Oxford University Press, 1946).
43. Professor John Kilcullen of Macquarie University, Australia, supplied some of this information (john.kilcullen@mq.edu.au).
44. Phyllis Parshall McDonald, *Managing Police Operations* (Belmont, CA: Wadsworth, 2002), 1.
45. Bennett and Hess, *Management and Supervision in Law Enforcement*, 187–188.
46. Stojkovic, Kalinich, and Klofas, *Criminal Justice Organizations*, 364. For a look at resistance, see Jack R. Greene and William V. Pelfrey, Jr., "Shifting the Balance of Power between Police and Community: Responsibility for Crime Control," in *Critical Issues in Policing: Contemporary Readings*, ed. Roger G. Dunham and Geoffrey P. Alpert (Prospect Heights, IL: Waveland, 2001), 435–465.
47. Walker and Katz, *Police in America*, 91.
48. David Osborne and Peter Plastrik, *Banishing Bureaucracy: The Five Strategies for Reinventing Government* (Reading, MA: Addison Wesley, 1997).
49. Eli B. Silverman, *NYPD Battles Crime: Innovative Strategies in Policing* (Boston: Northeastern University Press, 1999).
50. Innovation in Public Services, *Literature Review* (2005), 39–40. http://www.idea.gov.uk/idk/aio/1118552 (accessed March 18, 2008).
51. William Bratton and Peter Knobler, *The Turnaround: How America's Top Officer Reversed the Crime Epidemic* (New York: Random House, 1998), 201.
52. David Osborne, "Chief of Police, Chief of Reinvention: Bill Bratton Teaches Through Example," in the Osborne Letter (St. Paul, MN: The Public Strategies Group, 2006), p. 4. http://www.psgrp.com/resources/osbornebillbratton.html (accessed June 19, 2008).
53. Osborne, "Chief of Police, Chief of Reinvention: Bill Bratton Teaches Through Example," p. 5.
54. Osborne, "Chief of Police, Chief of Reinvention: Bill Bratton Teaches Through Example," p. 5.
55. Osborne and Plastrik, *Banishing Bureaucracy*.
56. Peter Drucker, *Management: Tasks, Responsibilities, Practices* (New York: HarperCollins, 1993).
57. Bennett and Hess, *Management and Supervision in Law Enforcement*, 11–14.
58. Rickey D. Lashley, *The Need for a Noble Character* (Westport CT: Praeger, 1995).
59. Pollock, "Ethics and Law Enforcement," 441.
60. Achieving and maintaining high ethical standards, *Police Chief* 69, no. 10 (2002): 64, 66, 68.
61. John Kleining, "Gratuities and Corruption," in *Policing: Key Readings*, ed. Tim Newburn (New York: Willan Publishing, 2001), 597–624.
62. Pollock, "Ethics and Law Enforcement," 357.
63. Pollock, "Ethics and Law Enforcement," 357. Also, consider this example: "Tom Delay and the GOP milking the system to live high on the hog. Tom DeLay saw a seat in Congress as

a way to live large at someone else's expense. From the time he arrived in Washington after the 1984 elections, DeLay started working the system to line his own pockets. 'I met Delay at the reception for freshmen members of Congress,' recalls retired lobbyist Jackson Russ. 'He walked up, looked at my name tag, introduced himself and asked how he could get some honorariums.' In 1984, honorariums were a quick way for members of Congress to line their own pockets. Special-interest groups would invite the Congressman to a get-together with executives of their company or top members of the organization and then pay that Congressman directly for the appearance. Congress banned honorariums in 1989 but that gave DeLay five years to become one of the top earner of fees for appearances on the Hill, adding an average of $27,000 a year to his Congressional salary. 'DeLay bugged everyone for honorariums,' says Roy Abrahams, who lobbied Capitol Hill for oil interests from 1975 through 1990. 'Others were subtle. He wasn't.'" [Doug Thompson, "CHB Investigates" (November 28, 2005), http://www.capitolhillblue.com/artman/publish/article_7709.shtml (accessed March 6, 2006).]

64. A. P. Cardarelli, J. McDevitt, and K. Baum, "The Rhetoric and Reality of Community Policing in Small and Medium Sized Cities and Towns," *Policing: An International Journal of Police Strategies and Management* 21, no. 3 (1998): 397–415.

65. Hans Toch, *Stress in Policing* (Washington, DC: American Psychological Association, 2001).

66. Walker and Katz, *Police in America*, 117.

67. Thomas O'Connor, "Police Organization" (2005), http://faculty.ncwc.edu/toconnor/205/205lect07.htm (accessed May 16, 2005).

68. Personal communication with a Boston police officer who was also a student of this writer (February 2005).

69. George L. Kelling, "How to Run a Police Department," *City Journal* (1995), http://www.city-journal.org/html/5_4_how_to_run.html (accessed December 1, 2006).

70. Walker and Katz, *Police in America*, 87.

71. Kelling, "How to Run a Police Department." Also see Noel Tichy with Eli Cohen, *The Leadership Engine: How Winning Companies Build Leaders at Every Level* (New York: Harper Business, 1997).

72. William Bratton and Peter Knobler, *The Turnaround: How America's Top Officer Reversed the Crime Epidemic* (New York: Random House, 1998).

73. Osborne and Plastrik, *Banishing Bureaucracy.*

74. David H. Bayley, *What Works in Policing* (New York: Oxford University Press, 1997).

75. Osborne and Plastrik, *Banishing Bureaucracy.*

76. Edward R. Maguire and Craig D. Uchida, "Measurement and Explanation in the Comparative Study of American Police Organizations" (2000), http://www.ncjrs.gov/criminal_justice2000/vol_4/04j.pdf (accessed December 2, 2006).

77. Mary Ann Viverette, President's Message: Prepare tomorrow's police leaders. International Association of Chiefs of Police (2006). http://www.theiacp.org/cpl2/index.cfm?fa=presMessage (accessed March 18, 2008).

78. Crank, "The Influence of Environmental and Organizational Factors on Police Style."

79. Willard Oliver, "Rural Police," in *Police Officer Stress: Sources and Resolutions*, ed. Dennis J. Stevens (Upper Saddle River, NJ: Prentice Hall, 2007), 241–250.

80. James Q. Wilson, *Varieties of Police Behavior* (New York: Atheneum, 1973), 7. Also see Bennett and Hess, *Management and Supervision in Law Enforcement.*

81. Bennett and Hess, *Management and Supervision in Law Enforcement.*

82. Richard T. Wortley, "Measuring Police Attitudes Toward Discretion," *Criminal Justice and Behavior* 30, no. 5 (2003): 538–558.

83. Walker and Katz, *Police in America*, 203.

84. William R. King and Thomas M. Dunn, "Dumping: Police-Initiated Transjurisdictional Transport of Troublesome Persons," *Police Quarterly* 7, no. 3 (2004): 339–358.

85. Maguire and Uchida, "Measurement and Explanation."

86. Joseph B. Kuhns III, Edward R. Maguire, and Stephen M. Cox, "Public-Safety Concerns Among Law Enforcement Agencies in Suburban and Rural America," *Police Quarterly* 10, no. 4 (2007): 429–454.

87. M. R. Haberfeld, *Theories of Police Leadership* (Upper Saddle River, NJ: Prentice Hall, 2006), 144.

88. Richard B. Ericson and Kevin D. Haggerty, *Policing the Risk Society* (Toronto: University of Toronto Press, 1997).

89. Steven Mastrofski and Richard Ritti, "Making Sense of Community Policing: A Theoretical Perspective," *Police Practice and Research Journal* 1, no. 2 (2000): 183–210.

90. Kuhns III, Maguire, and Cox, "Public-Safety Concerns Among Law Enforcement Agencies."

91. Charles Katz, Ed Maguire, and D. W. Roncek, "The Creation of Specialized Police Gang Units: A Macro-Level Analysis of Contingency, Social Threat and Resource Dependency Explanations," *Policing: An International Journal of Police Strategies and Management* 25, no. 3 (2002): 472–506.

92. Katz, Maguire, and Roncek, "The Creation of Specialized Police Gang Units."

93. Peter Manning, *Police Work: The Social Origin of Policing* (Mt. Prospect, IL: Waveland, 1997).

94. Ericson and Haggerty, *Policing the Risk Society.*

95. Richard B. Ericson and Kevin D. Haggerty, *The New Politics of Surveillance and Visibility* (Toronto: University of Toronto Press, 2006), 4.

96. Bureau of Justice Statistics, (2004, March). Law Enforcement Management and Administrative Statistics, 2000: Data for Individual State and Local Agencies with 100 or More Officers (NCJ 203350) (Washington, DC: U.S. Department of Justice, March 2004), http://www.ojp.gov/bjs/pub/pdf/lemas00.pdf (accessed October 15, 2006).

97. Jesse J. Holland, "House: All Right, Responders Can Unionize," *Associated Press* (July 17, 2007), http://www.thefreelibrary.com/House%3a+All+first+responders+can+unionize-a01610726804 (accessed March 19, 2008).

98. William B. Berger, "Vote of No Confidence," *Police Chief* (June 2002): 6.

99. Tom Jackman, "Police Chief Assailed in Vote by Va. Union: Gun Malfunctions Cited in Falls Church," *Washington Post* (August 31, 2005): B04.

100. David H. Bayley, *What Works in Policing* (New York: Oxford University Press, 1997), 66.

101. Peter Hutchinson, "Compliance: Getting 'Em to Understand That Carrying Out the Garbage Just Isn't What It Used to Be!" (2005), http://www.psgrp.com/resources/compliancegetting.html (accessed December 3, 2006).

102. John P. Crank, *Understanding Police Culture* (Cincinnati, OH: Anderson, 2004).

Police Management and Leadership

To the person who only has a hammer in the toolkit, every problem looks like a nail.
—Abraham Maslow

▸ ▸ LEARNING OBJECTIVES

When you finish reading this chapter, you will be better prepared to

- Describe police organizational accreditation
- Identify police officer ranks and their range of authority
- Characterize authority and delegation of authority among managers
- Describe management by objectives and total quality management
- Identify the components of Theory X, Theory Y, and Theory Z
- Describe Abraham Maslow's hierarchy of needs
- Characterize situational and transformational leadership styles
- Define leadership and characterize its primary techniques
- Identify the seven habits of highly effective police leaders

▸ ▸ CASE STUDY: YOU ARE THE POLICE OFFICER

You've just been promoted to a patrol sergeant. You're proud that your accomplishments have been recognized by your supervisors and the officers in your unit. You studied hard for the sergeant's exam, had two interviews including a psychology test, and know that some of the officers in your unit were interviewed on your behalf. Your directives are to take charge of the officers who supported you, but you realize you can't accomplish this task without repercussions. The difficulty lies in how to gain control over those officers who are your friends and supporters without alienating them.

Your sergeant tells you that it is an important part of the job to compromise and to understand that the officers have needs that must be addressed. One theory asserts that the officers have a need to feel safe when they're providing police services and free from the threat of attack by aggressors. If you become an aggressive force in an attempt to take charge, the officers' lower-level needs of safety will not be met. That would cause their higher-level needs to decrease, which might in turn negatively influence the quality of police services.

You have to find a balance between these two apparently opposing forces in police organizations: You need to take control without appearing to be an aggressor so that officers don't feel under attack, yet keep their motivation centered on higher-level need issues. The best result, you conclude, would be quality police services provided by the officers who have a positive reaction to your leadership.

1. Your directives are to take charge of the officers who supported your promotion to sergeant. What might be the repercussions of your actions as a sergeant when you attempt to follow those directives?
2. In what way do police officers (like almost everyone else) respond to relationships and supervisors if they feel threatened?
3. Why is this balance an important objective for a new sergeant?

Introduction

Chapter 7 describes contemporary police management models and leadership styles. Yet, before venturing into descriptions of management initiatives, one point should be understood: Police professionals with a high level of expertise in a specific intervention area inevitably see every problem as an opportunity to ply their trade.[1] Nevertheless, to say that a police organization in lower Manhattan, New York, has developed a more productive managerial style than a police agency in Manhattan, Mississippi, can be difficult. Every agency is unique, and what works in one Manhattan might not work in another Manhattan—or anywhere else, for that matter. Management must adapt a professional style that works best for each particular agency given its personnel, budget, regulations, laws, bargaining agents, and so on.[2]

Factors Affecting Police Management Styles

Police supervisors—similar to all supervisors—face a sometimes disappointing reality: Personnel work up to only 33 percent of their capabilities, 10 percent are high achievers, 10 percent are low achievers, and low achievers tend to cause 90 percent of the problems in most organizations, including police agencies.[3] The quote by Abraham Maslow advises us that if a police manager has limitations, similar to the carpenter whose toolbox contains only a hammer, than he or she will attack every problem in the same way. This is not always wise, of course. There are a number of methods available that can enhance the performance and effectiveness of police supervisors and top managers; those options are one focus of this chapter.

The selection of the best tool for the job, which ultimately determines managerial effectiveness, is greatly affected by many factors. For example, police managers must take into account their superiors as well as their subordinates, the nuances of the police organizational structure (see Chapter 6), politics, and economics. Other key factors include the unique characteristics of civil service, the role of collective bargaining units such as police unions (see Chapter 6), and accreditation, which affects both the organization and it managers.

Accreditation

Police **accreditation** is the receipt of a certificate formally recognizing the police agency as conforming to some specific body of regulations and standards. The most commonly sought certificate is from the Commission on Accreditation for Law Enforcement Agencies, Inc. (CALEA). This certification comes only after an internal audit (essentially a compliance audit)[4] that can initially take a year or more to complete; the audit is both time-consuming and labor-intensive. Many states also have their own certification boards that award accreditation. Police agencies tend to request an audit from both the state and CALEA should they wish to become certified.

"Becoming certified" demonstrates a department's professional standing. The accreditation has an expiration date, but the department can apply for recertification, which entails completion of a new audit.

History of Accreditation

The idea of police accreditation first emerged around 1979. Today, approximately 600 police departments have been accredited—that is, certified to be in compliance with CALEA's standards. CALEA is a nonprofit organization that started out as an innovative idea conceived by the International Association of Chiefs of Police (IACP). The CALEA commission is a unique blend of civilians (university professors, business leaders, politicians) and professionals appointed by the executive boards of IACP, the National Organization of Black Law Enforcement Executives (NOBLE), the National Sheriffs Association (NSA), and the Police Executive Research Forum (PERF).

Standards are usually classified as "mandatory," "essential," or "recommended" on the basis of the language used by the writers of the standards. If the standard reads, "Agencies *must* have Crown Victoria vehicles that are no less than a year old," then the *must* statement makes this standard "mandatory"; thus the agency must be in 100 percent compliance. If the standard reads, "Agencies *should* have personnel policies for the hiring of 4-year college graduates," then the *should* statement makes this standard "recommended"; in such a case, it is up to the agency to determine whether it wants to meet this standard.

The reading and classification of standards is more of an art than a science; it is more complex than the simple dichotomy of *must* and *should* statements. Nevertheless, it is often explained in this simple manner so as to involve employees in the self-assessment part of the process.

With regard to compliance, all assessments rate the agency, on any given standard, as being in "full compliance," "partial compliance," or "not in compliance." Agencies are allowed to explain in writing why they are not in compliance or only in partial compliance on some standards, and CALEA makes the final determination by using both quantitative and qualitative decision-making.

Benefits and Disadvantages of Accreditation

Accreditation has benefits and disadvantages.[5] The benefits include the following:

- Can mean that an agency formally recognizes the existence of professional standards
- Increases community status and support
- Enhances the confidence of police personnel
- Enhances the confidence of police policymakers
- Achieving accreditation can be described as being "state-of-the-art."
- Requires that the agency develop clearly articulated policies, procedures, and manuals
- Can lead to reductions in the agency's insurance premiums
- Serves as a deterrent to liability litigation
- Leads to improved communication with other community agencies
- Makes updated information about policing available
- Can be cited in applications for grant funding
- Can attract better qualified candidates during the recruitment process

Drawbacks associated with accreditation can include:

- Standardization might lead to a nationalized police force with a single mindset.
- Some standards for accreditation may be set too low or be too elastic, and an agency can always say the standard goes beyond what local laws or conditions merit.
- Some police chiefs resent the implication that their rules and policies are somehow inferior or not up to par.
- Accreditation may have a financial backlash, as some politicians may see the agency's ability to get accreditation as the ability to do more with less.
- Accreditation may foster resistance. Some line officers and unions in particular may resent accreditation if it is used as a shield for poor management, if it leads to demands for higher education or advanced technology training for officers, or if it puts officers' job security at risk.
- Accreditation may cause policymakers to lose control over segments of police organization.

Rank within Police Organizations

Police organizations use rank as a way to distinguish various levels of authority. The rank system includes both sworn police personnel and administrative authority—patrol officers, corporals, sergeants, captains, inspectors, deputy chiefs, majors, chiefs, sheriffs, and constables (in Texas), as well as civilians such as police commissioners, attorney generals, state prosecutors, and state and local public safety directors. Not all managers are sworn officers (i.e., officers who have detainment power and authority to use force), and not all managers have a similar amount of authority and power. For example, a captain in Santa Fe, New Mexico, might possess more authority than a captain in Boston, Massachusetts, because of differences in the organizational structure, resources, and policies of the department.

Three levels of supervisors are typically identified:

1. Front-line supervisors (sergeant)
2. Middle managers (major, captain, lieutenant)
3. Top management (deputy or assistant chief, chief, constable, sheriff, superintendent)

Regardless of their rank, police executives perform four essential functions:[6]

1. Planners: Tactical planning focuses on the short term, while strategic planning is future oriented.
2. Facilitators: Executives assist and motivate others toward performance that brings the agency closer to its objectives, goals, and work plans.
3. Interfacers: Managers interact with all levels of personnel, the public, and public officials, including personnel and officials from other local, state, and federal agencies.
4. Interactors: Executives must be capable of interacting effectively as a member of a group and as a commander when necessary. Police executives interact with the press, public officials, civic (business and religious) leaders, and other executives from other organizations such as labor unions and other state and federal agencies.

Depending on the size of the department, ranks can vary. For instance, the New York City Police Department (NYPD), which is the largest police department in the United States, has 11 uniformed ranks:

- Police Officer (The title "Patrolman" was phased out in the early 1970s.)
- Detective (considered to be the same rank as a Police Officer)
- Sergeant (symbol of rank: three chevrons)
- Lieutenant (symbol of rank: one gold bar)
- Captain (symbol of rank: two gold bars)
- Deputy Inspector (symbol of rank: gold oak leaf)
- Inspector (symbol of rank: gold eagle)
- Deputy Chief (symbol of rank: one gold star)
- Assistant Chief (symbol of rank: two gold stars)
- Bureau Chief (symbol of rank: three gold stars)
- Chief of Department (symbol of rank: four gold stars)

There are two ranks that are not uniformed and are appointed by the police commissioner: administrators who supersede the Chief of Department and Deputy Commissioner. These administrators may specialize in counterterrorism, training, and community affairs. Despite their rank, as civilian executive specialists, they are prohibited from taking operational control of a police situation (with the exception of the First Deputy Commissioner). The symbol of rank for the Deputy Commissioner is three gold stars.

First-Line Supervisors

Most often, first-line supervisors are sergeants who report to a designated supervisor of rank above them. According to one management expert, sergeants are the most powerful individuals in the police agency because they form "the tendons and sinews of an organization. They provide the articulation. Without them, no joint can move."[7]

Patrol officers (in the NYPD, police officers) are continually manipulated by their assigned sergeants.[8] For instance, one officer says that her sergeant assigned her to the worst beat in Boston and checked on her every 20 minutes during her entire graveyard shift. The officer adds that "there was little question in my head who was the boss. I hated that kind of intimidation but had to take it if I wanted to keep my job."[9] Sergeants make the decisions about which officers are deployed to which assignments, can divert citizen complaints and problematic officer behavior issues, and can provide the steps toward official recognition and promotion of patrol officers.

Most sergeants are highly efficient. When things go wrong in the field, these first-line supervisors spend their energies on discovering what went wrong, how to fix it, and how to make sure it doesn't happen again.[10]

Sergeant Chad Gann of the Arlington (Texas) Police Department (APD) began his career in 2000 and was promoted to sergeant four years later:

> Since then, my operational assignments and responsibilities have led to my continued development as a supervisor and, most importantly, as a leader. I believe that the position of patrol sergeant is the best position a person can attain in any police agency. For it is through this rank that I have the opportunity to influence people in a positive way to effectively accomplish the mission and goals of my department. Thus, my greatest day-to-day challenge is maintaining the right relationship with officers under my command as we work though operational, departmental, and even political concerns. . . . As a patrol sergeant, I must balance the relationships between those I supervise and those who supervise me.[11]

In 2001, the Baltimore Maryland Police Department (BPD) published the results of a study concerning its own first-line supervisors.[12] Study conclusions included the idea that exemplary sergeants can be distinguished from less competent sergeants on the following dimensions: moral reasoning ability; identification of moral exemplars; patterns and amounts of sick leave; and the fact that after working with exemplary sergeants, lieutenants selected exemplary sergeants as being preferable in times of crises. Nine vital traits were identified as being especially desirable among first-line supervisors:

1. Character and integrity
2. Knowledge of the job
3. Management skills
4. Communication skills
5. Interpersonal skills
6. Ability to develop entry-level officers
7. Problem solving and critical thinking skills
8. Effectiveness as a role model and as a disciplinarian
9. Ability to be proactive among subordinates

The BPD's study concluded that first-line training and performance evaluations require paying special attention to sergeants because their fundamental responsibilities are so crucial to police effectiveness.

First-Line Supervisors' Fundamental Responsibilities

One way to describe the work of first-line supervisors or sergeants is to say that they oversee the actual work being accomplished by patrol officers. Their task is to ensure that the routine day-to-day performance of officers is accomplished effectively and legally. The fundamental job of a duty sergeant can be parsed into the following tasks:[13]

1. Manage patrol officers in field
2. Supervise patrol activities
3. Conduct inspections
4. Maintain discipline
5. Enforce rules and regulations
6. Conduct roll call
7. Manage field operations
8. Report anomalies up the managerial ladder
9. Prepare reports

Supervisor Styles

First-line supervision styles vary. Chances are that personal experience, common sense, and intuition would confirm this thought. One study identified four distinctive styles of supervision among 64 first-line supervisors.[14]

Traditional supervisors are characterized by expectations of aggressive enforcement by subordinates rather than engagement in community-oriented activities or policing of minor disorders. Similar to micro-managers, they are task oriented and expect subordinates to produce measurable outcomes—particularly arrests and citations—and to complete extensive paperwork and documentation on their activities. (See the discussion of Theory X managers later in this chapter.)

Innovative supervisors are characterized by a tendency to form relationships (i.e., they consider officers to be friends), a low level of task orientation, and positive views of subordinates. (See the discussion of Theory Y managers later in this chapter.)

Supportive supervisors are characterized by supportive styles toward subordinates, protecting them from discipline or punishment perceived as "unfair" tactics and providing inspirational motivation. These supervisors often serve as a buffer between officers and management by protecting officers from criticism and discipline. They are less concerned with enforcing rules and regulations, dealing with paperwork, or ensuring that officers do their work. Supportive supervisors encourage officers through praise and recognition, act as counselors, or display concern for subordinates' personal and professional well-being. They see officers as members of a team and ask for their input. (See the discussion of Theory Z managers later in this chapter.)

Active supervisors embrace a philosophy of leading by example. These hands-on managers' want is to be involved in the field alongside patrol officers while controlling subordinates' behavior, thus performing the dual function of street officer and supervisor. Almost all active supervisors (95 percent) in this study reported that they often went to incidents and dealt directly with street issues.[15]

In the same study, the 239 patrol officers who participated reported that the number of arrests they made and the rate at which they issued citations had nothing to do with the style of their sergeant.[16] By contrast, patrol officers managed by traditional supervisors tended to favor the use of force more often than those officers managed by active supervisors. Traditional supervisors were the most respected among the officers largely because these supervisors provided the most training and believed that enforcing the laws was the most important responsibility for police officers (even to the extent of using force to gain compliance among constituents). Patrol officers with active supervisors spent more time per shift engaging in proactive (self-initiated) activities than did officers managed by supervisors using other methods of supervision styles. Also, patrol officers with active supervisors spent significantly less time per shift on administrative tasks. Overall, officers spent 16 percent of their time on personal business regardless of the style of their supervisor.

Middle Management

Middle managers usually refer to majors, captains, and lieutenants. Majors are rare in most departments; when this rank is present, the individual is usually involved in a specialized unit such as forensics or is the director of the police academy.

More commonly, captains operate districts within a department and have authority over all officers below top command: chief, sheriff, or constable (Texas). Captains, who are usually responsible only to top command, typically have authority over sergeants and all officers within their assigned responsibility. Captains and lieutenants perform the following functions:

- Manage specific command functions (e.g., administration, investigation, patrol, traffic, services, emergency preparation)
- Direct, control, supervise, plan, coordinate, and perform designated tasks within a specific command assignment
- Participate in the personnel selection process, department training, internal investigations review, and disciplinary procedures and review
- Perform inspection of an assigned operation, unit, or division
- Make recommendations
- Assist and inform complainants and other individuals who are seeking information about the department or who file complaints with the department
- Initiate investigations in accordance with the requirements of law or department procedure
- Perform staff functions as directed by the chief and serve on the advisory staff of the chief of police
- Develop plans, apply for grants, and oversee task forces (e.g., drugs, guns, stings)
- Prepare and supervise work schedules
- Implement plans

- Evaluate implemented plans
- Oversee equipment, confiscated items, and impounded evidence and vehicles

In large departments, captains usually manage units (investigations), divisions (District 3), and specialized units (e.g., hate crimes). Often those positions are difficult to manage because essentially the captain is the chief (of the unit, division, or specialized unit) but lacks the authority of the chief and his or her status depends on the authority provided by the chief. One of the captain's most critical responsibilities is to provide leadership for the community, civic leaders, and the officers under his authority, including officers from other departments who are engaged in a joint effort or task force. One final point, rank is often linked to salary-range; captains have greater authority, responsibility, and experience than lieutenants and therefore the thinking goes, captains earn a greater salary than lieutenants.

Top Command

One characteristic that appears to be typical of most successful top commanders or chiefs, sheriffs, and constables (Texas) is professional leadership competence (more on this later the chapter). This quality is needed because such police executives are both legally and morally responsible for everything that occurs in the department—every bit of training, interaction between personnel and outsiders, policy, and so on.[17]

The police executive officer faces demanding expectations in regard to both quality and rapidness of his or her performance and is exposed to constant intricate and comprehensive policing situations.[18] Members of top command must deal with both information technology, such as data recording and evaluation, and research work on the basic legal fundaments before the decision-making process is started. In essence, the function of the chief is to professionally facilitate constant change. Whether in large or small police departments, the primary function of the chief of police relates to leadership competence in every area of the department including the conduct of all police personnel, whether they are sworn officers, civilian employees, or volunteers.

Profile of a Police Chief: Jupiter, Florida

To provide a practical view of the responsibilities of a police chief, consider the situation in Jupiter, Florida, which had a population of approximately 45,000 in 2005. The Jupiter Police Department employed 100 sworn officers and 50 civilian personnel in that year, making it a typical-sized department in a typical-population jurisdiction.

When the town of Jupiter sought a chief of police, the Town Council provided a list of essential functions for the job.[19] According to this job description, the Jupiter police chief is expected to do the following:

- Supervise, direct, and evaluate assigned staff
- Process employee concerns and problems
- Direct employees' work
- Counsel, discipline, and complete performance appraisals for employees
- Coordinate employees' daily work activities
- Organize, prioritize, and assign work
- Monitor the status of work activities
- Consult with assigned staff

- Assist with problem situations
- Provide technical expertise
- Conduct interviews and make hiring decisions

The chief of Jupiter also directs operations and activities of the police department. These functions include planning, organizing, and directing all agency activities; directing and coordinating department staffing levels; and providing assistance to police officers when needed.

The functions of the Jupiter police chief also include interpretation and enforcement of all applicable (including traffic, criminal, and civil) codes, ordinances, laws, and regulations so as to protect life and property, prevent crime, and promote security. The chief ensures departmental compliance with general orders, police contracts, personnel codes, and all other applicable codes, laws, rules, regulations, standards, general orders, policies, and procedures. Likewise, he or she must ensure adherence to established safety procedures; monitor the work environment and use of safety equipment to ensure safety of employees and other individuals; and initiate any actions necessary to correct deviations or violations. In addition, the chief communicates with the Town Manager, Mayor, Town Council, and other individuals to review department operations and activities, review and resolve problems, receive advice and direction, and provide recommendations. In essence, the police chief serves as a liaison between the Jupiter Police Department and the Town Council.

The list goes on and on—but it should already be clear that the functions of a police chief are pretty incredible. Incidentally, the Jupiter police chief job paid less than $95,000 per year.

Reflections on Life as a Police Chief

What is like being chief of police? A chief of police from a typical Midwestern city says this:

> Similar to a sheriff's job, mine is that of a public figure who has to implement change. I walk softly and carry a big stick. Aside from strategic planning and the duties of the job, one of the hardest things to accomplish is to develop a "my town" attitude, whether you came up through the ranks or you're an import. People tell you what you want to hear, not what I need to know.[20]

Chief Francis D'Ambra at Manteo, North Carolina, describes his priorities as change and a lot of sophistication.[21] His department serves an affluent community where cost is never an issue when the agency needs new officers, new equipment, and training. But that generosity creates a different set of issues and expectations, which the chief describes as "phantom menace" problems. For instance, when the department requires new equipment, the chief and his deputy chief do not have to beg for cash or seek out the least expensive product, but they do have to develop an elaborate account of the item and be ready to explain all of its attributes. Town council likes to "discuss matters" and needs to be convinced that the desired item will benefit the department.

Given this proclivity for discussion, D'Ambra had to enhance his communication skills to deal with town hall. His approach is markedly different from that employed by chiefs in hard-pressed, crime-ridden communities who are searching for funds to control crime. But then, each chief's job is as unique as his or her community. "Most chiefs are in charge simply because they asked for the job without knowing what they were getting themselves into," says D'Ambra. Ultimately, what it takes to live up to the responsibilities of the job will never really be understood until an individual actually becomes police chief.

Female Police Mobility

Chief Nan Hegerty of Milwaukee, Wisconsin, says that males do not respect women who try to be one of the guys.[22] "They understand that women officers are essential, but they really want you to be a woman," the chief suggests. "On the other hand, they don't want you to be afraid to mix it up [fight] when the situation calls for it."

The profile of the typical police department is changing (as discussed in Chapter 1), and those changes include greater mobility for female police executives. For instance, Chief Beverly Harvard (Atlanta), former Commissioner Kathleen O'Toole (Boston), Chief Ella Bully-Cummings (Detroit), Nan Hegerty (Milwaukee), Chief Cathy Lanier (Metropolitan Washington, DC), Chief Heather Fong (San Francisco), and acting Chief Suzanne Devlin (Fairfax, Virginia) all belong to a unique sorority: top women executives of urban police departments.

Approximately 11 percent of all U.S. police officers are female, according to Bureau of Justice statistics.[23] The National Center for Women and Policing, a division of the Feminist Majority Foundation in Arlington, Virginia, disagrees with that estimate, stating that the percentage is actually 13 percent.[24] There are approximately 200 women chiefs in the United States. "We're not talking a lot of progress here, just to put it in some perspective. But we are talking high visibility, and all eyes will be on them," said Margaret Moore, who retired as the highest-ranking woman in the Federal Bureau of Alcohol, Tobacco and Firearms.[25]

Jim Kouri, vice president of the National Association of Chiefs of Police, is not surprised to see women reaching top positions in increasing numbers, noting that women have been working their way up the police hierarchy for several decades.[26] Kouri suggests that differences in academy training over the last 30 years may have contributed to this trend; such training is not necessarily "softer," but is certainly less paramilitary in nature.

Authority and Power

A good starting point in understanding management and leadership is to distinguish between authority and power.[27]

Authority is the legal right to accomplish goals through others by directing their behavior.

Power is the ability to accomplish goals through others, with or without a legal right to do so, to the extent of their compliance.

Authority is granted by law or regulation, whereas power is managerial influence that is exerted without the benefit of the law or regulation.[28] Managers possess the authority to direct specific personnel, but true leadership might require different skill levels of power. Both authority and power help managers compel subordinates toward compliance with organizational goals, yet authority's centerpiece is the law while power relies on the leadership style of the police manager. Think of it this way, says Stephen R. Covey: Managers work *in* the system, while leaders work *on* the system.[29]

Police managers delegate authority to their subordinates in an effort to enhance the decision-making process (especially because situations arise quickly and appropriate decisions can affect their resolutions). Of course, delegating authority to others is always risky business, because a manager must be a pretty good judge of character to transfer his or her authority to someone else, advises Peter F. Drucker.[30] Police chiefs and sheriffs alike are hesitant to delegate authority largely because the chief can be held legally responsible for the decisions of subordinates.

Managers who believe in the idea that "If you want something done, better do it yourself" tend to delegate less authority to others. How often these kinds of micromanagers succeed in police departments is difficult to say, especially given that delivery of police services (including emergency care) is a 24/7 business. Even so, most orders in the police hierarchy come from the top down, whereas reports start at the bottom and travel up to command personnel. Thus delegation of authority in police organizations is less likely to become a daily routine simply because of the nature of the hierarchal bureaucratic structure.

■ Officer Criticism of Police Executives

Police executives are not always viewed favorably by their subordinates. In two studies, patrol officers imply that police executives have forgotten or never knew what policing is about.[31] One officer explained:

> The people in this building [headquarters] are not policemen. Most wouldn't know how to be a policeman, and I don't say that bitterly. It's just how it is. Some are good book guys—you know, studied and quickly moved up the ladder if they were liked by the brass—but because they didn't like what they were doing, they wanted to be big bosses, and so they are.[32]

Many patrol officers have similar feelings for supervisors and top managers, who they believe tend to manage through intimidation models.[33]

A prominent police instructor who has conducted more than 500 ethics' leadership seminars reports the following:[34]

- Officers are routinely treated with disrespect by arrogant, insecure top managers.
- Promotions and discipline are ruled by favoritism.
- Officers who should have been fired, arrested, or decertified are quietly asked to resign.

One complaint of many patrol officers is that new policies and regulations are often inconsistent with existing policies and regulations. One officer says, "Policy is something the public and legislators see—kind of like the previews of a movie. But on the street, the reality of policy can only slap you in the ass if you go against the boss." Patrol officers may be obliged to carry through on a policy even if it is ineffective (or illegal). Indeed, patrol officers are often ordered to compromise on a policy and do so to accommodate the wishes of their supervisors and commanders.[35] Also patrol officers tend to face an enormous amount of uncertainty and stress resulting from mismanagement[36]—and often that mismanagement flows from the style of control exercised by middle managers and top command.[37]

In previous studies, a consensus arose among the participates that "they did not believe the department's management was honest or had integrity" and that it "systematically ignores their views and interest." [38]

In a typical response from the officers interviewed, one reported that "command is out for themselves and screw street officers." Also, it was suggested that patrol officers increasingly rate command and the police bureaucracy, "along with criminals," as being on the enemies list.[39] Other researchers note that "supervisors within the department were perceived to be a principal source of stress." [40]

Poor Supervisor Support

New supervisors eventually learn that the officers in the field are only as good as the supervisor's performance. Weak supervisors spend most of their time covering their own negative performance as well as the poor performance of those reporting to them.[41] A constant theme echoed by patrol officers is the belief that their supervisors have little regard for officers' well-being whether on or off duty.[42] This thought is consistent with a "moral blindness" as described in Chapter 6. To clarify the point, the conduct of police managers from the perspective of rank-and-file officers, fundamentally represents ethical behavior of their managers who commit unethical blunders "because they are blind to the implications of their conduct." [43]

One distinctive characteristic of police is that virtually everyone in the organization "from station house broom to chief, shares the common experience of having worked the street as a patrol officer." [44] Moving down the chain of command, lateral entry is unknown because supervisor positions are filled only from the pool of candidates directly below the vacant slot. One result of this system is that most supervisors have street experience—but that does not always mean that they were compatible with such duty.[43]

The competition in most police agencies for sergeant positions is keen and the results are generally uncertain, suggesting that becoming a sergeant is anything but an orderly, well-defined passage up the career ladder.[45] Sergeants learn their job by doing it (often taking risks along the way), but it is an individual and solitary process (see the discussion of the Peter Principle in Chapters 6 and 13).

Patrol Officer Criticism of First Line Supervisors

In the author's research into the Boston Police Department, almost two thirds of 329 patrol officers employed in several Boston police districts reported that duty-sergeants were rarely objective when dealing with patrol officers. Seventy-six percent of the participants in this study reported that their sergeants used favoritism to determine promotions, make desirable assignments (e.g., crime gang units), and defuse citizen complaints. The participants also reported that sergeants demonstrated inappropriate behavior and unprofessional responses to officers whom the sergeants had not respected when those officers complained about their duty assignments (such as assigning them to a midnight shift or in a rough part of town), equipment allocations (such as poorly operating police vehicles), and lack of back-up during volatile service calls.

More than half of the officers reported that their supervisors provided inadequate information that had endangered the officer's safety while on duty. In one instance, during roll call patrol officers were informed about a gang presence at a certain address—but the information was a week old. In another case, patrol officers were called to a scene and were not told that shots had been fired. In a third case, officers were ordered not to shoot at any moving target. "So what the heck does that mean? If a suspect is running and firing his weapon at innocent bystanders, I can't stop him?" an interviewed officer asked.

More than half of the study participants also reported that sergeants had used intimidation models to control officers. For example, one officer was attempting to secure a family violent crime scene, when she concluded—based on 12 years of experience—that a sexual assault had actually occurred regardless of what the victim had reported.[46] When the detective arrived, he "laughed off" the officer's speculation. When the patrol officer checked out at the station, she mentioned her thoughts to her sergeant on duty. The sergeant's reply: "Hey, that ain't your job [to gather evidence or develop a theory of a crime]. Follow the plan, Kojak [a name the sergeant gave to officers who thought they were detectives], or you'll be smoothing bunny ears [arresting prostitutes] in Chinatown [during the 11:00 P.M. to 7:00 A.M. shift]."

Police Management Theory

> **Management** can be described as an occupational position that legally administers organizational policy and legally controls its affairs or a particular sector of the organization.[47]

Positions such as police commanders, chiefs, sheriffs, and supervisors control the degree to which rules and policies are formalized and determine the extent of the police department's centralization (i.e., the degree to which the decision-making capacity within an organization is concentrated in a single individual or small select group).[48] In short, police management includes personnel who possess the responsibility of creating, maintaining, protecting, and perpetuating its system.

> Management is the process of planning, organizing, and controlling personnel and resources in the service of preventing crime, protecting victims, apprehending criminals, recovering stolen property, providing emergency care, and performing regulatory services.

Management involves much more than the official job description that a position might suggest. Managers are the role models most often emulated by personnel. Consequently, a police manager should always exhibit professional behavior.[49] Honesty and integrity must be part of their day-to-day routines. In fact, studies have shown that chronic police corruption among police officers is less likely when police managers are professional and righteous.

Public servants typically have authority to make decisions that affect others, including decisions that influence the public good.[50] "Ethical mentoring and role modeling should be consistent, frequent, and visible" at all times if managers expect ethical behavior from their subordinates.[51] Unfortunately, a "moral blindness" (as discussed in Chapter 6) often exists in police organizations, whereby fundamentally ethical people commit unethical blunders because they are blind to the implications of their conduct. Sometimes moral blindness may result from not knowing that an ethical dilemma exists; at other times, it may arise from a variety of mental defense mechanisms. How often do supervisors look the other way concerning unethical or outright corrupt behavior among their peers and their subordinates? Many view public servants as enjoying a "double standard" rather than adhering to a "higher standard," argues Jocelyn Pollock.[52]

For instance, in New York City, two detectives fired 42 of 50 rounds at a suspect during a bachelor party they attended and could face more than half a century behind bars after their indictment in the death of the unarmed groom.[53] Their supervisor, Lt. Gary Napoli, hid behind a car when the gunfire erupted and was placed on modified duty after the incident.

Given that managers are the guardians of organizational directives toward fulfilling organizational objectives, most personnel—regardless of the type of organization—rarely influence final decisions made by those managers.[54] "It's a fact of life," says one commander, "officers better do as they're told or I'll kick their lowly ass out of here."[55] Consequently, it has been insinuated that police managers know secrets about their organizations and the personnel in their command. As this line of thinking goes, the higher up the command ladder, the greater number of secrets managers hold, implies a division commander with the NYPD.[56] Managers also often use their leadership skills to move subordinates toward performance that the officers might not do otherwise; in turn, because of the manipulation practiced by skilled managers, their subordinates might "passionately" pursue a vague and often redundant objective of the organization. Consistent with this notion, the intensity and frequency of manipulation provided by a police manager are enhanced the higher up the chain of command that manager resides. If there is some truth to this thought, then a flattening of command might enhance public legitimacy and ultimately increase both organizational and public compliance, advise Nancy E. Marion and Willard M. Oliver.[57]

Nonetheless, there are several managerial models worth noting, although it is unlikely that any of the "pure models" highlighted here in real-world operation. Undoubtedly, police managers and policy makers shape the management model in place in their organization to fit their own experiences, circumstances, budget, personnel, history, skill levels, and agenda of the department.

Management by Objectives

Management by objectives (MBO) aims to enhance organizational performance by aligning goals and subordinate objectives throughout the organization. That is, police management and patrol officers agree on goals and objectives within a police organization in order for patrol officers to work toward those goals and objectives from an informed perspective because patrol officers have participated in developing these goals and objectives. MBO includes ongoing tracking and feedback in the process to reach objectives. MBO was first popularized by Peter Drucker in 1954 in his book *The Practice of Management.*[58]

Ideally, all personnel regardless of their rank, help to identify the department's objectives and develop timelines for completion. In the traditional police hierarchical bureaucratic structure, police commanders establish policy without the influence or input from middle managers (such as lieutenants and sergeants) and middle managers set objectives and timelines for patrol officers, simply because it is faster to make unilateral decisions. The MBO process requires commanders, middle managers, and patrol officers to collaborate toward reaching an agreement on tasks and goals. This perspective suggests that agreement might be one avenue toward increased compliance and greater solidarity within a police agency upon all personnel. Thus MBO can aid in police personnel unity and at the same time comply with agreed upon tasks and objectives.

As an example, suppose that, whatever else a commander and patrol officers may discuss and agree to in their regular discussions, they feel that it will be sensible to introduce a key performance indicator to show how a policy of proactive arrests benefits quality-of-life experiences. The manager and employees then need to discuss what is being planned, what the time schedule is, and what the indicator might or might not be. Unfortunately, MBO works only when everyone knows the objectives, according to Drucker. Druker implies that 90 percent of the personnel are not aware of the organizational objectives. Therefore, it is vital to gather input as a manager and disseminate it in an apporpriate manner to all personnel including adminstrative personnel.[58a]

MBO principles include the following ideas:

1. Cascading of organizational goals and objectives
2. Specific objectives for each member
3. Participative decision making
4. An explicit time period for meeting the goals and objectives
5. Performance evaluation and feedback
6. Objectives should meet the following criteria:
   - Specific
   - Measurable
   - Achievable
   - Realistic
   - Time-related

For example, the Greenbelt (Maryland) Police Department's MBO program has the following objectives:

- Maintain compliance and reaccredidation of accreditation with CALEA (The Commission on Accreditation for Law Enforcement Agencies).
- Continue working with the city of Hyattsville and other municipal jurisdictions on developing a joint dispatching operation along with a records management system. Include analysis of the current dispatch situation and the potential impact of the proposed solution.
- Upgrade all video cameras in police cruisers.

- Continue working with Prince George's County Police and other municipal jurisdictions on developing a 700 MHz radio system that will allow interoperability with all Council of Government members in the National Capitol region.
- Operate the inplace high-tech initiatives to locate missing children.

Chief James Craze adds that the Greenbelt PD is recognized as a contemporary, award winning agency with dedicated employees responsive to the community who constantly strive to make Greenbelt a better and safer place for those who live, work, or visit.[59]

Total Quality Management

Total quality management (**TQM**) is a management strategy aimed at raising awareness of quality in all organizational processes. In this sense, TQM is goal driven. It provides an umbrella under which personnel can strive to create constituent satisfaction. This personnel- and constituent-focused management system aims to continually increase satisfaction through the use of quality circles, where the increased satisfaction translates into continually lower real costs and greater compliance. **Quality circles** refer to groups of personnel who routinely come together to discuss policy, share information with other groups, and report on their progress to management. TQM is utilized in a variety of fields, including manufacturing, education, and service industries such as policing.

Abrahamson argues that trends in management such as quality circles follow a life cycle that takes the form of a bell curve.[60] Use of TQM, for example, is said to have peaked between 1992 and 1996.[61] TQM and quality circles are hardly new models for management of police agencies. For instance, David L. Carter and Louis A. Radelet have advocated the use of TQM in police organizations for the past two decades.

There are four reasons why police organizations should consider adapting TQM as a managerial style:[62]

- Police have their critics, some of whom suggest that the authority of police managers at all levels is affected by a host of unwieldy regulations and other obstacles advanced by politicians, community leaders, and organizational leaders in both the public and private sectors.
- The U.S. public has a litigious nature, and the trend toward bringing civil suits against police departments and police management at all levels (which have reached record numbers) often influences day-to-day, serve-and-protect decisions.
- Community policing strategies are at the core of future police services.
- Community relations initiatives such as problem-solving orientations and TQM have similar ideals and can act synergistically.

Within the TQM format, all personnel in the organization are goal directed and participate in goal attainment as opposed to fulfilling their own personal goals. Thus this model would appear to be a managerial tool that complements community relations initiatives, including the quality-of-life strategies described in Chapter 4. TQM (or an abbreviated form of it) can only help police organizations and enhance personnel professionalism. Nevertheless, the goal-driven management style linked to quality-circle reports is a process that, in the end, cannot function as well within a bureaucratic hierarchical structure. Authority within the structure needs to be flattened or decentralized for best effects.

Theory X and Theory Y

Two distinctive yet competing human motivation theories developed by Douglas McGregor at Massachusetts Institute of Technology in the 1960s are Theory X and Theory Y.[63] McGregor's perspective associated with Theory X and Theory Y relate to a humanistic approach to management and has been used in management, organizational behavior, and organizational development since their inception. These theories describe two very different attitudes toward the motivation levels of personnel.[6, p. 34] McGregor believed that organizations follow either one or the other approach.[63]

Theory X

Theory X refers to the perspective held by police supervisors who believe that police officers work for money and security, and that an officer's discretion should be limited and require close supervision. According to this view, officers are not paid to think, but rather to perform.

Theory X supervisors believe the average patrol officer in their command has the following characteristics:[64]

1. Dislikes the job and the public the officer serves
2. Has little ambition, wants less responsibility, and would rather follow than lead
3. Is self-centered and doesn't care about agency goals or mandates
4. Resists change
5. Is gullible, yet listens to peers more than superiors
6. Is not particularly intelligent
7. Took the job of police officer because it was the only job the subordinate could find
8. Stays on the job because the officer is lazy, and hides while working

One way to describe Theory X supervisors is to say that they practice an "in your face" form of supervision. Of course, it is unlikely that any police manager believes every item listed above, and it is equally unlikely that he or she would exercise that view to an extreme.

Theory X perspectives work well in a police hierarchical structure, however, where they encourage a narrow span of control at each level. As a result of this line of thought, Theory X managers adopt a more authoritarian perspective (see Chapter 13 for a discussion of the authoritarian personality) about management that relies heavily on the threat of punishment. **Authoritarian management** refers to a style that favors strict rules, established authority, obedience to the chain of command, and swift punishment. Another aspect of authoritarian principles is the demand that officers adhere to the parameters of their job descriptions and do nothing more, similar to the view espoused by the traditional sergeant in the Baltimore Police Department study described earlier in this chapter.

One criticism of Theory X practices is that this model can negatively affect arrest and conviction rates. This management style is also seen as enhancing officer resistance to change, including new police strategies. In essence, this authoritarian management system exercises a form of social control characterized by strict obedience to the authority of the organization, and maintenance and enforcement of control through the use of oppressive measures including intimation and ridicule of personnel.

Theory X's hard-line approach is grounded in coercion, implicit threats and intimidation, close supervision, and tight command and control. Such an approach typically results in hostility, purposely low output, and hard-line union demands. In contrast, a softer approach might produce an ever-increasing request for more rewards and ever-decreasing work output.

Theory Y

Theory Y is a theory of management that assumes personnel are ambitious, self-motivated, anxious to accept greater responsibility, and able to exercise self-control and self-direction. Subscribers to this theory believe that employees enjoy their mental and physical work activities, and that officers have the desire to be imaginative and creative in their jobs if given a chance. Managers who advocate this perspective suggest that greater productivity can result from giving patrol officers the freedom to perform their best.

Theory Y supervisors believe that, given the right conditions, most personnel under their command want to do well at performing their assignments and that there is a pool of unused creativity in the workforce. They can be described as employing optimistic supervisor practices, which are based on the following general assumptions:

- Police work can be as natural as play and rest.
- Officers can be self-directed to meet their work objectives.
- Officers will be committed to police organizational objectives if rewards are in place that address their higher-level needs, such as the need for self-fulfillment.
- Under these conditions, officers will accept the responsibility and seriousness of their job.
- Most officers can handle responsibility because they rank high in creativity and ingenuity.
- The personal goals of most officers are similar to the objectives of the police organization.
- Under Theory Y supervision, officers would depend less on the police subculture for guidance and support.

Theory Y holds that police organizations can take many steps to make full use of the motivational energy of their officers:

- *Decentralization and delegation.* If the police organization decentralizes control and limits the number of supervisors, then each supervisor will have more subordinates and will be forced to delegate more responsibility to each officer. This, in turn, may enhance morale and improve delivery of police services.
- *Job enlargement.* Broadening the scope of an officer's job adds responsibility and offers new opportunities to satisfy the officer's ego needs.
- *Participative management.* Consulting personnel in the decision-making process taps into officers' creative capacity and provides them with some measure of control over their work environment.

- *Performance appraisals.* Having officers set objectives and participate in the process of evaluating how well they meet those objectives aids in their performance towards agreed upon goals.

Theory Y managers believe that the satisfaction of doing a good job is a strong motivation in and of itself. Such a manager will attempt to identify and remove the obstacles preventing officers from fully actualizing themselves. For instance, as paperwork mounts, supervisors will institute more professional methods of fulfilling paperwork requirements, such as arranging for use of take-home computers, scheduling classes to learn about new computer software, and finding methods to reduce paperwork requirements.

In a Theory Y management scheme, a greater number of officers report to a single supervisor, so the span of control is wider and change is more likely. When Theory Y management is properly implemented, such an environment would produce a high level of motivation among officers to work toward satisfying their higher-level personal needs resulting in fewer citizen complaints, fewer civil liability suits, less corruption, and greater retention of officers.

Application of Theory X and Theory Y

Theory X and Theory Y could hardly be applied as perfect models in the real world. Indeed, they cannot be accepted too literally because they represent a dichotomy of unrealistic extremes. Most officers and their superiors fall somewhere in between these two poles. McGregor was well aware of the heuristic (as opposed to literal) way in which such distinctions are useful. Nevertheless, Theory X and Theory Y remain valid guiding principles that offer a positive approach to understanding management and motivation.

Abraham Maslow's Hierarchy of Needs

Douglas McGregor relied on **Abraham Maslow's hierarchy of needs** perspective to explain his managerial perspective (i.e., Theory X and Theory Y). One reason why Maslow's work is discussed in this chapter is that his perspective seems to be visible in most personnel behavioral dynamics as well as in the behavioral dynamics of the public. Maslow was a psychologist who developed a number of motivational models intended to help managers better understand how to motivate personnel.[65] Maslow's hierarchy of needs continues to be widely used in one form or another.

Maslow felt that personnel could be best managed by understanding their needs, which can be organized into a hierarchy that can be illustrated as a pyramid (Figure 7–1). His theory requires that an individual's personal needs must be satisfied at the lower levels before the individual can progress toward addressing his or her higher-level, more complex needs. When lower-level needs are satisfied, individuals are no longer motivated by them and instead progress to higher-level motivators. All needs are always present, however.

One assumption made when a manager utilizes Maslow's hierarchy of needs is that personnel want to succeed and to move toward higher-level needs, at least conceptu-

Figure 7–1 Maslow's Hierarchy of Needs

ally. Maslow calls the desired goal "self-actualization." A self-actualized person has the following characteristics:

- Is realistically oriented
- Accepts other people for what they are
- Is spontaneous in thinking, emotions, and behavior
- Is problem-centered rather than self-centered
- Needs privacy
- Is autonomous, independent, and able to remain true to himself or herself in the face of rejection or unpopularity
- Has a continuous freshness of appreciation
- Has mystic or oceanic experiences (although these experiences are not necessarily religious)
- Identifies with humankind
- Has deep, meaningful relationships with a few people
- Has a democratic structure and judges people as individuals
- Has highly developed ethics
- Resists total conformity to culture

These characteristics, coupled with the motivational needs described by Maslow, provide the tools for managing police personnel. For instance, if an officer asks for a specific shift and specific days, and automatically is assigned to that schedule, she or he would be less motivated to provide quality police services the next time a desired goal is sought. Once a need is obtained, whatever that need had been, it is now less capable of motivating an officer. The idea is to provide a percentage of the request to the officer—such as the shift while holding out on the days requested—until the officer demonstrates the quality of police services desired by the supervisor.

Most officers have little need to concern themselves with their physiological, security, and safety needs, so these basic, low-level needs might no longer motivate their actions (that is one reason why officers might take chances with their own safety), although these needs are always present. When officers are underpaid and deep in debt, however, lower-level needs can motivate their decisions, which ultimately can be seen in their behavior.

Most officers are motivated primarily by social, esteem, and self-actualizing needs. Most of us, for instance, require love relationships, acceptance by others (especially our supervisors and coworkers), and appreciation. Officers may join social, religious, fraternal, and educational organizations to fulfill these psychological needs. Even more often, officers form their own police subculture (see Chapter 13), which allows them to derive some comfort from the acceptance and appreciation that come from belonging to such a group. For that reason, it is advised that police organizations actively seek to fulfill the needs of their officers as opposed to leaving this task to the police subculture: Sometimes this subculture has detrimental effects by opening the door to corruption and approval of police behavior that is neither appropriate nor legal. Police experts report that when the organization is unstable, officers turn to one another for support more often; in some circumstances, these relationships can encourage police behavior that is contrary to the mission of police.[66]

Self-esteem needs are a step higher than social needs in Maslow's hierarchy. Beyond being merely accepted and belonging, officers want to be heard and to be publicly appreciated and rewarded for their services, which includes the good deeds they do that can lead to higher status within the community. Good managers know how to attract positive attention of the press and the community on behalf of their officers.

At the highest level of Maslow's perspective are self-actualizing needs. Officers seek to achieve their highest potential through professional, philanthropic, political, educational, and artistic channels. Obviously, police managers who work hard on their behalf are rewarded with quality police services, fewer constituent complaints, and fewer officers focus on their personal goals (see Chapter 13).

In summary, one aim of police organizations is to build unity or cohesion among officers. However, when managers lack the skill to forge departmental unity, officers tend to merge into a police subculture that can open the door to weaker police services and corruption, or so the theory suggests. It comes as no surprise that McGregor's work grouped Maslow's hierarchy into "lower-order" (Theory X) needs and "higher-order" (Theory Y) needs. McGregor suggested that management could use either set of needs to motivate and encourage police personnel toward delivering quality police services.

Theory Z

Theory Z was developed by William Ouchi, who applied this perspective to Japanese industrialization.[67] At the time, employment in Japan was long term, organizational authority was flat, promotions took a long time and few employees moved up the chain, workers were multitalented and willing to learn new skills, and workers were judged on objective performance. (Some of these characteristics of the Japanese economy has since changed.) Theory Z promotes development of a flexible, holistic, and cost-efficient police department that emphasizes the following concepts:

- Police supervisors must organize for work based on a human resources perspective.
- Officer participation, quality circles, leaner leadership, and worker empowerment are key aspects of managing the organization.
- Team work is at the core of police policy.
- Recognition of the global economy (i.e., an economy that crosses national lines) and emergence of new technologies necessitate new approaches for organizing and managing work—namely, empowering a more productive officer.

Mature Employee Theory

Mature employee theory is a managerial model developed by Chris Argyris, who examined issues related to worker apathy and lack of personnel effort.[68] His work, which has application to police organizations, reveals that poor delivery of police services is not simply the result of officer laziness or corruption, but rather is a product of managerial practices that prevent officers from maturing. In part, Argyris implies, the hierarchical bureaucratic structure of police command is responsible for officer immaturity because it provides officers with a minimal amount of control over their work environment. Officers are encouraged to be passive, dependent, and subordinate and, therefore, they behave immaturely.

Argyris argues that police officers are expected to behave in ways that demonstrate their immaturity because police organizations are usually created to achieve goals or objectives that can best be met collectively as opposed to individually. In this sense, the individual is fitted to the job and the design of police work centers on four concepts of scientific management originally explained by Fredrick Taylor:[69]

- Task specialization
- Chain of command
- Unity of direction
- Span of control

Each of these concepts has been explained earlier in this book, with the exception of task specialization.

Task Specialization

In policing, the concept of task specialization can be demonstrated by task forces. **Task forces** are used as an alterative managerial strategy that can operate within an existing formal structure. They are designed to target a particular criminal activity in a particular geographical area.

Generally, a task force consists of officers of different ranks within the same department or the same agency. Interagency task forces recognize that officers at the bottom of the hierarchical police bureaucracy can aid officers at the top of managerial pyramid.

Multiagency task forces include officers from different departments whose objectives and authority are clearly defined. For example, in a multi-county drug task force, officers from state, county, and municipal departments engage in drug interdiction initiatives. Sometimes multiagency task forces include officers and specialists from federal enforcement agencies such as the DEA and from state agencies such as probation officers and state troopers.

It is estimated that there are approximately 1,100 multiagency task forces funded by the federal government operating in the United States. Also, many major cases are the result of a collaborative effort between federal, state, and local police organizations, reports the U.S. Attorney's Office in Pennsylvania.[70]

Disadvantages and Advantages of Multiagency Task Forces

One potential disadvantage of task specialization is its potential for resulting in the oversimplification of a job, making the job repetitive, routine, and, to some officers, unchallenging. More typically, task specialization in the form of task forces enhances the checks-and-balances system of power and authority in a democratic system, leaving less opportunity for error and corruption. There are five major advantages to using multiagency task forces.[71]

First, multiagency task forces eliminate officer duplication of services in surrounding jurisdictions. For example, the Dane County (Wisconsin) Narcotics and Gang Task Force is staffed by officers from 14 different agencies, who were previously duplicating one another's investigations.[72]

Second, multiagency task forces can deliver to smaller agencies services that they might not otherwise have. For example, when the huge Mount Airy Lodge Resort and Casino was being built in January 2007, it became clear to the residents of the small village hidden in the Pennsylvania mountains that they needed a police force. Their options consisted of forming or joining a municipal or regional police department. In the end, the township employed a task force consisting of members of surrounding police departments and state troopers to serve its constituents.

Third, multiagency task forces enable shared resource management. In 1998, the Portsmouth and Keene, New Hemisphere, Police Departments, along with the Education Development Center, Inc., of Newton, Massachusetts, were awarded a $300,000 Justice Department grant to combat Internet-based sexual exploitation of children. The grant was used for undercover operations, the education of police officers, and the development of prevention materials. The Regional Task Force on Internet Crimes Against Children now serves Maine, New Hampshire, and Vermont. Formation of this task force has permitted agencies in these three states to identify and obtain sources of technological and investigative expertise to enhance proactive and reactive investigations of Internet-based crimes against children across New England.

Fourth, multiagency task forces extend authority for officers in felony pursuits into other jurisdictions. For example, some Fayetteville, North Carolina, municipal officers hold a dual "shield" with DEA (federal) authority. As a result, they can legally pursue drug traders across state lines and conduct probable cause arrests in other states' jurisdictions.

Fifth, multiagency task forces enhance communication among police agencies and among officers. For example, in early 2007, Tywan Lamont Williams was captured by members of the U.S. Marshals' Southeast Regional Fugitive Task Force and the Cobb County Police Department. Williams was wanted in Durham, North Carolina, for a 1997 murder and by the Cobb County Sheriff's Department for a felony probation violation.

Bridging the Gap

It has been suggested that police organizations adapt a managerial strategy that fosters perspectives that relate to "working smarter" rather than "working harder."[73] That is, police management needs to institutionalize officers' experiences and learn from those experiences—in essence, become a **learning organization**. Police departments should develop the capacity to reflect on organizational needs, plan toward fulfilling those needs, and implement changes to meet their objectives. In other words, police management needs to make informed decisions about the future.

One performance model for achieving this goal is called **Bridging the Gap (BTG)**. This initiative, which was originally developed by the Royal Canadian Mounted Police (RCMP), relies on human performance technology (HPT) to find, assess, and remove performance barriers for police personnel.[74] Whereas most managerial tools address management issues, performance improvement models differ in that they use a systematic methodology to find the root obstacles to quality performance and to fix the problems related to a specific performance deficit (such as arrest rates). In essence, this approach works backward relative to other management models. The broader issue is to clear or neutralize barriers that disrupt effective police practices (e.g., too much paperwork—provide formatted software and computers), but first the obstacles have to be discovered.[75]

In concert with BTG ideals, an assessment asked U.S. patrol officers about obstacles in the performance of a probable cause arrest. In response, 658 officers from 21 different departments ranked-ordered the following obstacles:[76]

- Paperwork
- Training/skills (The officer wasn't sure how to perform the task required.)
- Warrants/citations (No warrant or citation was available.)
- Minor crime (It was an insignificant act, in the officer's estimation.)
- Civil liabilities
- Complaints
- Supervisor/judge/prosecutor (These parties were incompetent and disrespectful of officers' past performance, so why provide them with another opportunity to insult officers?)
- Lack of staff (Personnel were not available to provide aid in processing, transporting, or securing a suspect.)
- Laziness
- Failure to understand the policy or a change in the policy

Once such a list of barriers has been constructed, the process shifts from analysis to problem solving. Similar to many managerial initiatives, efforts to solve patrol officer problems can be traced back to Fredrick Taylor's classical school of management, but BTG's primary form of problem solving was orchestrated by Thomas F. Gilbert[77] and stewarded by the International Society for Performance Improvement (ISPI) for purposes of establishing professional standards. A similar process to BTG is used by managers at the U.S. Navy, the U.S. Coast Guard, Motorola, and General Motors, to name a few.

Resistance to New Policies

There is a tendency among police managers and officers alike to resist new police policy, including showing a positive response to managerial initiatives and directives.[78] Officers, regardless of their rank, are not anxious to make changes.[79] They have likely seen many strategies come and go during the course of their employment. Consequently, there is always a lot of cynicism among police personnel. In addition, because policy is rarely developed at the police officer or even police middle-manager level, looking favorably upon policy and the changes it dictates adds fuel to the "we" and "them" split. This split between officers and everybody else is pervasive throughout most police departments and between most officers and top police managers.[80] Some refer to this split as a "siege mentality" in which the public is also increasingly seen as "them."

One expert identifies six belief perspectives that are held by police first-line supervisors and officers and that often prevent police management from implementing new strategies (of all types):[81]

- Research is impractical and "ivory tower" in its orientation.
- Outside researchers—and especially academic researchers—should be regarded with mistrust.
- Experiences with previous research and evaluations wasn't always positive.
- Findings from other jurisdictions do not apply to them; that is, each jurisdiction is unique unto itself. ("What do they know, anyhow?")
- Critical thinking exercises among officers will undermined discipline. ("They aren't supposed to think. They are supposed to do.")
- "Thinking inhibits doing."

The leadership qualities of police commanders and supervisors can make a major difference in how officers relate to changes.[82] It is the style or leadership qualities of a police manager that determines how successfully change will be implemented.

Police Leadership

The effectiveness of police service models often seems to depend on the managerial skill levels of police supervisors—that is, on leadership.[83] Are police managers professionally prepared to meet organizational challenges and implement needed changes so as to better serve their constituents and to manage police personnel? Naturally, there is not—and probably never will be—one single best way to manage and supervise police officers, including the tasks of developing policies, performance standards, productivity expectations, human relations, and implementation schemes.[84] Recall that every agency is different because of its location, personnel, history, budget, regulations, and public expectations, just to name a few critical factors. Likewise, most managers have their unique style.

Police management as a distinct entity has developed less extensively as compared to other organizational management entities owing in part to the organizational variations from police department to police department and in part to variations in the diversity of the constituents those departments serve. Nevertheless, there are universal characteristics that can be utilized by police supervisors to professionalize their management techniques.[85]

Leadership characteristics, it could be argued, help bring about appropriate police strategies without compromising the integrity of quality police service.[86] It can also be argued that police managers should possess proficient leadership skills if they hope to deliver contemporary police service.[87] For instance, management's failure to reward officers for excellent police performance can lead to cynicism, burnout, and reduced levels of effort by senior officers.[88]

Leadership is the process of directing and influencing the actions of others to bring a police agency closer to its objectives. Leadership is the style of a manager.[89]

Leadership deals with people, whereas management deals with things. You manage things and lead people.[90] This notion is consistent with the position of those who argue that leadership skills are typical of efficient management regardless of the service or product produced by an organization.[91] Leadership qualities encompass the individual's professional skills, characteristics, and style of managing others.

Police leadership can be seen as a process that effectively accomplishes organizational objectives but depends on how a supervisor interacts with other domains of leadership outside the department and with his or her subordinates.[92]

Leadership Techniques

Today's police executive should possess an extraordinary range of capacities.[93] Leadership characteristics among police supervisors are seen as crucial because these personnel must work closely both with individuals who provide police service (i.e., their own subordinates) and with officers over whom they have little control (i.e., local, state, and federal personnel from other law enforcement agencies).[94] They must also work closely with community members, business and social agency executives, and civic officials who can vigorously influence police service, yet these commanders seldom have the authority to make the final decision that will determine the outcome in most of the enterprises those individuals represent.[95] In addition, police managers are held accountable to courts and disciplinary committees populated with individuals whom police service affects, yet commanders have little influence in directing those courts or committees to action. Nonetheless, it can be argued that should police managers lack specific leadership skills, tension among their subordinates can be affected by their shortcomings.

Specific police leadership styles, which are discussed next, have their critics. Some of these critics suggest that the authority of both top management and middle management is affected by a host of regulations and obstructions advanced by politicians, community leaders, and organizational leaders in both public and private sectors.[96]

Situational leadership involves an interplay between three factors: the amount of direction (task behavior) a leader provides, the relationship behavior a leader displays, and the readiness level that personnel exhibit toward a specific task.[97] That is, as personnel become task-ready, a supervisor reduces his or her task directives and increases (task) delegation among the participating personnel. Quality leaders will assess each task (and know the limitations of their subordinates) and be prepared to reduce supervision directives while increasing personnel participation and authority. As personnel accept "ownership" of a task, it is expected that the quality of their individual and group input will enhance the output—that is, the results of the task.

The leadership strategy known as transformational leadership takes situational leadership a step further. **Transformational leadership** is a type of employee-centered leadership that focuses on providing empowerment to personnel.[98] In a sense, where the situational leader stops, the transformational leader takes up the baton. That is, transformational leaders empower personnel (and community members in community relations programs) to provide their fullest possible contributions. To do so, however, police personnel require training, resources, and authority to fulfill specific tasks. Transformational leadership is at the core of problem-solving strategies and community policing initiatives.

Some police leaders object to the idea of delegating authority to police personnel, and some of those objections are justified. Those who object to delegation of authority often prefer the authoritarian leadership style, a sobriquet that implies a micro-manager or "my way or the highway" approach. By contrast, one goal of many professional police managers is to develop and enhance the contributions made by officers. These managers prefer to act as mentors and facilitators among officers and constituents alike.

Characteristics of Good Police Leaders

A number of leadership skills are required of modern police managers, many which are congruent with corporate and other private-sector perspectives.[99] The 14 characteristics identified in this section, when objectively engaged to motivate police personnel to optimal performance, move a department closer to its mission, reduce the fear of crime among constituents, and enhance community and personnel quality-of-life experiences.[100]

Ability to Manage Organizational Change

Police leaders must employ change as part of a continuum to fulfill modern police strategies.[101] For instance, they may try out many variations of community relations initiatives, until the initiative is working appropriately for personnel and community members alike. This necessitates having change provisions and evaluation methods be spelled out as part of a new initiative.

Planning and Organizing Competencies

To initiate change, police executives must exercise a basic skill of managers—namely, the ability to plan and organize the department's competencies (or resources), including construction of a chronological timeline for change to occur and a list of players involved in the change.[102] Unfortunately, failure to plan professionally (objectively) is commonplace among police executives.

Problem-Solving Competencies

Problem-solving abilities are vital to leadership styles. Problem-solving competencies encompass three essential elements: defining public safety problems more precisely, analyzing each problem thoroughly, and finding more effective solutions to each problem.[103] Understanding the community's problems from the perspective of the community members, much like knowing the problems faced by the officers under the manager's command, and resolving those problems will go a long way toward winning respect for the leader and compliance with the proposed change.

Evidence for the need for problem-solving competencies comes from studies of community relations initiatives that include the strategy known as SARA (scanning, analysis, response, and assessment). The SARA model, which addresses crime in particular areas of a community, helps police reduce both the actual crime rate and the fear of crime among citizens.[104]

Creative Ability

Commanders must develop a mindset that allows them to see creative opportunities and different ways to deliver police services.[105] Rigid thinking is a thing of the past. Looking to officers and community members as partners and soliciting their thoughts about projects might be one way to add to the creative mix.

Toughness

Leaders should be decisive when making difficult choices or changes that are necessary for the health of the department and the well-being of the community.[106] For example, when a program such as Drug Abuse Resistance Education (DARE) isn't performing as well as another program that reaches out to youngsters, the leader must be prepared to cancel it (after professional evaluation) and not back down in the face of criticism for doing so. Of course, if the program is cancelled in a professional manner, there should not be much opposition to this change (see the section "Creative Ability").

Subordinate Trust and Public Trust

Empowering community policing officers and community members will enable them to mutually solve community problems and move the community closer to the ideal of public safety.[107] Once empowerment is provided (preferably limited, and in writing), an effective leader will support the decisions made by those officers and committees.

Delegation of Responsibility

Trust-based initiatives cannot be accomplished within the environment of a paramilitary hierarchy of command. For that reason, decentralization and delegation of responsibility are options that can move the organization closer to its mission.[108] Once authority is provided to others, the leader's own motivational level (assuming the leader has professional communication skills) can motivate officers toward change in an expeditious manner. (See the section "Subordinate Trust and Public Trust.")

Ability to Make Informed Decisions

Making informed decisions about issues such as deployment, department use of force limits, and police disciplinary and promotional prerogatives requires input from many sources.[109] When both the community and officers have input into the decision-making process, it is more likely that compliance will become forthcoming among those individuals and the individuals they influence.

Willingness to Take Action

Actuation/implementation competencies include the idea that leaders need to take action as facilitators of the appropriate managerial model versus playing the traditional authoritarian role of law enforcer.[110] Facilitating takes more time in the beginning, but in the long run it saves grief and aggravation when changing initiatives and expecting compliance. (See the section "Ability to Manage Organizational Change.")

Communication Skills

Discovering what is being experienced by rank-and-file officers and by community members is one of the most important assets of top management. Change must be communicated to everyone to improve the quality of police service.[111] Professional leaders must communicate and listen continually. The more they talk, the less time is available for them to listen. (See the section "Ability to Make Informed Decisions.")

Sharing Command

Evolutionary methods of sharing command represent a way of doing police business that is both inevitable in a democratic society and required to accommodate the changing demographics and expectations of U.S. society.[112] Commanders are politicians in the sense that they must reach out and share command with individuals who are affected by that command.

Visionary Orientation

Supervisors should be able to develop a mental picture of where the department is going and describe that vision to others, thereby ensuring that everyone is headed in a similar direction.[113] Drawing pictures for others is consistent with the thought that "A picture is worth a thousand words."

Integrity

Leaders should exemplify the boundaries of moral, ethical, and legal behavior, showing an inner strength that demonstrates their integrity to both police personnel and constituents. A leader must hold the confidence of many different types of individuals, yet still understand the lure of unprofessional behavior, even among offenders.[114] To truly be effective, officers must operate collaboratively with internal and external stakeholders in a constantly changing environment.[115]

Commitment

Leaders must demonstrate a commitment in today's climate of uncertainty to those individuals whom they serve, including both constituents and police personnel.[116] Police must become facilitators who keep an eye on quality police service, with the goal of enhancing quality-of-life experiences.[117]

Research into Police Leadership

In one study, researchers examined the 14 leadership qualities by looking at 97 police supervisors employed at 28 different police agencies across the United States.[118] All but three of the participants supervised an average of 60 officers each, and those three were responsible for an average of 701 officers each. The findings were indecisive, but reveal that the leadership characteristics of the police managers surveyed were typical of traditional police supervisors of the past, as opposed to being characteristics required of managers in the future.

One implication of these results is that most police managers operate from an "antiquated control, arrest, and command hierarchy with top-down dictates about deployment, tactical and use of force limits, and constituent conduct."[119] These thoughts are consistent with earlier observations that police mid-level managers need to develop team-building skills to create coalitions, manage task forces, establish linkages between departments and other units of the organization, and build relationships with consumers of police services, argue George L. Kelling and William J. Bratton.[120] Assuming that these findings are consistent across departments, one way to aid police leaders at all levels might be to emphasize the "seven habits of highly effective police leaders" program.

Seven Habits of Highly Effective Police Leaders

A program offered at the Federal Law Enforcement Training Center (FLETC) is based on the best-selling book written by Stephen R. Covey, *The Seven Habits of Highly Effective People.*[121] It provides a unique opportunity for law enforcement professionals to develop and refine their supervisory and leadership skills. The FLETC Management Institute and other law enforcement professionals have partnered together to design, develop, and deliver the "Seven Habits" workshop to meet the specific needs of police managers from local and other jurisdictions.

During the FLETC program, students are taught how to apply the "seven habits of highly effective people" both in their personal lives and in their police careers:

1. **Be proactive.** Promotes courage to solve problems, accept responsibility, and improve accountability so as to achieve goals.
2. **Begin with the end in mind.** Provides a common purpose and direction for police personnel and unites the department and community under a shared vision and mission.
3. **Put first things first.** Emphasizes prioritization and individual action in accomplishing the most important things first.
4. **Think win-win.** Enhances conflict resolution and helps individuals seek mutually beneficial solutions, thereby increasing community momentum.
5. **Seek first to understand, then to be understood.** Fosters deeper understanding and clearer communication through listening skills, resulting in heightened trust and timely solutions of problems.
6. **Synergize.** Encourages greater buy-in from agency personnel and citizens and takes advantage of diverse perspectives and ideas to discover new options.
7. **Sharpen the saw.** Promotes balance and continuous improvement while safeguarding against burnout and ineffectiveness through social, emotional, physical, and intellectual renewal.

RIPPED from the Headlines

Baltimore Police Chief Resigns

Police Commissioner Leonard D. Hamm's resignation leaves the Baltimore Police Department without a leader at an especially challenging time—less than two months before the mayoral primary election, as the city struggles with a surge in homicides and shootings.

Mayor Sheila Dixon is expected to name Frederick H. Bealefeld III to temporarily take over the 3,000-officer department. Bealefeld, a 26-year veteran, was tapped to be deputy commissioner of operations. As the No. 2 man in the department, Bealefeld is frank, at times emotional, often expressing outrage at the seemingly senseless crime in the city.

"This is exactly what's happening in the city, all over the place," said Bealefeld in June when a 4-year-old boy was hit by a stray bullet. "Just mindless, crazy shooting incidents that are occurring right now."

The interim chief will take over at a critical time. The city has had 177 homicides this year, putting Baltimore on pace to exceed 300 homicides for the first time since 1999. Nonfatal shootings are up more than 30 percent. Police morale is low, and the Fraternal Order of Police is set to begin contract negotiations soon.

Furthermore, the department is still adjusting to a drastic shift in policing strategies that took place when Dixon was elevated to mayor in January after Martin O'Malley became governor.

Dixon moved away from O'Malley's zero-tolerance policy, which had led to mass arrests for sometimes minor crimes, and moved to more foot patrols, community outreach, and targeted enforcement of the city's most violent offenders.

Bealefeld comes from a long line of police officers. His great-grandfather, grandfather, and great-uncle all worked for the department. And his younger brother, Charles, is a homicide detective.

"This is our eighth commissioner in the last eight years," said Blair, referring to both permanent and interim commissioners. "I want to see who is in charge. You put someone in there who really wants to work with the rank and file and is concerned about morale and getting back on the right track, that could be a great move. If it's someone who wants to go with the status quo until after the election, we probably won't see any change."

Source: Sumathi Reddy and Gus G. Sentementes, "Baltimore Police Chief Resigns," *The Baltimore Sun* (July 19, 2007), http://www.officer.com/web/online/Top-News-Stories/Baltimore-Police-Chief-Resigns/1$36962 (accessed May 14, 2008).

Summary

Management perspectives and leadership styles of police managers are key to the effectiveness of any police organizations. Police managers include sworn police personnel and those with administrative authority, from the rank of sergeant to captain, chief, sheriff, or constable (in Texas), as well as civilians such as police commissioners, attorney generals, state prosecutors, and state and local public safety directors. Because sergeants assign officers to tasks (among other things), they tend to be among the most powerful police managers.

To demonstrate their effectiveness, police organizations often seek local and national accreditation. This process has some drawbacks, however, including loss of control by policymakers over segments of police procedure.

Authority is the legal right to accomplish goals through others by directing their behavior. In contrast, power is the ability to accomplish goals through others, with or without a legal right to do so, to the extent of their compliance. Police management is defined as the process of planning, organizing, and controlling personnel and resources so as to prevent crime and satisfy various regulatory mandates.

Several managerial models have been developed that are applicable to police organizations, but it is unlikely that any "pure model" is in operation in the real world because of factors such as circumstances, budget, personnel, history, and skill levels and agenda of the department. For instance, management by objectives is defined as a process of consensus in which management and officers agree to work toward specific goals. Total quality management is a strategy aimed at incorporating awareness of quality into all organizational processes and is goal driven. This strategy often relies on the use of quality circles or groups of personnel who routinely come together to discuss policy.

Theory X and Theory Y are theories of human motivation that describe two very different attitudes toward the motivation levels of personnel. Theory X assumes that officers work for money and security and, therefore, should not be trusted. Theory Y assumes that given the right conditions, most personnel want to do well at performing their assignments and that there is a pool of untapped creativity in the workforce. In contrast, Theory Z suggests that organizational authority should be flat, promotions should take a long time, workers are multitalented and willing to learn new skills, and officers should be judged by objective performance. These theories and many managerial strategies rely upon Maslow's hierarchy of needs as a basis for managing personnel.

Mature employee theory is a managerial model that examines the issue of worker apathy and lack of personnel effort. It suggests that poor officer performance is a product of police managerial practices that prevent officers from maturing. Central to police **performance strategies** is the idea of "working smarter, not harder." Management needs to institutionalize its experiences and learn from those experiences, thereby becoming a learning organization.

Bridging the Gap is a performance strategy associated with human performance technology; it seeks to find, assess, and remove performance barriers for police personnel.

No matter which managerial strategy is employed, there is a tendency among police personnel to resist any new police policies. Some officers become cynical over time and are not anxious to make changes. To counteract this problem, police managers must develop leadership styles that help personnel perform at expected levels of output. Leadership is the process of directing and influencing the actions of others to bring a police agency closer to its objectives.

Characteristics of good managers include a willingness to mentor others and facilitate their development. Good managers also try to motivate police personnel to achieve optimal performance. Other preferred leadership characteristics include planning and organizing competencies, problem-solving competencies, and commitment. The "seven habits of highly effective people" are personal measures that can aid police managers in developing quality leadership styles.

Key Words

Accreditation
Authoritarian management
Authority
Bridging the Gap (BTG)
Leadership
Learning organization
Management
Management by objectives (MBO)
Abraham Maslow's hierarchy of needs
Mature employee theory
Performance strategy
Power
Quality circles
Situational leadership
Task forces
Theory X
Theory Y
Theory Z
Total quality management (TQM)
Transformational leadership

Discussion Questions

1. Describe organizational accreditation, including its advantages and its disadvantages.
2. Identify police officer ranks and describe the range of each rank's authority.
3. Describe the characteristics and styles of sergeants. Which supervisory style might encourage the use of force among patrol officers, and why?
4. Describe the characteristics and management styles of middle police managers.
5. Describe the characteristics and expectations of top police managers.
6. Characterize authority, delegation of authority, and power among managers.
7. Find an article about patrol officer criticism about supervisors. In what way do you agree or disagree with the criticisms provided by patrol officers in your article?
8. Characterize the primary components of police management and describe in what way the typical police manager might strengthen and weaken a typical patrol officer's delivery of police services.
9. Articulate the primary components and objectives of management by objectives (MBO) and total quality management (TQM).
10. Compare and contrast Theory X and Theory Y.
11. Describe the primary components of Theory Z.
12. Characterize the relevance of Abraham Maslow's hierarchy of needs as it relates to police management.
13. Characterize the primary components, process, and objectives of performance strategies for police managers.
14. Describe how police resistance to new policies affects the outcome of those policies, the police organization, and the community.
15. Characterize situational and transformational leadership styles, and explain in what way they differ from leadership styles focusing on facilitating and mentoring personnel.
16. Define leadership, and characterize the various leadership techniques employed by experienced managers.
17. Describe the seven habits of highly effective police leaders.

Notes

1. "Introduction to Performance Improvement" (2007), http://www.prime2.org/sst/intro.html (accessed January 14, 2007).
2. William A. Geller, "Suppose We Were Really Serious about Police Departments Becoming Learning Organizations?", *National Institute of Justice Journal* 234 (1997): 2–8.
3. Douglas McGregor, "Human Relations Contributors" (no date), http://www.accel-team.com/human_relations/hrels_03_mcgregor.html (accessed July 11, 2007).
4. Thomas O'Connor, "Police Professionalism and Accreditation" (January 4, 2004), http://faculty.ncwc.edu/TOCONNOR/417/417lect08.htm (accessed July 8, 2007).
5. O'Connor, "Police Professionalism and Accreditation."
6. Wayne W. Bennett and Karen M. Hess, *Management and Supervision in Law Enforcement*, 5th ed. (Belmont, CA: Wadsworth Thomson, 2007), 40.
7. Peter Drucker, *The Effective Executives* (New York: HarperCollins, 1993).
8. Hans Toch, *Stress in Policing* (Washington, DC: American Psychological Association, 2001), 46.
9. Boston police officer, personal communication with the author about her job (spring 2005).
10. Jim Weiss and Mickey Davis, "Empowerment or Finger Pointing," *Law and Order* (October 2004): 70–73.
11. Sergeant Chad Gann, Arlington Texas Police Department, a statement written especially for this chapter and edited by the author (2007).
12. Baltimore Police Department, *Identifying Characteristics of Exemplary Baltimore Police Department First Line Supervisors, Final Technical Report* (NIJ 189732) (August 20, 2001), http://www.ncjrs.gov/pdffiles1/nij/grants/189732.pdf (accessed July 9, 2007).
13. Bennett and Hess, *Management and Supervision in Law Enforcement*, 35.
14. Robin Shepard Engel, *How Police Supervisor Styles Influence Patrol Officers* (NIJ 194078) (June 2003), http://www.ncjrs.gov/pdffiles1/nij/194078.pdf (accessed July 9, 2007).
15. Engel, *How Police Supervisor Styles Influence Patrol Officers.*
16. Engel, *How Police Supervisor Styles Influence Patrol Officers.*
17. Rainer Schulte, "Which Challenges Will Police Managers Have to Meet in the Future: Policing in Central and Eastern Europe" (1996), http://www.ncjrs.gov/policing/which9.htm (accessed July 7, 2007).
18. Schulte, "Which Challenges Will Police Managers Have to Meet in the Future."
19. Town of Jupiter, Florida (2006), http://www.jupiter.fl.us/HumanResources/upload/Police%20Chief.pdf (accessed July 8, 2007).
20. Personal communication between the writer and the chief of police of a typical-sized Midwestern city (July 8, 2007).
21. Francis D'Ambra, "Community Partnerships in Affluent Communities," in *Policing and Community Partnerships*, ed. Dennis J. Stevens (Upper Saddle River, NJ: Prentice Hall, 2002), 17–27.
22. Karen Testa, "Women Rise to Top of Police Ranks in Several Major U.S. Cities," *Boston Globe* (May 24, 2004), http://www.policeone.com/policeone/frontend/parser.cfm?object=NewDivisions&rel=46210&operation=full_article&id=87594 (accessed July 4, 2004).
23. Bureau of Justice Statistics, "Local Police Departments, 2003" (Washington, DC: U.S. Department of Justice, 2007), http://www.ojp.usdoj.gov/bjs/pub/pdf/lpd03.pdf (accessed March 20, 2008).
24. National Center for Women and Policing, "Equality Denied. Status of Women in Policing Reports" (April 2002), http://www.womenandpolicing.org/publications.asp (accessed March 20, 2008).

25. National Center for Women and Policing, "Equality Denied. Status of Women in Policing Reports" (April 2002), http://www.womenandpolicing.org/aboutus.asp (accessed March 21, 2008).
26. Jim Kouri, Vice President of the National Association of Chiefs of Police, http://mensnewsdaily.com/blog/kouri/2005/04/minuteman-project-in-eyes-of-us-border.html (accessed June 12, 2005).
27. Dictionary.com, http://dictionary.reference.com/search?q=power (accessed May 15, 2005).
28. Bennett and Hess, *Management and Supervision in Law Enforcement*, 11–14.
29. Stephen R. Covey, *Living the 7 Habits*, 7th ed. (New York: Simon and Schuster, 2007).
30. Peter F. Drucker, *Management: Tasks, Responsibilities, Practices* (NY: HarperCollins, 1993).
31. Toch, *Stress in Policing*. Also see Dennis J. Stevens, *Police Officer Stress: Sources and Solutions* (Upper Saddle River, NJ: Prentice Hall, 2008).
32. Toch, *Stress in Policing*, 48.
33. Dennis J. Stevens, "Police Training and Management Impact Sexual Assault Conviction Rates in Boston," *Police Journal* 79, no. 2 (2007): 125–151.
34. Neal Trautman, *Corruption to Increase Dramatically* (National Institute of Ethics, 2004), http://www.ethicstrainers.com/pub.htm (accessed March 20, 2008).
35. David H. Bayley and Egon Bittner, "Learning the Skills of Policing," in *Critical Issues in Policing*, ed. Roger G. Dunham and Geoffrey P. Alpert (Prospect Heights, IL: Waveland Press, 2001), 82–106; Richard E. Kelly, "Psychological Care of the Police Wounded," in *Treating Police Stress*, ed. John M. Madonna, Jr. and Richard E. Kelly (Springfield, IL: Charles C. Thomas, 2002), 15–32.
36. Katherine W. Ellison, *Stress and the Police Officer* (Springfield, IL: Thomas, 2004).
37. International Association of Chiefs of Police, *Law Enforcement Code of Ethics*, revised (Institute for Criminal Justice Ethics, 1991), www.lib.jjay.cuny.edu/cje/html/lece-r.html (accessed July 20, 2003)
38. Toch, *Stress in Policing*, 89.
39. Toch, *Stress in Policing*, 89.
40. John P. Crank and Michael Caldero, "The Production of Occupational Stress Among Police Officers: A Survey of Eight Municipal Police Organizations in Illinois," *Journal of Criminal Justice* 19 (1991): 19–24.
41. Weiss and Davis, "Empowerment or Finger Pointing."
42. Crank and Caldero, "The Production of Occupational Stress Among Police Officers."
43. Stevens, *Police Officer Stress*, 97.
44. John Van Maanen, "Making Rank: Becoming an American Police Sergeant," in *Critical Issues in Policing: Contemporary Readings*, 4th ed., ed. Roger Dunham and Geoffrey P. Alpert (Prospect Heights, IL: Waveland, 2001), 132–148.
45. Van Maanen, "Making Rank," 136.
46. Wesley G. Skogan, Susan M. Hartnett, Jill DuBois, Jennifer T. Comey, Marianne Kaiser, and Justice H. Lovig, *On the Beat* (Boulder, CO: Westview Press, 1999).
47. Boston police officer, personal communication with the author.
48. Drucker, *Management: Tasks, Responsibilities, Practices.*
49. Bennett and Hess, *Management and Supervision in Law Enforcement*, 11–14.
50. Rickey D. Lashley, *The Need for a Noble Character* (Westport CT: Praeger, 1995).
51. Joycelyn M. Pollock, "Ethics and Law Enforcement," in *Critical Issues in Policing: Contemporary Readings*, ed. Roger G. Dunham and Geoffrey P. Alpert (Prospect Heights, IL: Waveland, 2001), 441.
52. Anonymous, "Achieving and Maintaining High Ethical Standards: IACP's Four Universal Ethics Documents," *Police Chief* (October 2002): 64, 66.
53. Pollock, "Ethics and Law Enforcement," 357.
54. Ikimulisa Livingston, Leonard Greene, and Andy Geller, "Point Blank Perspective," *New York Post* (March 3, 2007), http://www.policeone.com/investigations/articles/1230405/ (accessed March 23, 2007).

55. A. P. Cardarelli, J. McDevitt, and K. Baum, "The Rhetoric and Reality of Community Policing in Small and Medium Sized Cities and Towns," *Policing: An International Journal of Police Strategies and Management* 21, no. 3 (1998): 397–415.

56. Police commander in Newark, New Jersey, personal communication with the author (December 12, 2006).

57. Nancy E. Marion and Willard M. Oliver, *The Public Policy of Crime and Criminal Justice* (Upper Saddle River, NJ: Prentice Hall, 2006), p. 252.

58. Peter Drucker, *The Practice of Management* (New York: HarperCollins, 1954).

58a. Peter F. Drucker, *People and Performance: The Best of Peter Drucker on Management* (Cambridge, MA: Harvard Business School Press, 2007), p. 39–40.

59. James Craze, Greenbelt, MD, police department, http://www.greenbeltmd.gov/police (accessed May 13, 2008).

60. E. Abrahamson, "Management Fashion," *Academy of Management Review* 21, no. 1 (1996): 254–285.

61. Leonard J. Ponzi and Michael Koenig, "Knowledge Management: Another Management Fad?", *Information Research* 8, no. 1 (2002), http://informationr.net/ir/8-1/paper145.html (accessed April 10, 2007).

62. Dennis J. Stevens, *Applied Community Policing in the 21st Century* (Upper Saddle River, NJ: Prentice Hall, 2003), 114–116.

63. Douglas McGregor, *The Human Side of Enterprise* (Boston: McGraw Hill, 1960).

64. Guided by Business Knowledge Center, "Theory X and Theory Y" (no date), http://www.netmba.com/mgmt/ob/motivation/mcgregor/ (accessed December 19, 2006).

65. A. H. Maslow, "A Theory of Human Motivation", *Psychological Review* 50 (1943): 370–396.

66. Crank and Caldero, "The Production of Occupational Stress Among Police Officers."

67. William Ouchi, *Theory Z: How American Business Can Keep Up with the Japanese Challenge* (Reading, MA: Addison Wesley, 1981).

68. Chris Argyris, "Learning and Teaching: A Theory of Action Perspective," *Journal of Management Education* 21, no. 1 (1997): 9–27. Also see Chris Argyris, *Organizational Learning II: Theory, Method, and Practice* (Reading, MA: Addison-Wesley Longman, 1996).

69. In 1911, Frederick Winslow Taylor published *The Principles of Scientific Management* in which he described how the application of the scientific method to the management of workers greatly could improve productivity (see the Taylor website at http://www.netmba.com/mgmt/scientific/). Scientific management methods called for optimizing the way in which tasks were performed and simplifying the jobs enough so that workers could be trained to perform their specialized sequence of motions in the one "best" way. Prior to scientific management, work was performed by skilled craftsmen who had learned their jobs by completing lengthy apprenticeships. These craftsmen made their own decisions about how their jobs were to be performed. Scientific management took away much of this autonomy and converted skilled crafts into a series of simplified jobs that could be performed by unskilled workers, who easily could be trained for the tasks.

70. U.S. Attorney's Office, Western District of Pennsylvania, http://www.usdoj.gov/usao/paw/task_forces.html (accessed January 12, 2007).

71. Peter W. Phillips, "Special Units in Policing: An Overview," in *Policing and Special Units*, ed. Peter W. Phillips (Upper Saddle River, NJ: Prentice Hall, 2005), 2–32.

72. U.S. Attorney's Office, Western District of Pennsylvania, http://www.usdoj.gov/usao/paw/task_forces.html (accessed January 12, 2007).

73. William A. Geller, "Suppose We Were Really Serious about Police Departments Becoming Learning Organizations?", *National Institute of Justice Journal* 234 (1997): 2–8. Also see William A. Geller, *Police Leadership in America: Crisis and Opportunity* (Washington, DC: American Bar Association, 1985).

74. William Pullen, Yvon De Champlain, and Graham Muir, "Improving Police Performance with Human Performance Technology (HPT): Watch One, Do One, Teach One," *Journal of Police* 79, no. 2 (2006): 152–168.

75. Pullen, De Champlain, and Muir, "Improving Police Performance with Human Performance Technology."

76. Dennis J. Stevens, "Civil Liabilities and Arrest Decisions," in *Policing and the Law*, ed. Jeffrey T. Walker (Upper Saddle River, NJ: Prentice Hall, 2002), 53–70.

77. Thomas F. Gilbert, *Human Competence: Engineering Work Performance* (New York: Pfeiffer, 1996), 121–137.

78. For a closer look at the causes of officer resistance, see T. J. Dicker, "Tension on the Thin Blue Line: Police Officer Resistance to Community Oriented Policing," *American Journal of Criminal Justice* 23, no. 1 (1998): 59–82. Dicker implies that supervisor trust, satisfaction with the amount of control over one's work environment, rank, organizational trust, and level of pride in the department play the largest role in determining police officer resistance. Also see Robert Trojanowicz, Victor E. Kappeler, Larry K. Gaines, and Bonnie Bucqueroux, *Community Policing: A Contemporary Perspective*, 2nd ed. (Cincinnati: Anderson, 1998), 273.

79. John P. Crank, *Understanding Police Culture* (Cincinnati: Anderson, 2004), 3–4. Crank indicates that organizational traditions are customary ways of doing things, and they take on common-sense value that cannot be changed easily or frivolously.

80. Crank, *Understanding Police Culture*, 119.

81. Geller, "Suppose We Were Really Serious."

82. Daniel C. Ganster, Milan Pagon, and Michelle Duffy, "Organizational and Interpersonal Sources of Stress in the Slovenian Police Force," *College of Police and Security Studies*, National Criminal Justice Reference Service (1996), http://www.ncjrs.org/policing/org425.htm (accessed May 11, 2005). Also see Van Maanen, "Making Rank."

83. For a detailed review of police management changes, see George L. Kelling and Catherine M. Coles, *Fixing Broken Windows* (New York: Free Press, 1996), 70–74, 109–114. Also see Roy R. Roberg, Jack Kuykendall, and Kenneth Novak, *Police Management*, 3rd ed. (Los Angeles: Roxbury, 2002), 3–5, 384–91. For a look at occupational change among police, see Van Maanen, "Making Rank."

84. Dennis E. Nowicki, "Mixed Messages," in *Community Policing*, ed. Geoffrey Alpert and Alex Piquero (Prospect Heights, IL: Waveland Press, 1998), 265–274.

85. Bennett and Hess, *Management and Supervision in Law Enforcement*; David L. Carter and Louis A. Radelet, *The Police and the Community*, 7th ed. (Upper Saddle River, NJ: Prentice Hall, 2001), 51; James G. Houston, *Correctional Management: Functions, Skills, and Systems*, 2nd ed. (Chicago: Nelson Hall, 1999), 158; Carl Klockars, *The Idea of Police* (Beverly Hills, CA: Sage, 1999); Peter K. Manning, *Police Work: The Social Organization of Policing* (Prospect Heights, IL: Waveland Press, 1997); Nowicki, "Mixed Messages."

86. S. M. Cox, "Policing into the 21st Century," *Police Studies* 13, no. 4 (1990): 168–177.

87. Dennis J. Stevens, "Community Policing and Managerial Techniques: Total Quality Management Techniques," *Police Journal* 74, no. 1 (2001): 26–41.

88. Samuel Walker and Charles M. Katz, *The Police in America: An Introduction*, 6th ed. (New York: McGraw Hill, 2008), 117.

89. Bennett and Hess, *Management and Supervision in Law Enforcement*, 38–40; Roberg, Kuykendall, and Novak, *Police Management*; Drucker, *Management: Tasks, Responsibilities, Practices.*

90. Stephen R. Covey, *Principle Centered Leadership* (New York: Fireside, 1992), 34.
91. Norman M. Scarborough and Thomas W. Zimmerer, *Effective Small Business Management* (Upper Saddle River, NJ: Prentice Hall, 2006), 740–782; Cox, "Policing into the 21st Century"; Roberg, Kuykendall, and Novak, *Police Management*, 143.
92. Stan Stojkovic, David Kalinich, and John Klofas, *Criminal Justice Organizations: Administration and Management*, 4th ed. (Belmont, CA: Wadsworth, 2007). H. L. Tosi, J. R. Rizzo, and S. J. Carroll, *Managing Organizational Behaviour* (Marshfield, MA: Pitman, 1986).
93. International Association of Chiefs of Police, "Recommendations from the President's First Leadership Conference: Police Leadership in the 21st Century: Achieving and Sustaining Executive Success" (May 1999), http://www.theiacp.org/documents/pdfs/Publications/policeleadership%2Epdf (accessed May 13, 2004).
94. Robert Trojanowicz and S. Dixon, *Criminal Justice and the Community* (Englewood Cliffs, NJ: Prentice Hall, 1974). Also see Trojanowicz, Kappeler, Gaines, and Bucqueroux, *Community Policing*, 270–273.
95. James Q. Wilson and George L. Kelling, "Making Neighborhoods Safe," in *Community Policing: Contemporary Readings*, ed. Geoffrey P. Alpert and Alex Piquero (Prospect Heights, IL: Waveland, 1998), 35–44.
96. John Wiggins and Dennis J. Stevens, "The Effects of Police Management," *The Journal: The Police Official Journal* 6, no. 2 (2000): 14–17.
97. Bennett and Hess, *Management and Supervision in Law Enforcement*, 48–50.
98. Bennett and Hess, *Management and Supervision in Law Enforcement*, 50.
99. The magazine *Management Review* conducted a study to determine the attributes that leaders would need for the year 2000 and beyond. See also Dennis J. Stevens, "Community Policing and Police Leadership," in *Policing and Community Partnerships*, ed. Dennis J. Stevens (Upper Saddle River, NJ: Prentice Hall, 2002), 163–177. Stevens added three more attributes: organizational change, sharing command, and police decisions. The purpose of these additions is in keeping with community policing strategies. Also see Scarborough and Zimmerer, *Effective Small Business Management.*
100. Dennis J. Stevens, "Police Stress, Before and After 9/11," *Police Journal* 77, no. 2 (2004): 145–72. Also see Dennis J. Stevens, "Police Officer Stress and Occupational Stressors," in *Policing and Strategies*, ed. Heith Copes (Upper Saddle River, NJ: Prentice Hall, 2005), 1–24.
101. Kelling and Coles, *Fixing Broken Windows*, 70–71.
102. This particular trait was not included in the original list provided by Dennis J. Stevens in "Police Stress, Before and After 9/11." After additional research, this item was added, influenced by the following work: Rick Michelson, "Leadership Issues: Managing Change" (2006), http://www.hitechcj.com/chiefslist/id14.html (accessed March 15, 2006).
103. Lawrence Sherman and Heather Stang, "Policing Domestic Violence: The Problem-Solving Paradigm" (1996), http://www.aic.gov.au/rjustice/rise/sherman-strang.html (accessed April 12, 2006).
104. Loreen Wolfer, Thomas E. Baker, and Ralph Zezza, "Problem-Solving Policing Eliminating Hot Spots," *FBI Law Enforcement Bulletin* (1999), http://www.findarticles.com/p/articles/mi_m2194/is_11_68/ai_58177902 (accessed April 12, 2006).
105. Nowicki, "Mixed Messages."
106. Bureau of Justice Assistance, *Understanding Community Policing: A Framework of Action* (NCJ 148457) (Washington, DC: U.S. Department of Justice, Office of Justice Programs, August 1994). Also see Police Executive Research Forum, *Themes and Variations in Community Policing: Case Studies in Community Policing* (Washington, DC: Author, 1996).
107. Stevens, *Applied Community Policing in the 21st Century*, 270.
108. Carter and Radelet, *The Police and the Community*, 51.

109. Cardarelli, McDevitt, and Baum, "The Rhetoric and Reality of Community Policing."
110. Jill DuBois and Susan M. Hartnett, "Making the Community Side of Community Policing Work: What Needs to Be Done," in *Policing and Community Partnerships*, ed. Dennis J. Stevens (Upper Saddle River, NJ: Prentice Hall, 2002), 1–18.
111. Trojanowicz, Kappeler, Gaines, and Bucqueroux, *Community Policing*, 270–73.
112. Robert Trojanowicz and David L. Carter, *The Philosophy and Role of Community Policing* (East Lansing, MI: National Neighborhood Foot Patrol Center, Michigan State University, 1988).
113. Carter and Radelet, *The Police and the Community*, 52.
114. Bennett and Hess, *Management and Supervision in Law Enforcement*, 280–288.
115. Michelson, "Leadership Issues."

116. M. Moore, "The Pursuit of Integrity," *Law Enforcement Journal* (Winter 1998): 36–96.

117. Stevens, *Applied Community Policing in the 21st Century*, 280–281.

118. Stevens, "Community Policing and Police Leadership."

119. Dennis J. Stevens, "Civil Liabilities and Arrest Decisions," in *Policing and the Law*, ed. Jeffery T. Walker (Upper Saddle River, NJ: Prentice Hall, 2002), 71.

120. George L. Kelling and William J. Bratton, "Implementing Community Policing: The Administrative Problem" (1993), http://www.ksg.harvard.edu/criminaljustice/publications/pop17.pdf (accessed November 21, 2007).

121. Stephen Covery, *The 7 Habits of Highly Effective People: 15th Anniversary Edition*, 7th ed. (NY: Simon and Schuster, 2004); Stephen Covery, *The 8th Habit: From Effectiveness to Greatness* (NY: Simon and Schuster 2004).

ACCIDENT

Hiring and Training of Police Officers

As for the future, your task is not to foresee it, but to enable it.

—Antoine de Saint-Exupéry

LEARNING OBJECTIVES

When you finish reading this chapter, you will be better prepared to:

- Describe the responsibilities of the police in the twenty-first century
- Identify reasons why people would never apply for a police job
- Identify the general requirements to become a police officer
- Describe the process of applying for an officer's job
- Define education and training, and explain how learning is accomplished
- Characterize the objectives of training
- Describe what is meant by failure to train and deliberate indifference
- Identify the reasons why officers leave policing
- Summarize the federal qualifications for employment

CASE STUDY: YOU ARE THE POLICE OFFICER

You're one of many U.S. Border Patrol officers deployed at the United States–Mexico border, which stretches almost 2,000 miles through San Diego, California, to Brownsville, Texas. In between those cities, the terrain consists of urban cities, yet sand, rivers, and deserts are everywhere.

Today is your first day on the job. It wasn't easy getting here. You weren't sure what you wanted to do with your criminal justice degree, but you disliked the business concerns promoted by your dad, and you weren't sure about other disciplines such as counseling encouraged by your sister; there was just something exciting about your criminal justice professors, who were likeable, approachable, and experienced. The time in class when your professor talked about federal law enforcement opportunities, especially with the Department of Homeland Security, you couldn't sleep at night thinking about the possibilities.

You learned from your "Introduction to Police" college course two years ago that the requirements for federal law enforcement employment were especially high. For example, high academic standards, physical endurance, and recommendations from your professors were needed to work for the U.S. Customs and Border Patrol (CBP) as a Border Patrol officer. It was an electrifying thought that the CBP's priority mission is anti-terrorism and that you would be trained to detect and prevent terrorists and terrorist weapons from entering your country. You liked the CBP's perspective of "one face at the border" and wanted to contribute to quality-of-life standards of others. Initially, you thought about becoming a police officer, but you wanted more action. You thought about joining the military, but had no desire to be deployed in a foreign country.

Eventually, you went online to cbp.gov to see when the Border Patrol examination would be given in your community, and you completed a preliminary online application. Days turned into weeks, and finally—two months before graduation—you had a couple of interviews with agents, plus drug and mental health tests. You didn't know it at the time, but there was an investigation into your life that included interviews with your references, including your favorite professor, which led to your placement into a preliminary training group in Maryland after college graduation.

During training, your knowledge, skills, and abilities (KSAs) were assessed as they applied to your performance as a Border Patrol officer. The evaluators said that your KSAs are a product of your qualifying experiences, education, and training.

You also spent four months of intense training at the Federal Law Enforcement Training Center at Glynco, Georgia, where you learned about everything from high-speed chase tactics, as instructed by experienced DEA agents, to gaining a better command of Spanish. More tactical training at a military base followed, and finally all roads took you to an assignment at the Texas border. You feel prepared, happy you made the decision to follow your dreams, and ready to make a difference.

1. Explain the importance of policing the U.S.–Mexican border.
2. In what way are federal Border Patrol officers civil law enforcement officers?
3. Explain the relevance of KSAs to the training process of federal enforcement officers.

Introduction

Hiring and training of local police officers and federal law enforcement agents is the focus of Chapter 8. The emphasis here is on hiring assessments because ineffective officers are not necessarily a product of the job; instead, potentially poor officers and can be diverted at the candidate or training stage through professional assessment and training initiatives. Recruiting, training, and retaining quality personnel are some of the most important responsibilities faced by police managers in the twenty-first century.[1]

Predicting the future behavior of others is a task, yet enabling the future (as Antoine de Saint-Exupéry suggests) is more dynamic. That is especially true in policing, because police must deliver services and emergency care to the law-abiding public and suspects alike, which suggests the difficulty and complexity of the issues covered in this chapter.

■ Choosing a Career in Policing

Traditionally, police jobs have been occupied by white males from the working class. Most of those officers viewed policing as an upwardly mobile profession, allowing them to achieve a higher standard of living than was available to their parents. The challenge now is to attract educated middle-class applicants, persons of color, females, and members of diverse ethnic groups. This emphasis relates to the changes in the American society described in Chapter 4, including the challenges posed by a society that demands—actually, more than demands, litigates—due process rights. Police agencies compete with one another and with employers in the private sector for the same quality recruits, and the employer with the greatest resources enjoys a distinct advantage.

Considerations for Job Applicants

Before anyone applies for a job as an officer, that person should do some homework. What are the promotional and developmental opportunities at this department? Is the department technologically advanced, or does it supply its officers with equipment, vehicles, and weapons from the Stone Age? A little web surfing can provide up-to-date information even to a person with little experience in policing (try policeone.com). Is there an open-door policy for promotion to management positions, or does the department adhere to the "good ole boy" system? In what way is its officers trained? What about tuition reimbursement and other opportunities to advance in police work?[2]

At some point in every department's rank structure, promotion becomes a purely political process, advises a veteran of the Baltimore Police Department. This lieutenant adds, "You want a department with a civil service promotion process which gets you as far up the rank structure as possible. Most departments should include the rank of lieutenant under civil service. If the department you're looking at doesn't, trash it, and go on to the next one."[3]

Another method of examining a department's promotional process is to check with the civil service commission in the jurisdiction. Most often, officers become eligible for promotion after a probationary period ranging from 6 months to 3 years (see Chapter 6). In a large department, promotion may enable an officer to become a detective or to specialize in one type of police work, such as working with juveniles (see Chapter 11). Promotions to corporal, sergeant, lieutenant, and captain usually are made according to a candidate's position on a promotion list, as determined by scores on a **written examination** and on-the-job performance (as discussed later in this chapter).

Employment of police and detectives is expected to increase about as fast as the average for all occupations in the United States through 2014.[4] Many people are attracted to police work because of its challenging nature or because of the sensationalized view of policing presented by the media. Competent persons with college training in or who have military experience, or both, should have the best chance of finding a job in policing. College degrees can produce higher incomes, greater status, and more opportunities for managerial advancement. On average, police agencies hire only 4 percent of those who apply.[5]

Factors Working Against Police Recruiters

Here are nine major reasons why some people would never apply for a police job.

Cultural Issues

Some African Americans and members of other groups hold the police in low esteem because of past experiences as victims of or witnesses to police brutality and would be less likely to apply for jobs as police officers.[6] Stories of police brutality are actually observable by black residents and others who experience the "white policing syndrome" that began at least a century ago and has been carried over to modern policing.[7] This syndrome involves routinely negative white (working-class) police perceptions of, and treatment of, black local community members.[8] In general, white officers display dominant attitudes toward Americans of color. Although this experience is changing rapidly, the persistence of racism within police ranks becomes clear as officers relate stories to others about racial barriers within the departments where they work.[9] The essential, invariant structure of discrimination as experienced by African Americans is identified as fear, frustration, depression, and anger.[10] Becoming a police officer may be perceived as "selling out" for African Americans.

Macho Macho-Man Image

There is a tendency for police officers to compare and compete for performing the "toughest job." [11] Perhaps female and minority applicants just don't feel like they want to "break and enter" to fit in.[12] Masculine officers talk about shootings, gun battles, and high-speed chases.[13] In police academies, stories on every aspect of combat are told by students and instructors.[14] These stories take center stage and are focal points for coffee-break discussions and station-house roll calls. These thoughts are consistent with those expressed by experts who report on the "Dirty Harry" problem and by advocates who argue that street justice is a defensible behavior linked to social order requirements.[15]

In reality, and in spite of the media's portrayal of macho officers who deliver their own style of street justice, most officers will reach retirement without killing a single person or, for that matter, firing a weapon at a suspect,[16] Those officers who do "fire their weapons will not reload, finish their shift, and have to have their memory jogged to remind them of the event." [17]

Public Scandals

The "police scandal syndrome," as one researcher referred to it, has "shocked public conscience, . . . produced public outcry, and . . . diminished public confidence in those to whom it looks for protection." [18] Young people may be excited about what they see on the television dramas, but when they read the newspapers or become victims themselves of police brutality or corruption, the idea of working for a corrupt department seems less attractive.

More Prestigious Job Offers

Talented individuals have fewer road blocks in finding jobs in the private sector that can provide superior rewards relative to jobs in policing. Even private police agencies such as Wackenhut Services (http://www.wsihq.com) offers better benefits and career opportunities than jobs in the public sector (see Chapter 4).

Rigid Paramilitary Hierarchy

The independent-minded, educated youth of today are less tolerant of the rigid paramilitary hierarchy to which most police agencies adhere. Many individuals prefer to work in an organizational structure that is characterized by flexibility and network-type communication because mobility is greater and the chances of moving up do not depend on seniority.[19] Young people used to "live to work," but the current generation "works to live": Many choose to work late or work on weekends so that they can make money and retire early.[20] The complexity of a rigid bureaucracy is not a welcome sight compared to the flexibility offered by a private-sector firm such as the Nordstrom Company, whose one-page policy manual instructs personnel, "Use your own best judgment at all times." On the other hand, Nordstrom is not in the business of protecting, arresting, and using force.[21]

Dislike for Meeting Others Head-On

It comes as no surprise that most people dislike public speaking. In fact, public speaking is a major fear of many individuals, but particularly of young people who haven't experienced a lot of public speaking.[22] It might look easy for those who confront others as a routine practice but it takes training, experience, and a positive attitude to become confident in such person-to-person interactions with strangers.

High School Dropouts

Among those who avoid pursing a police career are individuals who drop from high school. Some reports indicate that one third of all high school–aged young people in the United States leave school before graduating.[23] Dropout rates are related to a variety of individual and family demographic and socioeconomic characteristics that are at odds with policing.[24]

Earlier Loss of a Police or Military Parent

As police officer and military deaths mount, many children are growing up with one parent, wondering if someone is about to take their remaining parent.[25] Look for a police job? "No way," might be an answer provided by a student who lost a father to the profession.

Fear of Discovery

Some individuals focus on their own personal weakness and are fearful of discovery of those flaws.[26] Interviewing for a police officer's job places an individual around others who can discover any kind of weakness—officers. Weaknesses might include criminal thinking or lifestyles, or being an alcoholic, addict, child beater, or bully. Individuals who might be hard-headed and uncompromising might also fit into this category, as would individuals who cannot or will not follow employment instructions. A fear of being discovered or a fear of vulnerability can set limits for relationships, lifestyles, and occupations.[27]

Supply of Qualified Candidates

The supply of good police applicants has been down for some time.[28] Most agencies report recent vacancies because of a lack of qualified candidates, officers leaving, and retirements. During the 12-month period ending June 30, 2003, 21 percent of local police departments had full-time sworn personnel called-up as full-time military reservists, accounting for an estimated 7,500 officers.[29] Agencies that had difficulty filling open positions had roughly one unfilled vacancy for every three that were filled. In the South, especially in Mississippi, Louisiana, and Texas, the destruction from Hurricanes Katrina and Rita sent approximately 25 percent of all officers packing, never to return to the region. In the West, out-of-control fires, civil rights litigation, and immigration encounters have caused many officers to leave in search of a better life.[30] "We're not able to find as many qualified applicants as we've had in the past," lamented Montgomery County (Ohio) Sheriff David Vore, who was looking to fill 12 deputy and 8 correctional jobs. "I can't explain it. It just doesn't appear people want to come into law enforcement like they did."[31]

Recruiting

Competition for quality candidates among agencies is fierce. Recruiters for the New York Police Department (NYPD) hand out coffee mugs on college campuses. In Los Angeles, the Los Angeles Police Department (LAPD) offers baseball jerseys with the words "Join Our Team."[32] "Walk-ins accepted for immediate testing!" says an advertisement at the LAPD, which sends recruiters to Florida to troll for prospective officers among college students lying on the beach during spring break. Police recruiters in Oakland, California, use a mobile home to travel to out-of-state military bases and colleges to administer tests to would-be officers.

Jason Abend, the Executive Director of the National Law Enforcement Recruiters Association, says that some police recruiters are now placing ads in movie theaters, where they run on-screen before films start. Some jurisdictions tell their veteran officers to help them find candidates; if their friends are hired, the officers will receive a bonus of 40 hours of extra vacation time.[33] In a competitive move, other agencies announce: "Put a star in your future—now offering a signing bonus of up to $5,000."[34]

Recruitment: Five Case Studies

In the wake of rising crime rates of the late 1980s and early 1990s, Congress passed the Violent Crime Control and Law Enforcement Act of 1994 (see Chapter 4). This legislation included the Public Safety Partnership and Community Policing Act (establishing the "COPS" program; http://www.officers.usdoj.gov/), which provided funding to hire 100,000 additional local police officers.[35] Approximately 61,000 of these officers were funded through hiring grants made available by the COPS program to state and local police agencies.[36]

Sacramento (California) Police Department

The Sacramento Police Department (SPD) wanted to attract police candidates who had strong histories of giving back to their communities and who were interested in community volunteer activities. As part of this effort, SPD wanted to ensure greater diversity in its applicant pool, so it invited a focus group of citizens to guide the project. The input from this group was then used to develop a recruiting plan, forms to be used, and an agreed upon hiring process. Lessons learned from these strategies are below.

Burlington (Iowa) Police Department

To transform its recruitment process, the Burlington Police Department (BPD) focused on five goals:

- Create a model-officer psychometric profile
- Change the selection system by supplementing state-sponsored selection tools
- Engage in an outreach effort to increase diversity through the use of marketing (radio and television) tactics and community involvement
- Provide professional development for recruits and mentoring
- Provide English tutoring and online entry-level practice tests

Hillsborough County (Florida) Sheriff's Office

The Hillsborough County Sheriff's office (HSO) serves Tampa Bay, Florida. HSO wanted to implement a model recruitment strategy and selection system that would recruit and train the best qualified candidates available. To accomplish this aim, the department had to develop procedures, including testing, to determine which personal characteristics were essential in service-oriented deputies within Hillsborough County. Items consistently identified by successful deputies included the following characteristics:

- Willingness to admit shortcomings
- Communication skills
- Frustration tolerance
- Avoiding procrastination
- Strong work habits

Detroit (Michigan) Police Department

Some of the unique accomplishments of the Detroit Police Department (DPD) were related to the department's community outreach programs. Three objectives were key to Detroit's strategic approach:[37]

- Expansion of the recruiting ambassadors program
- Communications strategy—media summit
- Involvement of the community in Compstat

Recruiting Ambassadors Program

Intensifying an existing recruiting program by urging the community to act as recruiting ambassadors demonstrates how a department can focus on better using its existing internal capacity rather than developing totally new initiatives. The DPD charged the community and its own personnel with seeking out recruit candidates who embodied the spirit of service. The department illustrated the seriousness of its intent by providing incentives for police personnel and community representatives who could bring in recruits who completed the process successfully.

Communications Strategy: Media Summit

Devising a comprehensive communications strategy, which included convening a highly successful media summit to introduce the program, proved to be an effective way to get ahead of issues concerning recruitment and hiring in Detroit. The summit drew attention to how the DPD was changing its recruitment and hiring practices and provided an opportunity for the city to define the issues rather than allowing newspapers or magazines to do so. The mayor was the focal point of the summit, showing that the city government supported this initiative.

Involvement of the Community in Compstat

Bringing community representatives into the Compstat process signals that the community role extends far beyond recruitment and hiring. In Detroit, the community has a role in providing input that affects how policing will be conducted. Who knows the community better than the citizens, and who can give the commanders better advice about where crime is occurring and why?

King County (Washington) Sheriff's Office

Project goals for the King County Sheriff's Office (KSO) included recruitment strategies, marketing strategies, occupational screening, clinic psychologist testing, and involvement of the community. Outreach included community members in partnership with police hiring professionals.

Lessons Learned

The experiences of these five police departments share a common theme: Community outreach helped attract more candidates, simpler hiring forms, and an established (and known) hiring process. Some lessons learned were consistent with the results of the community relations movement, which explored many issues related to policing, including recruitment.[38] Police departments that succeed in initiating community participation in recruitment have an entirely different relationship to the public that they serve. Yet questions arise about addressing the good of many or the agenda of the few committee members who serve in the hiring processes.[39]

However, as police policymakers have learned from previous community policing studies, what community members see as a crime deterrent and social order initiate is not consistent with what the police see as priorities toward the same goals.[40]

For example, community members maintain that officers should make arrests of juveniles gathering on street corners as long as one of them wasn't a local youth or selective enforcement.

In particular, committee supported hiring committees see what they want to see about a candidate and their views are not necessarily focused upon a candidate's attributes nor are community members concerned with how well a candidate would fit into the training classes and department as a person and an officer. Although there is a wide range of opinions about what type of person is best suited to handle the rigors of a police job, three factors are considered vital for diminishing the prospects of violence between the police and the community:

- Departments should have a ratio of employees of color and national origin that reflects the diversity of the community served.
- Continued emphasis should be placed on bringing into policing people who studied a variety of college disciplines.
- Individuals should be psychologically suited to handle the requirements of the job.

Alas, when the realities of budget constraints enter the picture, issues such as recruitment sometimes take second chair. For instance, the Detroit Police Department ran out of money and was confronted with an ugly reality: It was forced to lay off approximately 150 recruits and officers.[41]

Some measures to fill police personnel shortages might include these:[42]

- Step up recruiting practices on college campuses
- Recruit managers from other professions
- Improve pay and benefits
- Enhance recruiting officer skills
- Change job roles to enhance police officers' satisfaction
- Improve career development
- Change residency requirements
- Create incentives for retirement-eligible officers to remain with the agency
- Increase the use of technology and civilians in policing
- Enhance the managerial styles and accountability of command
- Provide more control over police agencies by command (qualified) personnel
- Disconnect political influence from the hiring, training, and promotion processes

Qualifications for a Police Officer's Job

General Requirements

Most police officers must be U.S. citizens, although the Austin, Texas, and New York City police departments, among others, accept naturalized citizens.[43] Candidates must be healthy and strong and of good moral character. To get a job, a person must pass a written test, be at least a high school graduate, possess a valid driver's license, and have some work experience.[44]

A minimum age of 20 or 21 is preferred by most jurisdictions, although in some agencies, such as the Minot (Nebraska) Police Department, the minimum age requirement is 18. Many police departments, such as the Chicago Police Department (CPD), have a maximum age for an entry-level applicant—40 years of age. Federal agencies' age minimum for candidates is 23, and their age maximum is 37 years.

Education Requirements

Patrol officers can provide quality police services without a college degree; experience and training can clearly help an officer make informed decisions about appropriate police action.[45] So why work hard to finish a college degree and then apply for the job?

College-educated officers are happier personally and feel a sense of achievement, and the decisions they make at home among family and friends are often centered in some of the things they learned in school. Another reason why a college degree or some college experience is sought for an entry-level police officer's job is that it adds to "the character of the applicant," according a police recruiter at the CPD, and "applicants need to make themselves as competitive as possible because most often, we hire for longevity and eventual supervisor's job through patrol officer positions."[46] This thought is consistent with an NYPD study that revealed marginally educated officers were significantly more likely than their better-educated colleagues to end their NYPD careers with involuntary separations[47]; conversely, better-educated officers were more likely than their less-educated colleagues to advance in ranks (see Figure 8–1). Thus more advanced academic credentials are tied to fewer involuntary separations (being fired) and more promotions.

College degrees offer more than just a competitive edge in policing. In an analysis of disciplinary cases brought against Florida officers from 1997 to 2002, the International Association of Chiefs of Police found that officers with only high school educations were the subjects of 75 percent of all disciplinary actions. Officers with four-year college degrees accounted for 11 percent of such actions. "An average patrol officer spends most of the time on dispute resolution," says Louis Mayo, executive director of the Police Association for College Education (PACE; http://www.police-association.org). Mayo adds that a degree "gives (officers) a broad perspective that makes them much more effective."[48]

Finally, applicants with degrees are more than simply competitive: They get to decide which police department they would like to work for, they are more likely to get

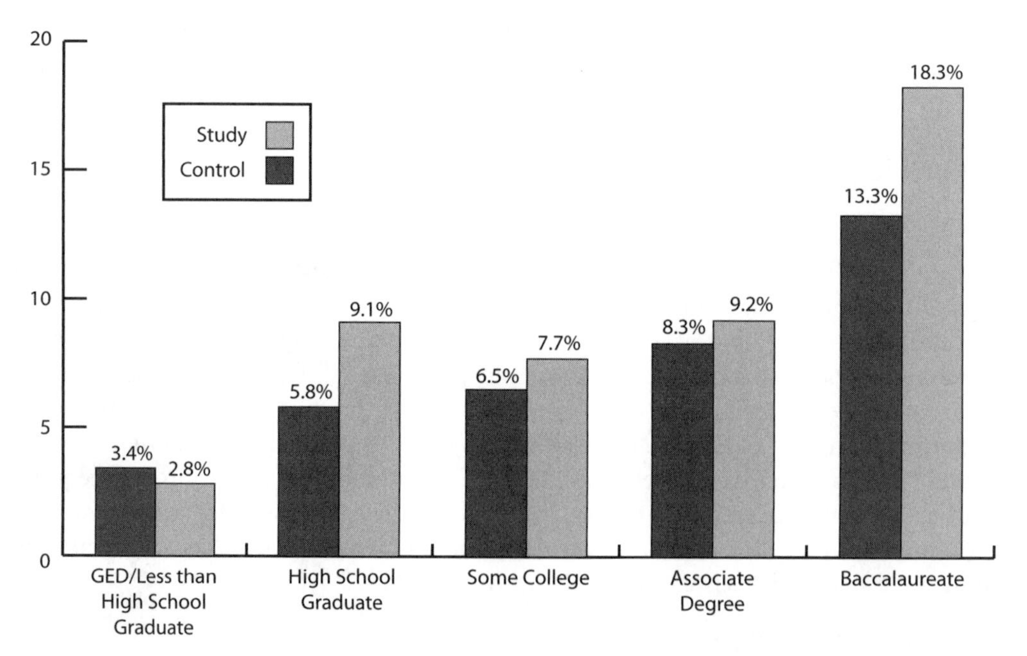

Figure 8–1 Percentage of officers with five years of service in supervisory or command ranks, by educational level at entry into the New York Police Department. Source: James J. Fyfe and Robert Kane, "Bad Cops: A Study of Career-Ending Misconduct Among New York City Police Officers" (September 2006), http://www.ncjrs.gov/pdffiles1/nij/grants/215795.pdf (accessed December 20, 2007).

the job when they apply, and they are more likely to receive greater remuneration—more bucks. The Chicago Police Department's starting salary plan gives college graduates $4,000 more per year than their less-educated counterparts; over a career, officers with a college degree should expect to receive $100,000 more than officers without a degree.

Other departments offer incentives that cover officers' tuition costs after they are hired. For instance, Gulfport (Mississippi) Police Department officers receive full tuition and textbook costs while attending college; once they graduate from an approved four-year institution of higher learning, they receive a raise amounting to almost $2,000 per year. Most agencies continue 100 percent tuition reimbursement through graduate school. The Boston Police Department also offers merit pay for each additional year of schooling accomplished. There is a move toward university representation at police academies, whereby police-students could attend university classes while on the clock.[49]

The enhanced self-confidence that results from earning a degree can help someone be happy, and there is a significant relationship between being happy and being successful.[50] In light of the knowledge that college educated officers tend to receive fewer constituent complaints and use or threaten to use force less often than other officers.[51] A study should be conducted to determine if college educated officers are injured or killed less on the job than other officers.

> Officers with college degrees are happier with themselves and possess a feeling of achievement; they are more competitive; have longer careers; are promoted more often; have a lower rate of constituent complaints and disciplinary actions against them; are paid better even from the get-go; and are provided amenities to advance themselves—adding up to an excellent and secure career that college-education officers can share with family and friends.[52]

State Educational Requirements

About one in three state agencies has a college requirement for new officers, with 12 percent requiring a two-year degree (from a community college, for example) and 2 percent requiring a four-year degree (although PACE reports a higher percentage).[53] For instance, the Louisiana State Police has set the following educational requirements: Applicants must have a minimum of (1) 60 semester hours from an accredited college or university or (2) four years of full-time law enforcement experience (which may include full-time Military Occupation Speciality M.O.S.-qualified military police duties) or (3) three years of military service or two years of full-time law enforcement experience plus 30 semester hours from an accredited college or university.[54]

Local Educational Requirements

About one in four municipal and county police departments (e.g., Chicago, Houston, and Boston) has an educational requirement of two years of college from an accredited institution of higher learning as a minimum. Approximately one in ten

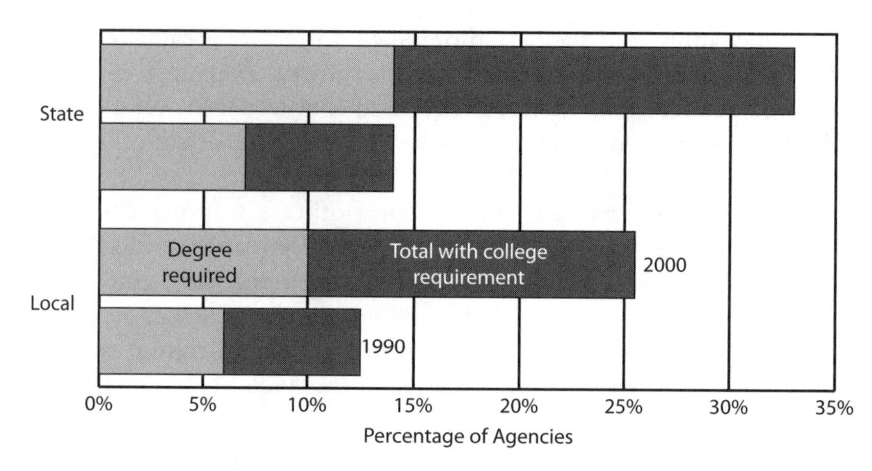

Figure 8–2 State and local law enforcement agencies with a college requirement for new officers, 1990 and 2000. Source: Bureau of Justice Statistics, *Local Police Departments, 2003* (NCJ 210118) (Washington, DC: U.S. Department of Justice, 2006), http://www.ojp.usdoj.gov/bjs/pub/pdf/lpd03.pdf (accessed June 20, 2007).

departments requires a degree (e.g., Arlington, Texas; Charleston, South Carolina; and Tulsa, Oklahoma).[55] Figure 8–2 shows these trends in graphic form, and PACE's website (http://www.police-association.org) provides more information.

In 2000, approximately 6 percent of county (sheriff) police departments and 2 percent of municipal police departments required officer candidates to have a four-year degree. About one in seven sheriff's offices had a college requirement, including 6 percent that required a two-year degree.

Overall, larger police agencies were about twice as likely to have a college requirement in 2000 as in 1990.

Federal Educational Requirements

All federal police agencies require a college degree. Advanced degrees such as a master's degree, law degree (with accounting emphasis), and doctoral degree in behavior sciences and statistics may also be required at the federal level depending on the agency within the U.S. Department of Justice. Specific requirements can be found on the web; for example, visit https://www.fbijobs.gov to learn about available opportunities as a special agent with the FBI.

The Hiring Process

Civil Service Regulations

Civil service regulations govern the appointment of police, state troopers, and detectives in most states, large municipalities, and special police agencies, as well as in many smaller jurisdictions.[56] Before an individual becomes a candidate for a police job, he or she must pass the civil service test by obtaining at least a minimum score (as determined by the civil service board, suggesting that a competitive score could be different in different jurisdictions). Notice that a civil service exam given on a certain date is communicated to the public by many means; for example, the information may be posted on websites, college campuses, and public buildings. Civil service exam schedules can also be found at departmental websites.

The majority of police agencies must follow state civil service regulations.[57] For example, the New York State Civil Service Commission administers the preliminary police officer exam and then reports the results to departments that might be supervising subsequent stages of selection and assessment within the regulations set by the Civil Service Commission.

Civil service regulations are the rules and regulations that govern classified employment linked to specific occupations in the state, such as police officers. Rules have the force and effect of law. Regulations implement the rules issued by an official commission.

One reason why civil service controls were established was to ensure that personnel decisions would be based on objective criteria as opposed to favoritism, biases, and political influence. In many jurisdictions, a civil service commission or board is established and its members appointed by city officials, which could include the mayor or town manager. The civil service board becomes the ultimate authority over police and other public personnel procedures, such as hiring requirements, pay scales, training requirements, promotion criteria, and disciplinary actions.

The use of a civil service board limits the power of police command in the hiring, promotion, and firing process, because command cannot change job descriptions without approval from the civil service committee. It also limits the financial incentives and promotional opportunities for officers, because each item is spelled out in the civil service criteria and there is little room for incentives and promotions that do not fit the criteria. Police command often argues that civil service criteria limit their power in disciplining police personnel. There may be some truth to that argument, but it is clear that officers and managerial authority have abused officers in the past and that civil service boards represent one method of controlling those abuses.

The civil service exam takes around two hours and is administered and graded by the civil service committee.

Background Investigations, Checks, and Interviews

The **selection process** varies from agency to agency even within the same state and jurisdiction. Generally, state agencies (94 percent) and county police departments (90 percent) are more likely to use credit history checks than are municipal police (79 percent) and sheriff's offices (73 percent). County police (97 percent) are the most likely to use drug tests and state agencies (76 percent) the least likely. State agencies are the most likely to use written aptitude tests (92 percent) and physical agility tests (90 percent); sheriff's offices are the least likely to use such tests (65 percent and 59 percent, respectively).

In 2003, all police departments required criminal record investigations (lasting from 8 to 45 weeks), and most required background, driving, and medical checks (Figure 8–3). Personal interviews, psychological interviews and evaluations (verbal and

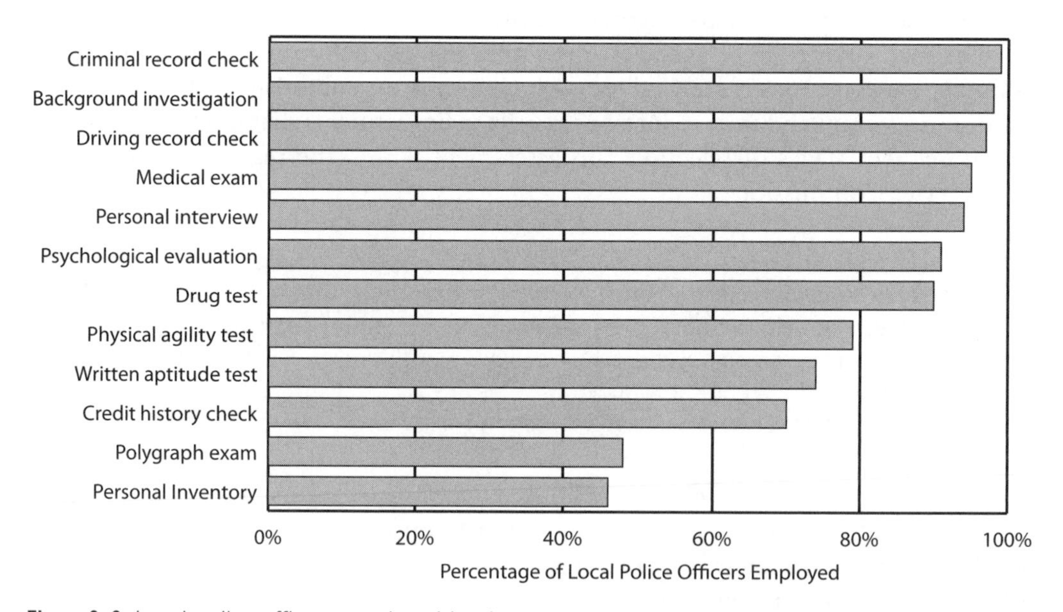

Figure 8–3 Local police officers employed by departments using various recruit screening methods, 2003. Source: Bureau of Justice Statistics, *Local Police Departments, 2003* (NCJ 210118) (Washington, DC: U.S. Department of Justice, 2006), http://www.ojp.usdoj.gov/bjs/pub/pdf/lpd03.pdf (accessed June 20, 2007).

written tests such as Minnesota Multiphasic Personality Inventory [MMPI]), and drug tests were also required by most hiring agencies. Physical agility tests were required by an estimated 80 percent of the agencies, and more than 50 percent required a written aptitude test (in addition to passing the civil service exam). Approximately 40 percent required a polygraph test and personal inventory check. An average of 23 percent of the police departments utilized an assessment center to evaluate candidates.[58]

Physical Fitness

Physical fitness evaluations can consist of timed push-ups, timed sit-ups, flexibility, bench press, leg press, and a 1.5-mile run/walk. Fitness tests usually take two hours or so. Applicants must meet the minimum guidelines in each area to proceed to the written examination.[59]

Written Exam

Many departments have developed a standardized written examination that is administered after a candidate passes the physical endurance tests. Examinations of this variety tend to measure reading comprehension, ability to learn written information, spelling and grammar, and related issues. Applicants are sometimes asked to write a short essay on a topic provided to them. The essays are evaluated based on the ability to communicate thoughts in writing, basic sentence structure, basic grammar, and spelling. Successful applicants may then be invited to an oral interview.

Oral Interview

Oral interviews may be conducted with one individual, such as the chief or chairperson of the hiring board, or with the entire board, which often consists of at least three officers and a person from the academic community such as a professor of criminal justice or psychology. Sometimes this interview is a get-to-know-the-candidate meeting. At other times, applicants are asked questions that include scenarios, logic questions, and/or analysis of information provided. Each applicant's performance in the oral interview is rated to determine whether the applicant meets the needs of the police department.

Background Investigation

Background investigations can take as long as two months and often include a complete review of the candidate's criminal history, employment history, references, education, driving history, military records, illicit drug use, and (in some jurisdictions) credit history. The candidate's fingerprints are sent to the FBI to check on prior participation in the military, passport status, and arrests and convictions.

Simulated Testing

Simulated testing may include mock crime scenes, simulated traffic stops, shoot/don't shoot decisions, "what would you do?" situations, leaderless group discussions, or role-playing scenarios. This part of the candidate evaluation process is designed to assess a dimension necessary for a police officer's work—namely, whether the applicant can display a respectful ability to communicate with the public, an attitude promoting teamwork, and the skills needed to resolve altercations that could lead to force during an assessment.[60] During the hiring process, 48 percent of departments use some type of real-life, simulated testing.[61]

Outcomes of the Hiring Process

Recommendations from background investigators are sent to top command. Sometimes a civil service representative or a member of the department's own office of economic opportunity might review recommendations to ensure that equal-opportunity policies have been followed.

After a brief interview (or sometimes not), a conditional offer of employment could be extended. It is typical to require a training program and a probationary period of from 4 to 12 months before a recruit actually becomes a sworn officer. Applicants accepting the conditional offer must complete a psychological examination, medical examination, illicit drug testing, and polygraph in some cases, or they may undergo an assessment process that could include all or part of the assessment center approach (used in 26 percent of departments and discussed later in this chapter). Conditional employees who fail one or more of these steps could have their offer of conditional employment withdrawn. Applicants who have favorable results from all of the conducted tests and who wish to pursue policing as a career are referred to the training office to schedule appropriate training (including a basic police academy, if needed).

Hiring Precautions

Police management might consider it necessary to seek expert assistance to protect the reputation of the police policymakers and their department in the hiring process. Steps recommended to ensure that the hiring process is effective might include the following:

- Establish standards and rules for the hiring process.
- Have individuals who input data and information sign off on their contributions.
- Document and review everything.
- Provide a clear paper trail.
- Include a police psychologist in the hiring process. Such a professional can be a major asset in the selection, fitness-for-duty examinations and training of officers, and other staff as well as in demonstrating to the courts and others that management took reasonable steps to ensure public safety.[62]

The Assessment Center Approach

Police administrators have come to realize that there is a difference between an assessment center and an assessment center approach.[63] An assessment center is a place where a series of events or exercises will occur.

> The **assessment center approach** is a method of supplementing traditional selection procedures with situational exercises designed to simulate actual officer responsibilities and working conditions.

First used in its basic form by the Cincinnati (Ohio) Police Department in 1961, the objective of the assessment center approach relates to both the need to determine the predictability of a candidate and the need for sound legal defensibility of the hiring process (should eliminated candidates wish to exercise their rights to contest their dismissal). Working against more widespread use of the assessment center approach is the relatively high cost to implement this strategy. Some critics suggest that a candidate is not expected to have knowledge of police procedures, which is one of the areas tested. Yet, to have candidates participate in a series of five to eight exercises, each designed to assess a particular "dimension" necessary for police work, could prove meritorious in the long run.

Experts say that police administrators who implement an assessment approach as part of their department's hiring process should strive to achieve the following goals:

- Standardize the model
- Develop relevant and realistic scenarios
- Have several alternative solutions
- Ensure that the model engages the candidate
- Ensure that the model elicits a number of possible emotional responses
- Not require specialized abilities
- Have assessors rate the performance of each candidate

Factors Contributing to Police Performance

Applicants might demonstrate indicators of future problematic behavior during the assessment process; ferreting out those indicators is one mission of the assessment exercise. Contributing factors that influence receptiveness or vulnerability toward problematic behavior (and **resilience**—the ability to recover from or adjust easily to misfortune or change) include personality, cognitive, group, and organizational factors.

Personality Factors

A person's vulnerability to critical incident stressors is an important consideration for most emergency care personnel, who are trained to enhance certain personality characteristics:[64]

- A need to be in control
- Obsessive/perfectionist tendencies
- Compulsive/traditional values—wanting things to remain unchanged
- High levels of internal motivation
- Action orientation
- High need for stimulation and excitement (easily bored)
- High need for immediate gratification
- Tendency to take risks
- High level of dedication
- Investment in the job owing to months of training and preparation
- View of the job as a lifelong career
- Strong identification with the role as a police officer
- High need to be needed

These personality characteristics contribute to quality police performance as well as to the resilience of an officer. They can be classified as belonging to one of three categories:[65]

- *Biological factors:* genetically based predispositions (e.g., heightened autonomic and physiological reactivity) and changes in physiological reactivity because of prior traumatic exposure.[66]
- *Historical antecedents:* socioeconomic status and preexisting psychopathology.
- *Psychological vulnerability:* learned behavior (e.g., avoidance of threat situations, hypervigilance toward threat-related cues), social skills deficits leading to problems in obtaining and using social support, inadequate problem-solving behavior, and drug and alcohol abuse.

Cognitive Factors

Cognitive factors that contribute to vulnerability include the differences between routine police work and training, and the significance of the incident.[67] In cognitive models, if performance expectations required to control a disaster are beyond the reach of an officer mentally, then the officer is more vulnerable to stress (a phenomenon seen with many New Orleans officers after much of the city was destroyed by Hurricane Katrina). This could also mean that an officer wasn't personally prepared to deal with the significance of the aftermath of events such as the terrorist attacks of September 11, 2001.[68]

Group Factors

Group factors that contribute to vulnerability of an officer suggest that officers should usually work in teams to control most incidents.[69] Personal vulnerability may be enhanced by inadequate coordination and intergroup conflict between officers. When various officers are in conflict with other officers, sharing of information and proactive response management are less likely, adding to the uncertainty among officers, which ultimately translates to officers making decisions based on limited information.

Organizational Factors

Organizational factors that affect officer vulnerability might include the level of trust and empowerment influences that tend to be found in hierarchical bureaucratic police organizations.[70] Development of managerial capacity as a resilience component is more likely to occur in organizational structures where police managers are supportive of personnel and manage through values, for instance, as opposed to the discipline perspective that is so often found in typical police organizations.[71] (See the discussion of attrition later in this chapter.)

Research into the Validity of the Assessment Center Approach

Studies have shown that there is a certain amount of validity in the predictions of an assessment arising from the selection process for entry-level officer candidates, particularly beyond the results of the cognitive ability tests.[72] One sample included 712 participants who underwent personality and cognitive ability testing; the researchers then compared these individuals' performance during training with their on-the-job performance. The findings from this study and others suggest that the assessment process can provide a unique insight into the candidate's suitability for life as a police officer, which subsequently can mean the delivery of quality police services, elimination of constituent complaints, and fewer disciplinary actions against an officer.[73]

It could be argued that a college graduate might score higher in each of these areas because she or he, as a former student, possesses relevant practical experience from the classroom, regardless of the subject actually studied. That is, universities communicate the need for social skills, patience, and tolerance.[74]

Rationale for Use of the Assessment Center Approach in the Hiring Process

The rationale supporting the assessment center approach is linked to the lessons learned about police deviance, police misconduct, and police corruption (see Chapter 13). Problematic officers are not entirely developed through the experiences of policing: Bad officers tend to demonstrate problematic behavior prior to becoming a police officer. The best way to assure that officers' careers do not end in disgrace is "to hire good people with clean histories and good educations."[75] Once hired, the agency must supervise officers carefully, taking note of and acting to see that their lesser scrapes with the agency's internal disciplinary system do not escalate into career-ending misconduct. James J. Fyfe and Robert Kane convey this point eloquently:

> Officers with criminal histories—indicated by arrests for violent, property, or public order crimes—prior employment disciplinary and reliability problems,

> or high average rates of annual complaints while employed by the NYPD were at higher risk than other officers for engaging in career ending police misconduct. By contrast, officers who appeared committed to the NYPD organization—as evidenced by promotion to supervisory ranks and increased years on the job—were significantly less likely than other officers to engage in career ending deviance.[76]

Some experts believe that more or improved training will sufficiently manage the risks associated with police officers' wrongful acts. To decrease the dangers associated with felony pursuit on the highways or on foot, for example, agencies must increase training and ensure that they have clear pursuit policies.[77] Unfortunately, departments rarely make improvements in the selection process of candidates prior to training.[78]

Basic Training

The transition from being a civilian to being a sworn officer begins at the basic law enforcement training centers located across the United States. Training is designed to change a candidate's behavior, provide alternative solutions in problems or confrontations, and persuade a candidate to assume values and ideals of the department that employs him or her.[79] There are distinct differences between training and education.

> **Training** is the practical instruction in how to use knowledge. **Education** is sharing knowledge.

In part, training occurs when officers apply the knowledge they learned in the classroom. One thought is that an officer (or anyone) has not learned anything until it "has changed or influenced you."[80] In other words, knowledge needs to become second nature.

Most educators, including those from past generations such as Jean Piaget,[81] suggest that, in its simplest form, learning is most often accomplished through observation and role-playing. Little is considered truly "learned" until the expected behavior becomes institutionalized, automatic, and routine. In the course of their duties, officers need to rely on their training (not their opinions) to perform at expected levels.

Length of Basic Training

Once hired, candidates attend a training center from 3 to 12 months, sometimes followed by further training in the field, for an average of 42 weeks of instruction.[82] Before receiving their first assignments, officers usually go through 12 to 14 additional weeks of training from their department after they leave the basic training center.

Training Instruction

Training includes classroom instruction in constitutional law and civil rights, state laws and local ordinances, and accident investigation. Recruits also receive training and supervised experience in patrol, traffic control, use of firearms, self-defense, first aid, and emergency response.

Police departments in some large cities hire high school graduates who are still in their teens as police cadets or trainees. These candidates do clerical work and attend classes, usually for 1 to 2 years, at which point they reach the minimum age requirement and may be appointed to the regular service. First, however, they must attend basic law enforcement training.

Training Data

Among basic law enforcement academy classes that completed training during 2002, an estimated 61,354 recruits started training and 53,302 (87 percent) successfully completed or graduated from the training program. An estimated 17 percent of recruits who completed training in 2002 were female, and 27 percent were members of a racial or ethnic minority.

General Characteristic Highlights (as of year-end 2002)[83]

- A total of 626 state and local police academies operated in the United States and offered basic law enforcement training to individuals recruited or seeking to become police officers.
- In addition to basic recruit training, 88 percent of the academies provided in-service training for active-duty, certified officers.

Academy Personnel Highlights

- In 2002, academies employed approximately 12,200 full-time and 25,700 part-time trainers or instructors.
- Three fourths of academies employed fewer than 50 full-time equivalent (FTE) training personnel.

Core Curriculum Highlights

- The median number of hours spent in basic recruit training—excluding any field training component—was 720 hours. The median number of hours above any state requirement was 100 hours.
- In those academies that conducted field training, recruits spent an average of 180 hours in field training.
- Those field hours were spent on the following topics: firearms skills (median 60 hours), health and fitness (50 hours), investigations (45 hours), self-defense (44 hours), criminal law (40 hours), patrol procedures and techniques (40 hours), emergency vehicle operations (36 hours), and basic first-aid/CPR (24 hours).

Force and Defensive Tactic Highlights

- Nearly all academies (99 percent) used semi-automatic pistols in basic firearms training.
- The median qualification score was 75 percent.

Training Concepts: How Versus When to Act

A Police Foundation study on the use of deadly force notes: "In the course of this study police chiefs and administrators were asked what steps they would consider most likely to bring about a reduction in unnecessary shootings by police officers." One common response was to recommend a tight firearms policy coupled with an effective training program.[84] Taking both the type and the approach of training practices into consideration, Herman Goldstein made several pertinent observations on police entry-level training.[85] One of his findings is that training programs sometimes fail to achieve the minimal goal of orienting a new employee to his or her job; a failure to equip officers to understand the built-in stresses of their job means that officers are left to discover on their own the binds in which society places them.

Goldstein's observations explain the limitations of automatically turning to training to solve a situation. According to the Community Relations Service (CRS), "perhaps it also suggests why some training programs are associated with a higher rate of police justifiable homicides."[86] Other researchers have explored Goldstein's perspective further.[87] The use of deadly force is used less often than expected and the actual occurrence of police hitting constituents is rare. Research shows that the average officer fires his or her weapon once every nine years. If the average officer works 200 days a year and handles eight assignments during a shift, then the chances of firing a shot on any particular assignment are 14,400 to 1.[87]

According to the CRS, most training tends to focus on one or possibly two isolated competencies. For example, shooting simulators attempt to train police officers to quickly identify threats.[88] Some crisis intervention training approaches focus almost exclusively on the verbal skills useful in dealing with a limited range of disputes. Effective training should reduce the aggregate number of police shootings and should focus on multiple psychological dimensions, emphasizing those capacities that influence police behavior in a wide range of armed confrontations. Such training should be conducted in environments simulating the "complex, and often bewildering, conditions in which deadly force episodes usually take place," according to the CRS. Ideally, for every hour agencies spend on training officers *how* to shoot, they should also spend several hours teaching *when* to shoot. In addition, training should cover tactics, policy, and liability.

Other examples include pursuit-driving techniques behind the wheel without accompanying classroom training, which teach officers *how* to pursue but not *when* to pursue. Based on the CRS's observations, this kind of training can actually increase the risks of pursuit-related injuries. According to the CRS, "the advantages to be gained from training will not be realized until programs go beyond teaching a single response to complex situations."[89]

> The objective of training should be to develop a "thinking police officer" who analyzes situations and responds in a defensible manner based on the desired value system.

Costs of Basic Training

Costs of training officers vary depending on the region of the country because of varying standard of living schedules. In 2003, these costs ranged from a high of $36,000 per trainee in New York City to a low of $5,400 in Montana.[90] Training an Arizona state trooper cost $15,600, while training a Texas state trooper cost $14,700. These amounts are initial costs, which do not include in-service training. The average expenditure per basic trainee in the United States is approximately $13,100.[91]

In fiscal year 2006, the average cost to train a Border Patrol agent at the federally run academy was approximately $14,700.[92] While differences in programs make direct comparisons difficult, it appears that the Border Patrol's average cost per trainee is consistent with the average cost of training programs that cover similar subjects and prepare officers for operations in similar geographic areas. For example, the estimated average cost per trainee for a Bureau of Indian Affairs police officer is approximately $15,300.

In-Service Training

In-service training is required of officers who are already on the job. For instance, most states require officers to qualify in weapons use each year. These sessions are usually conducted at a state or local police academy. Sometimes private agencies provide officer training in specific areas of certification at qualified shooting ranges or university classrooms depending on the topic (e.g., management).

Eighty-eight percent of police academies provide in-service training for active-duty, certified officers. Likewise, 84 percent provide specialized training of some type (e.g., K-9 and SWAT). More than two thirds (70 percent) provide first-line or higher supervisor training, and more than half (54 percent) provide training for field training instructors. Academies train most local, state, and county criminal justice personnel, including police officers, deputies, correctional officers, and probation officers.

Training and Civil Liability

Because of the increased litigation involving police agencies and an increase in the willingness of juries to award large verdicts against police officers, police departments, and training centers, police managers need to accept the idea that no action by any department can completely prevent that department from being sued.[93] For instance, the U.S. Supreme Court, in its ruling in *Monell v. Department of Social Services of the City of New York* (436 U.S. 658 [1978]), concluded that local governing bodies/municipalities can be held liable when a plaintiff proves "that official policy is responsible for a deprivation of rights protected by the Constitution."[94] A number of courts have since imposed liability on police supervisors and municipalities that do not guard against officer misconduct and do not provide adequate training for their police officers (see *City of Canton, Ohio v. Harris*, 489 U.S. 378 [1989]). According to the courts:

> A person's civil rights have been violated when the **failure to train** amounts to deliberate indifference to the rights of persons with whom the police come to contact, and the indifference subsequently is part of the officer's routine.[95]

Quality training programs increase the effectiveness and safety of police officers, and reduce the potential for liability of those officers, their supervisors, and their agencies.[96] Chapter 14 discusses officer accountability in more depth.

Defensible behavior requirements for officers mean that training institutions must document their training policy and the actions of the officers they train, because a commonly used tactic in civil liability litigation is to link an officer's behavior to a lack of department-sponsored training in an attempt to find an agency responsible for the officer's conduct. It is also possible that some officers might say they did not receive training in a particular area in an attempt to avoid blame. Such attempts and allegations necessitate examination, says Kenneth Flynn, "of the quantity and quality of the training, including assessments regarding the qualifications of the instructor."[97] Instructors need to keep records.

Municipal Liability for Failure to Recruit Properly

In *Board of the County Commissioners of Bryan County, Oklahoma v. Brown* (1997), the Supreme Court addressed administration inadequacy in the hiring of police officers.[98] In this case, Jill Brown was riding in a vehicle when she saw a police checkpoint and turned around. Bryan County Deputy Robert Morrison and Reserve Deputy Stacy Burns saw the turnaround and pursued the Browns. The deputies claimed that the pursuit reached speeds of 100 miles per hour and ended four miles from the checkpoint. The Browns said that they were not aware a deputy was in pursuit.

Once the couple stopped, Morrison pointed his gun at the Browns' vehicle and ordered them to raise their hands. Burns ordered Jill Brown to step out of the vehicle. When she refused, Burns used an "arm bar" technique and pulled her from the vehicle to the ground. The arrest resulted in a severe injury to Brown, who subsequently had to have knee replacement surgery.

Jill Brown "brought a damages action against Bryan County alleging that its Deputy Stacy Burns had arrested her with excessive force, and that it was liable for her injuries because Sheriff B. J. Moore had hired Burns without adequately reviewing his background."[99] In essence, the Browns sued Bryan County under 42 U.S.C. Section 1983, claiming that the county was liable for Burns's use of excessive force. It was Sheriff Moore's decision to hire Burns, who was the son of his nephew. The lawsuit alleged that Moore had failed to consider Burns's background, which included numerous driving violations and a variety of misdemeanors (e.g., assault and battery, resisting arrest, and public drunkenness).

In his defense, Moore said that while he had possession of Burns's criminal record, he had not reviewed it closely when making his hiring decision. The county argued that a single hiring decision by a municipal policymaker could not give rise to municipal liability under 42 U.S.C. Section 1983.

The court concluded that Burns effected an arrest of Jill Brown without probable cause and used excessive force. Also, the hiring policy of Bryan County was "so inadequate as to amount to deliberate indifference to Jill Brown's constitutional rights." The Court of Appeals affirmed the judgment against the county, but a jury did not find that Moore's assessment of Burns was inadequate and, therefore, was not enough to establish deliberate indifference.[100]

According to one expert, due in part to the ruling in the Brown case, litigation related to inadequate police training is likely to "skyrocket."[101] Plaintiffs' counsel will probably take great advantage of the U.S. Supreme Court's decision in *Brown*, which suggests that municipalities bear liability for failure to train a single officer, and failure to train will become the theory of choice in liability claims against governmental entities.

Municipal Liability for Failure to Train Properly

A common form of liability among municipal policymakers derives not from their actions, but rather from their failure to act. Allegations of inadequate or improper training are frequently the bases of "failure to act" claims filed under 42 U.S.C. Section 1983.

This issue was addressed in the U.S. Supreme Court's ruling in *City of Canton v. Harris* (1989).[102] In this case, a woman named Harris was arrested and brought to the police department.[103] She fell down several times and was incoherent while custody, but the officers did not summon any medical assistance for her. After her release, Harris was diagnosed as suffering from several emotional ailments requiring hospitalization and subsequent outpatient treatment. Some time later, she filed suit against the police department, alleging failure to provide necessary medical attention while she was in their custody.

The jury ruled in Harris's favor on this claim, based on evidence indicating that a city regulation gave shift commanders the sole discretion to determine whether a detainee required medical care, and suggesting that commanders were not provided with any special training to make a determination as to when to summon such care for an injured detainee. When the case was appealed to the U.S. Supreme Court, the Court's ruling held that failure to train can be the basis of liability under 42 U.S.C. Section 1983. In this case, the Court said, a "failure to train" was apparent in the "deliberate indifference" that affected the rights of Harris. According the Court, the "city's failure to provide training to municipal employees resulted in the constitutional deprivation . . . [and the] city's failure to train reflects deliberate indifference to the constitutional rights of its inhabitants."[104]

Deliberate indifference standard refers to a municipal "policy" or "custom" as the moving force behind the constitutional violation, and a failure to train reflects a "deliberate" or "conscious" choice by the municipality.[105]

To clarify the matter, the Supreme Court employed an example that lends itself to a better understanding of the latitude of the *Harris* verdict:

> City policymakers know to a moral certainty that their police officers will be required to arrest fleeing felons. The city has armed its officers with firearms, in part to allow them to accomplish this task. Thus, the need to train officers in the constitutional limitations on the use of deadly force (see *Tennessee v. Garner*)[106] can be said to be "so obvious" that the failure to do so could properly be characterized as "deliberate indifference" to constitutional rights.[107]

■ Why Some Officers Leave Policing

Across the United States, approximately 5 percent of all police officers leave their jobs each year. This attrition rate appears to have remained stable since the 1960s.[108] Reasons why officers leave include retirement, death, dismissal, voluntary resignation, and layoffs resulting from financial constraints.

Sixty-one percent of police departments had officer separations during the 12-month period ending June 30, 2003 (see Table 8–1). Large agencies serving populations of 1 million or more people lost an average of 284 officers (and hired an average of 359 officers). Overall, about 32,100 officers separated (of which 5,300 were lateral [from other departments] transfers), and 34,500 officers were hired. The reasons for these officers' departures included resignation (16,100; 49 percent), retirement (9,400; 29 percent), dismissal (2,600; 8 percent), medical/disability retirement (1,800; 6 percent), and probationary rejections (1,300; 4 percent).[109] On average, large agencies have a lower turnover rate (5 percent) than do small agencies (7 percent).[110] Many officers leaving their agencies in 1999 had served only a few years.[111]

Lateral Moves

Approximately 45 percent of officers who left small agencies and 24 percent who left large agencies continued in police work. About 5,300 officers transferred to other agencies.[112] These numbers and percentages are large enough to reinforce the idea that competition is getting stronger for both recruits and experienced officers.[113]

TABLE 8–1 Officer Separations and New Officer Hires in Local Police Departments, by Population Served, 2003

Separations and New Hires during the 12-Month Period Ending June 30, 2003

Population Served	Percentage of Agencies Having Separations	Average Number of Separations*	Percentage of Agencies Hiring New Officers	Average Number of New Hires*
All sizes	61	4	60	5
1,000,000 or more	100	284	94	359
500,000–999,999	97	67	97	67
250,000–499,999	100	41	90	54
100,000–249,999	100	19	94	22
50,000–99,999	96	8	90	9
25,000–49,999	89	4	86	5
10,000–24,999	76	3	77	3
2,500–9,999	65	2	63	2
Less than 2,500	44	2	45	2

*Excludes agencies without separations or new hires.

Source: Bureau of Justice Statistics, *Local Police Departments, 2003* (NCJ 210118) (Washington, DC: U.S. Department of Justice, 2006), http://www.ojp.usdoj.gov/bjs/pub/pdf/lpd03.pdf (accessed June 24, 2007).

Dismissals

In one study over 22 years that examined an estimated 78,000 NYPD careers, researchers found that 1,543 officers (0.1 percent) left involuntarily.[114] Reasons for their dismissal reflected a wide range of improper behaviors. Most often, officers were fired because they refused to undergo drug testing or were caught possessing or trafficking drugs, failed to follow administrative rules, committed profit-motivated crime (bribes and grand larceny), committed off-duty crimes or obstruction of justice, committed off-duty public-order crimes, committed on-duty abuses, or had poor conduct while on personnel probation. (See Chapter 13 for a discussion on police wrongful acts.)

Women Leaving Policing

In the same NYPD study mentioned earlier, the rate per 1,000 officers at which female officers were separated from the department indicated that female officers were charged with on-duty abuse nearly twice as often as male officers. During an 11-year period, 44 men (4 per year) and 12 women (1.1 per year) were separated from the NYPD on charges that included on-duty abuse. These data suggest that women and men are not as different in this regard as popular images would have it.

When attrition rates for female officers were studied across the United States, however, researchers found that female officers leave police work at higher rates (6.3 percent annually) than do male officers (4.6 percent annually).[115] Female officers are more likely to resign voluntarily than male officers (4.3 percent versus 3.0 percent, respectively) and are terminated involuntarily more often than their male counterparts (1.2 percent versus .6 percent, respectively). "Women officers experience a more hostile work environment" than do males.[116] Female officers who are single parents are also confronted with different problems than males in the workplace, and are treated differently than male officers by their commanders as well.

Little research has focused on the reasons why officers resign their jobs voluntarily. We do know that women officers and minorities resign at higher rates than do "mainstream" officers. This finding might reflect that, while much is done in the recruitment process to hire women and minorities, little attempt is made to maintain those individuals once they are trained and working in the department.[117]

Another study suggested that the higher female turnover rate may arise because female officers tend to have higher educational levels than traditional male officers, and college-educated personnel are more likely to grow disenchanted with routine beat duties.[118] That study assumed that education is the primary concern, however, while ignoring the notion that gender is an issue in itself.[119]

A Memphis study showed that dissatisfaction is a necessary but not a sufficient condition propelling an officer toward resignation.[120] There are various turning points in the path leading to voluntary resignation of female police officers, listed here in their order of importance:

- Perception of a stagnated career: "I just can't see any future in being a police officer."[121] One way to interpret this finding is to say that females tend to think more about the future than do males.
- A particularly intense experience that brought accumulated frustration to a head: a gunfight or the mangled bloody remains of a child.

- Lack of a sense of fulfillment on the job: "That job just doesn't fit me anymore."
- Family considerations: "I want to be there when my kids are in first grade and third grade. I'm missing too much."
- The conduct of coworkers: "It's so degrading how the male officers treat me—especially the older ones, who treat me like I'm their little daughter."
- Particular department policy or policies: "The regs [regulations] say cap our speed during felony pursuit at 65 and get approval for increased speed; by the time we do, the creep's gone."
- New employment opportunities: "Wackenhut [private police] offered me a stable shift, more pay, automated report writing, uniform allowance, take-home vehicle privileges, and day care."

Other Reasons for Attrition

Attrition refers to loss of police personnel. One assumption is that pay contributes to attrition among officers, although evidence suggests that starting salary is not directly related to attrition.[122] In other words, police departments with high attrition rates do not have significantly lower entry-level salaries than agencies with lower attrition rates.

Instead, a major factor driving attrition rates is organizational budget restrictions. Officers are often either forced to leave through layoffs caused by budgetary issues or voluntarily leave around the three-year employment mark owing to a lack of promotional opportunities and accompanying increases in salaries.[123] This relationship is consistent with findings related to private-sector attrition.[124] For instance, the three primary factors that influence a decision for personnel (including middle and top management) to leave an organization such as Ford Motor Company or Walgreens are as follows:

- Financial health and stability of the company
- Health of the company culture, including the work environment
- Manager–employee relationship

Police personnel want to feel as though they are part of something and can make a difference.[125] Unfortunately, too many patrol personnel feel excluded from the decision-making process and rarely contribute to police policy.[126]

The Relationship Between Length of Service and Attrition

Researchers have found that two thirds of the officers who leave both large and small agencies have five or less years of service.[127] Officers with intermediate lengths of service (i.e., 6 to 14 years of service) are the least likely to leave from both large and small agencies. Thus the officers with the least amount of time invested in the profession leave more often than others.

Low pay and benefits, and departure of officers for larger and more prestigious agencies (which can include federal jobs), account for many early-career departures.[128] Other factors driving attrition include growing discontent with the criminal justice system after arresting the same offenders over and over, only to see them repeatedly be released[129]; poor management and inadequate leadership; and rigid hieratical organizational structure.[130] In addition, as officers get older, they might have secret concerns for their own safety, and their families might be demanding their attention such that working rotating shifts is not desirable.[131]

Deploying Training Resources during Emergencies

In emergency situations, there is little time to think about right answers: That's when professional training kicks in. For instance, once Hurricane Katrina left the Mississippi Gulf Coast and the city of New Orleans, the massive destruction sent the population into desperation, as it spared few homes, businesses, schools, or household pets.[132] Shortly thereafter, Hurricane Rita leveled parts of cities missed by Katrina, including Houston, which had been a refuge for many individuals originally displaced by Katrina. In the immediate aftermath of Katrina's and Rita's wake, residents faced a dismaying reality. The life rafts of food, electricity, and public safety were months away. Should a resident require pharmaceuticals, legal or otherwise, desires went unfilled. Toilets never flushed. Children, if you could find them (more than 3,000 children were separated from their parents/caretakers, and many remain missing), cried out for care.[133] Three months later on the Mississippi Gulf Coast, cell phones and running water were available, but life and sanity depended on emergency care providers, including patrol officers and religious organizations that operated from trailers and abandoned warehouses.[134]

Incredible lessons were learned as a result of these and other experiences that can aid police policymakers in preparing for the collateral damage that might arise from terrorism, national calamities, and epidemics in twenty-first-century America. Much of this hard-learned knowledge will aid communities that are confronted by some very ruthless people who demonstrate a disregard for the rule of law and the human rights of others. For instance, looting and gang activity were so widespread in areas hit by Hurricanes Katrina and Rita (the latter damage occurring in Houston) that patrol officers were ordered to stop search-and-rescue efforts and to concentrate on the besieged streets in an effort to stop violence.[135] Police command was woefully unprepared for the damage inflicted by Katrina, but when it tried to rally through its antiquated chain of command, patrol officers had few choices other than to rely on their own discretion (see Chapter 13).[136]

Administrators face the unenviable task of reacting to disasters with an eye toward the immediate needs of the situation, but also with the knowledge that the consequences of their actions can affect future events for good or ill. To help handle such life-altering incidents, managers can employ a method that allows them to envision a desirable future, or end state, and then implement strategies to achieve a favorable resolution.[137] One method starts with gathering all the information from previous situations followed by brainstorming a variety of possibilities, from best- to worst-case scenarios, and training officers in "what-if" scenarios.[138] Thus appropriate training of police officers is a key component of any emergency planning.

The Federal Hiring and Employment Process

Nearly 50 percent of the United States' 2,800,000 federal employees are at or near retirement. Typically there are 19,000 job vacancies posted on just one federal jobs recruitment site, and hiring occurs in all sectors owing to the aging federal employee population. The federal government employs more than 188,000 law enforcement personnel in more than 40 job series. As mentioned in Chapter 4, the creation of the Department of Homeland Security brought policing functions from a variety of federal agencies—the Department of the Treasury, Justice, Health and Human Services, Defense, FBI, Secret Service, General Services Administration, Energy, Agriculture, Transportation, and the U.S. Coast Guard—under one roof. Perhaps the easiest way to find federal jobs is to access the federal.jobs.net site, which includes an application and tips about résumé writing.

Federal Qualifications

To be considered for appointment as an FBI agent, an applicant must be a graduate of an accredited law school or a college graduate with one of the following qualifications: a major in accounting, electrical engineering, or information technology; fluency in a foreign language; or three years of related full-time work experience. All new agents undergo 18 weeks of training at the FBI Academy on the U.S. Marine Corps base in Quantico, Virginia.

Applicants for special agent jobs with the U.S. Secret Service and the Bureau of Alcohol, Tobacco, and Firearms must have a bachelor's degree, a minimum of three years' related work experience, or a combination of education and experience. Prospective special agents undergo 11 weeks of initial criminal investigation training at the **Federal Law Enforcement Training Center** (FLETC) in Glynco, Georgia, and another 17 weeks of specialized training with their particular agencies.

Applicants for special agent jobs with the DEA must have a college degree with at least a 2.95 grade-point average or specialized skills or work experience, such as foreign language fluency, technical skills, law enforcement experience, or accounting experience. DEA special agents undergo 14 weeks of specialized training at the FBI Academy in Quantico, Virginia.

U.S. Border Patrol agents must be U.S. citizens, be younger than 37 years of age at the time of appointment, possess a valid driver's license, and pass a three-part examination on reasoning and language skills. A bachelor's degree or previous work experience that demonstrates the ability to handle stressful situations, make decisions, and take charge is required for a position as a Border Patrol agent. Applicants may qualify through a combination of education and work experience.

Federal Evaluation Criteria

Specific evaluation criteria or special qualifications are often indicated as part of the description of federal law enforcement positions.[139]

Knowledge, Skills, and Abilities

Knowledge, skills, and abilities (KSAs) are attributes required to perform a job. They can be demonstrated through qualifying experiences, education, and training.

Quality Ranking Factors

Quality ranking factors are KSAs that can be expected to significantly enhance performance in a position, but are not essential for satisfactory performance. Applicants who possess such KSAs may be ranked above those who do not, yet no one may be rated ineligible solely for failure to possess such KSAs.

Selective Placement Factors

Selective placement factors are KSAs or special qualifications that are required in addition to the minimum requirements in a qualified standard. They are essential and mandatory to perform the duties and responsibilities of a specific position. Applicants who do not meet all selective placement factors are ineligible for the position.

Selection Criteria

Selection criteria are outlined for specific salary grade levels of vacant positions. For example, a posted announcement for a GS-1811 Criminal Investigator (Special Agent) position at the CS-7 level specified the following selection criteria under the heading "Knowledge, Skills, and Abilities":

- Knowledge of investigative techniques, principles, detection methods, and equipment
- Knowledge of criminal law and rules of criminal procedure
- Skill in both oral and written communication
- Skill in computer software such as GIS and OFFICERLINK
- Ability to analyze and evaluate facts, evidence, and related information and arrive at sound conclusions
- Ability to deal effectively with individuals and groups at all levels of government and the private sector

While the KSAs referenced here are typical of criminal investigator positions at the GS-7 level, vacancy announcements for the same job series and salary grade could vary from one federal agency to another depending on the nature of the position and agency requirements. While some vacancy announcements describe selection criteria in detail, others might request KSAs in less unambiguous terms. For instance, a GS-1811 criminal investigator position at the GS-7 level could simply describe requirements as follows:

- Skill in oral communication
- Skill in written communication
- Ability to investigate

Finally, when applying for a federal position, applicants are required to submit a Supplemental Qualifications Statement along with their application forms or a résumé that explains the applicant's education, experience, training, accomplishments, and awards as they relate to the selection criteria. Vacancy announcements usually include instructions pertaining to the format of these statements.

Federal Training

Among federal agencies, training varies from department to department. Most training for federal agencies is conducted at the Federal Law Enforcement Training Center (FLETC), which serves as an interagency law enforcement training organization for more than 80 federal agencies.[140] (For details, see FLETC's website at http://www.fletc.gov/.) FLETC also provides training services for state, local, and international police agencies. It is headquartered at Glynco, Georgia (halfway between Savannah, Georgia, and Jacksonville, Florida).

In addition to its Glynco facility, FLETC operates two other residential training sites in Artesia, New Mexico, and Charleston, South Carolina. It operates an in-service requalification training facility in Cheltenham, Maryland, for use by agencies with large concentrations of personnel in the Washington, D.C., area.

FLETC has oversight and program management responsibility for the International Law Enforcement Academy (ILEA), which has facilities in Gaborone, Botswana; San Salvador, El Salvador; and Lima, Peru. FLETC also supports training at other ILEA facilities in Hungary and Thailand.

The Glynco operation is situated on 1,500 acres and features modern facilities such as classrooms, dormitories, and administrative and logistical support structures, including a dining hall capable of serving more than 4,000 meals per day.[141] Additionally, the Glynco site has 18 firearms ranges, including a state-of-the-art indoor range complex with 146 separate firing points, and eight highly versatile semi-enclosed ranges with 200 additional firing points. Other training assets include a sprawling complex of driver training ranges, a physical techniques facility, an explosives range, a fully functional mock port of entry, and numerous other structures that support the entire training effort. Glynco hosts many university and high school tours.

RIPPED from the Headlines

New York Recruits Get Lesson in Diversity

About 1,100 police recruits in crisp new uniforms filled the famed Apollo Theater in New York's Harlem neighborhood for a lesson on race relations and cultural diversity. The recruits are the most diverse ever, with immigrants born in 60 different countries and they are all scheduled to graduate at Madison Square Garden.

"What we're looking for is a better sense of communication," said musician Wyclef Jean at a news conference in the Apollo's chandeliered lobby. The Haitian-born Brooklyn resident was joining a panel discussion on policing the city's diverse neighborhoods. The panel was part of a special four-day "comprehensive multicultural immersion course" for new officers set to graduate next week from the police academy.

Many officers "behave as if they're detached, as if they don't understand that they're living in the most diverse city in the world," said another panelist, the Rev. Calvin Butts of the Abyssinian Baptist Church in Harlem. "So something has to happen with the training. Not just sensitivity training—it's a broader understanding of the world in which we live."

Police officials said the rookies would examine community relations, racial bias, and police abuse. Planned activities include a screening of *Crash*, the Academy Award-winning film about racial tensions in Los Angeles, role playing that simulates racial conflicts on the street, and examining racial incidents that have sparked community outrage.

Source: Tom Hays, "New York Recruits Get Lesson in Diversity," *Associated Press* (June 21, 2007), http://www.officer.com/web/online/Top-News-Stories/New-York-Recruits-Get-Lesson-in-Diversity-/1$36608 (accessed June 22, 2007).

Summary

Recruiting, training, and retaining quality personnel are among some of the most important responsibilities faced by police managers in the twenty-first century. People choose policing as a career because of a desire to help others, the job security, the crime fighter role, the excitement of the job, and the prestige of the job. Many other people would never apply for a police job for reasons ranging from cultural issues to disdain for policing's traditionally rigid paramilitary hierarchy to being a high school dropout. Consequently, the supply of applicants is low, though community outreach efforts can sometimes help attract applicants.

The general entry qualifications for an applicant include being at least age 21 (and younger than age 37 for federal positions), having a valid driver's license, and being healthy, strong, and of good character. Educational requirements vary from having a general equivalent degree (GED) to having a degree from a four-year college. Civil service regulations govern the appointment of officers and were designed to ensure that personnel decisions would be based on objective criteria. Most often the hiring process includes background investigations, credit history checks, mental and health checks, and interviews.

The assessment center approach supplements the traditional assessment and selection procedures with situational exercises designed to simulate actual officer responsibilities. Applicants might demonstrate indicators of future problematic behavior during the assessment process, ruling them out as future police officers. Contributing factors that influence receptiveness or vulnerability toward problematic behavior include personality, cognitive, group, and organizational factors.

Most police candidates average 42 weeks in a training center before their first assignment and receive another 12 to 14 weeks of training from their department. The objective of training should be to develop a "thinking police officer" who analyzes situations and responds in the defensible appropriate manner based on the desired value system. A person's civil rights have been violated when "failure to train" results in officers showing deliberate indifference to the rights of persons with whom they come to contact.

The reasons why officers leave departments vary, but at the top of the list are resignations, retirements, dismissals, medical/disability retirements, and probationary rejections. Most often new officers leave departments within the first three years after they join.

Qualifications for employment in federal law enforcement vary, but a college degree (and sometimes even more education) is required. Federal evaluation criteria focus on the applicant's knowledge, skills, and abilities. Most federal training is conducted at the Federal Law Enforcement Training Center.

Key Words

Assessment center approach
Attrition
Background investigations
Civil service regulations
Defensible behavior
Deliberate indifference
Education
Failure to train
Federal Law Enforcement Training Center (FLETC)
Physical fitness
Resilience
Selection process
Simulated testing
Training
Written examination

Discussion Questions

1. Describe some important responsibilities of police agencies in the twenty-first century.
2. Characterize why someone might want to become a police officer.
3. Identify the nine major reasons why people would never apply for a police job. In what way do you agree with the items? In what way do you disagree with them?
4. Explain the lessons learned from the recruiting case studies presented in this chapter.
5. Characterize the general requirements—including age and education—to apply to a police officer's job. In what way would a college degree benefit an officer?
6. Describe how civil service influences affect the job of an officer.
7. Describe the process after a candidate passes the civil service test.
8. Explain the assessment center approach, its mission, and the factors that predict problematic behavior among officer candidates.
9. Characterize the rationale underlying use of the assessment center approach in the officer hiring process.
10. Distinguish between education and training, and explain how learning is accomplished.
11. Characterize the objectives of training.
12. Describe what is meant by "failure to train" and "deliberate indifference."
13. Identify reasons why officers leave policing. Suggest reasons other than those listed in this book.
14. Summarize federal qualifications for employment as a law enforcement officer.

Acknowledgment

Sergeant Luke Thompson, Gulfport (Mississippi) Police Department and graduate student at the University of Southern Mississippi, aided the writer with parts of this chapter. His help is greatly appreciated.

Notes

1. Roy R. Roberg and Jack L. Kuykendall, *Police and Society* (Belmont, CA: Wadsworth, 1994), 227–28.
2. Barry M. Baker, *Becoming a Police Officer* (New York: iUniverse, 2006).
3. Baker, *Becoming a Police Officer*, 141.
4. Bureau of Labor Statistics (2007), http://www.bls.gov/ (accessed November 16, 2007).
5. David A. Decisso, "Police Officer Candidate Assessment and Selection," *FBI Law Enforcement Bulletin* (2000), http://www.fbi.gov/publications/ (accessed June 20, 2007).
6. Robert J. Kaminski, "Police Minority Recruitment: Predicting Who Will Say Yes to an Offer for a Job as a Cop," *Journal of Criminal Justice* 21 (1996): 395–409.
7. Statement of Charles H. Ramsey, Chief of Police, Metropolitan Police Department, Before the Council of the District of Columbia, Committee on the Judiciary, Public Round Table on the Police Deployment Act of 1999.

8. Michael L. Birzer and Jackquice Smith-Mahdi, "Does Race Matter? The Phenomenology of Discrimination Experiences among African Americans," *Journal of African American Studies* 10, no. 2 (2006): 22–37.
9. Ken Bolton, Jr., and Joe R. Feagin, *Black in Blue: African American Police Officers and Racism* (New York: Routledge, 2004).
10. Birzer and Smith-Mahdi, "Does Race Matter?"
11. Richard Kelly, "Psychological Care of the Police Wounded," in *Treating Police Stress*, ed. John M. Madonna, Jr. and Richard E. Kelly (Springfield, IL: Charles C. Thomas, 2002), 15–30.
12. Susan Ehrlich Martin, *Breaking and Entering: Policewomen on Patrol* (Berkeley, CA: University of California Press, 1980).
13. Dennis J. Stevens, *Police Officer Stress: Sources and Solutions* (Upper Saddle River, NJ: Prentice Hall, 2008), 148.
14. Personal experiences of the author.
15. Carl B. Klockars, "The Dirty Harry Problem," in *The Police and Society*, 2nd ed., ed. Victor E. Kappeler (Prospect Heights, IL: Waveland, 1999), 368–87; Gary W. Sykes, "Street Justice: A Moral Defense of Order Maintenance Policing," in *The Police and Society*, 2nd ed., ed. Victor E. Kappeler (Prospect Heights, IL: Waveland, 1999), 134–49.
16. Kelly, "Psychological Care of the Police Wounded," 17. Also see Bill Lewinski, "Stress Reactions: Related to Lethal Force Encounters," *Police Marksman* (May/June 2002): 24–27.
17. Kelly, "Psychological Care of the Police Wounded," 17.
18. John E. Ingersoll, "The Police Scandal Syndrome," *Crime & Delinquency* 10, no. 3 (1964): 269–75.
19. Andrew J. Harvey, "Building an Organizational Foundation for the Future: Managing Law Enforcement Agencies," *FBI Law Enforcement Bulletin* (November 1996), http://www.fbi.gov/publications/leb/1996/nov963.txt (accessed June 23, 2007).
20. Timothy Egan, "Police Forces, Their Ranks Thin, Offer Bonuses, Bounties and More," *New York Times* (December 28, 2005), http://www.nytimes.com/2005/12/28/national/28police.html (accessed March 23, 2008).
21. Harvey, "Building an Organizational Foundation for the Future."
22. James W. Kalat, *Introduction to Psychology*, 8th ed. (Belmont, CA: Wadsworth, 2007).
23. Focus Adolescent Services (n.d.), http://www.focusas.com/Dropouts.html (accessed June 23, 2007).
24. "Defining Dropouts: A Statistical Portrait" (no date), http://www.ed.gov/pubs/ReachingGoals/Goal_2/Dropouts.html (accessed June 23, 2007).
25. Michelle Perin, "Is My Daddy Dead: Helping Children of Officers Cope with Trauma," *Officer.com* (April 10, 2007), http://www.officer.com/web/online/Police-Life/Is-My-Daddy-Dead/17$35604 (accessed June 23, 2007).
26. Marcus Buckingham and Donald O. Clifton, *Are You Afraid of Weaknesses? Building Your Strengths* (New York: Free Press, 2001).
27. Bruce K. Alexander, "Finding the Roots of Addiction," *Transition* 34, no. 2 (Summer 2004) http://www.vifamily.ca/library/transition/342/342.html#1 (accessed March 23, 2008).
28. Christopher S. Koper, *Hiring and Keeping Police Officers* (NCJ202289) (Washington, DC: U.S. Department of Justice, July 2004), http://www.ncjrs.gov/pdffiles1/nij/202289.pdf (accessed June 22, 2007).
29. James J. Fyfe and Robert Kane, "Bad Cops: A Study of Career-Ending Misconduct Among New York City Police Officers" (September 2006), http://www.ncjrs.gov/pdffiles1/nij/grants/215795.pdf (accessed December 20, 2007).
30. Stephen D. Mastrofski, "Police Organization and Management Issues for the Next Decade," unpublished paper presented at the National Institute of Justice (NIJ) Policing Research Workshop: Planning for the Future, Washington, DC, November 28, 2006.

31. James Hannah, "Police Recruiting Wars Give Rise to Product Placement," *Associated Press* (November 11, 2005), http://www.policeone.com/police-recruiting/articles/120747/ (accessed June 24, 2007).
32. Hannah, "Police Recruiting Wars."
33. Egan, "Police Forces."
34. Egan, "Police Forces."
35. Koper, *Hiring and Keeping Police Officers.*
36. Ellen Scrivner, "Innovations in Police Recruitment and Hiring: Hiring in the Spirit of Service," COPS: U.S. Department of Justice, n.d., http://www.cops.usdoj.gov/mime/open.pdf?Item=1655 (accessed December 18, 2007).
37. Scrivner, "Innovations in Police Recruitment and Hiring," 43.
38. Community Relations Service, *Principles of Good Policing: Avoiding Violence between Police and Citizens* (Washington, DC: U.S. Department of Justice, September 2003), http://www.usdoj.gov/crs/pubs/principlesofgoodpolicingfinal092003.htm#74 (accessed June 25, 2007).
39. Wesley G. Skogan and Susan M. Hartnett, *Community policing: Chicago Style* (New York: Oxford University Press, 1998), 236–58.
40. Dennis J. Stevens, *Case Studies in Applied Community Policing* (Boston: Allyn Bacon, 2003), 157–68.
41. Scrivner, "Innovations in Police Recruitment and Hiring."
42. Decisso, "Police Officer Candidate Assessment and Selection"; Fyfe and Kane, "Bad Cops"; Scrivner, "Innovations in Police Recruitment and Hiring"; Samuel Walker and Charles M. Katz, *Police in American: An Introduction*, 6th ed. (Boston, McGraw Hill, 2008), 140–41.
43. Austin Texas Police Department, http://www.ci.austin.tx.us/police/recruiting/hiring.htm (accessed June 20, 2007).
44. "Police" (n.d.), http://www.bls.gov/k12/law01.htm (accessed June 20, 2007).
45. Dennis J. Stevens, "College Educated Officers: Do They Provide Better Police Service?", *Law & Order* 47, no. 12 (1999): 37–41.
46. Chicago police recruiter, personal communication with the author, June 26, 2007. Also see CPD's website, http://egov.cityofchicago.org/.
47. Fyfe and Kane, "Bad Cops."
48. Cited in Kevin Johnson, "Agencies Find It Hard to Require Degrees," *USA Today* (September 19, 2006), http://www.usatoday.com/news/nation/2006-09-17-police-education_x.htm (accessed June 26, 2007).
49. In this regard, the author has instructed college courses at the North Carolina Justice Academy for criminal justice professionals.
50. Sonja Lyubomirsky, Laura King, and Ed Diener, "The Benefits of Frequent Positive Affect: Does Happiness Lead to Success?", *Psychology Bulletin* 131, no. 6 (2005): 803–55.
51. William C. Terrill and Stephen D. Mastrofski. Situational and officer-based determinants of police coercion, *Justice Quarterly* 19, no. 2, 215–248.
52. Stevens, "College Educated Officers."
53. Police Association for College Education, http://www.police-association.org (accessed June 26, 2007).
54. Louisiana State Police (2007), http://www.officer.com/jobs/LASP.jsp (accessed June 20, 2007).
55. Bureau of Justice, "Law Enforcement Management and Administrative Statistics, 2000: Data for Individual State and Local Agencies with 100 or More Officers" (2004), http://www.ojp.gov/bjs/pub/pdf/lemas00.pdf (accessed June 20, 2007).

56. Bureau of Labor Statistics, *Occupational Handbook* (Washington, DC: U.S. Department of Labor, 2006), http://www.bls.gov/oco/ocos160.htm#training (accessed October 8, 2006).
57. Decisso, "Police Officer Candidate Assessment and Selection."
58. Decisso, "Police Officer Candidate Assessment and Selection."
59. The model guiding this presentation belongs to the New Mexico State Police, http://www.nmsuparking.com/pages/hiring.pdf (accessed June 21, 2007).
60. J. Pynes and H. J. Bemardin, "Entry-Level Police Selection: The Assessment Center Is an Alternative," *Journal of Criminal Justice* 20 (1992): 41–52.
61. Decisso, "Police Officer Candidate Assessment and Selection."
62. Cary D. Rostow, Robert D. Davis, Judith P. Levy, and Sarah Brecknock, "Civil Liability and Psychological Services in Law Enforcement Administration," *Police Chief* (2001): 36–43.
63. Decisso, "Police Officer Candidate Assessment and Selection."
64. Daniel A. Goldfarb and Gary S. Aumiller, "The Heavy Badge" (2005), http://www.heavybadge.com/cisd.htm (accessed May 22, 2005).
65. J. R. Scotti, B. K. Beach, L. M. Northrop, C. A. Rode, and J. P. Forsyth, "The Psychological Impact of Accident Injury," in *Traumatic Stress: From Theory to Practice*, ed. J. R. Freedy and S. E. Hobfoll (New York: Plenum, 1995), 181–212.
66. Douglas Paton, "Critical Incidents and Police Officer Stress," in *Policing and Strategies*, ed. Heith L. Copes (Upper Saddle River, NJ: Prentice Hall, 2005), 25–40, p. 33.
67. Paton, "Critical Incidents and Police Officer Stress."
68. In part, this component relates to cognitive psychology, which is a subfield of psychology associated with information processing and the role it plays in emotion, behavior, and physiology. *Psychology Dictionary*, http://allpsych.com/dictionary/c.html (accessed June 2, 2005).
69. Paton, "Critical Incidents and Police Officer Stress."
70. Paton, "Critical Incidents and Police Officer Stress."
71. A. P. Cardarelli, J. McDevitt, and K. Baum, "The Rhetoric and Reality of Community Policing in Small and Medium Sized Cities and Towns," *International Journal of Police Strategies & Management* 21, no. 3 (1998): 397–415.
72. Kobi Dayan, Ronen Kasten, and Shaul Fox, "Entry-Level Police Candidate Assessment Center: An Efficient Tool or a Hammer to Kill a Fly?", *Personnel Psychology* 55, no. 4 (2002): 827–49.
73. Diana E. Krause, Martin Kersting, Eric D. Heggestad, and George C. Thornton III, "Incremental Validity of Assessment Center Ratings over Cognitive Ability Tests: A Study at the Executive Management Level," *International Journal of Selection and Assessment* 14, no. 4 (2006): 360–71.
74. Stevens, "College Educated Officers."
75. Fyfe and Kane, "Bad Cops."
76. Fyfe and Kane, "Bad Cops."
77. John Hill, "High Speed Police Pursuits," *Federal Law Enforcement Bulletin* (July 2002), http://www.fbi.gov/publications/leb/2002/july02leb.pdf (accessed June 22, 2007).
78. Decisso, "Police Officer Candidate Assessment and Selection."
79. Kenneth W. Flynn, "Training and Police Violence," in *Policing and Violence*, ed. Ronald G. Burns and Charles E. Crawford (Upper Saddle River, NJ: Prentice Hall, 2002), 126–45.
80. Flynn, "Training and Police Violence," 128.
81. Jean Piaget, *The Origins of Intelligence in Children* (London: Routledge and Kegan Paul, 1953).

82. Bureau of Justice Statistics, *State and Local Law Enforcement Training Academies, 2002* (Washington, DC: Department of Justice, 2006), http://www.ojp.usdoj.gov/bjs/abstract/slleta02.htm (accessed December 18, 2007).
83. Bureau of Justice Statistics, "State and Local Law Enforcement Training Academies, 2002" (NCJ 204030; November 2004), http://www.ojp.usdoj.gov/bjs/pub/ascii/slleta02.txt (accessed June 21, 2007).
84. A report from the International Association of Chiefs of Police, *A Balance of Forces*, recommended the following training tips:
    In-service crisis intervention training as opposed to preservice training was associated with a low justifiable homicide rate by police.
    Agencies with simulator, stress, and physical exertion firearms training experience a higher justifiable homicide rate by police than agencies without such training.
    Marksmanship awards given to officers for proficiency in firearms training are associated with a high justifiable homicide rate by the police.
    In-service training in the principles of "officer survival" is correlated with a high justifiable homicide rate by the police.
85. Herman Goldstein, *Policing in a Free Society* (New York: Ballinger, 1977).
86. Community Relations Service, *Principles of Good Policing.*
87. Peter Scharf and Arnold Binder, *The Badge and the Bullet: Police Use of Deadly Force* (New York: Praeger, 1983), 178.
88. Community Relations Service, *Principles of Good Policing.*
89. Community Relations Service, *Principles of Good Policing.*
90. Bureau of Justice Statistics, "State and Local Law Enforcement Training Academies, 2002."
91. Bureau of Labor Statistics, *Occupational Handbook.*
92. United States Government Accounting Office, http://www.gao.gov/docsearch/abstract.php?rptno=GAO-07-997T (accessed June 21, 2007).
93. Dale H. Close, "How Chiefs Should Prepare for Nine Liability Risks," *Police Chief* (June 2001): 16–27.
94. Community Relations Service, *Principles of Good Policing.*
95. Flynn, "Training and Police Violence," 133.
96. Diane M. Daane and James E. Hendricks, "Liability for Failure to Adequately Train," *Police Chief* 58, no. 11 (1991): 26–29.
97. Flynn, "Training and Police Violence," 134.
98. Victor E. Kappeler, *Critical Issues in Police Civil Liability* (Mt. Prospect, IL: Waveland Press, 2001), 53.
99. *Bryan County, Oklahoma v. Brown*, http://www.oyez.org/cases/1990-1999/1996/1996_95_1100/ (accessed June 25, 2007).
100. A municipality may not be held liable under § 1983 solely because it employs a tortfeasor (a person who commits a tort or a civil wrong, either intentionally or through negligence). See, for example, *Monell v. New York City Dept. of Social Servs.*, 436 U.S. 658, 692. Instead, the plaintiff must identify a municipal "policy" or "custom" that caused the injury. See, for example, *Pembaur v. Cincinnati*, 475 U.S. 469, 480–481. Contrary to the respondent's contention, a "policy" giving rise to liability cannot be established merely by identifying a policymaker's conduct that is properly attributable to the municipality. The plaintiff must also demonstrate that, through its *deliberate* conduct, the municipality was the "moving force" behind the injury alleged. See *Monell, supra*, at 694. That is, a plaintiff must show that the municipal action was taken with the requisite degree of culpability and must demonstrate a direct causal link between the municipal action and the deprivation of federal rights (pp. 402–404).

101. Elliot Spector, "Emerging Legal Standards for Failure to Train," *Law & Order* (October 2001): 73–77.

102. *Canton v. Harris*, 489 U.S. 378 (1989).

103. Flynn, "Training and Police Violence," 133.

104. *Canton v. Harris.*

105. Findlaw (2007), http://caselaw.lp.findlaw.com/cgi-bin/getcase.pl?court=us&vol=489&page=390 (accessed June 25, 2007).

106. *Tennessee v Garner* 471 U.S. 1, 105 S. Ct. 1694, 85 L.Ed.2d 1 (1985).

107. Findlaw.com, "Failure to train" (2007), http://library.findlaw.com/1999/Jan/1/128567.html (accessed June 25, 2007).

108. Samuel Walker, *Sense and Nonsense about Crime and Drugs* (Belmont, CA: Wadsworth, 2001), 348.

109. Bureau of Justice Statistics, *Local Police Departments, 2003* (NCJ 210118) (Washington, DC: U.S. Department of Justice, 2006), http://www.ojp.usdoj.gov/bjs/pub/pdf/lpd03.pdf (accessed June 20, 2007).

110. Christopher S. Koper, Edward R. Maguire, and Gretchen E. Moore, *Hiring and Retention Issues in Police Agencies: Readings on the Determinants of Police Strength, Hiring and Retention of Officers, and the Federal COPS Program* (NCJRS 193428) (Washington, DC: U.S. Department of Justice, March 2002), 46.

111. Koper, *Hiring and Keeping Police Officers.*

112. Bureau of Justice Statistics, *Local Police Departments, 2003.*

113. Koper, *Hiring and Keeping Police Officers.*

114. Fyfe and Kane, "Bad Cops."

115. Susan E. Martin. "Outside within the Station House: The Impact of Race and Gender on Black Women Police," *Social Problems* 41 (1994): 398.

116. Walker, *Sense and Nonsense about Crime and Drugs*, 348.

117. Jerry Sparger and David Giacopassi, "Swearing In and Swearing Off: A Comparison of Officers and Ex-Officers' Attitudes Toward the Workplace," in *Police and Law Enforcement*, ed. Daniel B. Kennedy and Robert J. Homat (New York: Ams, 1987), 35–54.

118. Stevens, *Police Officer Stress*, 184–88.

119. Michael Buerger, "Educating and Training the Future Police Officer," *Federal Law Enforcement Bulletin* 73, no. 1 (2004), http://www.fbi.gov/publications/leb/2004/jan2004/jan04leb.htm#page_27 (accessed March 11, 2006).

120. William G. Doerner, "Officer Retention Patterns: An Affirmative Action Concern for Police Agencies," *American Journal of Police* 14, no. 3/4 (1995): 197–210.

121. Each of these statements was articulated by female officers in written (questionnaire) or verbal (derived from interviews) forms during the study for this work, or the comments were derived from classroom participation among police officers.

122. Douglas L. Yearwood, "Analyzing Concerns among Police Administrators: Recruitment and Retention of Police Officers," *Police Chief* (2007), http://policechiefmagazine.org/ (accessed June 27, 2007).

123. Yearwood, "Analyzing Concerns among Police Administrators."

124. Tim Augustine, "Why Employees Really Decide to Leave," *Daily Resource for Entrepreneurs*, Inc.com (2007), http://hiring.inc.com/columns/taugustine/20050819.html (accessed June 27, 2007).

125. Augustine, "Why Employees Really Decide to Leave."

126. Flynn, "Training and Police Violence."

127. Koper, Maguire, and Moore, *Hiring and Retention Issues in Police Agencies*, 47.

128. Koper, Maguire, and Moore, *Hiring and Retention Issues in Police Agencies*, 47.

129. Stevens, *Police Officer Stress.*

130. Mastrofski, "Police Organization and Management Issues for the Next Decade."

131. Captain Robert P. Dunford, lecture at University of Massachusetts, Boston, to criminal justice students and faculty members, 2001, http://www.umb.edu/news/2001news/reporter/1101/dorchester.html (accessed June 27, 2007).

132. Written by the author, who lived in Pass Christian, Mississippi, August 2005. In the months that followed, the author conducted critical-incident debriefing sessions among New Orleans police; Jefferson Parish deputies; and municipal, county, and state troopers in Mississippi.

133. Mary Swerczek and Allen Powell II, "2,000 Kids Still Separated from Parents," *NOLA* (September 2005), http://www.nola.com/newslogs/tporleans/index.ssf?/mtlogs/nola_tporleans/archives/2005_09_17.html (accessed April 14, 2006).

134. In Louisiana, 658 bodies have been identified, and an additional 247 victims have not been identified. In Mississippi, 250 bodies have been identified and another 200 continue to be missing. See Nathan Burchfiel, "Statistics Suggest Race Not a Factor in Katrina Deaths," *CNSNews* (December 14, 2005), http://www.cnsnews.com/ViewNation.asp?Page= percent5CNation percent5Carchive percent5C200512 percent5CNAT 20051214b.html (accessed April 14, 2006).

135. Associated Press, "New Orleans Mayor Orders Looting Crackdown" (September 1, 2005), http://www.msnbc.msn.com/id/9063708/ (accessed April 14, 2006).

136. Joseph B. Treaster, "Law Officers, Overwhelmed, Are Quitting the Force." *New York Times* (September 4), http://www.nytimes.com/2005/09/04/national/nationalspecial/04police.html (accessed March 23, 2008).

137. Charles S. Heal, "The Manageable Future: Envisioning the End State," *FBI Law Enforcement Bulletin* 71, no. 1 (2002): 1–13.

138. Andrew J. Harvey, "Building an Organizational Foundation for the Future," *FBI Law Enforcement Bulletin* (1996), http://www.fbi (accessed June 24, 2007).

139. Thomas H. Ackerman, *Guides to Careers in Federal Law Enforcement* (East Lansing, MI: Hamilton Burrows Press, 2001).

140. Federal Law Enforcement Training Center (2007), http://www.fletc.gov/ (accessed June 25, 2007).

141. "Federal Law Enforcement Training at Glynco," http://www.fletc.gov/about-fletc/locations/glynco (accessed June 25, 2007).

Police Work

POLICE
DEPARTMENT
METRO

On the Job: Patrol and Patrol Units

Man creates problems. Government and bureaucrats magnify them 100 times.

—George Van Valkenburg

LEARNING OBJECTIVES

When you finish reading this chapter, you will be better prepared to:

- Describe the role of patrol officers
- Characterize the four functions of police patrols
- Describe deployment initiatives and duties of patrol officers
- Identify the general work experiences of patrol officers
- Articulate the outcome of fresh vehicle pursuits
- Describe the outcomes of traffic stops and searches, including checkpoints
- Characterize the primary factors affecting patrol service delivery
- Identify eight patrol studies and their relationship to evidence-based policing
- Characterize random, directed, and directed aggressive patrols
- Describe ways of improving traditional patrol
- Characterize 911 call experiences of police departments
- Explain the necessity of differential response initiatives
- Identify several patrol units in various police departments

CASE STUDY: YOU ARE THE POLICE OFFICER

It is 2:30 on a sunny afternoon, and you are a patrol officer who is observing Main Street. You observe a vehicle driven by a driver who is recklessly weaving in and out of the incoming traffic, probably at a speed greater than the posted limit of 35 miles per hour. When you first spotted the car, you were entering your police vehicle after completing a report provided by a family who reported extensive damage to their property, presumably by a person or persons unknown. You were unable to read the license plate number of the recklessly driven vehicle, but the plates appeared to be consistent with the state bordering yours.

Once in your vehicle, you radio in the location, speed, partial vehicle description, and reason for your chase. Frankly, there are so many controls and buttons to pull and push in your new police cruiser that you almost forgot to look at the oncoming traffic. Your training has taught you to wait to initiate your emergency strobe-heads until you are close to the vehicle in pursuit. You hope that the reckless driver will not strike another vehicle or pedestrian.

Dispatch advises that a robbery has taken place a few blocks from your last service call and identifies a suspect vehicle consistent with the vehicle almost in front of you: a dark blue, late-model Pontiac coup with out-of-state tags. You ignite your strobes and siren, and then hear the dispatcher say that other officers are in the vicinity and close to intercepting the vehicle. Your department has a restrictive pursuit driving policy but at roll call you were told that those policies changed. One thing is certain: It is not up to your discretion to make that call. You ask yourself, "Officers make 17 million traffic stops every year—why can't this be one of them?"

The Pontiac accelerates, and suddenly you are in pursuit of a fleeing felon, but your training has taught you to control the adrenaline and excitement in situations such as this one. "Be in control of yourself first"—you can still hear your training officer's words. You ease off the pedal, knowing that a school is nearby and that the suspect is heading for the major highway where fewer people will be at risk. Now you hear your sergeant's orders: "Break off the pursuit." You respond, "Roger that boss, 10-4." Dispatch adds that the subject is close to an elementary school whose day is ending soon, crowding the streets with many children.

In your rear-view mirror, you see another police cruiser driven by your old training officer. She gives you a thumbs-up sign, suggesting that she knew you backed off the chase before you heard the order.

1. When officers are on the job, why would they continually think about their training as often as this officer does?
2. In what way is it possible that technology is moving faster than training as experienced by this officer, who almost failed to initiate safety procedures in the police cruiser, such as observing traffic before pulling out into it?
3. Why was the chase called off? Could the robber have known that he wouldn't be pursued on a school day at 2:30 in afternoon?

Introduction

Patrol officers are the centerpiece of American policing. They are highly visible because they work in clearly marked police vehicles, wear uniforms, and often engage in a traffic stop or arrive at someone's residence with blue and red lights flashing. Uniformed patrol officers have general police duties, which include regular patrols and emergency call responses. They direct traffic at the scene of accidents, investigate burglaries, provide first-aid to accident victims, and support other officers and other police agencies during altercations, which can include calls that involve a fleeing felon or an injured constituent.[1] Patrol officers, their general work experiences, and specialized patrol units are the focus of Chapter 9. In addition, this chapter examines emergency call accounts related to the department and the officer. The heart of this chapter, however, focuses on evidence-based studies of police performance.

■ Role of Patrol Officers

The central role of a patrol officer is crime control, which consists of serving and protecting. Another potential role, depending on the jurisdictional policy, is to provide guidance to help people protect themselves, their homes, and their neighborhoods.[2] Viewed from this perspective, members of the public are the workers on whom the police depend to get the job done.[3] That is, the public has a responsibility for crime control, too.

Police departments assign more than 62 percent of their officers to patrol duty,[4] and the highest proportion of a patrol officer's time (depending on the jurisdiction) is typically devoted to emergency 911 calls.[5] For instance, in large police departments, patrol officers' responses to emergency calls take up 27 percent to 65 percent of their time on duty, and an estimated one fourth of those calls are actually genuine emergencies.[6] There are approximately 450,000 officers (men and women) patrolling the streets of America.

> In 2005, the average number of sworn officers in a jurisdiction was estimated at 108 (ranging from 5 officers to 34,000 officers in 2005), representing 1 officer for every 500 residents, at a per capita expenditure of about $155 for each officer, or an average of $8 million in each jurisdiction. Approximately 68 of the 108 sworn officers are assigned patrol and service call duties.[7]

In most jurisdictions, 70 to 75 percent of a patrol officer's time is unassigned.[8] These findings suggest patrol officers always have had, and continue to have, a lot of downtime available for restructuring.[9] It also means that patrol officers are supervised less often than expected and make many decisions on their own while on duty (see Chapter 13 for a discussion of police discretion). As local and state personnel who make significant decisions about people's lives, patrol officers have been called "**street-level bureaucrats**."[10] (See Chapter 10 for more detail.)

Additionally, patrol experience is the formative part of any police officer's career. Police agencies' promotions, duty assignments (and the better-running vehicles and equipment), and street training officers are often based on seniority.

■ Functions of Patrol

Crime will never be stopped. According to Emile Durkheim, "Crime [is] really a natural kind of social activity, an integral part of all healthy societies."[11] Crime can be controlled through the prevention and deterrence efforts of a professional police patrol. As noted in Chapter 2, Sir Robert Peel, upon establishing the London Metropolitan Police in 1829, proclaimed that "it should be understood at the outset, that the object to be obtained is the prevention of crime."[12] Of interest, while other professionals such as lawyers and doctors make their clients come to their offices, police officers are among the few professionals who continue to make house calls.[13]

In the United States, policymakers have identified four major functions of police patrol:

- Deter crime (crime cannot be stopped, but it can be controlled through professional police efforts)
- Reduce the fear of crime (see Chapter 3)
- Enhance quality-of-life experiences (see Chapter 3)
- Control wrongful officer behavior (see Chapter 14)

Patrol Geography

Police agencies usually organize their jurisdictions into geographic districts, with uniformed officers being assigned to patrol a specific area, such as part of the business district or school areas.[14] Patrol also includes more than patrol cars in some jurisdictions: It sometimes incorporates air patrol, foot patrol, mounted patrol, and other means.

Public college and university police departments, public school district police, and agencies serving transportation systems and facilities are examples of special police agencies that deploy patrol officers. For instance, Boston, Chicago, and New York all employ officers whose primary duty is to control crime on public transportation such as the subways and buses. Also, Section 8 housing and project housing often deploy their own police patrols. These agencies have special geographic jurisdictions and enforcement responsibilities in the United States. Most sworn personnel in such special agencies are uniformed officers with full police powers but limited jurisdictional assignments.

Types of Patrol Contacts

A key part of any patrol officer's duties is interaction with the public. Patrol officers have four types of contacts with the public:[15]

- Contacts with victims or witnesses of crimes and accidents
- Contacts with suspects
- Frequent casual contacts with the public
- A field interrogation (FI), sometimes called a field stop, field contact, or stop and frisk

Duties of Patrol Officers

The duties of a patrol officer can be described as patrolling or monitoring activities in whatever zone or area assigned in the organization's jurisdiction. Patrol officers focus on detection, identification, and apprehension of violators of federal, state, and local laws. They are involved in case investigations when and where necessary.

While on patrol, officers learn about their patrol area and stay alert for anything unusual or out-of-place. For instance, delivery trucks are not usually unloading or loading at midnight. Patrol officers investigate and note suspicious circumstances and hazards to public safety, and are dispatched to emergency calls for assistance. During their shift, they may identify, pursue, and arrest suspects; resolve problems within the community; enforce laws; and direct traffic.

Patrol officers respond to a variety of requests from the public for related services. For example, they may respond to accidents on the highways, at homes, and in businesses. They often administer first-aid and contact providers for other services required by the injured, mentally ill, and children (see Chapter 10 for more detail). In addition, they perform public relations duties and complete required reports and paperwork.

While on duty, patrol officers must adhere to departmental rules, regulations, policies, and procedures (see Chapter 13). They must also adhere to federal, state, and local laws especially during the performance of official duties and provide care for the safety of others. Patrol officers perform the duties assigned to them by their supervisors: They transport and process suspects to hospitals, aftercare units, and city jails; testify in court; and review paperwork with supervisors, prosecutors, and investigators. Finally, they attend in-service training to enhance their skills.

The Acworth, Georgia, Police Department gives the following general instructions to its patrol officers:

> Under general supervision and definite operating procedures, a person in this position patrols the city and responds to calls for service and initiates calls in an effort to detect, stop, and apprehend violators of laws and ordinances in order to protect life and property. When on the scene of a violation or disturbance, the officer must take the initiative to control the situation, and must exercise considerable judgment in interpreting laws, ordinances, policies, and procedures. In general contact with the public, the officer must achieve a balance between enforcement and maintaining good public relations for the police department. A person in this position does not exercise supervisory controls over other employees.[16]

General Work Experiences of Patrol Officers

General work descriptions are discussed in Chapter 15 in terms of how they are associated with sources of stress, but it is useful to review these categories here for a better understanding of officer work-related issues. Patrol and service calls are officers' primary activities, and many aspects of those activities merit attention.

The following findings were obtained from a study of regionalized groups of officers working in jurisdictions from New England to Florida to New Orleans; accordingly, they might not necessarily apply to officers in other parts of the country. In this study, 310 officers were asked to rate 14 general work assignments.[17] The question in the survey read as follows:

> Following is a list of experiences that law enforcement officers like yourself experience sometime in your careers. How often do these experiences create major stress of officers?
>
> What's your guess? 5 = Always, 4 = very often, 3 = often, 2 = seldom, and 1 = never.

The officers' responses are shown in Table 9–1 and discussed next.

TABLE 9–1 General Work Activities of Patrol Officers ($N = 310$)

Descriptions	Mean Score
1. Conflict with regulations	4.75
2. Domestic violence	4.46
3. Losing control	4.29
4. Child beaten/abused	4.18
5. Excessive paperwork	3.46
6. Public disrespect	3.18
7. Another officer injured	3.15
8. Lack of recognition	3.03
9. Poor supervisor support	2.62
10. Disrespect of the courts	2.34
11. Shift work	2.25
12. Death notification	1.92
13. Poor fringe benefits	1.84
14. Accidents in patrol vehicles	1.37

Missing cases and "other" and lesser reported categories are not shown.

Conflict with Regulations

Conflict with regulations (rated 4.75, almost always) can occur when patrol officers are instructed to perform in a specific way that is not consistent with their police subculture expectations (see Chapter 13) or an officer's own understanding.[18] Also, the goals and objectives of some police agencies are communicated ineffectively, resulting in an officer redefining his or her own understanding of a goal (see the discussion of administrative rulings in Chapter 13). Consequently, performance mirrors the redefinition of the procedure rather than the organizational or supervisor's intended message.[19] Conflict or a misunderstanding of policy can put officers and constituents at risk and can open the door to street justice, corruption, or poor service.[20] Finally, as one expert argues, "police agencies have long been notorious for urging rank-and-file officers to do one thing, while rewarding them for doing something else."[21]

Domestic Violence Stops

Domestic violence stops (rated 4.46, very often) are part of the duties of patrol officers who respond frequently to domestic and family violence emergency service calls.[22] Such calls represent both the service calls made most frequently by patrol officers and the calls in which officers are most likely to be assaulted.[23] (See Chapter 10 for more detail about domestic violence service calls.)

Losing Control

"Losing control" (rated 4.29, very often) refers to the perception that an officer has little control over his or her work conditions, among suspects or situations, and over measures of performance.[24] "Work conditions" can refer to assignments, specific equipment, shift, promotion, transfer, or partners. Losing control of a suspect or a situation, witnesses, and the public at large is a serious concern, as control is the primary principle guiding an officer's relationship with other persons, places, and circumstances. Losing control could mean placing others at risk—including the officer. Most often, it is emphasized that the control and proper use of force can prevent or reduce injury to the officer, a suspect, and others.[25]

The perception of loss of control can also reflect the methods employed to measure the performance of a patrol officer. Traditionally, police officer productivity has been measured by factors such as arrests, stops, traffic citations, the value of recovered property, and reduction in rates of crashes and crime.[26] Patrol officers have little control over any of these factors, with the possible exception of crashes, as they depend on a number of other variables.[27]

Child Beaten or Abused Calls

Child beaten or abused calls (rated 4.18, very often) refer to service calls in which the victim is an abused or injured child.[28] For instance, one officer recalls, "I was in an emergency room and I saw this other officer come rushing in, giving a baby mouth-to-mouth resuscitation. He was still trying to breathe life into the child who was dead, and this big cop fell to his knees and started to cry, right there. I never, never forgot that."[29] Another officer talks about the frustration he felt when he found a young girl hung by the neck. Children who are injured as a result of auto accidents, circumstances, or criminal activity (including sexual abuse or neglect) create a great deal of discomfort among most first responders such as patrol officers. An abused child is beyond their individual assistance and out of their control. The child is already harmed, and a patrol officer realizes there is little that can be done to "unharm" a child despite all the officer's bravery, power, and skill.

Excessive Paperwork

Excessive paperwork (rated 3.46, sometimes) entails the completion of forms that require information about arrests, stops, traffic citations, and other occurrences arising from patrol and service calls.[30] One source recommends that officers learn to "control the paper flood or it will control them."[31] The sheer volume of reports required of patrol officers make their completion both time-consuming and difficult to absorb. Some officers report feeling as though they are clerks rather than professionals.[32]

Although a "paper trail" is critical, all too often forms are obsolete and redundant, and officers are called upon to do paperwork that could easily and efficiently be accomplished by clerical staff or with advanced technology.[33] Video and other techniques are available that could enhance police intervention, further public safety, and translate video into words, if necessary. In 2000, patrol officers in approximately four fifths of larger state and local police agencies used in-field computers or terminals.[34] Officers had direct access via in-field computers to vehicle and driving records in more than three fifths of local agencies in 2000. Although such technology may help officers with their paperwork, it intensifies the documentation needed to keep up with legal concerns such as racial profiling. Among state agencies, approximately three in four kept computerized files on traffic accidents and traffic citations in 2000, whereas only a little more than half maintained computerized data on arrests, incident reports, calls for service, and criminal histories.[35]

Public Disrespect

Public disrespect (rated 3.18, sometimes) relates to the issue of public trust.[36] If the public does not trust authority, then one way to demonstrate those feelings is by showing disrespect toward officers.[37] Evidence indicating a lack of trust of officers includes the high percentage of unreported crimes, less budget funding for patrol operations including equipment and training, and less police power as a result of civil liability lawsuits (see Chapter 14).[38] Officers often feel that the public holds a negative view of them and that the public's perspective is reinforced by the media.[39]

Practices of public disrespect are part of the cultural norms of some groups. For instance, youngsters learn that disrespecting officers might add to their reputation. An urban officer says, "Kids are just running their mouths at you because you're a cop and you got to learn not to take it personal; otherwise, you'd be arresting everybody." Then, too, early police disrespect can turn to subsequent violence. For example, in the inner city, young children learn to play the game, which is fairly simple: When the officers roll by, juveniles do their best to "punk 'em out."[40] The tougher kids often have fathers, brothers, and maybe even mothers in jail, and those are the ones who are usually willing to go the furthest. They spit at patrol cars, they call the officers filthy names, and many wave one-fingered salutes and smile.[41]

> Recent studies report that particular types of suspects, situations, and officers tend to lead to greater **displays of resistance** during police–subject encounters and increase the potential for an arrest.[42]

Variables related to the protection of social identities (e.g., the presence of bystanders, the location of the encounter) and measures of suspects' propensity for aggression (e.g., gender, intoxication) may also predict suspect behavior.[43] Such proxy measures of constituents' perceptions of officers' legitimacy have a mixed influence over suspects' displays of resistance toward officers. Of particular importance is the finding that nonwhite suspects are significantly more likely to display noncompliance (i.e., passively resisting officers' authority by refusing to answer questions of any type) with officers.

Why youths disrespect patrol officers can be seen in this example: A Swampscott, Massachusetts, patrol officer stood in the courtroom, "attempting to swallow his rage and mash his mortification,"[44] as one youth—described by the officer as a virtual one-person crime wave—had his case dismissed by a judge. The street officer had apprehended the youth breaking into a home. The homeowner was a witness to the arrest and came to court numerous times to testify but the case was repeatedly delayed. On this day, the homeowner was told by her boss that she was risking her job if she went to court again to testify. She sat in court waiting for the case to be called; finally, in mid-afternoon, her boss said that if she didn't come to work that her job was at risk. The case was called at 4:00 P.M. The judged chewed the officer out for letting the witness leave and dismissed the charges. "I'll never forget how that kid and his mother stood there and laughed, laughed at me," recalls the officer. The jeering by the accused, the verbal humiliation by the judge, and the release of a budding criminal, guilty in fact but exonerated by the system, "certainly were stressful experiences for the law enforcement officer."[45]

Reports of Another Officer Injured

"Another officer reported injured" (rated 3.15, sometimes) refers to an officer who is harmed as a result of an altercation with a suspect.[46] Almost 59,000 officers in the United States (118 per 1,000) were assaulted and 48 were feloniously killed in 2006 while providing police services.[47]

Lack of Recognition

Lack of recognition (rated 3.03, sometimes) refers to failure to recognize officers' accomplishments and promote them, in some cases. Sometimes officers are honored: For instance, in Orange City, Florida, Officer Mike Johnson was presented with a Good Samaritan Award in recognition of his successful life-saving efforts performed while on duty in the community.[48] All too frequently, however, police supervisors ignore the good work of patrol officers.[49] Most often, officers report that recognition from police command happens only when serious events occur. For example, in New York City after the terrorist attacks of September 11, 2001, officers report that superiors who rarely came to any event gathered officers together and gave a congratulatory speech.[50] Some of the strongest feelings of frustration that are expressed by officers seem to relate to inequities in the system associated with recognition, particularly linked to promotions.[51]

Poor Supervisor Support

Poor supervisor support (rated 2.62, almost sometimes) is a persistent problem: Officers say that their supervisors sometimes show little regard for their well-being.[52] One distinctive characteristic of police organizations is that virtually everyone in the organization, "from a station house broom to chief, shares the common experience of having worked the street as a patrol officer."[53] Moving down the hierarchical ladder, lateral entry is unknown because supervisor positions are filled from the pool of candidates directly below the vacant slot.

Disrespect of the Courts

Disrespect of the courts (rated 2.34, seldom) refers to perceptions held by patrol officers about how judges and other members of the court respond to them while they are testifying. One officer says that "a judge told me off for doing my job in front of the suspects and his criminal friends."[54] Also, patrol officers believe that court rulings tend to be too lenient on violators and too restrictive on procedural issues such as evidence admission; as a consequence, many officers view the courts as an adversary.[55]

Shift Work

Shift work (rated 2.25, seldom) refers the practice of rotating assignment periods for officers. Few are capable of dodging the rotating bullet unless an officer learns how to stay off the supervisor's radar, rotating shift work is inevitable.[56] Because humans are equipped with a biological clock that follows a circadian rhythm, they have a preference for day-oriented routines.[57] The circadian system's job is to prepare the body for restful sleep at night and active wakefulness during the day. Work patterns that tend to change are met with resistance, which an individual often cannot control. When the body's circadian rhythm is negatively affected by night work, it results in desynchronization (the relation that exists when things occur at unrelated times—for instance, a deviation from the night–sleep, day–wake pattern).[58]

Tired officers manage people, make observations, and communicate with less effectiveness than officers who are not tired. The complex interactions between thinking, feeling, and acting are all affected by sleep deprivation.[59] Obviously, reduced skill levels as a result of fatigued officers and shift work affect the quality of police services.[60] Some experts say that officers are less able to perform efficiently when fatigued and that greater sleep disruption occurs as they age.[61]

Death Notification

Death notification (rated 1.92, almost seldom) occurs when an officer must tell family members or friends of an individual's death—for instance, as a victim of a motor vehicle crash or as a victim of violent crime. The stress presented by this duty is clear. As one officer said, "I investigated a double fatality recently in Peabody, Massachusetts. What a tough, tough thing, seeing needless death caused by speeding. And how do you tell a mother and father that the pride of their life, their beautiful 16-year-old daughter, is dead? You might as well be going to their home to cut their legs out from under them."[62] Likewise, victim notification among robbery and aggravated assault victims, especially in families residing in disadvantaged neighborhoods, is not a pleasant event for officers.[63] Officers who attempt to notify families can sometimes encounter unexpected results:

> I had to notify this family that their son was in the hospital in serious condition for a hit-and-run after school. His parents had no phone, so off I go. They threw a pot of hot water on me as I climbed the steps to their fifth-floor walk-up. I yelled that I was there to help about Jesse, their son. "He doesn't live here no more," the mother shouted back.[64]

Poor Fringe Benefits

Poor fringe benefits (rated 1.84, almost seldom) refers to a form of personnel remuneration for the performance of services to a recipient.[65] One source suggests that fringe benefits are rigidly structured by civil service procedures and union contracts. Ultimately, however, both pay and fringe benefits are tied to rank.[66]

Fringe benefits can include health insurance, life insurance, dental insurance, worker's compensation, unemployment compensation, retirement, Social Security, and funeral benefits. They may also include vehicles, tuition reimbursement, travel expenses for professional development, free in-service training, legal guidance, counselors, and financial advisers and lawyers for the employee and often family members of the employee. One source suggests that up to 25 percent of budgets for police personnel go to fringe benefits, as compared to 30 percent to 35 percent in private-sector companies.[67]

Accidents in Patrol Vehicles

Accidents in patrol vehicles (rated 1.37, never) include all types of accidents involving police vehicles.[68] Research shows that officers who are engaged in pursuit of criminals on highways experience a collision of some type in 32 percent of those pursuits; in 13 percent of those cases, officers are injured.[69]

■ Fresh Vehicle Pursuit (Hot Pursuit)

Most vehicle pursuits are conducted by patrol officers. Some observers like to refer to vehicle fresh pursuits as hot pursuits or, more simply, as chases.

Fresh pursuit is an active attempt by a patrol officer to apprehend one or more occupants of a motor vehicle when the driver of the vehicle is intentionally resisting the apprehension by maintaining or increasing speed or by ignoring the police officer's signals to stop.[70]

Police vehicle pursuits usually involve a fleeing suspect. One concern associated with high-speed police chases is whether the benefits of apprehension outweigh the risks to anyone who comes in contract with the chase: other officers, the public, and the suspect. Pursuits are both effective and dangerous. It is incumbent upon police departments' leadership to consider the factors influencing the outcome of vehicular pursuits and to provide the training necessary to enhance the safety of their officers. "Failure to train" is a plausible justification for litigation when pursuits go wrong (see Chapter 8).

There is a consensus among patrol officers that chases are dangerous, must be controlled, and increase the adrenaline and excitement in those who participate. Ideally, the burst of adrenaline and excitement highs will not distract officers from public safety practices. Pursuit policies range from a "total ban to allowing officers complete discretion."[71]

The National Highway Traffic Safety Administration (NHTSA) estimates that approximately 315 deaths occur each year in the United States as a result of police vehicle pursuits.[72] A traffic accident constitutes the most common terminating event in an urban pursuit. Police pursuit records show that the majority of pursuits involve a stop for a traffic violation and that 42 percent of those injured or killed in the course of such pursuits are innocent third parties.

To chase or not to chase can depend on officer discretion, policy, or the seriousness of the crime. The public has demanded that police discretion related to fresh pursuit be limited through prohibitions, supervision, and technology.[73]

For example, officers sometimes carry spiked strips (**stop sticks**) and deploy them in the path of a fleeing suspect to blow out the vehicle's tires. The Cincinnati (Ohio) Police Department successfully implemented use of stop sticks as a risk-reduction technique following a string of pursuit-related tragedies. A large number of police departments are employing stop sticks with similar success. Also, a warning system can be activated that sends a signal to any motorist with a radar detector of an approaching police pursuit.

Other technological advances include an ultrasonic device that shoots a burst of microwave energy at a fleeing suspect's vehicle, which causes the suspect vehicle's electronic system to fail. A robot-like cart that jettisons from the front of the primary police pursuit vehicle is also available. This cart attempts to overtake the fleeing vehicle and electronically "zap" the engine out of service. Researchers are also testing radio-technologic devices (similar to stolen-car tracking systems) that electronically disable a vehicle.

Some agencies have used helicopters with good results in pursuits, such as in parts of California and in cities such as Baltimore, Miami, and Philadelphia.

Chicago and Dallas have increased criminal penalties in the hopes of reducing chases and established a "balance test" for officers to determine if they should chase a suspect. Ideas to help officers find the appropriate "balance" include these justifications for pursuit:

- When the officer has probable cause to believe that a felony involving the use or threat of physical force or violence has been, or is about to be, committed
- When the officer reasonably believes that the immediate need to apprehend the offender outweighs the risk to any person of collision, injury, or death
- When the officer is called to assist another police agency that has initiated a pursuit under the same circumstances.

Before undertaking a chase, officers should consider such conditions as weather, traffic-control devices, character of the neighborhood (residential or business), traffic volume, and road and vehicle conditions.[74] In some jurisdictions, a patrol officer can initiate a pursuit and call dispatch, which then assigns a field supervisor who can guide the officer through the process or terminate the chase.

In many jurisdictions, the policy is to terminate or to control vehicle pursuits.[75] Such policy could change in light of a recent U.S. Supreme Court ruling, which has provided officers and their departments with significant protection from chase-related lawsuits.[76] In this case, Victor Harris was left a quadriplegic after a Deputy Thomas Scott rammed his car off the road. "The car chase that [Harris] initiated in this case posed substantial and immediate risk of serious physical injury to others," Justice Antonin Scalia wrote for the majority. "Scott's attempt to terminate the chase by forcing [Harris] off the road was reasonable."

Before *Harris v. Scott*, local police agencies were four times as likely to have a restrictive type (73 percent) of pursuit driving policy (using criteria such as offense type or maximum speed) as they were to have a judgmental type of policy (19 percent); 8 percent left pursuits up to the officer's discretion. State agencies were equally likely to have a judgmental pursuit policy (45 percent) as to have a restrictive one (51 percent).[77] Research shows that placing too many restraints on the police regarding pursuits can put the public at risk, whereas insufficient controls on police pursuit can lead to excessive accidents and injuries.[78] Also, officers make an estimated 17 million annual traffic stops;[79] speeding was the most common reason for those stops in 2005.

■ Patrol Officers' Handling of Drunk Drivers and Checkpoints

Driving under the influence (DUI—although many jurisdictions refer to driving drunk by some other term) arrests are estimated to snag 1 million drivers each year in the United States.[80] Most alcohol-impaired driving arrests are made by officers on routine patrol who discern signs of impairment after stopping a driver for an ordinary traffic violation. In some jurisdictions, specialized patrols work exclusively on alcohol-impaired driving enforcement. Police in the majority of states can stop drivers at roadblocks known as sobriety checkpoints.

The National Safety Council advises that every 30 minutes someone dies in an alcohol-related crash in the United States.[81] Annually, alcohol-related motor vehicle crashes kill nearly 17,000 people. Alcohol is a factor in 6 percent of all traffic crashes and more than 40 percent of all fatal crashes.

Impairment is determined by measuring the person's blood alcohol concentration (BAC), which expresses the amount of alcohol in a person's blood as weight of alcohol per unit of volume of blood. For example, 0.08 percent BAC indicates 80 milligrams of alcohol per 1 liter of blood—considered legally drunk in most jurisdictions.

Police may use checkpoints to stop drivers in an attempt to identify impaired drivers.[82] **Checkpoints** are a visible enforcement method intended to deter potential offenders and to catch violators. This threat is key to general deterrence.[83]

In 1990, the U.S. Supreme Court confirmed that, when properly conducted, sobriety checkpoints are legal. Most state courts that have addressed this issue have upheld the legality of checkpoints as well, although some have interpreted state law to prohibit checkpoints that measure alcohol per 100 milliliters of blood. For most legal purposes, however, a blood sample is not necessary to determine a person's BAC; it can be measured more simply by analyzing exhaled breath.

A major purpose of frequent checkpoints is to increase public awareness. One study in two neighboring jurisdictions shows how checkpoints can change public perceptions.[84] Fairfax County, Virginia, had a history of vigorous enforcement of alcohol-impaired driving laws and used unpublicized drinking-driver patrols that produced high arrest rates. Nearby Montgomery County, Maryland, had historically lower arrest rates but used well-publicized sobriety checkpoints in an effort to deter alcohol-impaired driving. Surveys revealed that public awareness of enforcement programs was far greater in Montgomery County than in Fairfax County. People in both counties incorrectly believed the probability of arrest was higher in Montgomery County, simply because checkpoints were conducted there.

Changed perceptions can lead to fewer crashes. For example, one study showed a 22 percent decline in fatal crashes resulted from roadblocks.[85] Also, research shows that small-scale checkpoints with only a few officers can be conducted successfully.[86]

Despite the clear advantages of sobriety checkpoints, however, 10 states prohibit them.

> Measured in arrests per person-hour, a dedicated police patrol is the most effective method of apprehending traffic violators, including drunk drivers.[87]

In part because of patrol efforts, the number of persons killed in alcohol-related crashes each year in the United States declined by 34 percent from 1982 (when more than 26,000 deaths occurred) to 1994 (when approximately 17,000 fatalities occurred).[88] Sobriety checkpoints have been criticized for producing fewer arrests per person-hour than do dedicated patrols, but some studies show that arrest rates can be increased greatly when passive alcohol sensors are used to help police detect drunk drivers.[89]

Arrest-Related Homicides

During the three-year period from 2003 to 2005, 47 states and the District of Columbia reported 2,002 arrest-related homicides.[90] Homicide by police officers accounted for approximately 55 percent (1,095) of those deaths that occurred during the performance of an arrest (see the discussion of "suicide-by-cop" in Chapter 10). Other causes of arrest-related deaths included intoxication (13 percent), suicide (12 percent), accidental injury (7 percent), illness/natural causes (6 percent), and other persons (0.05 percent). In 8 percent of the cases, the causes were unknown. Three fourths of the arrest-related deaths involved violent crime suspects. Also, 97 percent of all suspects killed were male and 70 percent were white. Of concern, conducted-energy devices (CEDs), such as stun guns or tasers, were involved in 36 arrest-related deaths reported during the same period.

Factors Affecting Patrol Service Delivery

When members of the public want more police protection, they often imply that they want more frequent patrols. Several factors affect the number of officers available for patrol:[91]

- Ratio of sworn officers to population policed (see Chapter 5)
- Number of employed sworn officers by the department assigned to patrol
- Budget available to hire and train more officers
- Distribution of officers in police cars (one or two officers per unit)
- Distribution of officers by day, time, and area
- Work styles of the officers (see Chapter 8)
- Previous experience of the community (see Chapter 2)
- History of the department (both management and patrol officer experiences)
- History of the community (some communities have torn up patrol officers so badly that finding candidates to work there is next to impossible; see Chapter 8)[92]

Typology of Patrol Officers

Typologies (classifications of types) of individual police officer styles have been proposed by several researchers. For a look at organizational styles of policing, see Chapter 6. Here, we focus on descriptions of patrol officers' personal styles. Several are briefly summarized on the next page.

Robert B. Coates's Typology

Coates describes the "big picture" officer as follows:[93]

- Legalistic—follow the letter of the law when making stops, and so on. Such an officer is more likely to issue speeding citations and to make an arrest than to listen to excuses.
- Task officer—the primary focus of his or her job is the task at hand, no distractions.
- Community service officers—community relations are important to his or her success.

Susan O. White's Typology

White describes a "robo-cop" as having the following characteristics:[94]

- Tough officer
- Problem solver
- Crime fighter
- Rule applier

William K. Muir's Typology

Muir's concern relates to the passion to use force. He distinguishes between professionals and enforcers, who use force, and reciprocators and avoiders, who avoid force:[95]

- Professionals—are objective, use force when required by law
- Enforcers—tend to be cynical, use of force is coercion oriented
- Reciprocators—see force as the person's "just desserts," criminals get what they deserve
- Avoiders—avoid the use of force

J. J. Broderick's Typology

According to Broderick, when the major concern is due process, officers can be seen as falling into four categories:[96]

- Enforcers—have little respect for due process
- Idealists—want to keep the peace but have respect for due process
- Optimists—emphasize due process
- Realists—do not seem to care about anything, much less due process

John Kleinig's Typology

Kleinig's typology could be also influenced by organizational policy. According to this researcher, officers tend to think of themselves in the following ways:[97]

- Crime fighters—using the military model to portray criminals as the enemy or bad guys and police as the good guys
- Emergency operators—using the fire fighter model to portray themselves as emergency-handling professionals who just happen to be competent at crime control
- Social enforcers—using the band-aid model to portray themselves as fixer-uppers who settle things once and for all by force if necessary

- Social peacekeepers—using the peacekeeping model to portray themselves as pacifiers of the populace, bringing peace and psychologically satisfying closure to social conflicts

It is unlikely that an officer will respond with the same style all the time or by demonstrating all of the previously described styles. Instead, officers typically change their styles to match the situation.

Gender-Based Differences in Patrol Work Experiences

Studies show that the gender of the officer has little bearing on the effectiveness or outcomes of stops, although one difference has been found during service calls:[98] Female officers responding to sexual assault complaints make more arrests and the suspects they arrest are more likely to be convicted than the suspects arrested by male officers.[99]

A Miami–Dade County deputy explains how her job demands influenced her family relationships:

> A typical shift for me? Try this. One day last week I was good-to-go home after my shift. My sergeant comes to me in the locker room and says he needs me to fill in for Sanchez, who took sick because of some raw fish. I worked two shifts and eventually went home.
>
> It was around 11 when I pulled into the drive. The house was dark, kids were asleep. I could hear Jay Leno on the tube coming from my bedroom. When I peeked into the room, my husband was asleep, too. I tiptoed in and clicked it off. Another night alone.
>
> I had several cups of coffee to stay awake during my work shifts, and there was no way I was tired; I think I was wired. About an hour later, I couldn't stand the silence anymore and ran over to Harry's, a cop hangout near the station for some conversation and a beer. Some of the guys were there talking about a guy I hooked-up last week. The conversation was comforting, but I knew it was time to roll home when the sun was beaming through the windows.
>
> Got home, slept for an hour, got the kids off to school, and told my husband that I'd be home for dinner. That was a joke! Didn't get home until 10, and he was already in bed watching the tube. I didn't want to sneak out for a beer, but I did as soon as he dozed off.
>
> It didn't take long after that. We're separated, and he's got the kids and the house. I got a little place of my own. I go to Harry's probably more than I should, but so do a lot of the guys I work with. We have much to share. I miss my kids but never really saw them.[100]

Male and female patrol officers share similar experiences. As suggested by the Miami–Dade County deputy's narrative, merging into a police subculture might be an easy process no matter your gender. Nevertheless, female officers experience more hardships than do male officers simply because of their gender. For some, this thought may bring up memories of patronizing remarks from officers and offenders alike, but it is also as simple a truth as realizing that female officers must unbelt while visiting a restroom. The hurdles female officers jump over every day are the same ones male officers jump over—but then that entire set of hurdles was designed for men. (Also see Chapter 15 for more accounts of the experiences of female officers.)

In one evidenced-based study of 133 female officers, their responses were similar to those of male officers to questions about job stress.[101] Conversely, significant differences emerged in female officers' answers concerning their relations with other police officers, supervisors, the courts, and the public. A comment by one female street officer, Boston Officer Maria Gonzalez, is typical of the way in which many female officers responded:

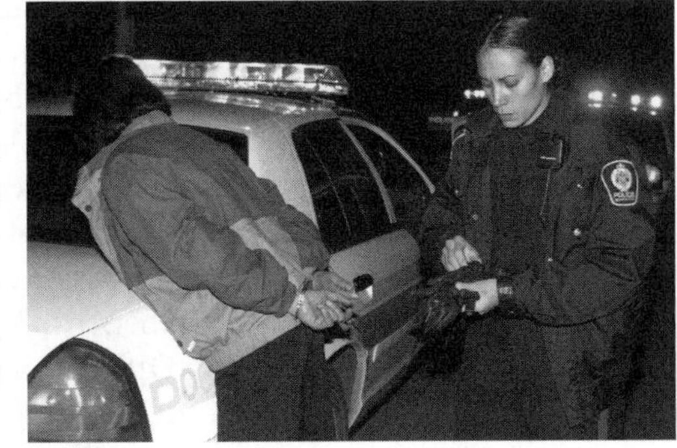

> Those of us who really like being out there can do the job as effectively as a guy. But we want to be respected as an equal. We're still women, and I expect that when it comes to the badge, that respect is there; at other times, I am a woman and I expect others to be mindful of that. Mostly, when guys find out that I'm a Boston copper, they're intimidated by me. Other times, other officers, including my supervisor, just deal with me because they don't want a lawsuit. . . . I just want to do my job, and I expect to be respected as a cop, a woman, and Puerto Rican women at that.[102]

Women officers experience other pressures, too. The director (retired) of the Massachusetts state trooper stress unit says, "There has been more difficulty with the assimilation of female police officers than other groups."[103] He thinks that some of these problems occur because of the perception that women receive preferred treatment, that their attributes of courage and strength are not similar to those of males, and that they are not able to meet the same physical standards as their male counterparts. Another mental health practitioner adds, "When a woman takes on a male macho job, she takes on the male macho disease. She feels she has to prove herself better than the men. And a woman has been raised to show her emotions, and now she has to start covering up those feelings."[104]

"That's typical macho crap," Massachusetts State Trooper Lisa Cesso says about the former director's comment.[105] She goes on to say:

> In my generation, guys grew up knowing that females are as good as, if not better [than male officers]. It's those good ole boys that keep playing their macho [crap]. Since they're in charge, they keep thinking [women are] . . . their little daughters.
>
> Let me give you an example. Sometime after 9/11, we had machine gun practice. They handed me this outdated cannon and said, "Take your point" at the range. If I said no to the instructor, he'd tell my boss and the next thing you know, he'd cut me out of overtime at Logan [Boston's airport]. These good ole boys want to control their turf and that's their little way of controlling me. For young guys, they have another approach. We're still early in the evolution of that process.

■ Evidence-Based Studies on the Effectiveness of Patrol

Evidence-based studies have shown that "less than 3 percent of the street addresses and 3 percent of the population in a city produce over half the crime and arrests."[106] Once that idea is clarified, deploying officers is easier. "When dispatch says there's 211A [domestic violence call in Boston], . . . By the time the address is popped, I know the apartment number before it's given. I've been there like a hundred times," says a Boston officer.

Study Findings

Preventive police patrol was traditionally considered a great deterrent to criminal behavior.[107] Indeed, a visible police presence and rapid deployment of patrol officers to emergency calls were viewed as particularly effective police techniques. Research, however, both challenged and fine-tuned those perceptions of patrol. The most famous study in this regard was the **Kansas City (Missouri) Preventive Patrol Experiment**, a 1973 study that reported patrol techniques have little effect on crime patterns, and that patrol presence has little influence on reducing the fear of crime among community members.[108] (This study is discussed in Chapter 4 in more detail.)

Other studies have since refined the findings of the Kansas City experiment. Some of their findings are highlighted here.

San Diego, California, Field Interrogation Report (1975)[109]

- Officers were encouraged to stop, question, and sometimes search and field interrogate (FI) suspects.
- Officers were provided with extensive training in recognizing valid FI situations.
- "Suppressible crimes"—that is, robbery, burglary, grand theft, petty theft, auto theft, assault/battery, sex crimes, and malicious mischief/disturbances—were not significantly reduced during the time periods studied.

Directed Patrol in New Haven, Connecticut, and Pontiac, Michigan (1976)[110]

- The results of crime analysis were used to direct noncommitted patrol time to problem locations.
- Direction by dispatchers proved problematic.
- A reduction in criminal incidents was observed (though the study lacked control areas with which to compare those results).

Split Force Patrol in Wilmington, Delaware (1976)[111]

- Structured patrols concentrated on problem areas and follow-up.
- Patrol officers' arrest rates increased by 4 percent.
- Detective division clearances decreased substantially.
- The entire department clearance rate decreased by 28 percent.

Newark, New Jersey, and Flint, Michigan, Foot Patrols (1981)[112]

- The implementation of foot patrols had no impact on crime rates in Newark, but led to a 9 percent reduction in crime in Flint.
- A decrease in fear of crime was found in both cities.
- Members of the public had significantly improved satisfaction with police services in both cities: 33 percent of Flint residents knew the beat officer by name, and 50 percent of the remainder could recognize the beat officer.

Minneapolis Repeat Call Address Policing Unit (1981)[113]

- The **RECAP** program was originally designed to reduce the homicide rate in Minneapolis.
- RECAP units were assigned to the top 250 commercial and residential addresses in the city, as ranked by the number of emergency calls made to police.
- Each one of the addresses could have four or more apartments with chronic repeat calls for domestic violence.
- Three percent of 115,000 addresses accounted for 50 percent of all calls for service (CFS).
- After 6 months, target addresses had 15 percent fewer CFS, but this changed disappeared after one year.
- Sixty-four percent of the calls to the police came from 5 percent of the city's addresses.
- The policy of mandatory arrest for domestic violence reduced the prevalence of repeat violence against the same victim by half, from 20 percent to 10 percent.
- The RECAP program served as a catalyst for mandatory arrest in misdemeanor domestic cases across the United States.

Kansas City, Missouri, Gun Reduction Experiment (1993)[114]

- Beat 144 is an 8-block by 10-block area. Its homicide rate was 177 deaths per 100,000 population, or 20 times the national average at the time.
- The population in beat 144 is 92 percent nonwhite, and has 66 percent home ownership.
- Directed police patrols in "hot spots" increased (by 65%) seizures of illegally carried guns.
- Gun crimes declined by 49 percent.
- Drive-by shootings and homicides dropped significantly.
- Saturation (spot checks) generated 29 more guns seized, 83 fewer gun crimes, and 55 patrol hours invested per gun crime prevented. Traffic stops were most productive in this regard, with one gun being seized for every 28 stops.

Joliet, Illinois, Repairing Neighborhoods with Partnerships (2000)[115]

- Joliet faced a crime problem that was intensified by the presence of burgeoning gangs.
- Open-air drug markets were observed in some older neighborhoods (which were characterized by absentee landlords).
- Calls for service were assessed in 15 target neighborhoods, with 890 CFS to 297 different addresses being noted. Water bills showed that 700 of these calls were to residential property and 137 calls were to commercial properties.
- Police developed a formal abatement process for dealing with drug, weapons, and nuisances. Properties had to pass city maintenance inspections, which forced landlords to cooperate with police.
- Abatement cases had a positive effect on the community by reducing the number of CFS (by 31 percent to 62 percent), reducing crime, and increasing quality of life.

Criticism of Patrol Studies

Many argue that the evidence-based police studies conducted to date tend to share several characteristics:[116]

- Clearly defined intervention strategies
- Targeted at particular offenses
- Crimes committed by particular offenders
- Crimes committed at specific places
- Crimes committed at specific times
- Conduct of the studies by individuals with little police street experience

Given their narrow focus, some argue, the studies relate to only a small segment of actual police experiences and are not reliable. In practice, it was recognized that specific initiatives were not as efficient as expected. Nevertheless, the results of these studies, through trial and error, can aid police performance. "As one delves more deeply into the various factors that shape police functioning, one finds that laws, public expectations, and the realities of the tasks in which police are engaged require all kinds of compromises and often place the police in a no-win situation."[117]

Many of the early studies were subsequently repeated in some way; this practice, called replication, is sometimes used by other researchers in an attempt to verify the original findings. For instance, a subsequent evaluation of the RECAP program was conducted in Minneapolis.[118] This one-year randomized test examined the effects of increased patrol at 55 of 110 crime "hot spots," which were monitored by 7,542 hours of systematic observations. It became evident that most crime occurred at a few addresses, and stepped-up patrols in those areas resulted in reductions in total crime calls ranging from 6 percent to 13 percent. Observed disorder was only half as prevalent in experimental hot spots as in control hot spots. Researchers concluded that substantial increases in police patrol presence can cause modest reductions in crime and more impressive reductions in disorder within high-crime locations.

Evidence-based researchers with little street experience often have fewer preconceived notions about "right answers" than do "insiders" and can provide indicators and predictions about how goals might be achieved. Ultimately, department policymakers need to ascertain how those indicators can be used and at what level they should be adopted.[119] Criticism of academic researchers' lack of street experience may be unfounded, at best: In an upside-down analogy, do you need to be an experienced felon to predict the patterns of behavior?

■ Strategies to Increase Patrol Effectiveness

Random Patrol

One idea that arose from the previously mentioned patrol studies is the notion that preventive patrol can become a pattern of routine and is observable by criminals. As a consequence, **random patrol**, which can use a computer to select random numbers that correspond to numbers on a zoned map of the jurisdiction, is recommended.[120] Patrol officers are provided with an envelope before going out to patrol and patrol those numbered zoned areas. The basic premise of random patrol is that it puts patrol officers closer to any potential incident or request for service before it happens (data from Compstat or other crime-map illustrations may be used to determine those areas). Goals are to reduce response time, erase set patrol pattern habits, and provide a police presence in hot spots.

Random patrols have been criticized as not being an efficient use of police resources. Carl Klockars extends the criticism of random patrols by concluding that "it makes about as much sense to have police patrol routinely in cars to fight crime as it does to have firemen patrol routinely in fire trucks to fight fire." [121]

Directed Patrol

Directed patrol focuses resources on those times and places associated with the highest risk of serious crime (i.e., hot spots as identified though evidence-based policing). Rather than simply providing high police visibility, directed patrol concentrates police activity where the crime is. The theory behind **directed patrol** is that the more police presence in hot spots, the less likely crime will occur there because the presence of police increases the threat of detection and punishment for criminal activity.[122]

Research suggests that police can, in fact, reduce—or at least displace—crime using directed patrol.[123] An analysis of the Minneapolis Hot Spots Patrol Experiment, for example, revealed that the longer the police stayed at one hot spot, the lower the crime rate dipped—up to a point. More than 10 minutes of police presence in one hot spot produced diminishing returns. As a result, "the optimal way to use police visibility may be to have police travel from hot spot to hot spot, staying about ten minutes at each one." [124]

Random patrolling or driving aimlessly is not as advantageous in providing public safety as compared to directed aggressive patrols[125] (as learned through evidenced-based policing studies).

Directed aggressive patrol focuses on preventing and detecting crime by putting attention on problem areas and by investigating suspicious activity.

Through this kind of patrol, officers can build an intelligence-base of information about who lives and works on the beat, learn about daily routines such as children coming and going to school, and learn what the youths look like who are hanging on street corners. They learn about the streets and alleys, the dead ends and underpasses, and the highways in their patrol areas. Officers acknowledge that highways are often haunted by criminals who use them as escape routes, as places to find victims, and "as a way to traffic their own nefarious brand of commerce—drugs, guns, stolen property, and cash." [126]

Aggressive directed patrol works well with police and community relations problem-solving initiatives. It is also effective in tackling specific problems. For instance, the Indianapolis (Indiana) Police Department (IPD) implemented a directed aggressive patrol project that targeted two areas.[127] Unlike in earlier studies, in which patrol officers had removed illegal weapons from a high crime neighborhood, IPD officers focused on increased surveillance on targeted high-risk individuals in high-risk neighborhoods. The results of this project show that directed patrol in high-violent-crime locations can have a significant effect on violent crime: A 29 percent reduction in firearms-related crime occurred in the IPD-targeted areas, and homicides were reduced from seven to one.

Other Improvements to Traditional Patrol

Directed aggressive patrol with specific targets and clearly articulated aims is one answer in some jurisdictions.[128] Other jurisdictions have come up with their ways to improve patrol.

The Santa Ana (California) Police Department (SAPD), for example, has sought to deploy more patrol officers in the field and fewer officers in administrative jobs at the office. Officers in the field have ongoing contact with the public, which keeps them in better touch with problems in particular neighborhoods and enables patrol officers to observe problems firsthand, according to SAPD Chief Paul W. Walters.[129] Equally important, the SAPD has increased the number of patrol officers it employs and enhanced their efficiency through employment of technology, civilians, volunteers, and a more efficient scheduling system.

In the Oak Park (Illinois) Police Department's (OPPD) Resident Beat Officer Program, police officers work out of their homes in this fashionable Chicago suburb of 54,000. The OPPD advertises the beats or zones, names, phone numbers, and email addresses of each designated beat officer. As neighborhood residents, beat officers work directly with their neighborhoods to identify and address local problems. OPPD also employs civilians as community officers to take accident reports, issue citations, and monitor traffic.

Use of Civilians to Free Up Officers for Patrol Duties

Like many other departments, the Omaha (Nebraska) Police Department uses civilians to perform some duties previously carried out by sworn officers. For instance, the rangemaster was replaced with a civilian who has more than 20 years of firearms handling and training experience. This change freed up an officer to return to the field and added a firearms expert to the department. In some departments, several investigator positions were converted to civilian police investigative specialists; each of these changes enabled an officer to be reassigned to the field. With more officers in the field to handle calls, the average response time for all patrol officers improved.

Use of Technology to Enhance Patrol Effectiveness

Technological advances, including the use of artificial intelligence (AI) such as COPLINK (as discussed in Chapter 10), have the power to identify effective patrol outcomes and enhance officer competence. Police patrols are strongly correlated with public safety, and patrols clearly enhance crime prevention and deterrence, especially in urban areas.[130] Nevertheless, the specification of successful police patrol routes is not a trivial task, especially in large demographic areas. Some departments have sought to use artificial intelligence to make better choices in designing patrol patterns.

Tucson (Arizona) Police Department's Patrols and Artificial Intelligence

Research was conducted to aid the Tucson Police Department's patrol pursuits, and the results were used to create GAPatrol, "a novel evolutionary multiagent-based simulation tool devised to assist police managers in the design of effective police patrol route strategies."[131] One aspect of the system uses GAPatrol's facility to automatically discover crime hot spots (see Chapter 12) or targeted areas that need to be better covered by routine patrol surveillance. Even with the AI system, however, communication between victims, departments, and patrol officers plays a role in response time and, ultimately, in detecting and apprehending offenders.

Boston Police Department's Operation Red Zone

Designed to address firearms-related violence, Operation Red Zone (ORZ) relies on the Boston Regional Intelligence Center (BRIC) to identify particular geographical city neighborhoods/blocks termed "Red Zones" because of their high incidence of violent crime. The Boston Police Department (BPD) has embraced RECAP's findings suggesting that most crime comes from few addresses. In particular, the BPD identified ten of the most violent areas in the city, which then became the focus of "Red Zone Teams." The teams consist of patrol officers, special agents from the FBI and DEA, civilian partners, faith-based partners, business partners, and academic partners.[132] Since ORZ began, reported violent crime in those areas has declined by 20 percent from previous periods.

Calls for Service

Telephone Reporting Units

Telephone reporting units (TRU) are responsible for all minor crime reports that do not immediately require field investigation by patrol officers. These units are called into action when a member of the public calls the police department's non-emergency phone number or when 911 dispatch screens a call and diverts it to TRU. TRU are usually manned by injured officers or officers on light duty. The officer at the TRU can document those crimes lacking suspect information or evidence (e.g., fingerprints, video, witnesses), thereby eliminating the need to dispatch a patrol officer to the scene to take the report, which increases the patrol officer's field time.[133] For instance, a victim of a burglary who doesn't know the aggressor and says there is no evidence or witnesses may file a police report with the TRU officer, listing the circumstances and the property taken. In jurisdictions where TRU are in operation, as many as 30 percent of all 911 calls can be sent to TRU.

At the Santa Ana (California) Police Department, for example, the TRU is charged with handling all incoming crime reports that could be taken over the phone. This unit is staffed by civilians and cadets, relieves patrol officers from having to respond and take the reports in person. Since the inception of this program, the unit has handled an average of 4,700 reports per year, or more than half of what were previously considered emergency calls.

311 Non-emergency Calls

The public often needs access to a non-emergency police telephone number when a report is not serious, when the incident is not life-threatening, or when a crime is not currently in progress. The 311 system fills this need in many jurisdictions; it allows members of the public to transmit general information to the police department. By the end of this decade, 90 percent of all U.S. police departments are expected to have 311 non-emergency services available.

For example, the Metropolitan Washington, DC, Police Department (MDCP) has implemented a 311 system that handles the following types of calls:

- Property crimes that are not in progress and the offender is not on the scene: vandalism, thefts, graffiti, stolen autos, and garage burglaries
- Animal control problems
- Illegally parked vehicles or vehicles blocking alleys or driveways
- Minor vehicle crashes where there are no injuries and traffic is not blocked

MDCP also has TRU service. When those providers are busy, they will return calls to constituents, usually in less than an hour, and provide callers with a report number (CCN number) for insurance or other purposes.

When the Baltimore Police Department pioneered its 311 system in the latter part of 1990s, only those calls that required immediate police presence were routed to a 911 dispatcher; all others were dispatched to 311.

911 Emergency Numbers

In 2003, 92 percent of local police departments and 94 percent of sheriff's offices participated in an emergency 911 system.[134] Seventy-three percent of local police departments and 71 percent of sheriffs' offices had enhanced 911 systems, capable of automatically displaying information such as the caller's phone number, address, and special needs.

Some departments are very sophisticated when it comes to 911. For example, the Arlington (Texas) Police Department (APD), under the leadership of Chief Theron Bowman, has implemented a system in which 911 calls are webbed (placed on the Internet) at the time they are received and linked to a map of the city of Arlington, Texas. (The APD's website, http://www.arlingtonpd.org/CAD.asp, provides more details.) For instance, at 4:38 P.M. on June 16, 2007, APD looked like Figure 9–1 (except that there were 51 entries for the afternoon—only a few are shown here).

Other police departments have created their own enhanced 911 systems. In Boston, police developed COPLINK by using BPD data from 911 reports, booking sus-

Current time is: **6/16/2007 4:38:24 PM** — Updates every minute.

District: All | North | East | South | West

Beat	Nature of Call	Party	Date	Time	Call No.	* Location	Map	Status
470	OPEN/OPEN /DOOR WINDOW	1	6/16/2007	1620	071670538	2000 AVALON LN		ONSCENE
320	ACCMAJ/ACCIDENT—MAJOR	1	6/16/2007	1539	071670506	4400 LITTLE RD		ENROUTE

Figure 9–1 911 call log for the Arlington (Texas) Police Department

pect lists, and police records. The 911 reports continue to play an important role in COPLINK technology, which crosses and matches data sets (addresses, phone numbers, incident reports, and suspect data). The BPD received almost 400,000 emergency calls in 40 different languages in 2005.[135]

Not all 911 calls deal with crime, of course. For example, from February 1, 1999, to April 30, 2001, Miami–Dade County 911 dispatchers received almost 2.25 million calls, 56,321 of which were medical emergency calls triggering dual deployment.[136] Cardiac arrest was the reason for 420 of these calls, with police arriving first (i.e., before fire or EMS personnel) 56 percent of the time. The survival rate was 17 percent for the 163 victims who had ventricular fibrillation or pulseless ventricular tachycardia (irregular heart rhythms that require an electric shock to correct)—up from the survival rate of 9 percent noted during the 1.5 years just prior to establishing the police responder program.

Text Messaging

The Tacoma (Washington) Police Department encourages its constituents to use 911 to report crime and allows constituents to text message any witness reports to the department. Other departments are following Tacoma's lead. "The thing is," says a Tacoma resident, "if I call to say I saw something like someone I know committed a crime on my house phone, 911 has my address. If I call on my cell, they only have my voice. If I text it, they don't know who sent it."[137]

Differential Response to Emergency Calls

> **Differential response** to emergency calls involves classifying calls according to their seriousness.

Such a system is intended to prioritize calls. With a differential response, three responses are possible:

- An immediate response by a sworn officer
- A delayed response by a sworn officer
- No police response, with the report instead being taken by telephone, email, text messaging, mail, or a person coming into the police department

In the Greensboro (North Carolina) Police Department, for example, 53 percent of the calls coming into the police department require immediate patrol officer attention, 27 percent require a delayed response, and 20 percent require no police response and reports can be completed by individuals who answer the call.[138]

The Tucson (Arizona) Police Department (TPD) receives 989 emergency calls every 24 hours, or approximately 312,000 calls annually. Table 9–2 summarizes the TPD's calls for the 12-month period from February 2006 to January 2007.[139] Reviewing these calls provides an indicator of a patrol officer's experiences in Tucson, which are likely to be matched by other officers' experiences in similar jurisdictions. (Tucson has a population of 515,000 and is ranked as the thirty-second largest city in the United States.)

TABLE 9–2 Tucson Police Department Calls for Service, February 2006 to January 2007

Incident	Totals	Incident	Totals	Incident	Totals
Homicide	51	Disorderly conduct	6,685	Found	2,780
Sexual assault	364	Vagrancy	278	Public assist	21,645
Robbery	1,554	Other offenses	8,337	Civil matter	8,471
Aggravated assault	2,104	Arrests for other jurisdictions	4,726	False alarm	10,657
Burglary	5,344	Juvenile violations	371	Suspicious activity	32,379
Larceny	19,122	Runaway juveniles	2,848	Disturbance no action	43,462
Motor vehicle theft	8,591	Fatal vehicle accident	50	Unfounded	23,914
Arson	274	Person injured accident	3,988	Assist other agency	3,420
Assault minor injury	10,738	Property damage motor vehicle accident	11,208	Miscellaneous	31,722
Forgery	707	Motor vehicle nonaccident	2,320	Field interviews	6,238
Fraud	4,263	Other vehicle accidents	62	Offenses against family–child	956
Embezzlement	898	Traffic motor vehicle	12,636	DUI	2,756
Stolen property	43	Death	1,229	Disaster	6
Criminal damage	11,473	Mental case	2,615	Public hazard	3,976
Weapons	619	Sick cared for	2,112	Fire	383
Commercialized sex	306	Animal bites	117	Lost	2,818
Sex offenses	1,078	Firearms accidents	28	Intoxication/liquor law	2,251
Narcotic drug laws	6,689	Persons injured accident	123	Totals	331,776

Source: Data from Tuscon Arizona Police Department and Pima (Arizona) County Public Library.

One way to interpret TPD's patrol officer experiences is to examine the seriousness of the service calls received by the department's 911 system. Of 331,776 calls, 26 percent (86,504 calls) required no action from a sworn officer (unfounded, false alarm, disturbance—no action, and civil matter). In 6,107 calls (2 percent), an immediate police response was required (homicide, sexual assault, robbery, aggravated assault, sex offense, and offense against family–child as crimes of violence possibly against a person). This list is not complete, and observers might see other calls as requiring better attention, yet the point is clear: Most 911 calls are not linked to violent crime. Thus, if a department wants to enhance its response time, a differential response to incoming calls is essential.

Response Time

When a member of the public calls 911, he or she expects a rapid police response. It might appear that the shorter the time it takes for police to arrive at an incident (i.e., the **response time**), the less crime will occur. This is not necessarily so, because of several gaps in the response process:

- Gaps in discovery times: the interval between commission of the crime and its discovery

- Gaps in reporting times: the interval between discovery and the time when the constituent calls police
- Gaps in processing times: the interval between the call and the dispatch of a patrol vehicle
- Gaps in travel times: the length of time it takes patrol to reach the scene

Rapid response times, reactive arrest, and random patrol are the three main strategies used by modern police departments to minimize these gaps, where possible.[140] In theory, rapid response should prevent injury, increase arrest, and deter crime. In practice, however, rapid response produces few of these outcomes: During a crime, injury is most often inflicted within the first few seconds, and by the time the police arrive (regardless how short a time it took for their arrival) the aggressor has often left the scene of the crime. "This neither increases arrest nor deters future crime."[141] An arrest occurs during only 2.9 percent of all calls for serious crime, usually when the police patrols arrive while the crime is in progress, and most often victims wait more than 5 minutes before contacting police.[142]

How are police departments attempting to reduce their response times? The 911 dispatch system used by the Miami–Dade County Sheriff's Office was reconfigured so that both police and fire/rescue were automatically dispatched to certain medical emergency calls.[143] Using this dual-dispatch mode, the time from the call to first responder arrival was shortened to 4.88 minutes, compared to the historical response time of 7.64 minutes. With the dual-responder system, help arrived on the scene of a cardiac arrest in less than five minutes for 41 percent of calls, compared to 14 percent for the standard fire/rescue calls.

In the Tucson Police Department, the response time varies based on the level of the crime:[144]

Level 1: 4.30 minutes. Emergency Response—An incident posing an immediate threat to life where the threat is present and ongoing and/or an incident posing an immediate threat to life involving the actual use or threatened use of a weapon. The mere presence of a weapon alone, however, without any indication of use or threat of use does not support or justify a Level 1 call.

Level 2: 9.18 minutes. Critical Response—An incident involving a situation of imminent danger to life or a high potential for a threat to life to develop or escalate. This incident must be in progress or have occurred within the past 5 minutes.

Level 3: 18.34 minutes. Urgent Response—Crimes against persons or significant property crimes where a rapid response is needed *and* the incident is in progress, has occurred within the past 10 minutes, or is about to escalate to a more serious situation.

Level 4: 71.37 minutes. General Response—Other crimes or matters requiring police response, generally occurring more than 10 minutes prior to dispatch and having a complainant.

■ Case Studies

Patrol Divisions of the Jacksonville Sheriff's Office

While Chapter 5 discussed the other divisions of the Jacksonville (Florida) Sheriff's Office (JSO), this section provides an overview of JSO's largest division—the patrol division.[145] The patrol function at the JSO is divided between Patrol Division East

(Zones 1, 2, and 3) and Patrol Division West (Zones 4, 5, and 6). Most of these zones include the city of Jacksonville. Police officers assigned to the patrol divisions are first responders to all emergency calls for police service. JSO's patrol officers, like their counterparts at most other police agencies, make up the largest component of the organization and perform a variety of functions: take reports of crime, conduct traffic control and enforcement, perform community policing functions, and contribute to the investigative role through diligent follow-up. Each patrol division is directed by a chief.

Zone 1's crime statistics are provided in Table 9–3. As a review of reported crime for 2005 and 2006 makes clear, patrol officers in Zone 1—like their colleagues in the other zones in Jacksonville—are pretty busy.

Patrol Units within Various Police Departments

Across the United States, police departments have implemented numerous specialized patrol police units depending on their specific needs, geography, budget, jurisdiction, personnel, and political contributions. These specialized units are administered through local, state, and county authority, and usually their authority, budget, and management are provided by a specific police organization.[146]

Airport Patrol

In Santa Barbara, California, airport patrol officers are sworn peace officers in the state of California. They wear the same uniforms as members of the Santa Barbara Police Department (SBPD), although their badges and shoulder patches set them apart. Nevertheless, the airport patrol and the SBPD are totally separate police agencies. The airport patrol officers' primary duty is to patrol airport property, but they are empowered by state law to take action outside of their designated patrol area if there is an immediate danger to a person's safety or if they observe a crime in progress.

Across the United States, there are an estimated 31 airport patrol departments, each of which is different and distinct from the police force of the municipality that

TABLE 9–3 Zone 1 Crime Report

January–November	January–November	Percent	Change
Murder	16	18	12.5%
Rape	35	27	−22.9%
Other forcible sex	31	17	−45.2%
Robbery	308	254	−17.5%
Aggravated assault	583	510	−12.5%
Total violent crime	**973**	**826**	**−15.1%**
Burglary	784	770	−1.8%
Larceny	2,713	2,611	−3.8%
Vehicle theft	458	411	−10.3%
Arson	16	19	18.8%
Total property crime	**3,971**	**3,811**	**−4.0%**
Total index crime	**4,944**	**4,637**	**−6.2%**

Source: Courtesy of Jacksonville Sheriff's Department, Sheriff John H. Rutherford, Jacksonville, Florida: Duval Florida County.

embraces the airport. For instance, the Charleston County Aviation Authority Police Department in Charleston, South Carolina, is a bike patrol department whose authority is limited to the regional airport.

Aircraft Patrol

The aircraft patrol operated by the Oregon State Police is an effective tool for apprehending aggressive drivers; enforcing speed laws; identifying "following too close" and other hazardous violations; and seeking out stranded motorists. The aircraft has also been used by the Fish and Wildlife Division to fly missions when the latter agency's aircraft were not available and by the Criminal Division for aerial photographs during its investigations. The aircraft supports all divisions including SWAT, and has provided a unique perspective for all police partners.

Bike Patrol

The Des Moines (Iowa) Police Department operates a bike patrol that is deployed in specific neighborhoods. This patrol starts in April and runs through September. Two officers work bicycle patrol four hours per day, in addition to working their regular eight-hour tour of duty. Typically bike patrols are seasonally deployed in cold-weather jurisdictions, but are relied upon year-round in warm-weather jurisdictions, particularly in park settings, business areas, and school districts. Bike patrols are less likely to be deployed in gang-infested communities such as the Rampart district in Los Angeles.

There are more than 2,000 police bicycle units in the United States that average 9 officers each.[147] Officers often spend the majority of their workday "in the saddle" (5.4 hours per day was reported in one study).[148] Bikes can travel farther at faster rates than can foot patrol officers and are cost efficient.

Canine Unit

Although few data are available on canine units operating among police agencies, one estimate is that at least half of all local agencies in cities with populations of more than 250,000 use police dogs to some degree. Many smaller agencies also employ police dogs depending on the specific jurisdiction's experience, budget, and availability of personnel to deliver quality police services with the aid of trained dogs.

The Houston Police Department (HPD) uses two varieties of police dogs: scent dogs and all-purpose dogs. According to a Houston canine handler, with scent dogs, the dog's innate sense of smell is focused toward specific venues via training (drug dogs); with all-purpose dogs (patrol dogs), the dog must be trained to perform multiple duties, which includes the use of force.[149] Police dogs may be deployed in search and seizure of narcotics, for instance, or in the apprehension of violent suspects. Some have suggested that patrol dogs should be used more often in "search and locate" procedures—rescue missions, detection of explosives and accelerants, tracking suspects, and the recovery of evidence.

How those dogs have been trained can make the difference in their success and in the department's chances of winning civil rights liability lawsuits against the dogs, their handlers, the trainers, and the HPD. Some agencies employ the "bite and hold" initiatives for their dogs, but HPD prefers the "bark and hold" procedure. Most dogs are purchased from German or Belgium breeders so that offenders cannot redirect a dog's action in English or Spanish.

One study among more than 300 police agencies that deploy patrol dogs in Texas discovered that few or no legal standards existed within the state or the law enforcement agencies to govern the use of patrol dogs—or, for that matter, in most jurisdictions across the United States.[150] Should lawsuits arise, however, a canine officer with certification and training would tend to have more credibility in court than would an officer without appropriate certification or training. The use of dogs in police work remains highly controversial among police officers, though some consider the dogs to be essential police partners.

Drug Enforcement Units

About 9 in 10 local police agencies, and particularly those agencies that employ 97 percent of all local officers, regularly perform drug enforcement functions. Eighteen percent of departments had officers assigned full time to a special unit for drug enforcement, with approximately 12,000 officers assigned to such duty nationwide. Nearly one fourth of departments had officers assigned to a multiagency drug task force.[151] The average number of officers assigned to those units ranged from 233 officers in jurisdictions with 1 million or more residents to 1 officer in those jurisdictions with fewer than 2,500 residents.

The Montgomery (Alabama) Police Department has 18 sworn officers and the Bridgeport (Connecticut) Police Department has 26 sworn officers assigned full time to drug units. By comparison, the Metropolitan Washington, D.C., Police Department assigns 257 officers to this duty and Florida's Orange County Sheriff's Office assigns 120 sworn officers to drug enforcement.

Sometimes, jurisdictions pool their resources and operate a joint or multiagency drug task force. For instance, Georgia's DeKalb County Sheriff's Department assigned 10 full-time officers to such a force, and the Dearborn (Michigan) Police Department assigned 18 officers to a multiagency drug task force (the Dearborn Police Department also assigned 214 officers to a special drug unit in the municipality of Dearborn). Sometimes large municipalities such as Dallas, Texas, have both a local drug unit (consisting of 116 sworn officers in Dallas) and a multiagency drug task force (6 officers) enforcing drug laws within the same jurisdiction. One reason for operating two drug units within the same jurisdiction is to maintain a system of checks and balances within the department, as a high-ranking Dallas officer explained.[152]

Local officers are often enlisted to aid the federal government's drug interdiction efforts but retain their role (and payroll) from the local department. One advantage to this system is that dual-shield officers (i.e., those with both local and federal jurisdiction) can pursue drug offenders over state lines. For example, many Columbia, South Carolina, drug enforcers are also federal DEA agents and possess the authority to pursue drug offenders on Interstate 95 (which connects drug-entry gateway cities in Florida to major East Coast buyers' markets such as New York City and Boston).

Working with other agencies can enhance enforcement endeavors in other ways. For example, while drug offenders might be well aware of the faces of the officers in their community, new faces can aid in building a case or infiltrating drug markets with an eye toward identification and apprehension of drug offenders.

Foot Patrol

Foot patrols, such as those used by the Rehoboth Beach (Delaware) Police Department, can increase the perception of public safety and provide police services consistent with community needs. Seasonal police officers walk the beat 24 hours a day during the summer months. These officers have closer contact with people and can get a good sense of what's happening in the community by getting to know residents, business owners, and visitors on a more personal basis. Overall, six (or more) of every 10 U.S. police departments use foot patrol routinely.[153]

Harbor or River Patrol

The Jacksonville Sheriff's Office (described earlier in this chapter) operates a special unit called the Seaport Security Unit. This unit is staffed by 11 officers and two sergeants from the Jacksonville Sheriff's Office whose positions are funded by Jaxport under a Memorandum of Agreement, which is renewable annually. The Seaport Security Unit reports to a lieutenant in the Homeland Security and Narcotics/Vice Division.

The primary duty and purpose of the Seaport Security Unit is to provide a full-time police presence and quick response at port facilities. Vigilance against potential terrorist and other suspicious activity that could disrupt port operations is central to these personnel's duties.

Motorcycle Patrol

Larger municipal departments often deploy motorcycle patrols as part of their regular patrols. As you can imagine, departments in no-snow states tend to have year-round motorcycle patrols. Also, state police units across the United States use motorcycle patrols.

The Pasadena (California) Police Motorcycle Division's officers go through a two-week Basic Motorcycle Academy, where officers learn to perform high-speed and slow-speed maneuvers through intricate cone patterns and practice track driving, mountain driving, and freeway driving.[154] The course is so difficult that more than half of the people who attend the school fail on their first attempt. At the school, officers learn to ride in adverse conditions, how to safely pull drivers over, and how to conduct pursuit driving. Officers are also taught how to shoot their weapons while driving and how to use their motorcycle as cover in the event of a shootout.

As you might have noticed, motorcycle officers dress differently than other police officers. They wear large boots and special pants, both of which are designed to protect the officer in the event of a collision.

The Pasadena Police Department uses both ST1100 Honda Police Motorcycles and Kawasaki 1000 Police Motorcycles. The motorcycle unit is part of the department's Traffic Section and consists of a lieutenant, two sergeants, a corporal, 16 motorcycle officers, a community service officer, and an office manager. The Traffic Section is responsible for monitoring traffic safety in the city of Pasadena. Because of the mild California weather, the motorcycle unit is the primary traffic control unit within the department.

Mounted Patrol

Similar to the deployment of motorcycle patrols, mounted patrols tend to be used in much the same circumstances, depending on the jurisdiction's weather and the size of the department. For example, Boston has 14 mounts (horses) that patrol its waterways, parks, and historic district during the day in the late spring, summer, and early fall.

The Fort Lauderdale Mounted Police Unit[155] was formed in September 1983, and celebrated its 25th anniversary. The goal of the Mounted Unit is to assist in enhancing the image of the police department. The Unit was started with confiscated funds taken from the assets of drug dealers and other criminal enterprises; the department purchased a 12 stall CB constructed state of the art facility complete with locker rooms, showers, class room, tack room, laundry, feed room, and offices. The horses were donated by private citizens. The present Unit is allocated for 5 permanent positions including a sergeant, and trains up to 5 other certified officers to ride. Presently the Unit rides a modified McClellan Saddle, with a U.S. Army configured halter bridle system. The donated horses must be athletic and nimble. They vary from Thoroughbreds, Quarter Horses, Morgan's, Tennessee Walkers, Arabians, Warm Bloods, and everything 15-3″ hands or taller, but no Draft Horses.

Following is a set of duty assignments for the Fort Lauderdale Mounted Police:

- Duty assignments generally come from district commanders, shift captains, or Operations Support Captain John Bollinger. Schedules are made out one week in advance and posted on the schedule board in the officers area. Check the schedule for your riding assignment on a daily basis.
- The daily objectives of the FLPD Mounted Unit are to participate in fighting Part One Crimes and Community Policing. Officers must be prepared to confront everything from minor disturbances to in progress bank robberies.
- Special events/situations: Special events, holiday duty, parades, and so on require a change of our normal schedule from time to time. Emergency situations also crop up without warning, such as demonstrations or crowd control problems. Be prepared to adjust to these demands on short notice. For this reason they have an extensive horsemanship training and fitness program. They practice basic military cavalry movements. Their formations and commands are taken from the U.S. Army/Cavalry Manual.
- They work a 5 day 40 hour work week. Officers get 1 hour at the beginning and 1 hour at the end of shift for preparation and maintenance of horse and tack. Their goal is to ride 6 hours a day, 7 days a week, in rain or shine. For special events the saddle time may be extended as much as 16 hours. Special precautions and preparations are made for the horse and rider during these events.

Early in the Mounted Unit's history, most of the Fort Lauderdale police department managers and officers alike had little respect for the horses and their abilities. But that changed one night during spring break, when approximately 1,000 college students took over a two-block section of State Road A1A, totally blocking the road. With traffic at a standstill, police managers first tried using 50 patrol officers and 10 motor officers to clear the street. That effort failed, and the crowd got rowdier. The deputy chief, who was on the scene with a city commissioner, was considering using teargas when eight mounted officers arrived. Nearly one third of the crowd left the street as soon as the mounted officers lined up for the sweep.

Terrorist Units

Intelligence gathering and drills happen nearly every day in many police departments.[156] In the years since the September 11 attacks, the NYPD has emerged as an international leader in urban security and counterterrorism measures. Its special terrorist unit, which is called the Emergency Service Unit, employs more than 200 officers and has installed thousands of surveillance cameras.[157] On any given day, more than 1,000 patrol officers are tasked with ensuring that New York City (the world's number one terrorist target, according to analysts) is doing everything in its power to prevent another attack. The NYPD has developed a wide variety of tactics, from positioning detectives abroad to reaching out to the Muslim community at home. Other police departments are following the NYPD's lead. The city spends approximately $200 million per year on its counter-terrorism efforts.

***RIPPED* from the Headlines**

Police Chief in Pleasant Hill, Iowa, Is Suspended Over Ticket Quota Allegations

The Pleasant Hill, Iowa, city manager suspended the city's police chief on April 26 [2007] after evidence surfaced that he had instituted an illegal traffic ticket quota. According to local police union leader Ron Zimmerman, 33, officers were being told to issue between five and ten tickets each month. A sergeant chided Officer Zimmerman on "the low number of tickets" that he issued.

Although Chief William Hansen, 58, denies the existence of a quota, a *Des Moines Register* review of court documents shows the amount of ticket revenue has more than doubled under Hansen's watch.

In the year ending April 30, 2005, the city issued 1,136 tickets worth $124,933. In the year ending April 30, 2007, that figure jumped to 2,427 tickets worth $309,261. The number of citations issued in Pleasant Hill per capita is nearly double that of Des Moines, which generates $5.9 million from its ticketing program.

Source: "Iowa: Police Chief Suspended Over Ticket Quota" (2007, May 9), http://www.thenewspaper.com/news/17/1743.asp (accessed June 20, 2007).

Summary

Patrol officers are the backbone of American policing and represent approximately 62 percent of all sworn officers in the United States. The functions of patrol are to deter crime, reduce the fear of crime, enhance quality-of-life experiences, and control wrongful act opportunities of patrol officers.

Police agencies organize their jurisdictions into geographic districts, with uniformed officers being assigned to patrol a specific area with that jurisdiction. Patrol officers focus on detection, identification, and apprehension of violators of federal, state, and local laws. They are involved in case investigations when and where necessary. They usually perform traffic control and respond to service calls.

The general work experiences of patrol officers include activities ranging from domestic violence stops to excessive paperwork. Hot pursuit by patrol officers can be defined as an active attempt to apprehend one or more occupants of a motor vehicle when the driver of the vehicle is intentionally resisting the apprehension by maintaining or increasing speed or by ignoring the police officer's signals to stop.

The Kansas City Missouri Preventive Patrol Experiment (1973) reported that patrol techniques have little effect on crime patterns, and patrol presence has little influence on reducing the fear of crime among community members. This study and others led the way to evidenced-based policing initiatives. Through research, it was learned that directed patrols can focus resources on those times and places with the highest risk of serious crime (i.e., hot spots). Rather than simply providing high police visibility everywhere, directed patrol concentrates police visibility in hot spots. Directed aggressive patrol, which focuses on preventing and detecting crime by putting attention on problem areas and by investigating suspicious activity, seems to be an especially effective technique.

Telephone reporting units (TRU) are responsible for taking all minor crime reports that do not immediately require field investigation by patrol officers; these units are activated when a member of the public calls the police department's non-emergency phone number or when the 911 dispatch system screens a call and diverts it to the TRU. Calling a non-emergency police number is appropriate when a report is not serious, when the incident is not life-threatening, or when a crime is not currently in progress. Differential response to emergency calls involves classifying calls according to their seriousness.

Key Words

Checkpoints
Differential response
Directed aggressive patrol
Directed patrol
Displays of resistance
Fresh pursuit
Kansas City (Missouri) Preventive Patrol Experiment
Random patrol
RECAP
Response time
Stop sticks
Street-level bureaucrats

Discussion Questions

1. Describe the role of patrol officers.
2. Characterize the functions of police patrols. In what way do you agree with those functions? In what way do you think they are inappropriate?
3. Describe deployment initiatives and duties of patrol officers. If you were a patrol officer, whose initiatives and duties would you find the most attractive? Why?
4. Identify the general work experiences of patrol officers. Which three would be the most exciting for you as an officer? Why?
5. Articulate the outcome of fresh vehicle pursuits. Why are such pursuits necessary?
6. Describe the outcomes of traffic stops and searches, including checkpoints. In what way do you agree or disagree with the rationale underlying checkpoints?

7. Characterize the primary factors affecting patrol service delivery. Of these factors, which would be the most detrimental for a patrol officer?
8. Summarize some of the patrol studies discussed in this chapter and identify their relationship to evidence-based policing. What have you learned from these studies?
9. Characterize random, directed, and directed aggressive patrols. In what way do these initiatives influence current patrol policy?
10. Describe ways of improving traditional patrol.
11. Characterize 911 call experiences in police departments.
12. Explain the necessity of differential response initiatives to emergency calls.
13. Describe the debate about police response time and its relationship to arrest data. In what way do you agree with response time findings related to an arrest?
14. Identify several of the specialized patrol units found within police departments.

Notes

1. Bureau of Labor Statistics, "Occupational Outlook Handbook" (2006), http://www.bls.gov/oco/ocos160.htm (accessed June 10, 2007).
2. Patrick V. Murphy, "The Patrol Function," presentation at the Annual Conference of Mayors, Washington, DC, http://www.totse.com/en/law/justice_for_all/patrol.html (accessed November 9, 2007).
3. Carl B. Klockars, "The Legacy of Conservative Ideology and Police," in *The Police and Society: Touchstone Readings*, ed. Victor E. Kappeler (Mt. Prospect, IL: Waveland Press, 1995), 349–57.
4. Bureau of Justice Statistics, "Law Enforcement Management and Administrative Statistics, 2000" (2004), http://www.ojp.usdoj.gov/bjs/pub/pdf/lemas00.pdf (accessed June 10, 2007).
5. Murphy, "The Patrol Function."
6. David H. Bayley, *What Works in Policing* (New York: Oxford Press, 1998), 53.
7. "Sourcebook of Criminal Justice Statistics, 2005," Tables 1.62, 1.64, 1.66, http://www.albany.edu/sourcebook/tost_1.html#1_af (accessed June 10, 2007).
8. Christine N. Famega, James Frank, and Lorraine Mazerolle, "Managing Police Patrol Time: The Role of Supervisor Directives," *Justice Quarterly* 22, no. 4 (2005).
9. Christine N. Famega, "Variation of Officer Downtime: A Review of the Research," *Policing: An International Journal of Police Strategies and Management* 28, no. 3 (2005): 388–419.
10. Michael Lipsky, *Street-Level Bureaucracy* (New York: Russell Sage, 1980).
11. Cited in Kai T. Erikson, *Wayward Puritans: A Study in the Sociology of Deviance* (New York: John Wiley, 1966), 3–5: "Crime was really a natural kind of social activity, an integral part of all healthy societies."
12. Quoted in Mike Nash, "Managing Risk: Achieving Protection?", *International Journal of Public Sector Management* 11, no. 4 (1998): 252–61.
13. Albert Weiss, *The Police and the Public* (New Haven, CT: Yale University Press, 1971), 3.
14. Bureau of Labor Statistics, "Police and Detectives," U.S. Department of Labor (2007), http://stats.bls.gov/oco/ocos160.htm (accessed June 10, 2007).
15. San Diego Field Interrogation Field Report, "Executive Summary" (n.d.), http://www.soc.umn.edu/~samaha/cases/san_diego_field_interrogation.htm (accessed June 13, 2007)
16. Acworth, Georgia, Police Department website used as a guide, http://www.acworth.org/depart/admin/files/PatrolOfficer.pdf (accessed June 10, 2007).

17. Dennis J. Stevens, *Police Officer Stress: Sources and Solutions* (Upper Saddle River, NJ: Prentice Hall, 2008), 168–96.
18. R. M. Ayers, *Preventing Law Enforcement Stress: The Organization's Role* (Washington, DC: Bureau of Justice Assistance, U.S. Department of Justice, Office of Justice Programs, 1990). Also see Katherine W. Ellison, *Stress and the Police Officer* (Springfield, IL: Charles C. Thomas, 2004), 67.
19. Gary W. Sykes, "Street Justice: A Moral Defense of Order Maintenance Policing," in *The Police and Society*, 2nd ed., ed. Victor E. Kappeler (Prospect Heights, IL: Waveland, 1999), 134–49.
20. Carl B. Klockars, "The Dirty Harry Problem," in *The Police and Society*, 2nd ed., ed. Victor E. Kappeler (Prospect Heights, IL: Waveland, 1999), 368–87; Carl B. Klockars, "Street Justice: Some Micro-Moral Reservations," in *The Police and Society*, 2nd ed., ed. Victor E. Kappeler (Prospect Heights, IL: Waveland, 1999), 150–53.
21. Herbert Goldstein, *Problem-Oriented Policing* (New York: McGraw Hill, 1980), 183.
22. Judy Van Wyk, "Hidden Hazards of Responding to Domestic Disputes," in *Policing and Stress*, ed. Heith Copes (Upper Saddle River, NJ: Prentice Hall, 2005), 41–54.
23. Some believe that domestic violence calls are the most serious of police calls and that in 75 percent of the cases, evidence suggests that serious conflict had existed. See Denise K. Gosselin, *Heavy Hands: An Introduction to the Crimes of Domestic Violence*, 2nd ed. (Upper Saddle River, NJ: Prentice Hall, 2004). Also see A. Roberts, "The Police Response," in *Battered Women*, ed. L. Gerdes (San Diego: Greenhaven, 1999), 32–40.
24. Ellison, *Stress and the Police Officer*, 67.
25. Kenneth W. Flynn, "Training and Police Violence," in *Police and Violence*, ed. Ronald G. Burns and Charles E. Crawford (Upper Saddle River, NJ: Prentice Hall, 2002), 127–246.
26. Geoffrey P. Albert and Mark H. Moore, "Measuring Police Performance in the New Paradigm of Policing," in *Critical Issues in Policing: Contemporary Reading*, 4th ed., ed. Roger Dunham and Geoffrey P. Albert (Prospect Heights, IL: Waveland, 2001), 238–54.
27. Thomas V. Brady, *Measuring What Matters, Part One: Measures of Crime, Fear, and Disorder* (Washington, DC: U.S. Department of Justice, Office of Justice Programs, 1996), http://www.ncjrs.org/txtfiles/measure.txt (accessed June 7, 2005).
28. Patricia A. Kelly, "Stress: The Cop Killer," in *Treating Police Stress*, ed. John M. Madonna, Jr., and Richard E. Kelly (Springfield, IL: Charles C. Thomas, 2002), 33–54.
29. Kelly, "Stress," 43.
30. Peter Manning, *Police Work: The Sociological Organization of Policing* (Prospect Heights, IL: Waveland, 1997).
31. Wayne W. Bennett and Karen M. Hess, *Management and Supervision in Law Enforcement*, 4th ed. (Springfield, IL: Charles C. Thomas, 2004), 152.
32. Ellison, *Stress and the Police Officer*, 66.
33. Ellison, *Stress and the Police Officer*, 66.
34. Bureau of Justice Statistics, *Law Enforcement Management and Administrative Statistics, 2000: Data for Individual State and Local Agencies with 100 or More Officers* (NCJ 203350) (Washington, DC: U.S. Department of Justice, 2004), http://www.ojp.usdoj.gov/bjs/pub/pdf/lemas00.pdf (accessed June 11, 2007).
35. Bureau of Justice Statistics, *Law Enforcement Management.*
36. W. C. Terry, "Police Stress: The Empirical Evidence," *Journal of Police Science and Administration* 9, no. 1 (1981): 61–75. Also see H. L. Madanba, "The Relationship Between Stress and Marital Relationships of Police Officers," in *Psychological Services for Law Enforcement*, ed. J. Reese and H. Goldstein (Washington DC: American Psychological Association, 1986), 463–66.

37. Michael D. Reisig, John D. McCluskey, Stephen D. Mastrofski, and William Terrill, "Citizen Disrespect toward the Police," *Justice Quarterly* 21, no. 2 (2004): 241–68.
38. Federal Bureau of Investigation, *Reported Crime, 2003* (Washington, DC: U.S. Department of Justice), http://www.fbi.gov/ucr/cius_03/pdf/03sec2.pdf (accessed June 7, 2005).
39. Vivian B. Lord, "The Stress of Change: The Impact of Changing a Traditional Police Department to a Community Oriented Problem Solving Department," in *Policing and Stress*, ed. Heith Copes (Upper Saddle River, NJ: Prentice Hall, 2005), 55–72.
40. Mike Seate, "Disrespect Can Turn Violent," *Pittsburgh Tribune-Review* (January 24, 2003), 7.
41. Kimberly Belvedere, John L. Worrall, and Stephen G. Tibbetts, "Explaining Suspect Resistance in Police–Citizen Encounters," *Criminal Justice Review* 30 (2005): 30–44.
42. Belvedere, Worrall, and Tibbetts, "Explaining Suspect Resistance."
43. Robin Shepard Engel, "Explaining Suspects' Resistance and Disrespect toward Police," *Journal of Criminal Justice* 31 (2003): 475–92.
44. Kelly, "Stress," 39.
45. Kelly, "Stress," 41.
46. Richard Kelly, "Psychological Care of the Police Wounded," in *Treating Police Stress*, ed. John M. Madonna, Jr., and Richard E. Kelly (Springfield, IL: Charles C. Thomas, 2002), 15–32.
47. Federal Bureau of Investigation, "Law Enforcement Officers Killed and Assaulted, 2006" (2007), http://www.fbi.gov/ucr/killed/2006/index.html (accessed December 19, 2007).
48. City of Orange, Florida (2007), http://www.ci.orange-city.fl.us/ (accessed June 12, 2007).
49. O. Finn and J. E. Tomz, *Developing a Law Enforcement Stress Program for Officers and Their Families* (Washington, DC: U.S. Department of Justice, Office of Justice Programs, National Institute of Justice, 1997).
50. Ellison, *Stress and the Police Officer*, 140.
51. Hans Toch, *Stress in Policing* (Washington, DC: American Psychological Association, 2001), 49.
52. John P. Crank and Michael Caldero, "The Production of Occupational Stress in Medium-Sized Police Agencies: A Survey of Line Officers in Eight Municipal Departments," *Journal of Criminal Justice* 19 (1991): 339–50.
53. John Van Maanen, "Making Rank: Becoming an American Police Sergeant," in *Critical Issues in Policing: Contemporary Readings*, 4th ed., ed. Roger Dunham and Geoffrey P. Alpert (Prospect Heights, IL: Waveland, 2001), 132–48.
54. Personal communication with Boston officer, February 2005.
55. Lord, "The Stress of Change," 59.
56. Thomas J. Aveni, "Identifying the Significance of the Problem," *S&W Academy Newsletter* (1999), http://www.theppsc.org/Staff_Views/Aveni/Shift-Survival.htm (accessed June 11, 2007).
57. Bennett and Hess, *Management and Supervision in Law Enforcement*, 411–12.
58. Bennett and Hess, *Management and Supervision in Law Enforcement*, 411.
59. Bryan Vila, "Officers Learn You're A, B, Zzzzzzs," *Law Enforcement Trainer* (September/October, 2002): 44–47.
60. Also see Peter Finn, "Reducing Stress: An Organization Centered Approach," *FBI Law Enforcement Bulletin* (August 1997), http://www.fbi.gov/publications/leb/1997/aug975.htm (accessed May 20, 2005).
61. Bryan Vila and Dennis Jay Kenney, "Tired Cops: The Prevalence and Potential Consequences of Police Fatigue," *NIJ Journal* 248 (2002): NCJ 190634, http://www.ncjrs.org/pdffiles1/jr000248d.pdf (accessed March 26, 2004).
62. Kelly, "Stress," 43.
63. Eric P. Baumer, "Neighborhood Disadvantage and Police Notification by Victims of Violence," *Criminology* 40, no. 3 (2002): 579–616.

64. Officer Thomas Holly, Oak Park, Illinois, Police Department.

65. Internal Revenue Services, http://www.irs.gov/businesses/small/article/0,,id=101065,00.html (accessed June 6, 2005).

66. Samuel Walker and Charles M. Katz, *The Police in America: An Introduction*, 5th ed. (New York: McGraw Hill, 2005), 170.

67. Bureau of Labor Statistics (2004), http://www.bls.gov/ncs/ebs/sp/ebsm0002.pdf (accessed June 6, 2005).

68. Vila and Kenney, "Tired Cops."

69. Geoffrey P. Alpert, *Police Pursuit: Policies and Training* (NCJ 164831) (Washington, DC: U.S. Department of Justice, Office of Justice Programs, 1997), http://www.ncjrs.org/txtfiles/164831.txt (accessed June 6, 2005).

70. Henry M. Wrobleski and Karen M. Hess, *Introduction to Law Enforcement and Criminal Justice*, 8th ed. (Belmont, CA: Thomson-Learning, 2006).

71. Wrobleski and Hess, *Introduction to Law Enforcement and Criminal Justice*, 386.

72. John Hill, "High Speed Police Pursuits: Dangers, Dynamics, and Risk Reduction," *Law Enforcement Bulletin* (July 2002), http://www.realpolice.net/police-pursuits.shtml (accessed June 3, 2007).

73. Kay Falk, "To Chase or Not to Chase?", *Policeone.com* (October 2006), http://www.officer.com/article/article.jsp?id=33481&siteSection=16 (accessed June 3, 2007).

74. Hill, "High Speed Police Pursuits."

75. Hill, "High Speed Police Pursuits."

76. Bill Mears, "Court: High-Speed Chase Suspects Can't Sue Police," *CNN* (2007), http://www.cnn.com/2007/LAW/04/30/scotus.chase/index.html (accessed June 3, 2007).

77. Bureau of Justice Statistics, "Law Enforcement Management and Administrative Statistics, 2000" (2004), http://www.ojp.usdoj.gov/bjs/pub/pdf/lemas00.pdf (accessed June 10, 2007).

78. Hill, "High Speed Police Pursuits."

79. Bureau of Justice Statistics, *Characteristics of Drivers Stopped by the Police, 2002* (NCJ211471) (Washington, DC: U.S. Department of Justice, 2006), http://www.ojp.usdoj.gov/bjs/abstract/cdsp02.htm (accessed June 10, 2007).

80. "Estimated Number of Arrests," *Sourcebook of Criminal Justice Statistics, 2002* (2004), Table 4.1, http://www.albany.edu/sourcebook/pdf/t41.pdf (accessed June 15, 2007).

81. National Safety Council, "Drunk Driving" (2007), http://www.nsc.org/library/facts/drnkdriv.htm (accessed June 15, 2007).

82. Insurance Institute for Highway Safety, "A. & Q. Alcohol" (January 2007), http://www.iihs.org/research/qanda/alcohol_enforce.html#2 (accessed June 15, 2007).

83. Insurance Institute of Highway Safety, www.iihs.org (accessed July 12, 2007).

84. Allan F. Williams and Adrian K. Lund, "Deterrent Effects of Roadblocks on Drinking and Driving," *Traffic Safety Evaluation Research Review* 3 (1984): 7–18.

85. R. W. Elder, R. A. Schults, D. A. Sleet, J. L. Nichols, S. Zaza, and Robert A. Thompson, "Effectiveness of Sobriety Checkpoints for Reducing Alcohol-Involved Crashes," *Traffic Injury Prevention* 3 (2002): 266–74.

86. John H. Lacey, S. A. Ferguson, Tara Kelley-Baker, and Raamses P. Rider, "Low-Manpower Checkpoints: Can They Provide Effective DUI Enforcement for Small Communities?", *Injury Prevention* 7 (2006): 213–18.

87. Robert B. Voas, A. E. Rhodenizer, and C. Lynn, *Evaluation of Charlottesville Checkpoint Operation: Final Report* (Washington, DC: National Highway Traffic Safety Administration, 1985).

88. Douglas J. Beirness and Robert D. Foss, *Review of Louisiana Highway Safety Commission Alcohol-Impaired Driving Programs* (Baton Rouge, LA: State of Louisiana, September 2007).

89. John H. Lacey, Tara Kelley-Baker, Debra Fun-Holden, Katharine Brainard, and Christine Moore, "Pilot Test of New Roadside Survey Methodology for Impaired Driving," Pacific Institute for Research and Evaluation (2007), http://www.pire.org/detail.asp?core=38873 (accessed December 19, 2007).

90. Bureau of Justice Statistics, *Arrest-Related Deaths in the United States, 2003–2005* (Washington, DC: Department of Justice, 2006), http://www.ojp.usdoj.gov/bjs/pub/pdf/ardus05.pdf (accessed November 25, 2007).

91. Geoffrey P. Alpert and Mark H. Moore, "Measuring Police Performance in the New Paradigm of Policing," in *Performance Measure for the Criminal Justice System* (NCJ 143505) (Washington, DC: U.S. Department of Justice, 1993). Also see Samuel Walker and Charles M. Katz, *The Police in America*, 201.

92. Statement of Charles H. Ramsey, Chief of Police, Metropolitan Police Department, Before the Council of the District of Columbia, Committee on the Judiciary, Public Round Table on the Police Deployment Act of 1999.

93. Robert B. Coates, *The Dimensions of Police–Citizen Interaction: A Social Psychological Analysis*, Ph.D. dissertation, University of Maryland, 1972.

94. Susan O. White, "A Perspective on Police Professionalization," *Law and Society Review* 7, no. 1 (1972): 61–85.

95. W. K. Muir, Jr., *Police: Streetcorner Politicians* (Chicago: Chicago University Press, 1977).

96. J. J. Broderick, *Police in a Time of Change* (Morristown, NJ: General Learning Press, 1977).

97. John Kleinig, *The Ethics of Policing* (New York: Cambridge University Press, 1996).

98. Police Foundation (2007), http://www.policefoundation.org/docs/policewomen.html (accessed June 17, 2007).

99. Jill Nicholson-Crotty and Kenneth J. Meier Gender, "Representative Bureaucracy and Law Enforcement: The Case of Sexual Assault," presentation at the annual meeting of the American Political Science Association, Boston, August 29, 2002.

100. Personal communication with Miami Dade officer who wished to remain anonymous, July 2005.

101. Stevens, *Police Officer Stress.*

102. Personal communication with a female Boston police officer, January 12, 2005.

103. Kelly, "Psychological Care of the Police Wounded," 40.

104. Richard E. Kelly, "Critical Incident Debriefing," in *Treating Police Stress*, ed. John M. Madonna and Richard E. Kelly (Springfield, IL: Charles C. Thomas, 2002), 139–49.

105. Personal communication with Massachusetts State Trooper Lisa Cesso, and a former student and research assistant, November 7, 2003.

106. Lawrence W. Sherman, "Attacking Crime: Police and Crime Control," *Crime and Justice* 15 (1992): 159–230.

107. Larry Siegel, *Introduction to Criminal Justice* (Belmont, CA: Wadsworth, 2004), 180.

108. George L. Kelling, Tony Pate, Duane Dieckman, and Charles E. Brown, *The Kansas City Patrol Experiment* (Washington, DC: Police Foundation, 1974), http://www.policefoundation.org/docs/kansas.html (accessed September 30, 2006).

109. San Diego Field Interrogation Field Report.

110. Gary W. Cordner, "The Effects of Directed Patrol: A Natural Quasi-experiment in Pontiac," in *Contemporary Issues in Law Enforcement*, ed. James J. Fyre (Beverly Hills, CA: Sage, 1981), 37–58.

111. James M. Tien, James W. Simon, and Richard C. Larson, *An Evaluation Report of an Alternative Approach in Police Patrol: The Wilmington Split-Force Experiment* (Cambridge, MA: Public Systems Evaluation, 1977).

112. Police Foundation, *Newark Foot Patrol* (Washington, DC: Author, 1981).

113. Lawrence Sherman, "Preventing Homicide through Trial and Error" (1992), http://www.aic.gov.au/publications/proceedings/17/sherman.pdf (accessed June 14, 2007).

114. Lawrence W. Sherman, James W. Shaw, and Dennis P. Rogan, "The Kansas City Gun Experiment," (Washington, DC: U.S. Department of Justice, 1995), NIJ 150855, http://www.ncjrs.gov/pdffiles/kang.pdf (accessed March 23, 2008).

115. Police Executive Research Forum, "Excellence in Problem Oriented Policing," (Washington, DC: U.S. Department of Justice, 2001) 201–22, http://www.ncjrs.gov/pdffiles1/nij/185279.pdf (accessed June 17, 2007).

116. Bayley, *What Works in Policing*. Also see Robert A. Johnson, "Integrated Patrol: A Case Study of a Law Enforcement Experiment in Anne Arundel County, Maryland," *FBI Law Enforcement Bulletin* 66, no. 11 (1997): 6–7.

117. Herman Goldstein, *Problem Oriented Policing* (New York: McGraw Hill, 1973), 3.

118. Lawrence W. Sherman and David Weisburd, "General Deterrent Effects of Police Patrol in Crime "Hot Spots": A Randomized, Controlled Trial," *Justice Quarterly* 12, no. 4 (1995): 625–48.

119. Samuel D. Faulkner and Larry P. Danaher, "Controlling Subjects: Realistic Training vs. Magic Bullets," *FBI Law Bulletin* (1997), http://www.fbi.gov/publications/leb/1997/feb974.htmr (accessed November 19, 2007).

120. "OJJDP Model Programs Guide" (2006), http://www.dsgonline.com/mpg2.5/policing_prevention.htm (accessed June 18, 2007).

121. Carl B. Klockars, *Thinking about Police* (New York: McGraw–Hill, 1983).

122. Lawrence Sherman and D. A. Weisbrud, "Does Police Patrol Prevent Crime? The Minneapolis Hot Spots Experiment," presentation to the International Society on Criminology, Conference on Urban Crime Prevention, Tokyo, April 1992.

123. Edmund F. McGarrell, Steven Chermak, and Alexander Weiss, *Reducing Gun Violence: Evaluation of the Indianapolis Police Department's Directed Patrol Project* (Washington, DC: U.S. Department of Justice, National Institute of Justice, 2002).

124. Lawrence Sherman, "The Police," in *Crime*, ed. James Q. Wilson and Joan Petersilia (San Francisco: ICS Press, 1995), 327–48.

125. Bennett and Hess, *Management and Supervision in Law Enforcement*, 389.

126. Chuck Hustmyre, "Catching Criminals on the Highways," *Law and Order* (September 2003): 13–15.

127. Edmund F. McGarrell, Steven Chermak, and Alexander Weiss, *Reducing Firearms Violence through Directed Police Patrol: Final Report on the Evaluation of the Indianapolis Police Department's Directed Patrol Project* (Rockville, MD: National Criminal Justice Reference Service, May 2002), http://www.ncjrs.gov/pdffiles1/nij/grants/194207.pdf (accessed June 16, 2007).

128. Michelle Chernikoff Anderson and Howard Giles, "Fairness and Effectiveness in Policing: The Evidence," *Journal of Communication* 55, no. 4 (2005): 872–74.

129. Santa Ana Police Department (2007), http://www.chiefwalters.com/index.html (accessed June 16, 2007).

130. Danilo Reis, Andriano Melo, Andre L.V. Coelho, and Vasco Furtado, "Towards Optimal Police Patrol Routes with Genetic Algorithms," in *Proceedings of the Intelligence and Security Informatics: IEEE International Conference on Intelligence and Security Informatics* (ISI 2006), San Diego, CA, May 23–24, 2006, http://ai.eller.arizona.edu/COPLINK/publications/isitowardsoptimalpolicepatrol.pdf (accessed May 13, 2007).

131. Reis, Melo, Coelho, and Furtado, "Towards Optimal Police Patrol Routes."

132. City of Boston, "Boston Police Annual Report" (2006): 9, http://www.cityofboston.gov/police/pdfs/2005AnnualReport.pdf (accessed June 16, 2007).

133. Huntington Police Department, "TRU" (2007), http://www.hbpd.org/patrol_tru.htm (accessed June 15, 2007).

134. Bureau of Justice Statistics, "State and Local Law Enforcement Statistics" (2007), http://www.ojp.usdoj.gov/bjs/sandlle.htm#911 (accessed June 15, 2007).

135. City of Boston.

136. American Heart Organization (2007), http://www.americanheart.org/presenter.jhtml?identifier=3004253 (accessed June 18, 2007).

137. Personal communication between the writer and a Tacoma resident who wishes to remain anonymous, June 14, 2007.

138. J. Thomas McEwen, Edward F. Connors III, and Marcia I. Cohen, *Evaluation of the Differential Police Response Field Test* (Washington, DC: U.S. Government Printing Office, 1986).

139. Tucson Police Department (2007), http://tpinternet.tucsonaz.gov/ucr/cfsucr.htm/ (accessed June 18, 2007).

140. "OJJDP Model Programs Guide."

141. "OJJDP Model Programs Guide."

142. William Spelman and D. K. Brown, *Calling the Police: A Replication of the Citizen Reporting Component of the Kansas City Response Time Analysis* (Washington, DC: Police Executive Research Forum, 1981).

143. American Heart Organization.

144. Tucson Police Department (2007), http://tpdinternet.tucsonaz.gov/ (accessed June 18, 2007); levels were clarified by Dina Richardson of the Tucson Police Department.

145. Jacksonville Florida Sheriff's Office (2007), http://www.coj.net/Departments/Sheriffs+Office/default.htm (accessed June 10, 2007).

146. Bureau of Justice Statistics, *Law Enforcement Management and Administrative Statistics, 2000: Data for Individual State and Local Agencies with 100 or More Officers* (Washington, DC: U.S. Department of Justice, 2003).

147. International Police Mountain Bike Association, http://www.ipmba.org/fact-sheet.htm (accessed June 17, 2007).

148. Steven M. Schrader, Michael J. Breitenstein, John C. Clark, Brian D. Lowe, and Terry W. Turner, "Nocturnal Penile Tumescence and Rigidity Testing in Bicycling Patrol Officers," *Journal of Andrology* 23, no. 6 (2002), 927–34.

149. Bruce M. Stewart, *Texas Police Canines, Search and Seizure Standards and Compliance*, unpublished dissertation, University of Southern Mississippi, October 2006.

150. Stewart, *Texas Police Canines.*

151. Bureau of Justice Statistics, *Local Police Departments, 2003* (NCJ 210118) (Washington, DC: U.S. Department of Justice, 2006), http://www.ojp.usdoj.gov/bjs/pub/pdf/lpd03.pdf (accessed June 20, 2007).

152. Personal confidential communication between the author and the informant who wishes to remain anonymous, October 15, 2006.

153. Bureau of Justice Statistics, *Local Police Departments, 2003.*

154. Pasadena Police Department, http://www.ci.pasadena.ca.us/police/Traffic/Traffic_Section_Home.asp (accessed October 14, 2006).

155. Fort Lauderdale Police Department, http://ci.ftlaud.fl.us/police/horse_1.html (accessed October 14, 2006).

156. New York Police Department, http://www.nyc.gov/html/nypd/ (accessed July 12, 2007).

157. Ed Bradley, "Inside the Antiterrorist Fight," *CBS News* (July 7, 2007), http://www.cbsnews.com/stories/2006/03/17/60minutes/main1416824.shtml (accessed July 12, 2007).

POLICE
9-1-1

The Role of Police Officers

Wherever law ends, tyranny begins.
—John Locke

LEARNING OBJECTIVES

When you finish reading this chapter, you will be better prepared to:

- Characterize police deviance control and civil order control
- Identify the elements of a legal search, seizure, and arrest
- Characterize critical incidents
- Explain legal issues for local officers versus federal agents
- Characterize counterterrorism methods
- Identify suicide bomber procedures
- Identify domestic terrorist activity
- Characterize contemporary initiatives for domestic and family violence
- Define a school threat and its assessment elements
- Characterize the role of police when responding to incidents involving mentally ill individuals, juveniles, and registered sex offenders
- Identify ways to identify a potential "suicide-by-cop" situation
- Identify how artificial intelligence can aid policing objectives

CASE STUDY: YOU ARE THE POLICE OFFICER

You are the officer—or at least you will be in a few weeks when your training is complete. You'll walk across another stage and shake the hand of the academy commandant, just like you did when you shook the hand of the college president a few months ago. You'll be graduating with 12 other candidates from your department and 65 candidates representing numerous jurisdictions in the metropolitan area. The feelings of another accomplishment make you feel absolutely incredible.

You're ready to put aside the textbooks. You've learned a lot at the academy, kept up on all your assignments, and took so many notes that your fingers were numb. Your hopes and prayers are to follow all the procedures and regulations, because your goal is to be a good police officer and to make a difference in the community where you're deployed. You learned from your academy instructors that a good officer follows ethical and legal standards and does the right thing at the right time in the right way.

A few days ago, you awoke in the middle of night to review your notes because "the right thing" seemed too simple a perspective. "Good policing comes from good people," your instructor lectured, and yes, that is what you had written. "Good police officers create a sense of well-being in their communities and fairness," you read aloud to no one and everyone. You climbed back into bed, yet the word "fairness" bounced around your head for a while, preventing you from returning to sleep. Your training seemed to help, as the words "due process" ushered you off to dreamland.

1. What makes a good police officer?
2. Which criteria are used to define the right behavior?
3. In what way is due process associated with being fair?

Introduction

The role of the police can be described as protector and as intruder, the focus of this chapter. One difficult practice perfected by most police officers is the process of intruding into the privacy of an individual, while simultaneously protecting and serving. Some describe an officer as a mom or a dad (often it's both), a preacher or a sinner, a medic or a mechanic, a flatfoot or a sprinter, a no-mind or a scholar, compulsive or sympathetic, and at times a crime fighter or a star finder. Officers are first responders to service calls, emergencies (ranging from highway accidents to lost children), and catastrophes (such as wildfires and tornadoes). Ultimately, they are also the defenders of democracy. Americans want to be safe and feel secure in their individual everyday choices, without fear of arbitrary government intrusion or the intrusion of others.[1] As philosopher John Locke noted, wherever law ends, tyranny begins. Thus officers are both guardians of the rule of law and caretakers of human rights. American police provide the bridge between the **social contract** (in which people give up some rights to a government in exchange for social order) and order.[2] To maintain this bridge and accomplish police objectives, people enforce deviance and civil order control.

Deviance Control

Deviance control refers to the police practice of reinforcing community values and laws by arresting lawbreakers, curbing threatening behavior that interferes with the rights of others, and ensuring that constituents can move around freely and exercise their rights without fear, harassment, or unnecessary impediments.[3]

Members of the American public expect the police to provide an environment in which they can act upon their individual choices. What that means in practical terms is that some folks will be arrested and some won't. The justice community has the moral but thorny duty of providing professional services that some officers describe as a Catch-22 situation: You're damned if you do and damned if you don't. Deviance control conduct among police officers is guided by both a code of ethics (see Chapter 14) and the rule of law.

The public expects police officials to operate in an efficient and professional manner without expressing their personal views and emotions. To accomplish this task, police personnel must have a strict and unwavering adherence to a code of ethics and a code of conduct. One recent revision of a police code of ethics was developed by the International Association of Chiefs of Police (IACP). This code defines appropriate police behavior and attitudes of individuals sworn into law enforcement:

> My fundamental duty is to serve the community; to safeguard lives and property; to protect the innocent against deception, the weak against oppression and intimidation, and the peaceful against violence and disorder; and to respect the constitutional rights of all to liberty, equality, and justice.[4]

Control practices are also guided by litigation, oversight committees, and internal investigative units (see Chapter 14). Even with these parties monitoring police performance, the best guide to prompt officer performance is still the individual set of values held by each officer. For most officers, ethical standards are determined by doing the right thing at the right time in the right way.[5]

Civil Order Control

> **Civil order control** refers to management of civil disorders, which includes initiatives employed to control one or more forms of disturbances caused by a group of individuals.[6]

Civil disturbances are typically a symptom of, and a form of protest against, major sociopolitical problems, according to Harry R. Dammer and Erika Fairchild.[7] Typically, the severity of a disturbance coincides with the level of public outrage. Civil disorder examples include, but are not limited to, illegal parades, sit-ins and other forms of obstructions, sabotage, and other forms of crime. Civil disorder is intended to be a demonstration to the public and the government, but can escalate into general chaos. Therefore, one way to describe civil order control is to say that it is one method used to curb public displays of disorder.

Many people assume that the justice system—and especially the police—is the only social institution (a formal, recognized, established, and stabilized way of pursuing some activity in society) tasked with "legally" controlling civil order.[8] In fact, civil order control can be performed by specialized units (such as SWAT) or by social services agencies or public and privately owned agencies other than the police. This list of institutions that demand our attention and our conformity to their expectations highlights two points:[9]

- The police are just one of many institutions designed to control civil order.
- There is a vast amount of interdependence between the various institutions that attempt to control civil order.

To better understand the institutions that shape civil order, consider the informal side of democracy. That is, free people tend to express the idea that they, both collectively and individually, have the right to make "righteous" decisions in the best interest of their community (although this does not mean they "legally" possess that right). When official agencies are unable to control civil strife, Americans have and will take matters into their own hands. A description of unofficial civil order control can be readily provided by those who have experienced civil strife such as riots, fires, and hurricanes.

For example, a disturbing scene unfolded in uptown New Orleans, where looters attempted to break into a local hospital a day after Hurricane Katrina struck the Gulf Coast. The hospital personnel and director feared for the safety of the staff and the 87 injured children inside the isolated and locked facility. Looters, junkies, and probably sex offenders had gathered outside the facility and were trying gain access to the building. Because of the rising and nasty flood waters, neither police nor National Guard personnel were available, or perhaps neither was aware of the dangerous situation. Reportedly, several adults armed themselves with stolen weapons and stood in front of a sign at the hospital's entrance that read, "Loot—We Shoot."

The point of this example is a simple one: Throughout American history and probably well into the future, Americans have exercised civil order initiatives, sometimes with regrettable consequences and other times not.

What Is a Good Police Officer?

Good police officers are difficult to characterize because of the nature of the job. Taking a cue from ethics, a good officer might be defined as one who "does the right thing at the right time in the right way." Another expert provides the following definition:

> Good policing requires that we understand both what it is and why we do it. Good police officers create a sense of well-being in their communities. They protect everyone—citizens, victims, and criminals—and they serve the good of the community, not themselves. They are active in community affairs not because they have to, but because they want to get involved. Good policing comes from good people. Good police officers are mentors for others, officers and citizens, and they set positive examples. The mark of good police officers may not be what they do but what they are remembered for after they have moved on. Good police officers are good teachers; they think, analyze, and listen; they are objective; they instill confidence in others; and they leave behind a perception that they are knowledgeable and, above all, fair. Good policing is all about doing the right thing at the right time because it is the right thing to do.[10]

The ability of a police officer to satisfy these criteria reflects his or her understanding of probable cause[11] and the arrest process.

■ Probable Cause

A constitutional arrest is based on probable cause.

> **Probable cause** can be defined as a reasonable belief, based on facts and circumstances, that a person is committing, has committed, or is about to commit a crime.

Probable cause meets the following criteria:

- Probable cause is where "known facts and circumstances, of a reasonably trustworthy nature, are sufficient to justify a man of reasonable caution or prudence in the belief that a crime has been or is being committed" (*Draper v. United States*, 1959).
- Probable cause is what would lead a person of reasonable caution to believe that something connected with a crime is on the premises of a person or on persons themselves.
- Probable cause is the sum total of layers of information and synthesis of what police have heard, know, or observe as trained officers (*Smith v. United States*, 1949, which established the "experienced police officer" standard).

In terms of seizure of items, probable cause merely requires that the facts available to the officer warrant that a "man of reasonable caution" would conclude that certain items may be contraband or stolen property or useful as evidence of a crime.[12] The process employed in establishing an arrest must be rooted in due process[13] practices.

> **Due process** is the requirement that laws and regulations must be related to a legitimate government interest (such as crime prevention) and may not contain provisions that result in the unfair or arbitrary intrusion of an individual.

The definition of unfair or arbitrary intrusion into the lives of Americans continues to evolve as the war on terrorism escalates. As part of new government initiatives, for example, use of technology by local and federal police that electronically scans individuals, structures, and vehicles to identify hidden dangers is becoming more common.[14] Some might say that the new police sensory capabilities and mobile digital terminals are outside existing procedural standards. The boundaries of due process are not fixed, however, and are the subject of endless judicial interpretation and decision making.[15] Elected officials would be wise to develop policies that will aid officers in meeting the challenges of a new age while both satisfying probable cause standards and maintaining public safety.

Officers practice due process procedures that were established by the U.S. Constitution, which cites probable cause as one justification for arrest. Some might ask, "If probable cause is always present at the time of the arrest, why isn't every suspect found guilty?" The answer to this question relates to the nature of probable cause: Probable cause is more than reasonable suspicion but less than actual proof.[16] An officer builds probable cause through a step-by-step ascent from his or her reasonable suspicion, which can come from any of the following sources:[17]

- *Observation:* The officer obtains knowledge via his or her senses—sight, smell, and hearing. Observation also includes inferences made by an experienced officer based on a familiar pattern, such as circling the block twice around an armored car that is unloading money at a bank.
- *Expertise:* An officer is specially trained in gang awareness, recognition of burglar tools, the ability to read graffiti and tattoos, and various other techniques in the general direction of knowing when certain gestures, movements, or preparations tend to indicate impending criminal activity.
- *Circumstantial evidence:* The officer notes evidence that points the finger away from other suspects or an alibi, and employs a process of elimination to identify the true suspect.

The Arrest Process

The criminal justice process typically begins when a police officer places a suspect under arrest. An **arrest** occurs when an individual is taken into police custody and is no longer free to leave or move about. The use of physical restraint or handcuffs is not necessary, but remains an option for an officer. An arrest could be complete when an officer simply tells a suspect that he or she is "under arrest," and the suspect submits without the officer's use of any physical force.[18]

> The key to an arrest is the exercise of police authority over a person, and that person's voluntary or involuntary submission.

To determine whether an arrest has been made, the following objective standard must be met: "a reasonable person would have believed that he (or she) was not free to leave."[19] An officer can conduct an arrest when he or she observes a crime in progress, when an officer has probable cause, or when a warrant has been issued for a specific arrest.

An arrest warrant is a legal document issued by a judge or magistrate, usually after a police officer has submitted a sworn statement that sets out the basis for the arrest. When issued, an arrest warrant typically includes the following information:

- Identifies the crime(s) committed
- Identifies the individual suspected of committing the crime
- Specifies the location(s) where the individual may be found
- Gives a police officer permission to arrest the person(s) identified in the warrant

Search and Seizure and the Fourth Amendment

The Fourth Amendment to the U.S. Constitution protects personal privacy and specifies all constituents' right to be free from unreasonable government intrusion into their persons, homes, businesses, and property.[20] Lawmakers and the courts have established many legal safeguards to ensure that police officers interfere with individuals' Fourth Amendment rights only under limited circumstances, and through specific methods such as probable cause. When an officer acts outside the Fourth Amendment rights of a suspect, any evidence seized can be excluded under the provisions of the exclusionary rule.[21]

> The **exclusionary rule** is a legal principle holding that evidence collected or analyzed in violation of the Fourth Amendment is inadmissible for criminal prosecution.

In its 1961 ruling in *Mapp v. Ohio*, the U.S. Supreme Court held that the exclusionary rule should apply to every state.[22] It was "logically and constitutionally necessary," wrote Justice Clark for the majority, "that the exclusion doctrine—an essential part of the right to privacy—be also insisted upon as an essential ingredient of the right" to be secure from unreasonable searches and seizures.[23] "To hold otherwise is to grant the right but in reality to withhold its privilege and enjoyment."

The policy in most states reflects federal guidelines associated with the exclusionary rule. For instance, the Florida legislature has identified that one purpose of the exclusionary rule is to deter misconduct on the part of officers and agencies and, therefore, that all arrests must be conducted within probable cause standards.[24]

Most often, opponents of the exclusionary rule emphasize its vast limitations associated with crime control. "Conservatives often oppose the rule as not grounded in the Constitution, not a deterrent to police misconduct, and not helpful in the search for truth. Abolishing the exclusionary rule has been a high priority for conservatives for more than 30 years."[25]

Supporters of the exclusionary rule argue its relevance to the democratic process. "Because the exclusionary rule is the only effective tool the judiciary has for preserving the integrity of its warrant-issuing authority, any legislative attempt to abrogate the rule should be declared null and void by the Supreme Court."[26] The exclusionary rule persists because there is no credible alternative: Freeing the guilty is as unpalatable as providing police with infinite search and seizure powers.[27]

Responding to Critical Incidents

Critical incidents are part of a typical police officer's job description, job training, and obligations, and are discussed in detail in Chapter 15.[28] For now, what you need to know about the role officers play during critical incidents is that some encounter a number of critical incidents during their career, whereas others—even those in the

same jurisdiction—experience few.[29] Conceptually, a critical incident can refer to the apprehension of emotionally disturbed offenders, organized domestic terrorists, hostage-takers, barricaded subjects, riot control, high-risk warrant service, and/or sniper incidents.[30] One way to deal with critical incidents is through tactical units such as SWAT.

SWAT

Developed by LAPD commander Daryl Gates in 1966, special weapons and tactics (SWAT) units were designed to respond to critical incidents similar to those described previously. They were conceived of as an urban counterinsurgency bulwark. Since the 2001 terrorism attacks, almost 60 percent of U.S. police agencies have been supplied weapons of war by the U.S. military and selected members of those departments have been trained by active members of the U.S. military. Their mission is to aggressively respond to any altercation with the same technical know-how and weapons systems employed in military battles. Largely, the mission of police "tact units" is to tactfully control a serious incident without injury or death.

Most police agencies have either strategically trained officers ready for a tactical call-out among their own officers or trained teams drawn from other municipal police departments, sheriff departments, or state police agencies. A commander of a tact unit in a Midwestern city of 75,000 describes how his team cruises the streets in an armored personnel carrier:

> We stop anything that moves. We'll sometimes even surround suspicious homes and bring out the MP5s [an automatic weapon manufactured by gun manufacturer Heckler and Koch and favored by military special-forces teams]. We usually don't have any problems with crackheads cooperating.[31]

Often tact units appear to be similar to military units in training and weapons; for that reason, they can be called a command and control model, which includes the hierarchal chain of command.[32] Across the United States, tact units have evolved from emergency response teams into a standard part of everyday policing.[33] Raids are no longer confined to big cities or limited to critical incidents. In 1982, 59 percent of police departments that responded to one study had an active paramilitary police unit.[34] Fifteen years later, nearly 90 percent of those same agencies had an active paramilitary unit. The number of paramilitary police "call-outs" also quadrupled between 1980 and 1995.[35]

With proper equipment, first-responding units can control critical incidents and reduce the amount of both public and officer risk. The costs of "tact" weapons can be expensive, but the U.S. Department of Defense makes its surplus available to police agencies through the Law Enforcement Support Office 1033 program. For as little as $39 per weapon, a police agency can acquire surplus M-16A1 rifles, which can be converted from fully automatic to semi-automatic for use on the streets. Between 1995 and 1997, the Department of Defense gave local police 1.2 million pieces of military hardware, including more than 3,800 M-16 automatic assault rifles, 2,185 Rugar M-14 semi-automatic rifles, 73 M-79 grenade launchers, and 112 armored personnel carriers

(APCs). From 2001 to 2006, the number of military hardware doubled.[36] One tactical outfit calls its APC "mother"; another unit in Texas named its APCs "Bubba One" and "Bubba Two."

Tact units have increased in number, too. One estimate put the number of tactical units at 5,000 units across the United States.

Tactical unit officers are trained on the philosophy of "time versus opportunity" when handling hostage-rescue situations.[37] SWAT members can employ aggressive action and deadly force when directed by a SWAT commander in keeping with departmental guidelines and laws. Regular training maintains both the expertise and the cohesion of the group, which in turn enhances confidence and pride of the unit.

Active Shooter

When tact teams are called, its member gear up together, the team is briefed, and an active shooter is designated.[38] The active shooter is the officer who will employ deadly force, when so ordered. Ideally, a tactical response will be made with a precision-firing long gun, such as an AR-15/M-16 (.223 caliber), according to Chief Jeff Chudwin, president of the Tactical Officers Association.[39]

Events such as the Columbine High School shooting and the deadly shootout at the North Hollywood bank robbery are tragic altercations that justify the use of tact units.[40] More recently, however, tact units have been called out to perform relatively mundane police work, such as patrolling city streets and serving warrants.[41]

SWAT units are not without their critics. One writer refers to the priority placed on tactical units as a postmodernization effect that undermines the traditional identities and practices associated with police organizations.[42] Ultimately, whether someone has a positive opinion or a negative opinion about tactical teams, they are everywhere—and they can be effective when professionally managed.

■ Responding to the Call for Counterterrorism Measures

Federal officers and local officers are not equal, in that they do not operate by the same set of rules and regulations. In addition, federal officers face different legal risks than local officers. Municipalities and local officers are vulnerable to litigation for damages and prospective relief for civil rights violations (Title 42: U.S.C. 1983; see Chapter 14 for more detail). For instance, local police cannot employ the provisions of the USA Patriot Act as a general work guide. By comparison, federal officers are not as likely to be sued as local officers but can be criminally prosecuted.

Defensive **counterterrorism measures** are notoriously tricky to implement and can easily backfire.[43] For instance, the installation of metal detectors in airports may have produced a dramatic reduction in the number of airplane hijackings, but it has also resulted in a proportionally larger increase in bombings, assassinations, and hostage-taking incidents.[44] Target hardening of U.S. embassies and missions abroad has produced a transitory reduction in attacks on those sites, but an increase in assassinations.

Given this history, it would seem that a prudent initiative for the police would be to harden potential targets (e.g., subways, train stations, and utility centers) without triggering potential terrorist substitution effects associated with racial profiling.[45]

Some justification for questioning certain people comes from the Patriot Act, which allows federal law enforcement agents, while engaged in an open investigation, to detain and question suspected terrorists.[46]

Responding to Racial Profiling

Developing standardized procedure and policy for traffic stops, *Terry* stops, and other situations where an officer restricts the movements of a person is one of the most effective means of addressing **racial profiling** complaints.[47] Erie County–Buffalo, New York police and city officials formed a group called the Law Enforcement and Diversity Team (LEAD) to improve community relations between the police and community groups, as well as to address racial tension issues. Groups like these create a buffer between police and minority groups, which can help if a racially charged incident does occur.

Some want racial profiling to be renamed "biased-based profiling," because the targeting of individuals is not always based on their race.[48] Indeed, ethnicity, religion, sexual orientation, and other biased factors sometimes come into play. One study sought to examine this issue by comparing college students' and police officers' ideas about drivers and their vehicles.[49] There was reasonable agreement among college students as to the race/ethnicity, gender, age, and social status of the drivers of different vehicles. Both the college students and the police officers reported similar thoughts about the drivers based on vehicle cues, suggesting that "profiling" originates from the larger society and not from a police subculture, as some observers argue.[50]

Responding to Suspected Suicide Bombers

What should police do to prevent a suspected **suicide bomber** from detonating a bomb in a crowded subway or a crowded school bus? In London, 56 subway commuters were killed and 700 persons were injured in July 2006 when four bombs were exploded on three subway cars and a bus.[51] A day later, Brazilian Jean Charles Menezes left his London apartment. It was a very hot day, but Menezes was wearing a long heavy coat and his home was under surveillance as a possible terrorist hideout. When he was about to board a bus, the police tried to arrest him; when Menezes attempted to evade the officers, they killed him.[52]

In response to the terrorism attacks, the United Kingdom had suspended its constitution (something the United States cannot do). However, U.S. police officers and federal agents typically have been authorized to use deadly force if lives are in imminent danger. Since the September 11, 2001, terrorist attacks, the definition of "imminent danger" has changed, prompting officials to rethink the rules of engagement.

"There is not a responsible chief or head of a law enforcement agency in this country who isn't now pondering the dilemma a suicide bomber presents to their officers," says U.S. Capitol Police Chief Terrance W. Gainer. Gainer became the first police chief (although his employer is the federal government) in the United States to adopt a "shoot to kill" policy if his (federal) officers are confronted with a suicide bomber.

The International Association of Chiefs of Police has issued new guidelines saying that "officers who confront a suicide bomber should shoot the suspect in the head."[53] The head is the preferred target in such cases because a suicide bomber is probably wearing a "bomb" and firing at the center mass might detonate it. Conversely, an analysis by the Police Executive Research Forum (PERF), a Washington-based law enforcement advocacy group, urges officers to use lethal force only as a last option to deal with an "inevitable" threat. "Police officers in these situations [in which they are confronted by what appears to be a suicide bomber] are on the horns of a dilemma," PERF Executive Director Chuck Wexler says.[54] "On one hand, you don't want to ignore a situation. You also don't want to read something into a situation that doesn't exist." Wexler adds that some departments have been "cautious" on suicide bomber training.

Responding to Domestic Terrorism

> **Domestic terrorism** is the unlawful use, or threatened use, of violence by a group or individual based and operating entirely within the United States (or its territories) without foreign direction, committed against persons or property to intimidate or coerce a government, the civilian population, or any segment thereof, in furtherance of political or social objectives.[55]

Domestic terrorism groups range from white supremacists, anti-government types, and militia members to ecoterrorists and people who hate corporations. They include violent anti-abortionists who threaten to attack clinics and doctors; violent biker gangs that may be involved in organized crime; and black and brown nationalists who envision a separate state for blacks and Latinos. American terrorism has been remarkably diverse in terms of the causes and ideologies of the terrorists—and those terrorists have been busy. One researcher tallied more than 3,100 bombings, shootings, kidnappings, and robberies carried out for political or social objectives between 1954 and 2005.[56]

Ecoterrorists

According to the FBI, ecoterrorists commit acts of violence, sabotage, or property damage because they are motivated by concern for animals or the environment. Such ecoterrorists have committed more than 1,100 criminal acts and caused property damage estimated at least $110 million since 1976 in the United States.

In recent years, the Animal Liberation Front (ALF) has become one of the most active extremist elements in this country. Despite the destructive aspects of ALF's operations, its operational philosophy discourages acts that harm "any animal, human, and nonhuman."

The group called Earth First! has engaged in a series of protests and civil disobedience events in the United States. Twenty years ago or so, its members introduced "tree spiking" (insertion of metal or ceramic spikes in trees in an effort to damage saws) as a tactic to thwart logging.

In 1993, the Earth Liberation Front (ELF) was listed for the first time along with the ALF in a communiqué declaring solidarity in actions between both groups. It is not uncommon for the ALF and the ELF to post joint declarations of responsibility for criminal actions on their websites. In 1994, founders of the San Francisco branch of Earth First!, writing in *The Earth First! Journal*, published a recommendation that

Earth First! mainstream itself in the United States, leaving criminal acts other than unlawful protests to the ELF.

The ELF advocates "monkeywrenching," a euphemism for acts of sabotage and property destruction against industries and other entities perceived to be damaging to the natural environment.[57] "Monkeywrenching" includes tree spiking, arson, sabotage of logging or construction equipment, and other types of property destruction. The most destructive practice advocated by ALF/ELF is arson. ALF/ELF members consistently use improvised incendiary devices equipped with crude but effective timing mechanisms; these devices are often constructed based on instructions found on the ALF/ELF websites.

Another environmental terrorist group, Evan Mecham Eco-Terrorist International Conspiracy (EMETIC), was formed to engage in ecoterrorism against nuclear power plants and ski resorts.

Anti-abortion Terrorist Groups

The Army of God is an underground network of terrorists who believe that the use of violence is an appropriate tool for fighting against abortion. This group was responsible for the bombing of women's health clinics in Alabama and a gay night club. One of its members, Eric Randolph, was responsible for the bombing that occurred during the 1996 Summer Olympics in Atlanta.

Agroterrorism

Experts are concerned about the introduction of foot-and-mouth disease into the U.S. food supply, as this pathogen is 20 times more infectious than smallpox. The goal of (foreign and domestic) agroterrorists would likely be to damage the economic stability in the United States.

The Role of the Police in Domestic Terrorism

During the past decade, the United States has witnessed a dramatic change in the nature of the domestic terrorist activities. In the 1990s, right-wing extremism overtook left-wing terrorism as the most dangerous domestic terrorist threat to the country, says FBI's domestic terrorism chief.[58] Because extremist groups engage in activity that is often protected by constitutional guarantees of free speech and assembly, the police become involved when the volatile talk of these groups transgresses into unlawful action. The FBI estimates that the ALF/ELF group, for example, has committed more than 600 criminal acts since 1996, resulting in damages in excess of $43 million.

An FBI spokesperson has suggested that one role of local police is to identify threats of domestic terrorism and stop them before they happen.[59] According to the Police Executive Research Forum, local police have a critical role in responding to critical incidents that includes monitoring domestic terrorist activities, stabilizing the community after an incident, reducing community fear, and sharing information with other police agencies. Nevertheless, it is clear that protecting the United States against domestic (and foreign) terrorists will require a combined effort by industry and commerce, government, policing, and the academic and scientific communities, with all of these parties working together to minimize both the likelihood of an attack and the severity of its impact if such an attack does occur.[60]

Responding to Domestic or Family Violence

> **Domestic violence** involves one person dominating and controlling another person or persons by force, threats, or physical violence.

Family violence ranks high among all the crimes committed in the United States and remains both a community-level issue and a national problem because of the large number of individuals influenced by it. For instance, National Crime Victimization Survey (NCVS) highlights the following statistics:[61]

- Intimate partner crimes against females in the United States declined from 1993 to 2001.
- Despite the decline, intimate partner violence accounted for 20 percent of all nonfatal violent crime experienced by women in 2001.
- Some 1,247 women and 440 men were killed by an intimate partner in 2000.

Data from New York show that the rate of domestic violence in that state averages 85 criminal incidents per 10,000 residents, though the precise rate varies across New York counties.[62] By comparison, the rate of violent crime in New York is about 59 incidents per 10,000 residents, whereas the rate of property crime is about 272 incidents per 10,000 residents.

In the past, family violence was often ignored by police and frequently went unreported. A Washington, DC, evidence-based study identified the following barriers that keep women from seeking police assistance from intimate partner violence:[63]

- Predisposing characteristics—situational and personal factors
- Fears and negative experiences with police response
- Fears of repercussions

In addition to these emotional and economic reasons for not contacting police, researchers also found that the following police-related issues may prevent women from reporting intimate partner violence:[64]

- Requirement that physical proof of injuries must be present
- Humiliation of showing injuries of private parts to male officers
- Police using homophobia against the victim
- Batterer manipulation of the incident and bonding with officer
- No penalty for the batterer; retaliation for the victim

Cultural attitudes and language often play a role in determining whether emergency (911) calls are made to police. When researchers constructed a "wish list" related to police responses to domestic violence, it included requests such as quick response, more female officers, enforcement of protection orders, and follow-up visits with the victim.

Mandatory Arrest Policies

Mandatory arrest policies, under which officers must make an arrest when responding to a domestic violence call when probable cause is present, are in place in most jurisdictions in the United States. Such policies emphasize that responding domestic violence officers are authorized to conduct an arrest.

The state of New York advises that mandatory arrest policies have the following benefits:[65]

- Such a policy provides for the immediate safety for the victim and other members of the household.
- It creates a window of opportunity for the victim to identify options, develop a safety plan, gather resources, and find support from family, friends, and domestic violence advocates.
- It brings an alternative source of power into the situation.
- It sends an important message to children that violence and intimidation in the family are wrong and that someone cares about them and will act to protect them.
- It may reduce the amount of "unfinished business" the next shift has to handle.
- It sends a strong message to offenders that criminal acts will not be tolerated.
- It forms the first link in a chain of accountability that may include a combination of protective orders, civil and criminal penalties, treatment, and batterers' education programs.

All too often, however, offenders return to the same household after an arrest and repeat the process. The Women's Justice Center reveals claims that police frequently fail to take the following steps, which might prevent repeated acts of domestic violence:[66]

- Write a report.
- Get an adequate victim statement.
- Get a history of abuse.
- Ask about or properly record threats.
- Ask about sexual violence.
- Get witness statements.
- Deal with restraining order violations and stalking.
- Enforce custody orders, visitation orders, and other family court orders.
- Properly determine dominant aggressor.

Although most officers consider domestic violence to be "real crime" that warrants police intervention, some officers either see it as a private family matter or do not believe that official intervention reduces it.[67] Some officers struggle to understand domestic violence victims' actions and attitudes, sympathizing with their plight but questioning some of their behaviors and outlooks. These thoughts are consistent with a domestic violence arrest study involving a sheriff's department, which showed that less than 29 percent of the 1,870 domestic violence cases serviced in the community (over a 12-month period) ended in an arrest.[68]

One study of 329 police agencies showed that advocacy groups, victim shelters, medical professionals, counseling services, and treatment services headed the list of agencies that partnered with the police to respond to domestic violence calls.[69] It was also learned that officers were concerned with their own safety while responding to such calls, while some were irritated by the amount of time they knew the call would take. Still others were frustrated by their inability to make a difference. Equally important, a recent study comparing violence against male and female police officers reported that female officers are more likely to be assaulted during family conflict calls, especially when the assailant is intoxicated.[70]

Failure to Make Arrests in Domestic Violence Incidents

Why do police sometimes fail to make arrests when they respond to incidents involving domestic violence? The reasons vary, but most often officers who do not make an arrest despite probable cause can be characterized as follows:

- Officers see domestic violence as a private matter.
- Officers have learned that past arrests end with dismissals because victims do not testify against their abusers.
- Although victims placed an emergency call, when officers arrive, the victims ask police to perform some different task other than an arrest.
- An arrest increases an officer's workload—for example, by creating paperwork.
- Departments place a higher priority on other crimes.
- Officers may know the offender or the offender may be another officer. (If an officer is arrested on a domestic violence charge, he or she can lose the right to carry a weapon.)

Contemporary Models of Domestic Violence Response

Authorities have developed many models to deal with domestic violence. For example, the Office of Community Oriented Police Services tries to address the complexities of the problem; it focuses on the safety of the victim. This model provides victims with access to services more frequently and efficiently; provides officers with a better understanding of the domestic problem; and enables officers to share their legitimate frustrations over domestic violence with advocates and service providers who respect their role as law enforcers.

In Arlington, Texas, police have forged a partnership with The Women's Shelter to provide a coordinated on-scene response to domestic violence. Their aim is to reduce domestic violence incidents, educate domestic violence victims, aid children who witness domestic violence, and break the cycle of domestic violence and repeat victimization. The police department uses Victim Assistance staff paired with highly trained and specialized response team volunteers who are on duty during evenings and late nights and are dispatched at the request of patrol officers. The response team provides crisis intervention, informs victims of services available to them, and assists victims in requesting emergency protective orders. The response team also assists victims in making safety plans, helping with witness statements, and arranging or providing for transportation of victims and pets to shelters. By calling on the response team, police can concentrate on investigating the allegation of violence and on dealing with the offender. The partnership helps assure that everyone involved in a domestic violence case—whether it is identified through 911 or shelter hotline calls—works together throughout the process to provide the best possible service to the victim, using the strengths of each partner's role.

The Broward County (Florida) Sheriff's Office has partnered with Women in Distress (a battered-women's shelter) and Victim Assistance to provide a coordinated response to domestic violence incidents. This partnership seeks to better serve victims through a multidisciplinary method to increase the safety of victims and to provide counseling services and transportation. The Sheriff's Office formed a Special Victims

Unit specifically to target domestic violence. This unit consists of six in-house advocates who respond on-scene with deputies; it can also provide court advocacy for victims, review cases, and provide follow-up for victims. Deputies have special training and access to special evidence-collection kits (including a camera and other evidence-collection instruments) that help the state attorney's office build solid cases against perpetrators of domestic violence.

In Fort Smith, Arkansas, the police department has partnered with the Crisis Center for Women to present a coordinated response effort intended to reduce domestic violence incidents. The two main purposes of the collaboration are to assist women and families who experience abuse and to advocate for, and provide services to, rape victims. Goals include improving the quality of services to victims, appropriately evaluating successes and making necessary changes to the program when necessary, increasing officer awareness, and going beyond just arresting the batterer to having a victim-oriented response. The Crisis Center provides training to department personnel, victim information referral sheets, a 24-hour response van that arrives at the scene of disputes to provide shelter, assistance in filing emergency protective orders, and court advocacy and transportation. In addition, the Crisis Center has partnered with the police department in developing a Sexual Assault Nurse Examiner (SANE) program. When an officer responds to a domestic violence call, he or she may contact the dispatcher, who will call the Crisis Center for Women's 24-hour hotline. The Crisis Center then sends a response van to the scene and will transport the victim and children to its shelter.

In Huntsville, Alabama, the First Responder program is a partnership between Crisis Services of North Alabama and the Huntsville Police Department to provide on-scene response to victims of domestic violence. Crisis Services serves five counties and operates three shelters. Counseling at Crisis Services includes a trauma counseling program for children who witness violence and for victims of violence. The program coordinator, who is a victim advocate from Crisis Services, works at the police department and reviews reports filed at the department (not just domestic violence reports). The program coordinator works with the domestic violence investigators to assess the needs of domestic violence victims and help them access needed services. The second key feature of this partnership is that volunteer advocates ride along with patrol officers to respond to incidents involving domestic violence. Once the officer clears the advocate to go onto the property, the advocate will work with victims, provide crisis counseling, and offer information packets that let victims know what options they have and how they can make safety plans. The advocates also arrange shelter admittance and transportation, and ensure that children of victims also receive information packets.

■ Responding to School Violence

School violence is a staple of newspaper headlines across the United States. In Pearl, Mississippi, a 16-year-old boy allegedly killed his mother, then went to his high school and shot nine students, two fatally.[71] Three students were killed and five others were wounded in a high school in West Paducah, Kentucky; a 14-year-old student pleaded guilty to the crime. During a false fire alarm at a middle school in Jonesboro, Arkansas, four girls and a teacher were shot to death and 10 individuals were wounded when two boys, 11 and 13 years old, allegedly opened fire from the woods. A science teacher was shot to death in front of students at an eighth-grade dance in Edinboro, Pennsylvania; a 14-year-old awaits trial for that crime. Two teenagers were killed and more than 20 individuals were hurt when a 15-year-old boy allegedly opened fire at a high school in Springfield, Oregon.

On April 20, 1999, in Littleton, Colorado, two Columbine high school students, Eric Harris and Dylan Klebold, went on a shooting rampage. They killed 12 students, a teacher, and wounded 24 others. The two school resource officers on duty were unable to prevent the tragedy. The Columbine High School massacre is one of three of the deadliest incidents of American school violence. The 2007 Virginia Tech University massacre claimed 33 lives by a single killer and at the University of Texas at Austin there were 14 people and 31 wounded by Charles Whitman. He had fired from the observation deck of the University's 32 story administrative building in 1966 after killing his wife and mother. These tragic events have provoked debate regarding gun control laws and gun violence among youths, the availability of firearms in the United States, and the additional deployment of officers equipped with boarder arrest powers on high school and college campus. These events have also encouraged police agencies "show of force" such as zero tolerance strategies and police raids among school populations which can produce a negative affect.[72]

For example, on November 5, 2003, near Charleston, South Carolina a police drug raid with drug dogs at Stratford High School in Goose Creek, sparked incredible outcry from parents across the nation and civil liabilities suits. One parent reported: "I was shocked and outraged that the principal would let this happen," said Sharon Smalls, mother of Nathaniel Smalls, a ninth-grader who was forced to his knees with his hands behind his head while his socks, wallet, and pockets were searched. "When I saw the video on television I almost lost it. It looked like something from the war, not from my son's school."[73]

Statistics on School Violence

The report *Indicators of School Crime and Safety 2003*[74] indicates that in the latest school year evaluated, 32 school-aged youth were victims of school-associated violent death (8 were suicides). The National Crime Victimization Survey shows that students ages 12–18 were victims of about 2 million nonfatal crimes (theft plus violent crime) while they were at school and about 1.7 million crimes while they were away from school in 2002. These figures represent victimization rates of 73 crimes per 1,000 students at school.

In 2002, 17 percent of students in grades 9–12 reported they had carried a weapon everywhere, and about 6 percent reported they had carried a weapon on school property. Annually, over the five-year period from 1998 to 2002, teachers were the victims of approximately 234,000 total nonfatal crimes at school, including 144,000 thefts and 90,000 violent crimes (rape, sexual assault, robbery, aggravated assault, and simple assault).

Nonetheless, research continues to show that schools are relatively safe places for children: In fact, students were twice as likely to be victims of serious violent crime away from school than at school in 1998. Unfortunately, the perception of imminent danger in the school environment has become commonplace in many communities, leaving parents, students, and school personnel with, at best, a tenuous sense of security.

School Shooters: Myths and Reality

There are many accounts of school shooters, which have led to the following myths:[75]

- School violence is an epidemic
- All school shooters are alike
- The school shooter is always a loner
- School shootings are exclusively motivated by revenge
- Easy access to weapons is the single most significant risk factor in school shootings
- Unusual or aberrant behaviors, interests, and hobbies are hallmarks of the student destined to become violent

In reality, school shootings and other forms of school violence are problems involving schools, families, police, and the communities. A child arrives at school with cultural baggage and collective life experiences, both positive and negative, that are shaped by the environments of family, school, peers, and community. Part of this cultural baggage and collective experience includes values, prejudices, biases, emotions, and the student's responses to training, stress, and authority.[76] His or her behavior at school is affected by the entire range of experiences and influences. No single factor is decisive, and no single factor is without effect. Thus, when a student shows signs of potential violence, schools and other community institutions have the capacity—and the responsibility—to keep that potential from turning into reality. Unfortunately, it is often difficult to tell when a threat is real.

School Threat Assessment

> A **threat** is an expression of intent to do harm or act out violently against someone or something. A threat can be spoken, written, or symbolic—for example, motioning with one's hands as though shooting at another person.[77]

School threat assessment (STA) rests on three principles:[78]

- Threats and threateners are never equal
- Most threateners never carry out their threat
- All threats must be taken seriously and assessed (with police assistance)

Some threats represent a clear and present danger of a tragedy on the scale of the Columbine High School and Virginia Tech massacres. Others present little or no real risk to anyone's safety. Nevertheless, many school authorities tend to adopt a one-size-fits-all approach to any mention of violence. All school threats should be accessed in a timely manner, and decisions regarding how they are handled must be acted upon quickly. Threats should be discussed with school resource officers (SROs) or other appropriate police personnel. Authorities need to be cautious with STA because it is easy to exaggerate a threat's dangerousness—or lack of it, leading to potential underestimation of serious threats. Authorities must be mindful not to unfairly punish or stigmatize students who are not dangerous. Every threat and each threatener does not represent the same danger or require the same level of response.

Consider the following examples cited by the Justice Policy Institute:[79]

- In Houston, Eddie Evans, a sixth-grader, forgot that his Boy Scout knife (from a meeting that had taken place the night before) was in his coat pocket. After asking a friend what he should do, the boy decided to keep quiet and hide the knife in his school locker. But his friend mentioned the knife to a teacher, and school officials called police. Eddie was arrested and taken to a juvenile detention center, suspended from school for 45 days, and enrolled in an alternative school for juvenile offenders. Eddie's father says his son is a good student, a youth leader in his church, and a first-class Boy Scout, but now "he's so miserable he talks about suicide."
- A 17-year-old youth in Richmond, Illinois, shot a paper clip with a rubber band, missing his target but hitting a cafeteria worker instead. He was expelled.
- A 12-year-old child in Ponchatoula, Louisiana, diagnosed with a hyperactivity disorder, told others in a lunch line not to eat all the potatoes or "I'm going to get you." Turned in by the lunch monitor, he was referred by the principal to the police, who charged the boy with making "terroristic threats." The child spent several weeks in a juvenile detention center.
- A 13-year-old boy in Denton County, Texas, was assigned in class to write a "scary" Halloween story. He concocted one that involved shooting up a school, which got him a visit from police—and six days in jail before the courts confirmed that no crime had been committed.

"Zero tolerance has been implemented mindlessly," says Cecil Reynolds of Texas A&M University.[80] "Anytime you take something as complex as the way children behave and apply something simplistic to it, you can't be doing a good job."

Prevention of School Violence and Bullying

School safety issues take a spotlight especially among students harmed by school bullies. Bullying has two central components: repeated harmful acts and an imbalance of power. As many as 10 percent of students bully others with some regularity.[81] Jumping in, punitively or otherwise, when a student is in high school is less effective in preventing school violence. Instead, to prevent school violence, a broad-based effort guiding students at early ages and reinforcing these guides throughout their educational experiences has merit.[82]

Experts suggest that the police, parents, and educators band together and try to accomplish the following eight initiatives, which are better preventive measures against school violence:[83]

- Publicize the philosophy that a gang presence will not be tolerated.
- Institutionalize a code of conduct.
- Alert students and parents to school rules and punishments.
- Create alternative schools for those students who cannot function in a regular classroom.

- Train teachers, parents, and school staff to identify at-risk children.
- Develop community (facilitated through police community relations) initiatives focused on breaking the cycle of family violence, with programs focusing on conflict resolution, anger management, and substance abuse recovery.
- Establish peer counseling in schools to provide troubled youths with advisers.
- Provide an academically visible police presence to provide troubled youths with mentors (this includes officers qualified as certified teachers, who can lead anger management and other relevant courses).[84]

To reduce school violence, curricula should become student centered and focus on the following goals:[85]

- Promote academic attainment as opposed to lower standards; otherwise, students are being set up to fail.
- Reduce the risk of antisocial attitudes and behavior through conflict resolution and anger management courses.
- Prevent substance abuse through rewards and instilling positive values.
- Reduce conflict among students.
- Enhance the well-being of students.
- Include a strong positive police presence such that police are viewed not as adversaries, but rather as mentors.

The school resource officer (SRO) concept offers an approach to improving school security and alleviating community fears, given that the safety of children at school is an important national issue.[86] The basic SRO program model involves SROs in enforcing the law, teaching, and mentoring.[87] Forty-three percent of U.S. police departments had full-time school resource officers in June 2003; collectively, these agencies employed about 14,300 such officers.[88]

On average, most SROs divide their time as follows:

- 50 percent on policing activities
- 25 percent on counseling or mentoring
- 13 percent on teaching
- 12 percent on other activities (e.g., meetings)

At Lafayette (Louisiana) High School, one officer is assigned to an estimated 2,000 students. At Oak Park–River Forest (Illinois) High School, three SROs are assigned to an estimated 3,400 students. At North Quincy (Massachusetts) High School, there are two SRO and 3,500 students. This trend is typical across the nation. Clearly, each SRO, no matter where she or he is employed, carries a heavy responsibility.

Responding to Juveniles

Officers are not particularly comfortable when dealing with juveniles (often defined as persons younger than 17 years of age, depending on the jurisdiction) because juveniles are less predictable than adults, require more individual attention than adults, and fall under an entirely different set of jurisdictional regulations than adults.[89] In one sense, an officer has less control over the relationship when dealing with juveniles. The thoughts of most officers on this subject are neatly summarized by the following statement from Dale Ledford (formerly of the Chattanooga, Tennessee, Police Department): "Lord knows what they're thinking or if they're thinking at all. But I wouldn't bet my career or pension on taking a kid into custody unless I absolutely had to."[90]

Put simply, most officers dislike interacting with juvenile offenders and juvenile victims, especially sexual assault victims.[91] The role of a police officer among juveniles is filled with thorny issues. The fact that police are "gatekeepers" to the justice system highlights the importance of an officer's decision when dealing with juveniles. As a consequence, probable cause among juveniles often leads to an arrest, but not always.[92]

Uniform Juvenile Court Act

The Uniform Juvenile Court Act (UJCA) guides the process of taking a youth into custody. It includes the following provisions:

- Taking a youth in custody can occur pursuant to an order of the court.
- It can also occur pursuant to laws of arrest.
- A sworn officer with reasonable grounds to believe that a child is suffering from illness, is injured, or is in immediate danger may take the child into custody.
- Custody is not an arrest.

Because of the protective nature of the juvenile justice system, an officer does not require a warrant to take a juvenile into custody. Three forms of conduct typically lead to police involvement with a youth:

- Delinquent behavior
- Status offenses
- Neglect or dangerous behavior of care providers

Discretionary Detainment of Juveniles

Decisions about whether to take a juvenile into custody have come under intense review. Often, police discretion to detain a youth takes into account three factors:

- Legal: the seriousness of the crime, the frequency of the criminal behavior of the youth, and the youth's prior involvement with gang members
- Extralegal: race, gender, and socioeconomic factors
- Other factors: the youth's demeanor, his or her family situation, and the existence of a victim or citizen complaint

Most often, officers say that demeanor and disrespect are major factors influencing whether to detain a youth. For instance, one study involving 428 Cleveland, Ohio, police officers found that most officers, when disrespected by a juvenile during an altercation, would take that juvenile into custody.[93] Other reasons to take a juvenile into custody included breaking curfew, having a suspicious demeanor, and engaging in suspicious activity. Males were more likely to be detained than females.[94] Other factors that were considered by the Cleveland officers included a juvenile's hair style, type of clothing, and facial expressions.

The same study determined that veteran officers were less likely to arrest juveniles than were younger officers. It is unclear whether this difference is a consequence of mature parenting decisions or reflects the fact that the veteran officers just did not want to do the paperwork.

Follow-ups of Juvenile Custody

After taking a youth into custody, little follow-up information is given to the officer about the disposition of the case, the penalty assessed, and any treatment provided. Sometimes the only way the officer knows of the consequences of the arrest is if the officer happens to meet the offender at a later time, especially if the juvenile is apprehended again.

Numerous efforts have been made to decrease the amount of juvenile crime. Unfortunately, owing to these initiatives' disorganization and a lack of appropriate funding, it may take longer to find positive outcomes. One juvenile program that has adapted itself to the twenty-first century is the Drug Abuse Resistance Education (DARE) program. Its mission is to transform the program to meet the needs of the high-tech and fast-paced world of adults and kids alike. Instead of using the old lecture and podium method of teaching, DARE officers interact with juveniles, employ computer technology, and use practical exercises to teach reality-based issues. DARE officers are certified as trained SROs.

Responding to the Mentally Ill

A common misconception is that all mentally ill persons are tucked safely away in high-tech mental health facilities surrounded by lush parks and well-manicured lawns. The reality is that, since the trend toward deinstitutionalization of mentally ill patients (who were often sent home from mental hospitals and simply advised to take medication) began in the 1990s, most mentally ill persons have not been institutionalized. One estimate is that 1 in every 28 Americans suffers from a mental disorder.[95]

> Police officers are more likely to be killed by a person with mental illness than by an assailant with a prior arrest for assaulting police or resisting arrest. People with mental illnesses are killed by police in justifiable homicides at a rate nearly four times greater than members of the general public.[96]

An estimated 284,000 prisoners have some kind of mental illness, or 16 percent of all incarcerated prisoners. Many of those persons are eventually released through maxing out or parole. Almost 587,000 are currently on parole in the United States, though they violate the conditions of their parole more often than do individuals without mental illness. One implication of these data is that participants in the justice system must deal with mentally ill individuals on a daily basis.

When police officers respond to an incident that involves a mentally ill person, they simultaneously have to become a psychologist, a psychiatrist, a social worker, and an officer. Many officers have little training or patience with mentally ill persons, yet one expectation is that officers will process them in a similar manner as other offenders. Many mentally ill persons are unpredictable, and they can also be violent, as many are easily rattled and do not necessarily understand the consequences of their behavior. In many situations, an officer must make the decision of whether to process the mentally ill person into the criminal justice system by making an arrest or try to place the individual into a mental health facility, if beds are available. (Of course, knowing whether a bed is available is a whole other matter.)

In many situations, an officer might arrest an individual who exhibits behavior characteristic of mental illness because of convenience and practicality to the person; mentally ill patients who are confined are provided with medications and treatment.[97] Police trying to protect those persons who appear to have severe mental illnesses often use "mercy bookings" to get them off the streets. This is especially true for mentally ill women and teenagers, who are easily victimized, and often robbed and raped while on the streets. Indeed, mentally ill persons are more likely to be victimized than to be aggressors. The attitude of many officers is typified by the following statement: "I find myself arresting them to ensure their safety."[98]

Critical intervention training can change officers' perceptions about the mentally ill, so that officers are better able to meet the needs of these individuals. In one study, pre- and post- training surveys were given to participants who underwent such training; the results showed that a higher percentage of the officers said that they could respond more effectively after the training.[99]

For example, after a tragic confrontation between police and a person with severe mental illness, the Memphis (Tennessee) Police Department developed crisis intervention teams (CIT). CIT officers are "generalist–specialist" police officers who have 40 hours of training and experience in a special-duty assignment (responding to emotional disturbance crisis calls), in addition to making regular police services calls. This approach fosters a partnership between the police and the community. CIT officers learn to interact with people with mental illness who are in crisis in a way that de-escalates—rather than inflames—a tense situation. CIT officers can also divert a person to a mental health treatment facility rather than jail when appropriate.

CIT has been shown to reduce officer injury rates fivefold. In recent years, more cities have begun moving toward CIT training, including Portland, Oregon; Albuquerque, New Mexico; Seattle, Washington; Houston, Texas; San Jose, California; Salt Lake City, Utah; and Akron, Ohio.[100]

Almost one fourth of the 911 emergency calls to the New York City Police Department are linked to mentally ill offenders. In Florida, police officers respond to people with mental illnesses who are in crisis by having them assessed under the state's mental health treatment law, the Baker Act. In 2000, there were 34 percent more Baker Act cases (80,869) than DUI arrests (60,337).[101] Florida police officers initiate nearly 100 Baker Act cases each day, a volume comparable to the number of aggravated assault arrests for the state in 2000 (111 per day) and 40 percent more than the number of arrests for burglary (71 per day).

■ Responding to Registered Sex Offenders

Sexual assault offenders, including pedophiles, are a challenging experience for most officers. In many cases, sex offenders are assessed by their status as opposed to their behavior. Consider the following event, which occurred one evening in Portland, Maine.

A young man wandered into a church meeting room where church members were involved in an evening workshop. The young man, who was unknown to the church members, said that he had been physically attacked outside the church along his walk home from a retail store that was located between his apartment and the church. He said he needed help, as was obvious from the blood on his face. One member pointed to the restroom, and another called police about the attack.

When officers arrived, they asked for young man's identification and called it in. Almost immediately, the demeanor of the officers changed toward the young man, going from helpful to hostile. The officers had learned that he was a registered sex offender.

More officers arrived, and some checked out the restroom and handcuffed the bleeding young man. The sergeant interrogated the bleeding youth in front of the congregation. Eventually, his restraints were released. Another officer looked at his watched and proclaimed that the young man had violated curfew. "But I don't have a curfew," he said. "I'm not a parolee or anything."[102] With that, one of the officers slapped him in the face and told him to "shag your rotten ass home before I change my mind." The laughter among officers and church members relieved the tension in the room among everyone except the young man, who looked very sad.

Several months later, when one of the officers was asked about his own conduct during the incident, he said that he "wasn't sure what to do, so I followed the lead of the other officers. I sort of felt sorry for the kid because he really needed medical assistance. Later some of the other officers said they would have liked to take the kid outside and beat the shit out of him."[103]

In another case, Anthony Mullen was sentenced to life in prison for shooting to death two sexual offenders.[104] Mullen found his victims on Washington's online sex-offender list, posed as an FBI agent to enter their homes, and killed Victor Vazquez, 68, and Hank Eisses, 49. He said he let a third resident go because he showed remorse. During the trial, Mullen was comforted by several hundred emails sent to him, the prosecutor, and the judge saying that Mullen should receive a medal rather than a trial.[105]

How an officer behaves toward sexual offenders is not always an issue. As one officer said, "[At] the end of the day, an officer is responsible to himself or herself for bad behavior while serving and protecting. Who cares what other officers think when you're dealing with other people's lives? Do what's right at the right time and maybe an officer won't think about eating your service weapon."[106]

On a more prosaic level, it is the responsibility of the local police to supervise or participate in monitoring the sex-offender registry in their jurisdiction.[107]

Responding to Human Trafficking

Esperanza, a 20-year-old Mexican school teacher, left her three children with her mother and her home to journey to the United States.[108] Her hope was to secure a job, build a cash reserve, and have her family travel to America to live with her. A group of other women with similar goals flew from Mexico City to Tijuana and crossed over to San Diego. The travelers prepaid their expenses. They were told that their identification would not be returned until they worked off the balance of unexpected travel debts. They were forced to work in a sewing shop for 17 hours per day. They slept there, too. Esperanza was told that if she went to the police, her children would be harmed. Despite the outcome, these illegal immigrants were luckier than most: Some—both boys and girls—have been forced into prostitution, performing a minimum of 25 sexual acts per day to pay off their debts to the human smugglers.

Clandestine Nature of Human Trafficking

It is difficult to accurately track how many people have been trafficked into other countries, including the United States, because of the clandestine nature of human trafficking. The U.S. State Department estimates that 600,000 to 800,000 men, women, and children are trafficked across international borders each year; approximately 80 percent are women and girls and as many as 50 percent are minors.[109] The majority of transnational victims are trafficked into commercial sexual exploitation. Given their focus on transnational trafficking in persons, however, these data fail to account for the millions of victims around the world who are trafficked within their own national borders.

An estimated 17,000 to 20,000 women, men, and children are trafficked into the United States each year. Thus there may be as many as 100,000 to 200,000 people working as modern-day slaves in U.S. homes, sweatshops, brothels, agricultural fields, and restaurants.

Trafficking Victims Protection Act

The United States has criminalized "involuntary servitude" for more than 100 years. The Trafficking Victims Protection Act (TVPA) created new forced labor and sex trafficking criminal offenses for perpetrators and specified additional protective measures for victims.[110] As a recent court opinion interpreted the TVPA, it intended to define and expand the anti-slavery laws that would apply in trafficking situations, so as to reflect a modern understanding of victimization.[111] By more broadly encompassing the subtle means of coercion that traffickers use to bind their victims, these new criminal statutes make good on the promise made in the Thirteenth Amendment to the Constitution: that no person shall suffer slavery or involuntary servitude on American soil.

From 2001 to 2005, federal investigations of human trafficking quadrupled from 106 to 420, resulting in 95 prosecutions. Each year fewer than 10 percent of those trafficked in the United States (about 1,000 individuals) are liberated through the actions of policing.[112]

One way to control human trafficking is seen in the recent policy implemented by Montgomery County, Maryland.[113] In that jurisdiction, investigators go after pimps who smuggle people for forced prostitution, while extending a helping hand to their victims, the prostitutes. "The way we treat prostitution is completely different from when I first came up here," says vice squad Detective Thomas Stack.[114] "It was go out, pick them up, take them over to jail—that was it." But not anymore, he says. "We treat every person as a victim, and it's important that we should do that."

This trend might become common practice especially in light of the fact that local officers are now being asked to confirm the immigration status of non-citizens, a duty previously reserved for federal officers. At issue and as yet unresolved across the country is the controversy over local officers enforcing federal immigration laws.[115]

■ Responding to Suicide-by-Cop

Suicide-by-cop, also known as police-assisted suicide, refers to the situation in which an officer kills an individual who intended for the officer to kill him or her.[116] Of the almost 400 justifiable homicides (individuals killed to prevent imminent death or serious bodily injury to the officer or another person) that occur annually in the United States, an estimated 10 percent of those killed intended this result from the beginning of the interaction.[117] One study shows that many of the individuals who were ultimately killed by the police had attempted suicide prior to police arrival.[118] In some instances, the individual took his or her own life during the police encounter (see the discussion of arrest-related homicide in Chapter 9).

> Suicide-by-cop is a colloquial term used to describe a suicidal incident whereby the suicidal subject engages in a consciously, life-threatening behavior to the degree that it compels a police officer to respond with deadly force.[119]

In suicide-by-cop, the offender has the intention of provoking a lethal response from officers. For example, a suspect might point an unloaded or nonfunctioning weapon (such as a toy gun or starter's pistol) at an officer.

The following guidelines suggest ways to recognize a potential suicide-by-cop:[120]

- The person is barricaded and refuses to negotiate.
- The person has just killed someone, especially a close relative (e.g., parent, spouse, or child).
- The person says that he or she has a life-threatening illness.
- The person's demands to police do not include negotiations for escape or freedom.
- The person has undergone one or more traumatic life changes (e.g., death of a loved one, divorce, financial devastation).
- The person has given away all of his or her money or possessions.
- The person has a record of assaults.
- The person says he or she will only surrender to the person in charge.
- The person indicates that he or she has thought about planning his or her death.
- The person expresses an interest in wanting to die in a "macho" way.
- The person expresses interest in "going out in a big way."
- The person expresses feelings of hopelessness or helplessness.
- The person dictates a will to negotiators.
- The person demands to be killed.
- The person sets a deadline to be killed.

When one or more of these indicators are present, the officer might be dealing with a subject who wants to be killed and who may be willing to take any steps to reach that goal, including shooting the officer, advises Policeone.com experts. In these situations, tactical vigilance is critical, and officers should recognize that the subject is interested only in having the officer shoot him or her. Nevertheless, this knowledge should never cause the officer to hesitate to perform his or her job. When an officer feels his or her life or the lives of others are threatened, regulations dictate performance.

Some have argued that suicidal subjects may not be in a "knowing" state of mind during the event that enables them to fully appreciate the outcome.[121] Interestingly, few of those individuals killed in such encounters are intoxicated or mentally impaired.

Interviews of officers describe such individuals as being in a high emotional state. It appears that those killed did not care what the outcome between them and the police would be, suggesting they already knew the outcome—which is consistent with the idea that those killed intended to be killed by the police.

One concern is that suicide-by-cop incidents can leave an officer with feelings of guilt or anger about being tricked into using deadly force, feelings that often can be compounded by media accounts of the incident that present the deceased as the victim.[122] "Police find out after the fact that they didn't need to shoot to protect themselves. It angers them that they hurt or killed someone," says David Klinger, a former police officer and an associate professor of criminology and criminal justice at the University of Missouri at St. Louis. Another result of taking someone's life can be symptoms of post-traumatic stress disorder, such as hypervigilance, impulsiveness, and difficulty sleeping. (Chapter 15 covers stress in more detail.)

■ Responding to Police Suicide

During the days after the devastation of Hurricane Katrina, the New Orleans Police Department was feeling the strain of the disaster, and so were its officers. Not long afterward, two of New Orleans' finest were dead—but not from attacks by irate citizens or from weather-related danger. Instead, the officers committed suicide.[123] While police officer suicide does not usually make headlines, it affects those officers who learn about the incident in much the same way that a line-of-duty death does.

Factors Underlying Police Suicide

Many officers think that the job kills officers, rather than other factors. That is, the culprit is seen as not a suspect or a vehicle accident, but rather the nature of being a police officer. Because officers know the stress, the frustrations, and the all-around pressures of wearing the badge, they can understand to a certain extent what an officer may have been feeling when he or she decided to pull the trigger.

A police officer is two to three times more likely to die by suicide than to die on the job, a ratio confirmed by the National Police Suicide Foundation in Pasadena, Maryland. However, it is difficult to obtain accurate statistics about police suicide because officer deaths are often not reported as suicides, largely because officer death benefits would not be available to the individual's survivors and because officer suicide is seen as a weakness of the "fallen brother." One officer explains it this way:

> Officers are taught to take charge, be in control, and make decisions under high levels of stress. They do this on an almost daily basis and develop various officering mechanisms to deal with situations the general public couldn't and doesn't want to deal with.[124]

Although attitudes of police administrations are changing, the prevailing philosophy is that "police suicides are not an issue because the officer taking his life did so for reasons outside of the job, due to domestic situations, and many are just reported as accidental deaths."[125] A New York City Police Foundation study says that "people kill themselves because they don't know how to solve their problems." [126] The study identifies personal problems, substance abuse, and depression as direct causative factors in suicide, but not job stress and, in this case, police officer stress.

Another study involving almost 600 officers points to another factor in officer suicides:

> [Among police] officers, the risks of the streets are less traumatic than the organizational factors they face, which includes policy developed by non-police professionals, top-down bureaucratic practices, and policy initiatives linked to strategies developed as a law enforcement response to the war on crime, drugs, and terrorism.[127]

The attitude of most commanders is that police suicide is a personal result of poor living habits developed by the individual officer and, therefore, does not need to be addressed at the official level.[128] Even so, one website reports that a police suicide happens every 24 hours in the United States.[129]

Suicide Prevention

Preventing suicide becomes possible only when one understands the mindset of the suicidal officer. The first and foremost aspect driving suicidal impulses relates to feelings of lack of control and helplessness.[130] Feelings of a loss of control can culminate in the stressor event referred to as learned hopelessness: The sense of loss of control is complete and continued feelings of hopelessness grow until they create an insurmountable negative motivation toward suicide as an acceptable conclusion to the problem.

Generally, the officer who commits suicide is male, white, 35 years of age, and working patrol. Notable risk factors include abusing alcohol, separated or seeking a divorce, and experiencing a recent loss or disappointment. Typically, a domestic dispute is involved. Other risk factors include the following:

- Close relative (e.g., a parent) committed suicide
- Talked or joked about suicide
- Mood changes either gradually or suddenly happening recently
- Obvious changes in behavior
- Increase in citizen complaints about aggressiveness
- High number of off-duty accidents
- Making arrangements as though he or she is going away
- Noticeable gain or loss of weight
- Talking about himself or herself in the past tense
- Lack of interest in events previously holding a strong interest (e.g., sports, hobbies)

Suicide awareness training for new officers and in-service training should be mandatory at all law enforcement academies. In addition, supervisors should aid in identifying at-risk officers and refer them to appropriate and confidential services. Officers should know from the beginning that strength comes in many forms, and one of them is knowing when to ask for help.

■ Responding to Identity Theft

In 2005, 6.4 million households (5.5 percent of all U.S. households) discovered that at least one member had experienced one or more types of identity theft.[131] Identity theft can refer to identity fraud and can include any of the following:[132]

- Financial identity theft
- Criminal identity theft
- Identity cloning
- Business or commercial identity theft

Identity theft occurs when an individual uses the identifying information of another person (e.g., name, Social Security number, mother's maiden name, or other personal information) to commit fraud or engage in other unlawful activities, according to the Federal Trade Commission. Criminal identity theft occurs when a person uses another individual's information instead of his or her own when the person is stopped by police to avoid detection. In such cases, the offender may want to harm a third party for reasons known only to the offender, but revenge and jealousy are possible motives, too. Unfortunately, if an arrest warrant is issued, the person whose information was given—the victim of the criminal identity theft—will be named on the citation.

In identity cloning, an offender uses a victim's information to start a new life. The criminal can work and live as the victim. Illegal aliens often do this to start a life in the United States. Business or commercial identity theft refers to the situation in which an offender obtains checking accounts or credit cards in a business's name.[133]

There are many ways to obtain another person's identity:

- Change of address notices
- Eavesdropping (at businesses or schools)
- Hackers (on computers)
- Phishing (websites, credit card offers, emails)
- Scamming (fake court worker serving warrants or IRS agent—the FBI reports more than 200,000 computer complaints in 2006)[134]
- Skimming (clerks, service station attendants storage, devices on ATMs)
- Stealing (from mail boxes, dumpsters, unlocked vehicles, or gym lockers)

Successful public safety depends on ready data availability. Think about a patrol officer making a traffic stop or responding to a service call. The officer needs to know if the person being interviewed has been involved in the previous incidents, is a gang member, is an immigrant, is a fugitive, or is a terrorist. By contrast, a detective needs data to find out whether there is a verifiable crime trend in a neighborhood or whether a vehicle involved in one incident is linked to other incidents.

Expert systems and artificial intelligence technology play significant roles in the development of tools to support police performance. Such systems have successfully demonstrated their value in the areas of identify theft, breaking and entering, criminal profiling, and tracking serial criminals, among others.[135] Typically, police agencies have captured data on paper or have fed it into a database or a crime information system. Unfortunately, there is always more than one database, and often the various systems are incompatible. As a result, retrieving data can be difficult and time-consuming. In the meantime, criminals can go about their business undetected.

A group of researchers provided this example:[136] Suppose a crime has occurred at a given address in Tucson. One suspect is a person known by the alias "Baby Gangster." The responding police officer would like to know more about this person, including his real name, previous involvement with other crimes, and membership in any gangs, and, if possible, would like to see a picture of him. At the Omaha (Nebraska) Police Department, a person's mug shot (picture) and information such as involvement in a gang are stored in separate databases, which in turn are stored separately from incident records. The police officer would have to know how to search all of the information in these separate databases using different user interfaces to obtain the desired information. The problem may be further complicated if this suspect has lived in another jurisdiction. The police officer would have to call that jurisdiction (if he or she knew the facts) and ask officers there to search for this suspect in their own databases.

How extensive is this problem? In 2003, 55 percent of local police departments and 58 percent of sheriff's offices in the United States used paper reports as the primary means to transmit criminal incident field data to a central information system.[137]

■ Responding to Artificial Intelligence

Information and knowledge management relates to **artificial intelligence** for police officers, and it offers hints about what the future holds for policing. Artificial intelligence relies on multiple data sources, with each source typically having a different user interface. Efforts have been made to unify these various sources, thereby allowing for more efficient access to the data they hold.

For example, the **COPLINK Connect** network provides a single easy-to-use interface that integrates different data sources such as incident records, mug shots, and gang information. It allows diverse police agencies to share data easily, on a 24/7 basis.[138] COPLINK's cross-reference database enables information sharing, is the most efficient means of police communications available, and can expedite case investigations. An officer can provide little or partial information, such as an incomplete license plate number or an identifying tattoo, which the database then cross-references and processes. The system's cross-referencing capabilities enable the officer to access information that other law enforcement agencies may have compiled about the suspect if the suspect had had a previous run-in with the law in another county or jurisdiction.

COPLINK was designed by Hsinchun Chen of the University of Arizona and began in Boston by compiling all the information from three large databases: 911 reports, booked-suspect lists, and police reports.[139] COPLINK Connect is currently deployed at the Tucson (Arizona) Police Department (TPD). Although TPD reports that the network is a huge success, its implementation has necessitated additional officer training and led to a new, tougher set of hiring qualifications.

In the past, the police in Colorado's Grand Junction and Mesa County Sheriff's Office operated two separate database systems, even though both patrolled the same jurisdiction. The departments eventually invested in COPLINK in an effort to streamline their activities. "Combining the information into one system, which also is connected to agencies on the Front Range, could save law enforcement hundreds of hours in investigative time," says Deputy Chief Troy Smith of the Grand Junction Police Department. "Any number of cases could be solved much quicker than right now."[140]

Mesa County Sheriff Stan Hilkey concurs, noting that the COPLINK system is designed to help police make connections in cases and identify and track suspects more efficiently and more expeditiously. "This [COPLINK] is something that we absolutely want to do," says Hilkey. "I believe in any kind of system that allows us to access and share data. That takes the whole conversation to the next level."

Artificial intelligence initiatives have recently been developed for patrol officers as well (see Chapter 9). It should be acknowledged that new technology is never, by itself, the entire solution. Rather, the solution lies in providing effective policy and enforcement strategies for use of the technology. For instance, emerging weapons-detection technologies pose complex questions for the police, particularly in terms of the development of legally defensible protocols (capable of being protected from attack) for using them.[141]

One goal of the Homeland Security forces and local police who patrol the U.S. border is to create a "smart border" that provides "greater security through better intelligence and coordinated national efforts."[142] The Department of Homeland Security already monitors vehicles that enter and leave the country at land ports of entry. Some vehicles are targeted as part of the search for drugs and other contraband. U.S. Customs believes that vehicles involved in illegal activity operate in groups, and it is argued that "if the criminal links of one vehicle are known, then their border crossing patterns can be used to identify other partner vehicles."[143]

Are the extra expenditures required to install artificial intelligence systems worth it? General Meir Dagan, former head of the Bureau for Counterterrorism in the Israeli prime minister's office, explains that "investments in intelligence are invisible, whereas increased security is visible but often wasteful. The first priority must be placed on intelligence, then on counterterrorism operations, and finally on defense and protection."[144]

RIPPED from the Headlines

Fatal Shooting of Boston Man Called "Suicide-by-Cop"

Police shot and killed a 65-year-old Quincy [Massachusetts] man yesterday after he attacked officers with a knife as they responded to a report of an attempted suicide. Police and an emergency unit responded to a 911 call at about 10 A.M. and found James Hart with self-inflicted knife wounds inside a small shed. When officers tried to force Hart out of the small structure with a chemical spray, he ran out and attacked the officers with a 12-inch kitchen knife.

"Mr. Hart lunged at two of the officers while urging them to shoot him," a Quincy PD spokesperson said. "Two officers fired a total of three rounds, two of which struck Hart, who was transported to Boston Medical Center, where he was pronounced dead. Hart had a history of mental illness.

"The evidence clearly indicates the shooting was justified."

Source: Claire Cummings, "Fatal Shooting of Boston Man Called 'Suicide-by-Cop,'" *Boston Globe* (July 4, 2007), http://www.policeone.com/suicide-by-cop/articles/1286798/ (accessed July 8, 2007).

Summary

The role of the police is to provide both deviance control, which refers to the police practice of reinforcing community values and laws, and civic order control, which refers to the management of civil disorders. Good policing is all about doing the right thing at the right time because it is the right thing to do. Good officers are impartial **street-level bureaucrats**—that is, they are public employees who actually perform the actions that implement laws. Those laws include making arrests, where an arrest is defined as taking a suspect into custody based on probable cause. Probable cause is a reasonable belief, based on facts and circumstances, that a person has or is about to commit a crime.

Search and seizure is protected by the Fourth Amendment, which protects personal privacy. When an officer acts outside the Fourth Amendment, evidence can be excluded from litigation against a suspect.

A critical incident is any situation beyond the realm of an officer's usual experience that overwhelms his or her sense of vulnerability and lack of control over the situation. In dealing with such incidents, local officers (unlike federal officers) can be civilly sued by the public if they fail to meet appropriate standards. This difference in their relationship to the public means that local officers must be especially carefully when enforcing the provisions of the USA Patriot Act, implementing counterterrorism methods such as racial profiling, and dealing with terrorists of any stripe.

Contemporary policing initiatives related to domestic and family violence include policies of mandatory arrest when probable cause is present, promotion of domestic violence training for officers, community efforts to prevent domestic violence, a stronger official response to such violence, mapping of decision points in domestic violence incidents, review of prosecution functions in such cases, and partnerships between police and other agencies.

The role of police at school includes assisting with threat assessment. In this regard, police rely on three major principles: Threats and threateners are never equal; most threateners never carry out their threat; and all threats must be taken seriously and assessed. Many schools have applied zero-tolerance policies to school-aged children, exaggerated the threat, and responded too harshly to perceived threats, visiting harm upon both children and their parents.

The role of the officers when dealing with mentally ill individuals often evolves into that of caretaker, especially in the case of "mercy arrests." When officers deal with juveniles, they often have a tendency to pull back on police actions because of the enormous number of regulations surrounding juvenile intervention. When it comes to confronting registered sex offenders, however, officers tend to be guided by cultural and media view points.

Suicide-by-cop is particularly difficult for police, who are sometimes in doubt about what to do when confronted with a suspect who wants the police to kill him or her. Police officer suicide is another matter of concern: Officers kill themselves at a higher rate than do members of other professions. Usually administrators blame the individual officer for his or her weakness, when, in fact, the organizational structure helps shape the decision of the officer.

In identity theft, one person uses another's information instead of his or her own, such as when stopped by the police to avoid detection. Identify theft and other modern-day crimes can be reduced through artificial intelligence programs. Artificial intelligence refers to information and knowledge management systems that use multiple data sources having different user interfaces, such as COPLINK Connect. With this kind of network, officers can obtain reliable information around-the-clock about suspects, patrol hot spots, and border crossings.

Key Words

Arrest
Artificial intelligence
Civil order control
COPLINK Connect
Counterterrorism measures
Critical incidents
Deviance control
Domestic terrorism
Domestic violence
Due process
Exclusionary rule
Probable cause
Racial profiling
School threat assessment (STA)
School violence
Social contract
Street-level bureaucrats
Suicide bomber
Suicide-by-cop
Threat

Discussion Questions

1. Characterize police deviance control and civil order control perspectives.
2. Describe the characteristics of a good police officer.
3. Define an arrest and characterize the elements of a legal arrest.
4. Explain the relevance of the Fourth Amendment to search and seizure.
5. Describe critical incidents, and explain the importance of tact units.
6. Explain the legal issues that confront local officers versus federal agents.
7. Characterize counterterrorism methods, and describe their strengths and weaknesses.
8. Explain the controversial issues related to use of racial profiling.
9. Describe recommended procedures for dealing with suicide bombers.
10. Characterize contemporary initiatives for preventing domestic and family violence.
11. Define a school threat, and explain how it is assessed.
12. Characterize the role of police when responding to incidents involving mentally ill individuals.
13. Describe the role of police when responding to juveniles. Why do some officers hesitate when dealing with youthful offenders?
14. Articulate the role of officers when confronting registered sex offenders.
15. Describe 15 ways to identify a potential suicide-by-cop situation.
16. Characterize suicide prevention methods of police officers.
17. Explain what criminal identity theft is and why it is important to national security.
18. Identify the various artificial intelligence models used by the police.

Acknowledgment

This chapter was in part developed by Ph.D. graduate assistant, Jennifer Taylor at the University of Southern Mississippi. Taylor spent eight years as a sworn patrol officer in the Lafayette (Louisiana) Police Department. She has been employed by the Louisiana

Department of Correction, and has been an instructor at the University of Louisiana at Lafayette. Taylor was on duty during and after Hurricane Katrina in August 2005. Her dissertation is on female law enforcement experiences.

Notes

1. Harry R. Dammer and Erika Fairchild, *Comparative Criminal Justice Systems*, 3rd ed. (Belmont, CA: Wadsworth Thomson, 2008), 105–15.
2. Jean Jacques Rousseau, *The Social Contract or Principles of Political Right* (Paris: 1762).
3. Dammer and Fairchild, *Comparative Criminal Justice Systems*, 105–15.
4. International Association of Chiefs of Police, http://www.theiacp.org (accessed June 3, 2008).
5. Dennis J. Stevens, *Applied Community Policing in the 21st Century* (Boston: Allyn and Bacon, 2003), 23.
6. Dammer and Fairchild, *Comparative Criminal Justice Systems.*
7. Dammer and Fairchild, *Comparative Criminal Justice Systems.*
8. Steven E. Barkan, "Legal Control of the Southern Civil Rights Movement," *American Sociological Review* 49, no. 4 (1984): 552–65.
9. Dennis O'Neal, "Social Control Overview" (2006), http://anthro.palomar.edu/control/con_1.htm (accessed November 22, 2007).
10. Robert W. Peetz, "The Courage to Teach," *Federal Law Enforcement Bulletin* 75, no. 10 (2006): 10–11, http://www.fbi.gov/publications/leb/2006/oct06leb.pdf (accessed March 16, 2007).
11. *United States v. Puerta*, 982 F.2d 1297, 1300 (9th Cir. 1992).
12. *United States v. Dunn*, 946 F.2d 615, 619 (9th Cir. 1991), cert. Denied, 112 S. Ct. 401 (1992).
13. *United States v. Dunn:* The guarantee of due process is found in the Fifth Amendment to the Constitution, which states, "no person shall . . . be deprived of life, liberty, or property, without due process of law," and in the Fourteenth Amendment, which states, "nor shall any state deprive any person of life, liberty, or property without due process of law."
14. Samuel Nunn, "Seeking Tools for the War on Terror: A Critical Assessment of Emerging Technologies in Law Enforcement," *Policing: An International Journal of Police Strategies & Management* 26, no. 3 (2003): 454–72.
15. "U.S. Constitution" (2007), http://www.usconstitution.net/ (accessed November 21, 2007).
16. Edward M. Hendrie, "When an Informant's Tip Gives Officers Probable Cause to Arrest Drug Traffickers," *FBI Law Enforcement Bulletin* 72, no. 12 (2003), http://www.fbi.gov/publications/leb/2003/dec03leb.pdf (accessed December 19, 2007).
17. Anthony J. Pinizzotto, Edward F. Davis, and Charles E. Miller, "Intuitive Policing: Emotional/Rational Decision Making in Law Enforcement," *FBI Law Enforcement Bulletin* (2004), http://www.au.af.mil/au/awc/awcgate/fbi/intuitive.pdf (accessed December 19, 2007).
18. In either case, an arrest includes physical force (even a touch) or submission to the assertion of authority (*California v. Hodari*, 499 US 621, 1991).
19. *United States v. Mendenhall*, 446 US 544, 1870.
20. "U.S. Constitution." Also see Findlaw.com, http://criminal.findlaw.com/ (accessed May 11, 2007).

21. Edward M. Hendrie, "The Inevitable Discovery Exception to the Exclusionary Rule," *FBI Law Enforcement Bulletin* (1997), http://www.fbi.gov/publications/leb/1997/sept697.htm (accessed December 19, 2007).
22. *U.S. Supreme Court, Mapp v. Ohio, 367 U.S. 643.*
23. Findlaw.com, http://caselaw.lp.findlaw.com/data/constitution/amendment04/06.html (accessed July 13, 2007).
24. State of Florida, "2005 Status," https://taxlaw.state.fl.us/view_statute.aspx?req=322.202&file=pta_s05&ttype=2005%20Florida%20Statutes%20-%20Property%20Tax%20Administration (accessed November 21, 2007).
25. Thomas Lynch, *Police Analysis: In Defense of the Exclusionary Rule* (Washington, DC: Cato Institute, 2006), http://www.cato.org/pubs/pas/pa-319es.html (accessed November 22, 2007).
26. Lynch, *Police Analysis.*
27. "Landmark Cases: Supreme Court," Street Law & the Supreme Court Historical Society (2006), http://www.landmarkcases.org/mapp/exclusionary2.html (accessed November 22, 2007).
28. Dennis J. Stevens, *Police Officer Stress: Origins and Solutions* (Upper Saddle River, NJ: Prentice Hall, 2008), 200.
29. Michael S. McCampbell, "Field Training for Police Officers," in *Critical Issues in Policing: Contemporary Readings*, 4th ed., ed. Roger G. Dunham and Geoffrey P. Alpert (Prospects Heights, IL: Waveland, 2001), 107–16.
30. R. C. Davis, *SWAT Plots: A Practical Training Manual for Tactical Units* (Washington, DC: U.S. Government Printing Office, 1998).
31. Peter Cassidy, "Operation Ghetto Storm: The Rise in Paramilitary Policing," *Covert Action Quarterly* 62 (1997): 20–25.
32. Christian Parenti, *Lockdown America: Police and Prison in the Age of Crisis* (New York: Verso, 1999), 111. Parts of this book are available online at http://www.thirdworldtraveler.com/Prison_System/War_All_Seasons_LA.html (accessed June 7, 2004).
33. Samuel Walker, *Sense and Nonsense about Crime and Drugs* (Belmont, CA: Wadsworth, 2001), 8; Elliott Currie, "Reflections on Crime and Criminology at the Millennium," *Western Criminology Review* 2, no. 1 (1999), http://wcr.sonoma.edu/v2n1/currie.html (accessed June 6, 2004).
34. Peter Kraska and Victor Kappeler, "Militarizing American Police: The Rise and Normalization of Paramilitary Units," *Social Problems* 44, no. 1 (1997): 101–17.
35. Peter Kraska and Victor Kappeler, "Militarizing American Police." Also see Peter B. Kraska, "Enjoying Militarism: Political/Personal Dilemmas in Studying U.S. Paramilitary Units," *Justice Quarterly* 13, no. 3 (1996): 405–29.
36. Center Mass Inc. (2007), http://www.centermassinc.com/1143587.html (accessed July 13, 2007).
37. C. Remsberg, "Seize the Moment: Time vs. Opportunity in Hostage Rescue," *Law Officer Magazine* 3 (January 2007): 40–41.
38. Jeff Chudwin, "Active Shooters: Training and Equipment for First Responders," *Law Officer Magazine* 3 (February 2007): 60–63.
39. Center Mass Inc.

40. On February 28, 1997, at the Bank of America in North Hollywood, California, SWAT and patrol officers were confronted by two heavily armed robbers. The shootout resulted in the wounding of 11 people (nine police officers and two civilians) and the deaths of both bank robbers. Although only the suspects were killed, the sheer number of injuries made this one of the bloodiest single cases of violent crime in the 1990s and one of the most significant single bank robberies.

41. Kraska and Kappeler, "Militarizing American Police."

42. Jonathan Simon, "Paramilitary Features of Contemporary Penalty," *Journal of Political and Military Sociology* (1999), http://www.findarticles.com/p/articles/mi_qa3719/is_199901/ai_n8834565 (accessed July 5, 2005).

43. Barnard E. Harcourt (March 17, 2006), https://www.law.uchicago.edu/academics/publiclaw/123.pdf (accessed May 12, 2007).

44. Harcourt.

45. Harcourt.

46. Michael Smithkey, "Terrorism, Profiling, and Common Sense," *National Academy Associate* 8, no. 3 (2006): 14–17.

47. G. W. Schoenle, Jr., "Developing a Program to Address Racial Profiling: A Collaborative Approach," *National Academy Associate* 8, no. 1 (2006): 14–17, 24.

48. E. Nowicki, "Racial Profiling Problems and Solutions," *Law and Order* 50, no. 10 (2002): 16–18.

49. Jennifer N. Luken, Rhonda R. Dobbs, and Alejandro del Carmen, "Vehicle Cues and Perceptions of Driver Characteristics: A Comparative Analysis of Police Officers and College Students," *Southwest Journal of Criminal Justice* 3, no. 2 (2006): 127–47.

50. John P. Crank, *Understanding Police Culture* (New York: Anderson, 2004).

51. CBS News Online, "London on Edge" (2006), http://www.cbc.ca/news/background/london_bombing/ (accessed May 12, 2007).

52. Jed Babbin, "Shoot to Kill?" (August 3, 2005), http://www.californiarepublic.org/archives/Columns/Babbin/20050803BabbinShoot.html (accessed May 12, 2007).

53. Sari Horwitz, "Police Chiefs Group Bolsters Policy on Suicide Bombers," *Washington Post* (August 4, 2005), http://www.washingtonpost.com/wp-dyn/content/article/2005/08/03/AR2005080301867.html (accessed June 3, 2008).

54. Kevin Johnson, "U.S. Police Guidelines Differ on Suicide Bombers," *USA Today* (May 10, 2007), http://www.officer.com/ (accessed May 16, 2007).

55. James F. Jarobe, "The Threat of Eco-terrorism," testimony of James F. Jarobe, Domestic Terrorism Section, Counterterrorism Division, FBI, Before the House Resources Committee, Washington, DC, February 12, 2002, http://www.fbi.gov/congress/congress02/jarboe021202.htm (accessed November 25, 2007).

56. Christopher Hewitt, *Political Violence and Terrorism in Modern America: A Chronology* (CT: Greenwood Publishing, 2007).

57. Jarobe, "The Threat of Eco-terrorism."

58. Jarobe, "The Threat of Eco-terrorism."

59. Glenn R. Schmitt, "Agroterrorism: The Role of Law Enforcement" (NIJ 257) (June 2007), http://www.ojp.usdoj.gov/nij/journals/257/agroterrorism.html (accessed November 25, 2007).

60. Schmitt, "Agroterrorism."

61. Bureau of Justice Statistics, "National Crime Victimization Survey (NCVS)" (2006), http://www.ojp.gov/bjs/abstract/ipv01.htm (accessed November 22, 2007).

62. Division of Criminal Justice Services, "Comparison of Domestic Violence Reporting and Arrest Rates" (2006), http://criminaljustice.state.ny.us/crimnet/ojsa/domviol_rinr/index.htm (accessed June 18, 2007).

63. Marsha E. Wolf, Uyen Ly, Margaret A. Hobart, and Mary A. Kernic, "Barriers to Seeking Police Help for Intimate Partner Violence," *Journal of Family Violence* 18, no. 2 (2003): 121–29.

64. Wolf, Ly, Hobart, and Kernic, "Barriers to Seeking Police Help."

65. Office of Domestic Violence, State of New York (2007), http://www.opdv.state.ny.us/criminal_justice/police/index.html (accessed April 26, 2007).

66. Women's Justice Center (2007), http://www.justicewomen.com/handbook/part2_c.html#police_errors (accessed April 29, 2007).

67. "Domestic Violence Insight" (2005), http://www.asu.edu/news/stories/200512/20051212_morrisondv.htm (accessed April 29, 2007).

68. Sherrie Bourg and Harley V. Stock, "A Review of Domestic Violence Arrest Statistics in a Police Department Using a Pro-arrest Policy: Are Pro-arrest Policies Enough?", *Journal of Family Violence* 9, no. 2 (1994): 177–89.

69. Melissa Reuland, Melissa Schaefer Morabito, Camille Preston, and Jason Cheney, *Police-Community Partnerships to Address Domestic Violence* (Washington, DC: U.S. Department of Justice, COPS Office, 2006), http://www.cops.usdoj.gov/files/ric/Publications/domestic_violence_web3.pdf (accessed April 27, 2007).

70. Cara E. Rabe-Hemp and Amie M Schuck, "Violence against Police Officers: Are Female Officers at Greater Risk?", *Police Quarterly* 10, no. 4 (2007): 411–27.

71. Stephen R. Band and Joseph A. Harpold, "School Violence: Lessons Learned," *FBI Law Enforcement Bulletin* (1999), http://www.dps.mo.gov/HomelandSecurity/ReferenceDocuments/FBI%20Law%20Enforcement%20Bulletin%20-%20School%20Violence%20%20Lessons%20Learned.pdf (accessed December 19, 2007).

72. Daniel Macallair and Tim Roche, *Widening the Net in Juvenile Justice and the Dangers of Prevention and Early Intervention*, (Washington, DC: U.S. Department of Justice, The Justice Policy Institute United States, NCJ 192131, 2001). Also see Thomas G. Blomberg, "Widening the Net: An Anomaly in the Evaluation of Diversion Programs," http://www.criminology.fsu.edu/crimtheory/blomberg/netwidening.html (accessed June 20, 2008)

73. "South Carolina Students were Terrorized by Police Raid with Guns and Drug Dogs, ACLU Lawsuit Charges." American Civil Liberties Union, (2003), http://www.aclu.org/drugpolicy/gen/10672prs20031215.html (accessed June 21, 2008).

74. *Indicators of School Crime Survey* (Washington, DC: Bureau of Justice Statistics, U.S. Department of Education, 2004), http://nces.ed.gov/pubsearch/pubsinfo.asp?pubid=2005002 (accessed October 29, 2006).

75. Federal Bureau of Investigation, "The School Shooter: A Treat Assessment Perspective" (2007), http://www.fbi.gov/publications/school/school2.pdf (accessed May 14, 2007).

76. Federal Bureau of Investigation, "The School Shooter."

77. Federal Bureau of Investigation, "The School Shooter," 42.

78. Guided by Ronald F. Tunkel, "Bomb Threat Assessments: Focus on School Violence," *FBI Law Enforcement Bulletin* (2002), http://findarticles.com/p/articles/mi_m2194/is_10_71/ai_93915934 (accessed December 19, 2007).

79. Michael Crowley, "That's Outrageous: No Mercy, Kid," *Readers Digest* (2007), http://www.rd.com/content/thats-outrageous-no-mercy-kid/ (accessed May 17, 2007).

80. Crowley, "That's Outrageous."

81. Rana Sampson, *Bullying in Schools* (Washington, DC: U.S. Department of Justice, Office of Community Oriented Policing Services, March 2002).

82. Gene Marlin and Barbara Vogt, "Violence in Schools," *Police Chief* (April 1999): 169.

83. Ira Pollack and Carol Sundermann, "Creating Safe Schools: A Comprehensive Approach," *Journal of the Office of Juvenile Justice and Delinquency Prevention* 8, no. 1 (2001): 13–20.

84. This point has been strongly advised by this writer in several publications. The police require a strong presence in our schools not as adversaries, but rather in mentorship roles, which includes their work as teachers. See Stevens, *Applied Community Policing in the 21st Century.*

85. Office of Juvenile Justice and Delinquency Prevention, http://ojjdp.ncjrs.org/ (accessed May 18, 2007).

86. Office of Juvenile Justice and Delinquency Prevention, *School Resource Officer Training Program* (Washington, DC: U.S. Department of Justice, Office of Justice Programs, March 2001), http://www.ncjrs.gov/pdffiles1/ojjdp/fs200105.pdf (accessed October 28, 2006).

87. Peter Finn, Michael Shively, Jack McDevitt, William Lassiter, and Tom Rich, "Comparison of Program Activities and Lessons Learned among 19 School Resource Officer (SRO) Programs" (document no. 209272) (March 2005), http://www.ncdjjdp.org/cpsv/Acrobatfiles/SRO_Natl_Survey.pdf (accessed October 29, 2006).

88. "SRO" (2006), http://www.ojp.usdoj.gov/bjs/pub/pdf/lpd03.pdf (accessed July 13, 2007).

89. Terrance T. Allen, "Taking a Juvenile into Custody: Situational Factors That Influence Police Officers' Decisions," *Journal of Sociology and Social Welfare* 32 (2005): 121–29.

90. Dr. Dale Ledford, Divisional Chair at University of Southern Mississippi, personal communication, July 2007.

91. Stevens, *Police Officer Stress.*

92. Page 56 in Dennis J. Stevens, "Civil Liabilities and Arrest Decisions," in *Policing and the Law*, ed. Jeffery T. Walker (Upper Saddle River, NJ: Prentice Hall, 2002), 53–71. The evidence shows that officers are uncomfortable with arresting juveniles and often deliberately overlook probable cause, including the presence of illegal substances in some cases.

93. Allen, "Taking a Juvenile into Custody."

94. D. J. Conley, "Adding Color to a Black and White Picture: Using Qualitative Data to Explain Racial Disproportionality in the Juvenile Justice System," *Journal of Research in Crime and Delinquency* 31, no. 2 (1994): 135–48.

95. Joan Epstein, Peggy Barker, Michael Vorburger, and Christine Murtha, *Serious Mental Illness and Its Co-occurrence with Substance Use Disorder, 2002.* Department of Health and Human Services (SMA 04-3905, Analytic Series A-24) (Rockville, MD: Office of Applied Studies, 2004).

96. "Law Enforcement and People with Severe Mental Illnesses," briefing paper: Psychlaws (2005), http://www.psychlaws.org/BriefingPapers/BP16.htm (accessed May 15, 2007).

97. Randolph Dupont and Samuel Cochran, "Police Response to Mental Health Emergencies: Barriers to Change," *Journal of American Academy of Psychiatry and the Law* 28, no. 3 (2000): 338–44.

98. W. Wells and J. A. Schafer, "Officer Perceptions of Police Responses to Persons with a Mental Illness," *International Journal of Police Strategies and Management* 29 (2006): 578–601.

99. Wells and Schafer, "Officer Perceptions of Police Responses."

100. Dupont and Cochran, "Police Response to Mental Health Emergencies."

101. Florida Department of Law Enforcement, "Data and Statistics: UCR Arrest Data," http://www.fdle.state.fl.us/FSAC/data_statistics.asp#UCR20Arrest20Data (accessed January 8, 2002); Annette C. McGaha and Paul G. Stiles, Florida Mental Health Institute, "The Florida Mental Health Act (The Baker Act)," *2000 Annual Report* (2001): 9.

102. Personal confidential communication between a church member and the writer.

103. The officer was a member of the congregation and confidentially agreed to discuss the matter with the writer provided the officer's identity remained confidential.

104. John R. Ellement and Suzanne Smalley, "Maine Killings Refuel Debate over Registries," *Boston Globe* (April 18, 2006), http://ethics.tamucc.edu/article.pl?sid=06/04/18/2251200 (accessed July 13, 2007).

105. "Sex Offender Issues," http://sexoffenderissues.blogspot.com/search/label/Suicides=over (accessed May 15, 2007).

106. Confidential personal communication between the writer and a New Orleans, Louisiana, police officer, July 13, 2007.

107. All 50 states' and the District of Columbia's sexual registries and their laws pertaining to registration can be found at the following website: http://www.prevent-abuse-now.com/register.htm (accessed May 16, 2007).

108. Ellie Hidalgo, "Human Trafficking: A Harsh Reality," *Tidings News* (March 18, 2005), http://www.the-tidings.com/2005/0318/traffic.htm (accessed November 24, 2007).

109. U.S. State Department, "Trafficking in Persons Report: June 2005" (2006), http://www.state.gov/g/tip/rls/tiprpt/2005/46606.htm (accessed November 22, 2007).

110. Kevin Bales and Steven Lize, "Investigating Human Trafficking," *FBI Law Enforcement Bulletin* 76, no. 4 (2007), http://www.fbi.gov/publications/leb/2007/april2007/april2007leb.htm#page_24 (accessed December 19, 2007).

111. U.S. Department of State, "The Common Thread of Servitude" (2005), http://www.state.gov/g/tip/rls/tiprpt/2005/46606.htm (accessed November 22, 2007).

112. Bales and Lize, "Investigating Human Trafficking."

113. Lea Terhune, "U.S. Law Officers Take a New Approach to Combat Prostitution," International Information Programs, USINFO, June 2007, http://usinfo.state.gov/ (accessed November 22, 2007).

114. Terhune, "U.S. Law Officers Take a New Approach."

115. In 2003, the George W. Bush administration overturned a Department of Justice legal opinion and stated that the police have the "inherent authority" to enforce all federal immigration laws; it also mandated that the names of immigration violators be entered into a national criminal database. This policy is being challenged in the courts. Source: American Immigration Lawyers Association, http://www.immigrationforum.org/ (accessed November 22, 2007).

116. Hal Brown, "Suicide by Cop" (2003), http://www.geocities.com/~halbrown/suicide_by_cop_1.html (accessed May 14, 2007).

117. Brown, "Suicide by Cop." Also see Bureau of Justice Statistics, "Policing and Homicide, 1976–98: Justifiable Homicide by Police, Police Officers Murdered by Felons" (NCJ 180987) (March 2001), http://www.ojp.gov/bjs/pub/pdf/ph98.pdf (accessed May 15, 2007).

118. Rick Parent, *Aspects of Police Use of Deadly Force in North America: The Phenomenon of Victim-Precipitated Homicide*, published dissertation, Simon Fraser University, 2004.

119. Rebecca Stincelli, "Suicide by Cop: Victims from Both Sides of the Badge" (2002), http://www.suicidebycop.com/page3.html (accessed May 14, 2007).

120. Scott Buhrmaster, "Suicide by Cop: 15 Warning Signs That You Might Be Involved. A Police One Special Report" (May 26, 2006), http://www.policeone.com/writers/columnists/ScottBuhrmaster/articles/84176 (accessed May 15, 2007).

121. Vivian B. Lord. *Suicide by Cop—Inducing Officers to Shoot* (Flushing, NY: Looseleaf Law Publications, 2004).

122. Susan Weich, "MO: Suicide by Cop Takes Toll on Police," *St. Louis Post Dispatch* (February 7, 2006), http://www.stltoday.com (accessed March 29, 2008).

123. D. Grossi, "Emotional Survival: Police Suicides," *Law Officer Magazine* 3 (April 2007): 34–36.

124. Jay Quinlan, "Police Suicide: What You Need to Know and What Police Officers Need to Do," *Policeone* (February 7, 2006), http://www.policeone.com/writers/columnists/marksman/articles/123078 (accessed May 14, 2007).

125. Quinlan, "Police Suicide."

126. Gale Scott, "Job Not Guilty in Cop Suicides," *New York Newsday* (September 14, 1994), p. A23.

127. Stevens, *Police Officer Stress*, 10.

128. Quinlan, "Police Suicide," 2.

129. National Police Suicide Foundation, http://www.psf.org/ (accessed May 15, 2007).

130. Quinlan, "Police Suicide," 2.

131. Bureau of Justice Statistics, *National Crime Victimization Survey: Identity Theft, 2005* (NCJ 219411) (Washington, DC: Department of Justice, 2007). There were approximately 117,110,800 households in the United States in 2006 and approximately 1.2 billion credit cards.

132. Identity Theft Resource Center, "Scams & Consumer Alerts" (2007), http://www.idtheftcenter.org/alerts.shtml (accessed April 15, 2007).

133. Identity Theft Resource Center.

134. These complaints included many different types of fraud, such as auction fraud, nondelivery, and credit/debit card fraud, as well as nonfraudulent complaints, such as computer intrusions, spam/unsolicited email, and child pornography: http://www.ic3.gov/media/annualreport/2006_IC3Report.pdf (accessed May 14, 2007).

135. John W. Brahan, Kai P. Lam, Hilton Chan, and William Leung, "AICAMS Article: Intelligence Crime Analysis and Management System," *Knowledge-Based Systems* 11, no. 5–6 (1998): 355–61.

136. Hsinchun Chen, Jenny Schroeder, Roslin V. Hauck, Linda Ridgeway, Homa Atakakhsh, Harsh Gupta, Chris Boarman, Kevin Rasmussen, and Andy W. Clements, "COPLINK Connect: Information and Knowledge Management for Law Enforcement," Artificial Intelligence Lab, University of Arizona and Tucson Police Department (2002), http://ai.bpa.arizona.edu/go/intranet/papers/COPLINK-Connect-Special-Issue-DG.pdf (accessed May 13, 2007)

137. Bureau of Justice Statistics (2007), http://www.ojp.gov/bjs/sandlle.htm (accessed May 15, 2007)

138. Chen et al., "COPLINK Connect."

139. "Knowledge Computing," *Boston Globe* (2003), http://www.knowledgecc.com/bglobe703.htm (accessed June 16, 2007).

140. "Top News: Colorado Police Embrace New Communications Technology" (June 12, 2007), Policemag.com (accessed June 15, 2007).

141. Chris Tillery, "Detecting Concealed Weapons: Directions for the Future," *NIJ Journal* 258 (2007): NCJ 219608 (Washington, DC: Department of Justice).

142. Office of Homeland Security, "National Strategy for Homeland Security," http://www.whitehouse.gov/homeland/book/index.html (accessed May 14, 2007).

143. Siddharth Kaza, Yuan Wang, and Hsinchun Chen, "Suspect Vehicle Identification for Border Safety with Modified Mutual Information," in *Proceedings of the IEEE International Conference on Intelligence and Security Informatics* (ISI 2006), San Diego, CA, May 23–24, 2006, http://ai.eller.arizona.edu/COPLINK/publications/isisuspectvehicle.pdf (accessed May 13, 2007).

144. Derek Jinks, *Disaggregating "War,"* Public Law Working Paper 69 (Chicago: University of Chicago, July 2004).

NO PARKING
ANYTIME
STOP
LINE
Salon

Crime Scene Investigations

The Ripper case is not one to be conclusively solved by DNA or fingerprints . . . and in a way, this is good. Society has come to expect the wizardry of forensic science to solve all crimes, but without the human element of deductive skills, teamwork, very hard investigation, and smart prosecution, evidence means nothing.

—Patricia Cornwell

▸ ▸ LEARNING OBJECTIVES

When you finish reading this chapter, you will be better prepared to:

- Describe the criminal investigation process
- Define the chain of custody
- Characterize the myths associated with criminal investigations
- Describe and compare cold cases versus hot cases
- Compare an interview to an interrogation
- Explain ways to respond to a suspect's response, resistance, and lies
- Articulate the importance of a baseline
- Characterize burden of proof
- Describe strategies to help avoid interrogative failure
- Explain the tasks of a crime scene analyst
- Characterize police reluctance to use crime labs
- Describe how the literature supports police reluctance to use crime labs
- Describe the positive effects of the "*CSI* effect"

▸ ▸ CASE STUDY: YOU ARE THE POLICE OFFICER

One reason you worked hard at the police academy was to finish your police training so that you could become a crime scene analyst, much like the ones you saw on popular television dramas such as *CSI* and *Law & Order.* Although your college major was criminal justice, your minor was forensic science. Your GPA was over 3.4 in everything, but you did better in your minor courses such as microbiology and chemistry. Your work during a DNA course on trace evidence drew the attention of your professor, and he asked you to assist in the lab during the next semester for college credit.

During your last summer at college, you interned at the state crime lab, which was located 50 miles from the campus. The drive was okay because of the benefits it brought you, and your parents—especially your dad—encouraged you to accept the internship. Your mom, an emergency room nurse, hesitated at first because she knew some of the technicians who worked at the state lab.

You learned the first day at the state crime lab that most of the forensic science technicians or criminalists had little college education. Many of the technicians had come up through the police ranks and were trained on the job. They seemed to be men who were roughly your father's age with your father's XL physical appearance. You got the impression that many of them resented your education, the things you learned while a lab assistant, and the fact you're a girl. You now understood your mother's hesitation. Of course, that pushed you all the more.

1. Why did your mother hesitate to encourage you to accept the internship at the state crime lab?
2. In what way is a forensic science curriculum different from a criminal justice curriculum? Why?
3. Which skills would the training of a police officer aid a criminalist or crime scene analyst?

Introduction

Contemporary policing relies on three strategies to provide police services:[1]

- Motorized patrol (see Chapter 9)
- Rapid response to calls for service (see Chapter 10)
- Retrospective investigation of crimes (this chapter's focus)

This chapter describes crime scene investigations, which includes the investigation process and forensic or crime laboratory support. Detection is the discovery of a crime. Most often, a patrol officer detects a crime and a detective investigates it. Detectives are plainclothes investigators who can be employed by local, state, or federal agencies and usually specialize in a certain type of violation, such as homicide or fraud.[2] As Patricia Cornwell[3] implies, even forensic science has its limitations, despite the myth that conducting a successful investigation takes little skill and that forensics is a cure-all.

Criminal Investigation Units

Ninety percent of all local police departments have investigative units, which can include units with a focus on juvenile, robbery, homicide, arson, cybercrime, or hate crimes, to name a few possibilities.[4] Some units are designed as interagency task forces to combat specific crimes, such as a multi-county drug unit. An estimated 15 percent of all sworn officers across the United States are tasked with criminal investigative responsibilities.[5]

The federal government maintains a high profile in many areas of investigative law enforcement. For instance, special agents at the Federal Bureau of Investigation (FBI) are the government's principal investigators, responsible for investigating violations of more than 200 categories of federal law and conducting national security investigations.[6] Agents may conduct surveillance, monitor court-authorized wiretaps, examine business records, investigate white-collar crime, and participate in sensitive undercover assignments. The FBI investigates organized crime, public corruption, financial crime, fraud against the government, bribery, copyright infringement, civil rights violations, bank robbery, extortion, kidnapping, air piracy, terrorism, espionage, interstate criminal activity, drug trafficking, and other violations of federal statutes.

After a report of a crime is made or a probable cause observed by an officer, a case can be assigned to the investigative unit (if there is one), and the detective follows through with the investigation.[7] Of course, detecting a crime and investigating it does not automatically lead to an arrest (see the discussion of clearance rates in Chapter 1), and making an arrest does not automatically lead to prosecution and conviction.

Criminal Investigations

Once a criminal offense has been committed, there are three possible outcomes:[8]

- The crime can go undetected by victims, witnesses, or the police.
- A victim may not report the crime because it might lead to an increase in the person's insurance rate (auto, home, or a private business), or because a business owner is selling illegal products such as drugs or untaxed cigarettes, or because a rape victim might think the authorities might not bring a rapist to justice and fears retaliation. There are many reasons why crime goes unreported.
- Crime might come to the attention of police through observation, a witness, or a victim.

> An **investigation** is an information-gathering process that includes interviewing victims, witnesses, and suspects. It also includes collecting and analyzing evidence such as fingerprints, DNA, documents, computer hard drives, shell castings, and other potential clues. When collecting evidence in this way, it is crucial to document the chain of custody.

Purpose of Investigations

The second primary crime-control strategy of police organizations is the investigation of reported or known crimes.[9]

> The purpose of an investigation is to carry out the following steps both legally and professionally:
>
> - Gather and confirm evidence and facts of a crime (sometimes utilize forensic analysis).

- Secure and protect the chain of custody (evidence).
- Identify law breakers, witnesses, and victims.
- Arrest law breakers (assuming probable cause is available).
- Obtain an admission of guilt through interviews and an interrogation if necessary.

Patrol officers (see Chapter 9) are the first responders at most crime scenes (because of 911 calls), and it is the task of first responders to secure the crime scene while awaiting the arrival of investigators who will "process" the evidence.[10] Detective Mike McCleery of the New Orleans Police Department adds a detective needs to keep an account of the names and badge numbers of the patrol officers who are present at the scene in case questions arise about their roles as first responders at a later date.[11]

Process of an Investigation

Only when a case is detected or reported does the offense become an "official" concern to an investigator (otherwise, litigation or complaints are options for handling illegal intrusion). The preliminary investigation consists of the initial facts and circumstances that are collected and documented in the initial offense report completed by the primary officer who was first called to the scene. Any subsequent detective work on the case would then be considered a continuing investigation. Continuing the investigation is usually conducted by specific detectives who possess some degree of specialization (e.g., in homicide, fraud, or robbery). As part of the continuing investigation, detectives might also call upon the services of forensic analysts, who conduct laboratory analysis of evidence (in some jurisdictions their presence could be mandatory).

The process of investigations can only be described ideally because circumstance and events are different even among similar criminal activities. Nevertheless, investigations typically follow these steps:

Request medical services. First responders should make an immediate request for medical and other public health services (e.g., fire, Board of Health, medical examiner's office, child services) in instances where the victim or the suspect has sustained an injury.

Secure the crime scene (CS) while awaiting the arrival of detectives. Try not to destroy evidence while securing the CS.

Determine whether a crime was committed. Was a crime committed and, if so, what specific type of offense was it? Ruling out the notion that a crime has not occurred is possible in circumstances including suicide, death by natural means, or accidental causes leading to, for instance, the injury or attempted injury of others. Most complaints are made honestly, but an investigator must always be alert for indicators of a false complaint or concealment of an additional crime on the part of the victims. In some situations, for example, individuals may inflict injury upon themselves owing to a psychological dysfunction.

Make an arrest. An arrest can be conducted when probable cause (see Chapter 9) is present or when a suspect confesses. This option remains available during the entire investigative process. Of course, officers always try to build a strong legal case, even when a suspect is arrested at the CS, thereby ensuring a successful prosecution if the suspect is formally charged at a later date (see the discussion of burden of proof later in this chapter).

Build a case. An investigator might initiate a "hot" (immediate) search for evidence or a "warm" search, which might include the help of forensic investigators. The latter type of search may occur even after a suspect is apprehended because a prosecutor requires (legally) documented evidence to secure a conviction. It is unlikely that forensic analysts will be present at every crime scene because of the expense and given the circumstances of the case.

Maintain the chain of custody. The **chain of custody** is a chronological, written record of those individuals who have had custody of the evidence from its initial acquisition until its final disposition. Each person in the chain of custody must be identified. Chain of custody begins when an item of evidence is collected. Each individual, depending on the size of the agency, in the chain of custody is responsible for an item of evidence to include its care, safekeeping, and preservation while the evidence is under his or her control. The chain is maintained until the evidence is disposed of.[12] The chain of custody assures continuous accountability. If this accountability is not properly maintained, an item may be deemed inadmissible in court or may be lost or misused. Because of the sensitive nature of evidence, an evidence custodian typically assumes responsibility for the evidence when it is not in use by the investigating officer or another competent authority involved in the investigation. In some agencies, an evidence custodian is a sworn officer; in others, it is an enlisted person.

Contact the district attorney. If anyone might possibly have an expectation of privacy in any portion of the CS, gathering certain kinds of evidence might require a search warrant. The evidence that investigators recover is of little value if it is not admissible in court.

Conduct a walk-through. A walk-through aids in determining the scope of the CS. In this review, investigative personnel make note of anything moved and identify potential evidence on the scene.

Call specialists. Detectives might call in specialists if they are not certified in a particular area or if additional tools are needed to gather evidence. For example, blood spatter on the ceilings or maggot activity on a corpse requires specialists. It is difficult to deliver a section of the ceiling to the lab for blood spatter analysis, and maggot activity changes with each passing minute.

Bag and tag the evidence. Crime scene investigative personnel systematically collect all potential evidence, tag it, log it, and package it to ensure its safety. Most often specific evidence is delivered to a crime lab, suggesting that it is unlikely that evidence is analyzed at the CS.

Take photographs, draw sketches, and use other forms of documentation. Sometimes videos aid in documentation of the CS. Documentation is one of the most important responsibilities of crime scene personnel.

Create investigative records. All conversations with first responders, witnesses, and victims (if any) should be recorded electronically or in writing.

Eliminate "persons of interest." Find a suspect.

Interrogate the suspect. The goal is to gain a confession.

Myths Associated with Investigators and Investigations

James Bond Syndrome

James Bond (007) and many investigators featured on popular media dramas share the uncanny ability of knowing just about everything on this planet. For instance, they immediately know the exact location of an obscure park based on a droplet of debris scratched on the sole of a suspect's shoe found in Lower Manhattan at noon. An investigator who consistently solves many cases tends to possess an abundance of professional qualities, including a high degree of self-discipline, patience, and other qualities that will be characterized later in the chapter. Unfortunately, few of these characteristics resemble James Bond's image as depicted in the popular media—including his income.

Myths about Known Crime

Most crimes are never reported to the police. The literature documents that nearly all drug crimes, most property offenses, most acts of family violence including sexual assault, and many other crimes of violence including murder are vastly underreported.[13] This leads to a worrisome question: How many murders and sexual assault crimes actually occur in the United States?

One report shows that when a body is discovered, determining that the culpable agent for death was murder may not be as easy as expected. For instance, suicide or accidental death can be reasoned as a cause of death and sometimes natural diseases can masquerade as trauma.[14] As for sexual assaults, most rapes and sexual assaults are never reported.[15] Thirty-six percent of rapes, 34 percent of attempted rapes, and 26 percent of sexual assaults were not reported to police in 1992–2000 according to one estimate, though other researchers would put the number much higher—for example, 1 crime reported for every 100 crimes actually committed.[16]

Myths about Typical Crime Scenes

There are few, if any, typical crime scenes; there are few, if any, typical bodies of evidence (*corpus delicti*); and there are few, if any, typical investigative approaches, according to a member of the Colorado Bureau of Investigation.[17]

Myths about Crime Outcomes

Television dramas and films might suggest that all crimes inevitably end in an arrest and a conviction, but actual **clearance rates** paint a different picture (see Chapter 1 for clearance rate details). One way to interpret these clearance rates is to say that 6 out of 10 "known" murders end in an arrest, and 5 out of every 10 "known" forcible rapes end in an arrest. The non-conviction rate of those arrested and prosecuted for murder stands at around 30 percent; in cases where sexual assault violators are identified, changed, and prosecuted, only an estimated 5 percent are convicted (4 out of every 5 cases are thrown out for lack of evidence).[18] Thus many crimes go undetected, and those that are reported might not end in arrest. In fact, many crimes are never solved. The Bureau of Justice Statistics also reports that clearance rates are on the decline—that is, police are solving fewer violent crimes.

Myths about Prosecution and Solved Crimes

One way to look at the crime-solving skills of investigators is by reviewing clearance rates. That said, keep in mind that an arrest is not the same thing as a conviction. A comparison of annual arrest rates for crimes of violence (620,510 arrests were made in the United States in 2003)[19] with prosecution rates (less than half of those suspects arrested are prosecuted) reveals that investigators often face challenges beyond their control, although they tend to influence prosecutor discretion toward charging (or indicting) a suspect with a crime. What happens to a case after an investigator processes it depends on other justice practitioners.

Myths about Hate Crimes

Many people think that hate crimes are rare occurrences. In reality, hate crimes occur relatively frequently. As a result, many police departments have developed special investigative units to deal with those activities. The FBI reports that crimes of hatred and prejudice—"from lynchings to cross burnings to vandalism of synagogues—are a sad fact of American history."[20] Nevertheless, the term "hate crime" did not enter the FBI's vocabulary until the 1980s, when emerging hate groups such as skinheads launched a wave of bias-related crime. The FBI began investigating what we now call hate crimes as far back as the early 1920s, when the agency opened its first Ku Klux Kan case.[21]

The *Uniform Crime Report* (UCR) provides data about both single-bias and multiple-bias hate crimes. A single-bias incident is defined as an incident in which one or more offense types are motivated by the same bias. A multiple-bias incident is defined as an incident in which more than one offense type occurs and at least two offense types are motivated by different biases. Highlights available from the 2006 UCR include the following findings:[22]

- In total, 2,105 law enforcement agencies reported 7,722 hate crime incidents involving 9,080 offenses.
- The 7,720 single-bias incidents involved 9,076 offenses, 9,642 victims, and 7,324 offenders; only 2 multiple-bias incidents were reported.

An analysis of the 7,720 single-bias incidents reported in 2006 reveals the following:

- Fifty-two percent were racially motivated.
- Nineteen percent were motivated by religious bias.
- Sixteen percent resulted from sexual-orientation bias.
- Thirteen percent stemmed from ethnicity/national origin bias.
- One percent were prompted by disability bias.

In 2006, police agencies reported that 4,737 single-bias hate crime offenses were racially motivated. Of these offenses:

- Sixty-six percent were motivated by anti-black bias.
- Twenty-one percent were motivated by anti-white bias.
- Six percent were driven by bias against groups of individuals consisting of more than one race (anti-multiple races, group).

- Six percent resulted from anti-Asian/Pacific Islander bias.
- Two percent were motivated by anti-American Indian/Alaskan Native bias.

Typology of Hate Crime Offenders

One way to analyze criminal behavior is to construct an offender typology (categories). A widely adapted typology of hate crime offenders is based on an evidence-based study of 169 cases in Boston.[23] Researchers identified four major categories of hate-crime motivation:

- *Thrill-seeking:* offenders who are motivated by a desire for excitement (66 percent).
- *Defensive:* offenders who commit hate crime to protect their turf or resources in a situation that they consider threatening (25 percent).
- *Retaliatory:* offenders acting to avenge a perceived insult or assault (8 percent).
- *Mission:* offenders who are so strongly committed to bigotry that hate becomes their career (less than 1 percent).

Police Response to Hate Crimes

Many jurisdictions have established hate crime units in their police departments, and some regional task forces are dedicated to investigating hate crimes.[24] Some states have increased training on this topic and implemented prevention programs aimed at this type of crime. California and Massachusetts are notable for including these and other strategies in their efforts to combat these crimes. The Bureau of Justice Statistics reports that most hate crimes described by victims (84 percent) accompanied violent crimes such as a rape or other sexual assault, robbery, or assault.[25]

■ Cold Cases versus Hot Cases

Ongoing investigations may be classified by their temperature—that is, as "hot" or "cold."

Hot versus Cold Homicides

In homicide cases, the use of the **cold case** descriptor is controversial among homicide investigators because it can suggest that some cases are unworkable, are impossible to explain, and offer little hope for a solution. Think of the false impression sent to the family and friends of unsolved murder victims by this label, which suggests that the victim in a cold case is insignificant. "Unsolved crime" might be a more appropriate term rather than "cold case" in such circumstances.

By contrast, "hot" homicides are those murders that have been reported to the police and might follow this scenario:

- A murder or finding of a body is reported.
- Police arrive at the scene.
- Witnesses are located and interviewed.
- Forensic services are enlisted.
- The victim is identified.
- The cause and manner of death are established.
- Investigators seek to identify what happened and who did it.

- Investigators write reports and document their actions, "building the case."
- Investigators' actions result in the identification and arrest of a suspect.

If this series of steps does not ensue, the case may become an unsolved homicide.

In both hot and cold homicide scenarios, investigators seek to answer the following questions:

- What happened?
- How did it happen? (means)
- Who had the opportunity to commit the crime? (opportunity)
- Why was the victim killed? (intention/motive)
- Where did it happen?
- When did it happen?

In summary, investigators seek to learn why the victim was murdered and who had the *means*, *opportunity*, and *motive* (MOM) to do so. Since humans have been on this planet, motives have not changed. Individuals commit murder for love, for money, or for power, and in anger, in rage, or for retribution of a real or imagined wrong. Some persons also commit murder because of their cultural perspectives. Homicide investigations might also consider the following motives:

- Murder–suicide
- Thrill killing
- Self-defense
- Random killing
- Sex and sadism
- Love triangles
- Jealousy
- Drug connection
- Gang involvement
- During catastrophes

Evolution of Hot Cases into Cold Cases

The cases that most commonly turn into cold ones involve the following types of crimes:

- Missing persons
- Gang- or drug-related killings
- Crimes involving immigrants, transients, and homeless victims
- Crimes with unidentified victims
- Suicides
- Accidental deaths
- Murder of police officers

For investigators, the transition from a hot case to a cold one may be related to many other factors, including issues associated with control of the CS. Investigators may be able to anticipate and mitigate some of these factors, as discussed next.

Factors Influencing the Investigation Process

The following factors all have the potential to influence the course of an investigation:

- Time when the crime occurs (e.g., before lunch, after midnight, Christmas, on your birthday)
- Crime scene location (private versus public)
- Lack of physical evidence
- Lack of witnesses
- Skill of witnesses and victims in identifying and testifying against a defendant
- Inability to identify victim
- Weapons used
- Gang- or drug-related scenario
- Inability to narrow investigation to a few solid persons of interest
- Inability to identify a person of interest (a suspect)
- Inadequacy of technology to fully analyze the evidence
- Television or newspaper coverage of a crime
- Public perception of criminal act

In terms of the last factor, consider that in the Washington, DC, sniper case, authorities received 100,000 calls, and more than 500 investigators pursued 16,000 leads.[26] Clearly, the publicity accorded to this series of crimes encouraged greater public help in solving it.

Other factors associated with the police organization may become either obstacles or catalysts in solving cases. For example, the number of detectives originally assigned to the case (too few, too many, or from too many jurisdictions, of which some do not communicate with other agencies), the detectives' caseloads, and budgetary and financial concerns can all influence the outcome of the investigation.

Sometimes police departments do not have adequate resources to aid detectives in investigating cases, such as equipment, staff, and forensic analysis. For instance, Detective Lee Bollinger of the Dallas Police Department often spends as much as 30 minutes of his day looking for a police car.[27] That's time he knows he should be using to investigate the robbery cases assigned to him. At Dallas's Northeast patrol station, where Bollinger is assigned, there are not enough cars to go around for his unit's 18 detectives. "I have to beg, borrow, and steal," Bollinger says. "I've had to postpone interviews and showing line-ups because I didn't have available transportation."

Another reason a case may go unsolved is because of poor interview and interrogation strategies provided by detectives.[28] The next section focuses on these crucial skills.

Interviews and Interrogations

An **interview** is the "process of obtaining information from individuals who possess knowledge about a particular offense."[29] Interviewing can be a key part of the investigation process; investigators may conduct interviews with witnesses and even first responders to a crime scene who might have information about a crime or a suspect. Interviews answer the major investigative questions: who, what, where, when, how, and why.[30] An interrogation is conducted when an individual is identified as a suspect through the help of an interview process.

An **interrogation** is associated with someone whom the investigator thinks is "good for the crime" (a person of interest becomes a suspect) based on probable cause evidence. It is designed to match acquired information to a particular suspect in an attempt to secure a confession. That is, whereas an interview is designed to gather information, an interrogation is the process of testing that information and its application to a particular suspect for the express purpose of obtaining a confession (Table 11–1).

As Table 11–1 shows, interviews and interrogations differ in several ways, including their purposes and goals. The purpose of an interview is to gain information, whereas the purpose of an interrogation is to test that information and obtain a confession. The due process rights of an individual do not necessarily come into play when someone is being interviewed. By contrast, someone who is being interrogated must be made aware of his or her rights—in particular, the Miranda rights. If a confession is eventually obtained but the suspect's rights were violated during the process, that confession would not necessarily be admissible in court (see the discussion of the exclusionary rule in Chapter 10). Guarding of the suspect's rights and planning of the session are critical to the outcome of an interrogation. Interviews, however, often take place in much less formal settings.

It is unlikely that an investigator will be able to control the surrounding or environment where an interview is conducted because it is usually conducted spontaneously—that is, whenever an individual is available or discovered. An interrogation's environment or surrounding must be controlled because of the suspect's constitutional rights and the objectives of the interrogation, which should include absolute privacy. Often investigators seek to establish a rapport with individuals during interviews, but rapport is necessary in interrogations, too.

In both interviewing and interrogation, investigators should ask good and relevant questions about the case. Careful listening is helpful during the interview, but vital during the interrogation process. Listening includes observation of both verbal and

TABLE 11–1 Similarities between Interviews and Interrogrations

Interview	Interrogation
Purpose: gather information	Purpose: test information and obtain a confession
Due process rights may not apply	Due process rights must be documented
Planning is helpful	Planning is critical
Controlling surroundings is important	Controlling surroundings is a must
Privacy or semiprivate surroundings are desirable	Absolute privacy is essential
Establishing rapport aids the process	Establishing rapport aids the process
Asking good questions is important	Asking good questions is important
Careful listening	Careful listening and observation of nonverbal clues
Appropriate documentation (notes)	Appropriate documentation (notes or video)
Goal: to gather information	Goal: to obtain a confession

nonverbal communication. Research shows that as much as 70 percent of most communication is nonverbal in nature.[31]

The interrogation process typically has five objectives:

- Obtain legally valuable facts about the case, the suspects, the witnesses, and the victims
- Eliminate the innocent
- Identify the guilty
- Obtain a (legal) confession
- Reduce a suspect's oral admission to a permanent form, either written or video-taped

Most often, obtaining and documenting an admission of guilt requires that the first three of these common objectives be met. One way of accomplishing those objectives in the interrogation process is by establishing a rapport with the suspect.

Interrogation Rapport

Rapport allows for the use of approaches that are less obvious and that will minimize conflict between the suspect and the interrogator. Rapport can be characterized as building trust between an individual and the investigator. Don Rabon and Tanya Chapman of the North Carolina Justice Academy explain that rapport provides the opportunity to establish commonality, which can provide a means by which a feeling of trust is developed in the suspect's mind.[32]

"The more an individual comes to believe we understand him, the more he tends to trust us."[33] Rapport can be defined as being in "sync" with the individual with whom you are communicating.

A number of techniques are beneficial in building rapport:

- Matching your body language (e.g., posture, gestures)
- Maintaining eye contact; matching breathing rhythm
- Avoiding judgment about any of the suspect's thoughts or gestures

Rapport intervention techniques could allow an interrogator to enter the worldview of the suspect. From that vantage point, the suspect may be able to build a pathway toward an admission of the crime, which is the goal of the interrogation.[34]

Despite the best efforts of an interrogator, suspects are usually unwilling to admit their crimes.[35]

Nevertheless, interrogators should try to build rapport with the suspect for the purpose of changing a suspect from an unwilling suspect into a willing participant. In doing so, the investigator should consider the following characteristics of the suspect:

- Response (categories)
- Resistance (suspect's fears)
- Denials (predisposition toward denial)
- Lies (and sensory channels)

Suspect Response

There are at least 12 ways to describe a suspect's response, which can be made verbally, with gestures, or through silence:[36]

- Suspect resistance (described below)
- Confession of crime (full, partial, or false—some suspects will admit to anything regardless of whether they committed the act or not)
- An alibi (supported or not supported by witnesses)
- No comments (silence)
- Results of polygraph testing
- Self-written statements
- Self-drawn sketches
- Showing and talking about items at the scene of crime
- Reconstruction of the crime event (before, during, and after the incident)
- Denial of crime (active, passive, real, or false)
- Rude jests, comments, and body functions
- Threats to officers and their families and friends

Suspect Resistance

Confronting a suspect is a complicated process that can be described in terms of seven fears:[37]

- Fear of loss or termination of a job. This category includes fear of financial repercussions. A suspect may be reluctant to make an admission because of a fear of present and future employment opportunities and an inability to pay bills and other financial obligations.
- Fear of arrest or prosecution. This concern is usually greatest among suspects who have had little previous contact with police.
- Fear of embarrassment. This hurdle to a confession relates to a suspect's self-image. Suspects fear that they will shock family, friends, or coworkers and lose their respect.
- Fear of losing self-image. Some suspects are unable to face what they have done without destroying their own self-image. Being interrogated for sexual assault of a child, for instance, can easily complicate a suspect's responses.
- Fear of restitution. Some suspects may be resistant to confession because they could not compensate the victim for the damage or loss related to their actions.
- Fear of retaliation. A suspect's fear of his or her safety or that of family members can represent a huge hurdle for an interrogator. For example, imagine the thoughts of a suspect fearful of gang retaliation for a crime.
- Fear of admitting to the crime. A suspect may confess to a lesser crime than the one actually committed. By admitting to small crime, the more serious crime might come out.

Some suspects admit to a crime to relieve feelings of guilt; others admit to crimes because of investigator's skill. Most often, however, guilty suspects go through a progression of negative responses: anger, then depression, followed by denial, bargaining, and finally acceptance.[38] Many incarcerated felons say that their resistance toward a confession would have increased had their interrogator treated them disrespectfully, contrary to the perspective presented by popular television police dramas.[39]

Suspect Denials

Suspects can posses a predisposition to deny a crime, and the fears perceived by the suspect (as discussed earlier) may play a role in the decision not to confess.[40] Denial is a survival technique learned as a child to avoid punishment. This perspective suggests that suspects can be clever and manipulative. Arguing against this notion, noted criminologists point out that the "burglar typically walks to the scene of the crime; the robber victimizes available targets on the street; the embezzler steals from his own cash register; and the car thief drives away with keys left in the ignition. What planning does take place in burglary seems designed to minimize the momentary probability of detection and to minimize the effort required to complete the crime."[41]

The implication of this view is that most suspects are not overachievers. Few things are easier than telling a lie, and few things are harder than spotting a liar. Nonetheless, the skill of an interrogator, which includes conducting an interrogation in a manner that meets legal and ethical standards, plays a major role in moving the interrogation toward the ultimate goal (getting a confession). Because the suspect will often deny the act regardless of an interrogator's ability, the role of the interrogator should be to avoid "forcing the suspect into a position where he or she must deny" an act.[42] The decision to deny a crime is linked to the choices an interrogator offers and the strategies employed to engage the suspect.

Environment, suspect, and interrogator all influence the outcomes of the interrogation:

- Environment. The environment of the interrogation includes the timing, location, room setting, position of suspect versus interrogator, windows, and doors.
- Suspect. During the interrogation, the suspect is making decisions about what the interrogator knows about his or her involvement in the crime. The suspect reacts to the perceptions and strategies of the interrogator by either reducing or increasing resistance. Some suspects buy time through denial to assess the state of investigation and the skills of the investigator.
- Interrogator. An interrogator's confidence in the suspect's guilt, word usage, questions, statements, and plan of attack determine the probable response of the suspect

Suspect Lies

Most suspects lie[43]—and most chronic offenders lie about everything, all the time.[44] Some experts suggest that lie-detection technology can aid interrogators, but its use is often prohibited by its cost, accuracy concerns, and constitutional issues.[45] Polygraph advocates state the technology is at least 85 percent accurate in criminal investigations, but the National Research Council of the National Academy of Sciences recently dismissed the machines as useless. That perception might be inconsistent with the actual application of the polygraph, which is not necessarily utilized for accuracy. Instead, its results may be used to help build a case and can provide a focus for investigators during the integration process.

The best liars tend to be the least troubled by their dissembling and produce the fewest outward clues. However, most of us lie easily, but we don't lie well, especially when the truth could bring about criminal penalties. Fibbing can cause the heart to pound, breathing to accelerate, and sweating to increase, and a polygraph measures all those things. Sometimes the machines work well, but often the experience of being wired up to a piece of gadgetry and asked questions by a stranger can produce symptoms similar to that seen with a lie.

Observation of Suspects During Interrogation

So how can interrogators tell whether suspects are telling the truth? A number of strategies have been explored, including technology for analyzing facial expressions. Quicker and less expense methods that can be pressed into service by police investigators and federal agents include changing sensory channels during the interrogation process.

Poker Face (Micro-expressions)

Observation of micro-expressions is a strategy based on the assumption that the best poker players see tics and flutters in an opponent's face—the so-called poker tells—that can telegraph when a player is bluffing. Scientists agree that the face tells tales, even though some of us wish it didn't.

San Francisco psychologist Paul Ekman has codified 46 facial movements into more than 10,000 micro-expressions in what he calls the Facial Action Coding System (FACS). According to Ekman, this technique can detect deception with 76 percent accuracy. While similar behavioral screening has been used in British airports for several years, FACS is only now being rolled out as a terrorist-screening tool in a dozen U.S. airports.

Sensory Channels

Investigators acknowledge that if they remain vigilant, they can gain insight into the dominant sensory mode of a suspect through his or her vocabulary, articulation, and use of slang. Don Rabon argues that some suspects attempt to control their communication styles to prevent interrogators from gaining this knowledge.

One way to determine truth focuses on the "windows to the soul"—that is, a suspect's eyes.[46] There are corresponding eye movements and positions for each of the senses of sight, hearing, and touch (the five senses also include smell and feel). All the data and information we possess came into our brains through the senses, so, in theory, a suspect can be pursued through the senses, too. To do so, however, an interrogator has to be on the same "channel" as the suspect. That is, to develop rapport, if the suspect is more attuned to a visual, or hearing, or touch (feelings) orientation, the interrogator should change to the channel of the suspect.

For example, an interrogator might tell the seeing-sense suspect, "I see what you are showing me, and I want to see how this looks to you." To the hearing-sense suspect, the interrogator might say, "I hear what you're saying, and I want to determine how this sounds to you." To the suspect whose sense is tied more to touch or feelings, the interrogator might respond, "I understand your feelings, and I would like to know how this strikes you?"

Visual Eye Movements When a suspect recovers data stored in his or her visual memory, there are typically used vocabulary usage and eye movement patterns (**visual eye movements**) that correspond to visual processing. Consider these examples:

- Suppose the interrogator says, "Mary, let's take a look at last Saturday night. Can you show me what happened from the time you left Bill's house to the time you got home?" This type of an answer should produce (1) eyes moving upward and to the subject's left and (2) eyes looking straight ahead. When the suspect begins to speak, she looks straight ahead and then her eyes move upward and to her left. "Well, let's see," she says, "I got there . . ."
- When a suspect is constructing a conversation that could include lies, the eyes look up and to the right when the words, and sometimes false sounds (um, uh, mm, aaaa) roll off the lips. Sometimes a suspect provides a hedge or qualifier words such as "That's about it" or "Mostly we rode around." In these cases, the suspect looks at the narrative he or she wishes to share, and the eyes move from upward to the right and then straight ahead.

Auditory Eye Movements Certain eye movements reveal that a suspect is trying to remember. Eye movements for auditory memory (**auditory eye movements**) are seen in two separate positions:

- Eyes down and to the left: The suspect will refer to what he or she has heard (external sounds). For example, "Jenn said she would be at my apartment on Saturday night."
- Eyes to the left: The suspect will refer to something he or she thought or said.

Sensation Eye Movements The feeling or sensation channel is the most functional for including behavior. For instance, someone who has lost a loved one tends to look deep into the ground. Similarly, a suspect in an interrogation might continue to hang his head down and stare at the ground, but eventually might take a deep breath, hold it for a moment, and slowly exhale, while his head comes up. He might then agree with the interrogator: "You were there when the girl was killed, right?" The sensation or feeling channel is the decision-making channel, and an interrogator has to move a suspect to this channel before getting an admission of guilt.

The following eye patterns are associated with emotion or sensation:

- Eyes looking downward
- Eyes looking downward and to the right
- Eyes closing
- Eyes fluttering or blinking rapidly

Interrogators can read these **sensation eye movements** and move the suspect to the feeling channel to get a confession.

Interrogator's Task and Deception

Interrogators understand three things, says Don Rabon: "Deception is deception is deception."[47] Interrogators acknowledge that most suspects are deceptive and, therefore, one of their tasks is to identify the "nature of that deception, and of changing falsehoods, into truth."[48] An interrogator can say that he or she has information that he does not, in fact, have. The interrogator might say, "Why would someone contact me? Someone who has nothing to do with this situation, but would go to the trouble to call me and place you there that night? Why would someone do that?" Of course, in reality no one may have called.

> The primary task of an interrogator is to motivate a suspect to change from an unwilling participant to a willing participant and provide an admission to the crime.

Detectives are aware of "optimal moment in time" when a suspect is most susceptible to making a confession. A case can change for many reasons: Witnesses may relocate, evidence may be misplaced, and victims may change their feelings about prosecution. Interrogators also recognize the elements of the crime or violation, the varieties of enticement questions or bluffs that could legitimately be used, and the times when they should use special personnel or legal requirements to close the investigation.

Many factors must be considered in the preparation phase, including the strengths and weaknesses of using substantial evidence early or late in the interrogation and the decision to reveal evidence, if it is used at all. Revealing strong evidence too early in an investigation might promote several problems:

- The presentation of evidence may show the weaknesses of the case.
- Presenting evidence while a suspect is emotionally or physically stable can aid in his or her contradicting or destroying the significance of the evidence.
- In a small number of cases, an individual confesses to a crime that he or she has not committed. Once the evidence is revealed, the person could repeat what was said about it later in the proceedings, thereby leading to a conviction in error.

Techniques of Interrogation

One well-established formal process for interrogation is the **Reid technique**. Its core values are outlined in the *Essentials of the Reid Technique: Criminal Interrogation and Confessions* manual.[49]

More generally, during an initial conversation between detective and suspect, the detective observes the reactions of the suspect, making mental notes about both verbal and nonverbal actions.[50] One key factor in determining the success of an interrogation is the ability to establish a **baseline**—that is, a study of the suspect's normal responses so that deviations from them can be identified. This perspective is one of the most important parts of sound interrogation (or interviewing, for that matter).

> Establishing a **baseline** is the key to a successful interrogation, and entails establishing an understanding or a "reading" of the suspect.[51]

Even patrol officers tend to question individuals to get a "reading" on their responses before asking about the crime. An investigator will use the baseline understandings later as a comparison point.

One method of creating a baseline involves asking questions that cause the suspect to access different parts of the brain. To do so, the detective asks the following types of questions:[52]

- Non-threatening questions that require memory. Such a simple recall question might be, "Did you see a lot of traffic at the interaction before making your turn?" When the suspect is remembering something, his or her eyes will often move to the right.
- Questions that require thinking. Such a creativity-requiring question might be, "Did you have a chance to notice the dog in the car you passed on the highway?" When thinking about the answer, the eyes of the suspect might move upward or to the left, reflecting activation of the cognitive center.

The detective then asks about the crime and compares the suspect's reactions to the baseline. For example, if a suspect is asked where he was the night of the crime and he answers truthfully, he'll be remembering, so his eyes may move to the right; if he's making up an alibi, he's thinking, so his eyes will move to the left. If the interrogator determines that the suspect's reactions indicate deception, and all other evidence points to guilt, the interrogation of a guilty suspect begins.[53]

Once a baseline is established, the interrogator may pursue the following tactics in moving the interrogation toward the objectives of getting a confession from a suspect.[54]

Direct, Positive Confrontation

The detective presents the facts of the case and informs the suspect of the evidence against him or her. This evidence might be real, or it might be made up. The detective typically states in a confident manner that the suspect is involved in the crime. The interrogator needs to reduce anticipated resistance through rapport by conveying a strong belief that the suspect's guilt is known and that many people have faced similar circumstances and outcomes like those encountered by the suspect. To do so, the interrogator begins with a firm statement that the suspect is guilty. The suspect's stress level starts increasing, and the interrogator may move around the room and invade the suspect's personal space to increase the discomfort. If the suspect starts fidgeting, licking the lips, or grooming himself or herself (running a hand through the hair, for instance), the detective takes these as indicators of deception and knows he or she is on the right track.

Theme Development

The interrogator creates a story about why the suspect committed the crime. Theme development is about looking through the eyes of the suspect to figure out why he or she did it, why the suspect would like to think he or did it, and what type of excuse might make the suspect admit he or she did it. Does the suspect use any particular mode of reasoning more often than others? For example, does he or she seem willing to blame the victim? The detective lays out a theme, a story, that the suspect can latch on to either to excuse or justify his or her part in the crime, and the detective then observes the suspect to see if he or she likes the theme. Is the suspect paying closer attention than before? Nodding the head? If so, the detective continues to develop that theme; if not, the detective picks a new theme and starts over. When developing themes, the interrogator speaks in a soft, soothing voice to appear nonthreatening and to lull the suspect into a false sense of security.

Stopping Denials

Letting the suspect deny his or her guilt will increase the suspect's confidence, so the detective interrupts all denials, sometimes telling the suspect it will be his or her turn to talk in a moment, but right now, the suspect needs to listen. In addition to keeping the offender's confidence low, stopping denials also helps quiet the suspect so he or she doesn't have a chance to ask for a lawyer. If no denials are forthcoming during theme development, the detective takes this as a positive indicator of guilt. If initial attempts at denial slow down or stop during theme development, the interrogator knows that he or she has found a good theme and that the suspect is getting closer to confessing.

Overcoming Objections

Once the interrogator has fully developed a theme the suspect can relate to, the suspect may offer logic-based objections as opposed to simple denials: "I could never rape somebody—my sister was raped and I saw how much pain it caused. I would never do that to someone." The detective handles these objections differently than denials, because objections can give the detective information to turn around and use against the suspect. The interrogator might say something like, "See, that's good—you're telling me you would never plan this, that it was out of your control. You care about women like your sister—it was just a one-time mistake, not a recurring thing." If the detective does the job right, an objection ends up looking more like an admission of guilt. A sympathetic series of possibilities of how and why the crime took place keeps the suspect on the defensive.

Getting the Suspect's Attention

At this point, the suspect should be frustrated and unsure of himself or herself. He or she may be looking for someone to help the offender escape the situation. The interrogator tries to capitalize on that insecurity by pretending to be the suspect's ally and attempting to being more sincere in the theme development. At this point, the interrogator should get physically closer to the suspect to make it harder for the suspect to detach from the situation. The interrogator may offer physical gestures of camaraderie and concern, such as touching the suspect's shoulder or patting his or her back.

Suspect Loses Resolve: Passive Mood

If the suspect's body language indicates surrender—head in hands, elbows on knees, shoulders hunched—the interrogator seizes the opportunity to start leading the suspect into confession. At this point, the interrogation makes a transition from theme development to motive alternatives that force the suspect to choose a reason why he or she committed the crime. The interrogator makes every effort to establish eye contact with the suspect to increase the suspect's stress level and desire to escape. If, at this point, the suspect cries, the detective takes this as a positive indicator of guilt. Look for the "buy signs" that show a suspect is ready to confess.

Alternatives

The interrogator might offer two contrasting motives for some aspect of the crime, sometimes beginning with a minor aspect so it's less threatening. One alternative is a socially acceptable explanation ("You needed to do what any real man needs to do, I know; it was a crime of passion.") or a morally repugnant explanation ("You killed her for the bucks, is that right?"). The detective builds the contrast between the two alternatives until the suspect gives an indicator of choosing one, like a nod of the head or increased signs of surrender. Detectives should offer some persuasive arguments for telling the truth.

Bringing the Suspect into Conversation

When the suspect chooses an alternative, the confession has begun. The interrogator encourages the suspect to talk about the crime and arranges for at least two people to witness the confession. One may be the second detective in room, and another may be brought in for the purpose of forcing the suspect to confess to a new detective—having to confess to a new person increases the suspect's stress level and his or her desire to just sign a statement and get out of there. Bringing a new person into the room also forces the suspect to reassert his or her socially acceptable reason for the crime, reinforcing the idea that the confession is a done deal. Obtaining the admission by revising the roles of interrogator and suspect and explaining the benefits of clearing the suspect's mind of the crime is a good option, too. The noting of doing the right thing and emphasizing the benefits it would generate to victims, witnesses, and the suspect's family, friends, and co-workers may be rationalizations that allow the suspect to save face while building a persuasive argument in favor of admission.

The Confession

The final stage of an interrogation is getting a legally acceptable confession and its documentation. Moving a suspect toward admitting his or her guilt is the primary goal of the interrogation, but that admission is worthless unless it is appropriately documented (in writing or recorded) and can be used in court to charge the suspect (prosecutor's office). The suspect is usually willing to do anything at this point to escape the interrogation. The suspect confirms that his or her confession is voluntary, not coerced, and signs the statement in front of witnesses.

Criticism of Interrogation

An alleged flaw of the previously outlined interrogation process is the expectation that every criminal is motivated by guilt and that he or she lives in constant fear of discovery.[55] An interrogator helps suspects conclude that a confession will let them move past the act and get on with their lives. That might ring true for first-timers, but many practitioners in policing and corrections would disagree with this basic tenet underlying the interrogation process.[56] A police practitioner might protest, "Chronic offenders don't feel shame or guilt for their heinous acts. If they do, it's all an act."[57] Experienced interrogators realize the importance of this perspective and adjust their interrogation style to fit the experiences of the suspect.

Also, the police confessional room is a space where truth is produced by the interrogator's strategic use of narratives about guilt and innocence.[58] Sharply articulated confessions do not spring full-blown from the minds and mouths of criminal suspects. Indeed, many of them, as prison population studies report, are not that highly trained, especially in their ability to write a six-page confession. Rather, those confessions can emerge from a collaboration between investigator and criminal, in which the officer usually plays the role of lead author, who spins specific narrative plot lines in an effort to get the suspect talking and keep him or her talking until the offender talks himself or herself right into a prison cell.[59]

There is always the possibility of error on the part of investigators, and police organizational traps may sometimes lead to failure in criminal investigations owing to misconceptions and lack of knowledge about these probabilities.[60] The police organization may contribute to the failure of an investigation through its inability to change and adapt to match the ongoing investigation. One expert uses a gambling-related analogy: Anyone who has spent a few hours watching gamblers eventually realizes that probability is a difficult concept to understand. It is not uncommon for gamblers (and investigators) to use heuristics (problem-solving methods that incorporate trial and error). Because of investigators' (or patrol officers') experiences, definitive answers (final answers that will not be challenged; black and white) when solving crimes are preferred, rather than shades of gray. Probability errors are common among many types of officers, including investigators and forensic analysts, a former detective writes.[61]

Many detectives look for patterns in, or draw inferences from, a small number of incidents. For example, an analyst might examine the dates for a series of 15 street robberies and observes that none of the crimes occurred on a Monday. Is this pattern meaningful? Probably not. With only 15 crimes, chances are at least one day of the week will be free of robberies. Yet some might argue that the intentions of the investigator are as relevant to a successful investigation as is his or her training.

■ Strategies to Help Avoid Investigative Failure

One way to avoid investigative failure is to employ full-time professional investigators. These investigators should undergo appropriate pre-service and in-service training that focuses on the following issues:

- Rossmo's strategies[62]
- Enhancing the suspect's self-respect
- Changing the detective's mindset
- Professional training

Rossmo's Strategies

D. Kim Rossmo suggests the following idealized options to expand the investigator's arsenal of tools:[63]

- Encourage investigators to express alternative, even unpopular, points of view.
- Assign the role of devil's advocate to a strong team member.
- Employ subgroups for different tasks.
- Facilitate parallel but independent decision making.
- Recognize and delineate assumptions, inference chains, and points of uncertainty; always ask, "How do we know what we think we know?"

- Obtain expert opinions and external reviews at appropriate points in the investigation.
- Conduct routine systematic debriefings after major crime investigations and organize a full-scale "autopsy" after an investigative failure.
- Encourage and facilitate research into criminal investigative failures and how they might be prevented.

Enhancing the Suspect's Self-Respect

One technique that has proved effective in interrogation is enhancing the self-respect of the suspect by not forcing him or her to lie in the early part of the interrogation. One way to set up a liar is to ask a question early in the process that has a socially approved answer. For instance, an interrogator might ask a white suspect who attacked and killed a Mexican, "Do you hate Mexicans?" The suspect knows that it is best to say that he does not dislike Mexicans and, when asked another question later about Mexicans, will conform to his earlier lie. Some refer to this outcome as "accounts theory" (talk involves the giving and receiving of "accounts"—that is, statements made by a social actor to explain unanticipated or untoward behavior).[64] "Excuses (or lies) are socially or culturally approved vocabularies when suspect is questioned."[65]

Changing the Detective's Mindset

The mindset of the investigator plays a key role in determining the success of an interrogation, a relationship that can be better understood by examining an evidence-based scenario.[66] This study compared two groups: criminal investigators and undergraduate students. Individuals in both groups read a set of facts from the preliminary investigation of a homicide case. Participants' initial hypothesis regarding the crime was manipulated by providing background information that implied either (1) the prime suspect had a jealousy motive or (2) there might be an alternative perpetrator for the homicide. Students displayed a framing effect, such that guilt was ascribed to the prime suspect only when a potential motive was presented, whereas investigators did so regardless of the hypothesis presented.

In this study, investigators were less sensitive to alternative interpretations. This finding suggests that if detectives hold a mindset that a suspect is guilty, they are more likely to pursue that suspect as opposed to other potential leads. Investigators are less likely to acknowledge inconsistencies in the material when presented with a potential motive, but are more likely to do so when made aware of the possibility of an alternative perpetrator.[67] Based on this study, it appears that both mindset and flexibility shape the outcome of an investigation.

Professional Training

Training is an important issue in determining investigator success or failure. In one study consisting of 319 Boston police officers engaged in first response and sexual assault investigations, researchers found that a lack of professional training among those officers shaped sexual assault conviction rates.[68] When sexual assault training was available, it consisted of "war stories"; although the officers learned much from some of the sessions, much of what they learned had little to do with processing a crime scene, interviewing, and interrogation. The officers said that while they were grateful for the specialized training conducted by the state police and FBI, it supplied little new knowledge that might affect sexual assault convictions.

Professional training is available through a number of private and public sources. For example, the North Carolina Justice Academy (NCJA) offers professional training to investigators employed by both local and state police agencies. Any sworn North Carolina officer, regardless of his or her employer, who is tasked with investigative duties can participate. Usually fees are paid by the employer. The NCJA program includes the following schedule of training:

- Crime scene investigation: 80 hours
- Death investigation: 40 hours
- Legal issues and court room testimony: 16 hours
- Chemical processing and advanced photography: 40 hours

Its advanced crime scene investigation training consists of these courses:

- Blood Stain Pattern Analysis Investigation: 80 hours
- Shooting Reconstruction: 40 hours
- Buried and Scattered Remains Recovery: 24 hours
- Entomology Recovery: 16 hours
- Crime Scene Reconstruction: 40 hours

Many jurisdictions enroll their investigators to a state facility such as the NCJA, but often they also enroll investigators in courses offered at the Federal Law Enforcement Training Center (FLETC). As discussed in Chapter 8, FLETC serves as an interagency law enforcement training organization for more than 80 federal agencies. Investigative courses are offered by FLETC through the Forensics and Investigative Technologies Division (FIT). This division provides training to the agents and officers in the day-to-day, hands-on, technical aspects of law enforcement by providing them with contemporary, scientific, and technological skills necessary to perform their respective law enforcement functions. Courses instructed by FIT include Crime Scene Investigations, Crime Scene Preservation, Identity Documents, Crimes Against Persons, Fingerprints, Drugs of Abuse, and Clandestine Labs.

Qualifications for Detectives

All investigators are not created equal: They differ in terms of their expertise, caseloads, funding, support, and jurisdictions. Nevertheless, the best professional investigators share certain characteristics:[69]

- A lot of self-discipline, patience, and internalized control
- Knowledge of a systematic method of inquiry that is more science than art
- Excellent interview skills
- Initiative and resourcefulness (because investigations cannot be performed by rote application)
- Oral and written communication skills (witness, victims, and suspects should be addressed differently)
- Knowledge and experience in use of informants

- Strong deductive reasoning skills (begins with the formulation of an explanation of the crime, which is then tested against available information)
- Strong inductive reasoning skills (examination of the evidence and particulars of a case, and subsequent use of this information as a basis for formulating a unifying and internally consistent explanation of the crime)
- Good listening skills (when you're talking, it is hard to learn)
- Knowledge of crime scene reconstruction
- Experience in use of computers and report writing
- Knowledge of courtroom procedures and testimony
- Knowledge of department resources, procedures, and protocols
- Good health and stamina
- An analytical approach
- Motivation and enthusiasm

Even with the increasingly more important role played by evidence processing and forensic laboratory expertise in contemporary crime scene investigations, the information that suspects provide to police professional investigators remains the greatest driving force in obtaining an admission of guilt and ultimately prosecution and conviction of the suspect. But why all the fuss about investigators and their skills? Among other things, there are two central answers to this question: the task of evaluating a case and the burden of proof.

Evaluating the Case

An investigator creates a hypothesis around the physical evidence of a crime and creates a visual record that will allow a crime analyst, police investigator, and prosecutor to easily recreate an accurate view of the crime.[70] An investigator determines risk factors. Despite probable cause, many investigators do not automatically perform an arrest. Some detectives believe that conducting an arrest too early during an investigation might set up road blocks toward getting a conviction. Of course, if an arrest is not quickly performed, a suspect might flee or alter significant evidence. Sometimes dangerous suspects may harm others if they are not taken into custody. Many investigators balance the merits of the case and potential factors related to the suspect prior to making an arrest. Also, when evaluating a case, investigators realize where the burden of proof lies.

Burden of Proof

All of the skills, training, resources, and discretion of investigators, even after a confession is obtained, must be devoted to the desired outcome—namely, obtaining a conviction. In court, the prosecution has the "responsibility of affirmatively proving the allegations on which it has based its accusation."[71] This responsibility, which is supported by the U.S. Constitution, places the **burden of proof** on the government or the prosecutor. The prosecutor's task is to prove beyond and to the exclusion of every reasonable doubt that the defendant is guilty of the crime with which he or she is charged. Innocence is presumed. There are exceptions, but what is relevant to your understanding of a conviction (which includes a guilty plea) will depend on the performance of police, and particularly the performance of those officers who investigated the case and provided legal evidence to the prosecutor, who then indicted the suspect.

■ Forensics

An investigation might be carried out by both field investigators who gather evidence and laboratory personnel who perform forensic assessments upon the evidence. Services provided by forensic personnel, including field investigators, vary for three reasons: (1) variations in local laws, (2) different capabilities and functions of the organization to which a laboratory is attached, and (3) budgetary and staffing limitations.[72]

Tasks of a Crime Scene Analyst

The field investigator gathers the evidence, protects the chain of custody, and transfers evidence to laboratory (lab) personnel, who then assess the evidence. The lab worker who assesses the evidence can be called a crime scene analyst. The assessment or evaluation in the laboratory is often called forensic science. Depending on the size of the department, field and lab personnel may be different people, although sometimes their tasks overlap.

Here is how one analyst describes her job:

> Depending on what scientific examinations are needed or requested, I may be involved in the actual "bench work" once the evidence is submitted to the laboratory. I have expertise in blood pattern identification (blood spatter), trajectory determination, serology (blood and body fluids), and photography. I also have knowledge in many other areas (firearms, fingerprints, questioned documents . . .) that may assist me at the scene. As a primary crime scene responder at the CBI, my role at the scene may involve one or more of my particular disciplines. While I would not do a functionality test on a firearm here at the laboratory, my role at the crime scene would be to collect the gun and understand its potential evidentiary significance.[73]

The chief duty of a crime scene analyst is to arrange and collate numerous individual events, details, and observations in an order that may become a part of a comprehensive picture associated with the crime, victims, and suspects.[74]

Types of Crime and Analysis Laboratories

The many types of crime and analysis lab units vary depending on their expertise, funding, and policy. Some crime labs are operated by local and state governments and the federal government; others are privately owned. Which lab is employed to assess evidence might depend on police policy, available funds, and type of evidence. Some of those units can be described as follows:[75]

- Physical science: application of the principles/techniques of chemistry, physics, and geology to the identification and comparison of crime scene evidence.
- Biology unit: biochemists who identify and perform DNA profiling
- Firearms unit: discharged bullets, cartridge casing, and ammunition of all types
- Document examination unit: handwriting and typewriting experts

- Photography unit: photographic lab, which also examines and records physical evidence
- Toxicology unit: body fluids and organs
- Latent fingerprint unit
- Polygraph unit: lie detector
- Voiceprint analysis unit: telephone and tape-recorded messages

Police Reluctance to Use Crime Labs

Before evidence can be assessed by lab personnel, police must inform them of the crime. Many police are reluctant to use crime labs, however, because officers face a series of dilemmas related to their inclusion in the investigatory process. In particular, the vast number of unsolved crimes and turnaround issues related to evidence seem to create a greater police reluctance to utilize forensic services.

First, the United States has a massive backlog of unsolved rapes and homicides.[76] One estimate is that local police agencies have not submitted 994,000 cases to laboratories for analysis in 2002 consisting of more than 221,000 violent crimes, including an estimated 52,000 homicides and 169,000 rape cases that remain and unsolved. The number of property crime cases with possible biological evidence that local law enforcement agencies have not submitted to a laboratory for analysis exceeds 264,000. The number of unanalyzed DNA cases reported by state and local crime laboratories is more than 57,000.[76]

Second, the slow turnaround time of the analysis is a problem for police, because time is not on their side when they are investigating criminal activities. Lab managers estimated that 930 additional labs were needed to achieve a 30 day turnaround.[77] Apprehending a perpetrator becomes less likely as more time passes after a crime has been committed.

In a Michigan study among police officers, researchers found that they were generally reluctant to utilize crime labs for the following reasons:[78]

- Fifty-one percent said DNA was not considered a tool for criminal investigations.
- Thirty-one percent said no suspect had been identified.
- Twenty-four percent of the police agencies had no funds for forensics.
- Ten percent of the suspects stopped were not charged.
- Nine percent said prosecution had not required testing.

Other conclusions seem to agree with these findings. For instance, it is widely accepted that actuarial risk instruments among sexual offenders outperform clinical assessments and judgments.[79] How about expert forensic behavioral scientists? A Texas study reviewed 155 Texas capital cases in which prosecutors used behavioral experts to predict a defendant's future dangerousness; in 95 percent of the cases, those experts were wrong, reports the Texas Defender Service.[80] Another study reveals that first responders to CS in Boston do not adequately safeguard evidence, inadequately secure the CS, and withhold their own theories of the crime because of their inadequate training and often inappropriate supervisor manipulation of police personnel.[81]

Four other elements—officers' attitudes, institutional support, personnel, and network—guide decisions about whether the police utilize crime labs.[82] Police decisions to process CS evidence in forensic labs are also directly related to budgetary issues. Clearly, there is a need to reconcile the consequences of resource allocation if forensic science at the local police level is ever to succeed.[83]

Whether officers and detectives utilize crime labs or not can be a matter of discretion for them and their commanders, unless mandated. The final decision is often based on these individuals' own initiatives and interpretations, especially if they see less value in forensic assessment than in their own theory of the crime.[84] Other concerns also have to be satisfactorily addressed:

- Differences of forensic science opinions can derail investigative interviews among child abuse victims.[85]
- Forensic pathologists may produce vague results when investigating a variety of violent crimes, such as child homicide.[86]
- Natural disease sometimes masquerade as trauma: the fatal event is usually not witnessed by anyone other than the accused; a problem may arise in distinguishing accidental injury from assault; multiple injuries pose challenges to analysis, including the timing of different injuries; ascertaining the relationship between old injuries and death can be problematic; and the relationship between shaken baby syndrome and subdural hemorrhage can cause blurring of the lines between these two types of events.[87]
- Suicides as identified through CS initiatives are sometimes subsequently revealed to be homicide cases, as was discovered in Arizona when more than 10,000 bodies were reviewed by the five boards of certified forensic pathologists after receiving a pre-cremation examination over a two-year period.[88]
- Private (individually owned and operated) labs that perform DNA assessments have been criticized for their failure to adhere to appropriate and adequate standards that can aid prosecutors in winning convictions.[89]
- When dealing with trace evidence after a fire, the analysis arising from that evidence depends more on the interpretation of the field personnel than on the crime lab assessment.[90]
- When forensic experts testify and present their forensic assessments, their scientific expertise may not be reflected in their trial presentation, and vice versa. One study suggests that criminal defendants do not always benefit from those presentations, either.[91] Ultimately, developing even high-quality assessments might have less to do with trial outcomes than expected, depending on the witnesses' capacity and the expertise of trial delivery.
- Early goals of the National Institute of Justice research program to develop an organized body of knowledge about police matters, which includes CSI, has yet to be fulfilled.[92]

Many of the above problems impact the legal process in obtaining a conviction.[93]

Different jurisdictions also have different laws and regulations governing the trial process that could easily either validate or cancel forensic findings. Prosecutors have their own challenges when jury members' beliefs are shaped by the media and do not jive with real forensic circumstances.

Criticism of Forensic Science

In some respects, the difference between televised mythical CSI performance and real CSI performance places an extraordinary hardship upon the justice system, almost as if it were a runaway train racing down a wintry slope. "[If] you don't accept their [crime lab] findings as reliable, regardless of the validity of those findings, that train will derail a case faster than greased lightning," says one prosecutor.[94] "To challenge crime lab findings, it's almost as if you were unpatriotic," she adds.

In fact, "Reality and fiction have begun to blur with crime magazine television shows such as *48 Hours Mystery*, *American Justice*, and even, on occasion, *Dateline NBC*. These programs portray actual cases, but only after extensively editing the content and incorporating narration for dramatic effect," advises the Honorable Donald E. Shelton in 2008 for the National Institute of Justice.[95]

When a case comes to court, "Jurors expect it to be a lot more interesting and a lot more dynamic," says Barbara LaWall, the county prosecutor in Tucson, Arizona. "It puzzles the heck out of them when it's not."[96]

The truth is that the many of the nation's crime labs are underfunded, undercertified, and under attack because they cannot meet the level of public expectations or match up to the myths believed by the public. These concerns can also increase crime lab backlogs and lead to wrongful convictions.

Backlog of Crime Labs

Crime scene technicians who collect evidence are sometimes confronted with a reality worse than expected when they turn over their evidence to a forensic lab. The Bureau of Justice Statistics reports that the 50 largest labs in the United States began 2002 with about 117,000 backlogged requests for forensic services, and they ended the year with more than 93,000 backlogged cases, including about 270,000 requests for forensic services.[97]

Officers apprehending sexual offenders, for instance, have experienced processing times at crime laboratories that ultimately posed significant delays and placed cases at risk of being thrown out of court. State laboratories take an average of 23.9 weeks to process an unnamed-suspect rape kit, and local laboratories average 30.0 weeks for such tests.[98]

Wrongful Convictions

According to the Innocence Project, several hundred defendants are wrongfully convicted every year in the United States.[99] Many of those defendants are eventually exonerated when it is found that their convictions resulted from human error leading to faulty forensic assessments, despite all the advancements made in both the forensics field in general and forensics laboratories in particular. Other reasons for wrongful convictions may include eyewitness errors, overzealous/unethical police officers and criminal investigation personnel, lazy or incompetent prosecutors, and incompetent expert witnesses, which include forensic personnel.[100]

The version of the justice process presented in the media can affect the way prosecutors and defense lawyers prepare cases, as well as the expectations that police and the public have for real crime lab results.[101] Real crime scene investigators report that, because of the flamboyant and idealized versions offered up by the media, individuals often have unrealistic notions of what forensic science can deliver.[102]

Human Incompetence and Forensic Science

Real forensic science is seldom fast or as certain as the media portrays.[103] Too often, authorities say, the science is unproven, the analyses unsound, and the experts unreliable.[104] For instance, a crime analysis unit or fingerprinting unit at the Boston Police Department (BPD) was responsible for the wrongful conviction of Stephan Cowans of Roxbury, Massachusetts, who was charged in the shooting and wounding of a police officer in 1997 and awarded $3.2 million in damages following his wrongful conviction.[105] Cowans was released from prison in January 2004 after serving about 6 ½ years. Because of these earlier injustices, the BPD hired, trained, and assigned professional crime scene investigators to all its open cases. Will they fare better in the courtroom? "Probably not, because of the public's expectations," says one of the officers trained by the Henry C. Lee Institute of Forensic Science in Connecticut.[106]

In Chicago, Larry and Calvin Ollins, Omar Saunders, and Marcellius Bradford were convicted of the October 1986 kidnapping, rape, and murder of 23-year-old medical student Lori Roscetti.[107] At their trial, Bradford testified that the four teens abducted Roscetti and drove her to a remote location, where he and Saunders got out of the car to act as the lookouts while the two cousins assaulted her. When she tried to escape, according to Bradford, Larry Ollins crushed her head with a chunk of concrete; he then carried Roscetti back to the car, and he and Calvin raped her. A friend of Bradford's provided additional testimony, recounting a confession by Saunders with substantially similar facts. Bradford later recanted his statements, saying that police coerced him into falsely confessing and that he did so to avoid a life sentence.

Crime lab analyst Pamela Fish testified that semen found on the victim's body belonged to Larry and Calvin Ollins, but a recent examination of her notes by a DNA expert showed that none of the four men's blood types matched the CS samples. Fish's testimony helped convict many innocent persons. As a result of Fish's incompetence and because of the highly publicized postconviction DNA exonerations of more than 156 death row inmates, the governor of Illinois placed a moratorium on the use of the death penalty in that state.[108]

Nonetheless, numerous crimes are potentially preventable through better, more efficient use of forensic DNA analysis.[109] A review of specific cases in 19 states revealed more than 100 serious crimes that could have been prevented through either the inclusion of all convicted felons in the database or shorter DNA analysis processing times.[110]

Unfortunately, human incompetence among forensic personnel seems to be at an all-time high. Therefore, some conclusions can be drawn about television performances that might concern many.

■ The *CSI* Effect

In light of the information presented in the previous section, it is easy to understand why some observers suggest that popular television drama police shows miss the reality mark. Popular television dramas such as *CSI*, *CSI Miami*, and *Criminal Minds* portray the notion that all criminal cases can be solved through the employment of high-tech criminal investigations.[111] The ***CSI* effect** (i.e., the message delivered by fictionalized accounts of forensic analysis) offers glamour, certainty, self-discipline, objectivity, truth, and justice all rolled into one, and in doing so effortlessly accommodates much-heralded successes.[112] The *CSI* effect tends to raise crime victims' and jury members' real-world expectations of investigators and forensic science and, as a result, hinders the investigative process.[113]

Prospective students and others who overestimate the reality basis of shows such as *CSI* may develop unreasonable expectations of actual forensic practitioners. For instance, a crime laboratory that produces forensic science evidence, which includes DNA testing results, may be accepted as infallible regardless of the quality of the evidence. In fact, the literature advises that many CSI units that provide DNA testing and fingerprint analysis cannot standardize their operations, control their excessive objectives, or meet judicial responsibilities of their mission.[114] This thought is consistent with findings suggesting that while "practitioners have compelling research ideas, these often must take a back seat to the reality in our nation's crime labs, where shelves of evidence await testing and there is daily pressure from agencies and the communities they serve. Crime laboratory professionals may realize that research is the key to long-term solutions, but with limited resources and overwhelming caseloads, what can they do to move a great research idea from their heads to the laboratory bench?"[115]

On a more positive note, the media's focus on forensic science has had several beneficial outcomes. For instance, the courts have taken an interest in technological advancement through the *CSI* effect and have made judicial decisions shaping punitive decisions. For instance, in the U.S. Supreme Court's decision in *House v. Bell* (126 S. Ct. 2006), a Tennessee death-row inmate, Paul Gregory House, was provided with the right to a hearing in federal court because newly discovered DNA and other evidence allowed House to assert his claim that he received ineffective assistance of counsel at trial.[116]

Here are seven other benefits of the *CSI* effect:

- It stimulates public interest in rejuvenating unsolved cases.
- It provides understanding when policing designate resources to investigate cold cases.
- It encourages those with knowledge of unsolved homicides and other crimes to come forward.
- It has stimulated police agencies to become proactive, particularly in terms of reviewing cold cases.
- It reminds those who have so far gotten away with murder that the police do not give up and their cases have not been forgotten.
- It encourages police agencies to find new ways to serve and protect.
- It heightens the accountability of police officers to act as role models who demonstrate integrity and competence.

Qualifications for Forensic Personnel

One recent characteristic for new hires in forensics seems to be a college degree for entry-level personnel and a degree from graduate school (which includes a Ph.D.) for senior-level positions. Experience as an officer might be a second priority. For instance, the state of California's Department of Health seeks graduate hours for a forensic alcohol analyst and an undergraduate degree in chemistry, biochemistry, or other natural sciences degrees for an entry-level alcohol analyst.

When the Massachusetts State Police seek a Forensic Chemist 1 for their DNA Unit, the pay range is $1,506 to $1,992 biweekly. The minimum qualifications include at least two years of full-time professional or technical experience in the field of chemistry; a bachelor's or higher degree with a major in chemistry or biochemistry may be substituted for the required experience. In senior jobs, expert testimony experience is sought along with advanced degrees and experience, and income is similar to that found in the private sector.

***RIPPED* from the Headlines**

NYPD's CSI Head Demoted

The head of the NYPD's vaunted CSI-type units was dumped because federal and state authorities were kept in the dark about an embarrassing 2002 integrity test that caught two lab technicians lying about evidence.

Police Commissioner Ray Kelly transferred Deputy Chief Denis McCarthy, a 27-year veteran, from commander of the Forensic Investigations Division, which includes the Police Laboratory, the Crime Scene Unit, and the Bomb Squad, to a patrol spot in Manhattan. Sources said the move against the respected chief was meant to send a stern message that the top cop expects the highest standards at the NYPD's critical scientific and technological units.

Kelly is also expected to announce the creation of a "forensic science review committee" to evaluate the lab and the overall division, its quality controls, and its adherence to standards.

NYPD Internal Affairs records showed the two technicians flunked an annual integrity test in August 2002 where they were given drug evidence to analyze.

Sources said one "criminalist," identified as Elizabeth Mansour, 58, claimed 37 bags of drugs contained cocaine when only 34 did. The sources said she subsequently retired, but not before she suffered a heart attack at the lab on the day she was to be questioned by IAB, sources said.

The other civilian technician, Rameschandra Patel, 64, reported cocaine hits in six bags, when only five had the drug. Charges were filed against him, but he continues in a desk job answering phones outside FID while fighting the allegations.

Source: Murray Weiss, "NYPD's CSI Head Demoted," *The New York Post* (April 19, 2007) http://www.officer.com/web/online/Internal-Affairs-News/NYPDs-CSI-Head-Demoted/5$35777 (accessed July 14, 2008).

Summary

A criminal investigation is defined as an information-gathering process, which includes interviews among victims, witnesses, and suspects. The purpose of an investigation is to legally gather and confirm the evidence; secure the chain of custody; and identify, arrest, and obtain a confession from lawbreakers. Cold cases comprise unsolved cases, whereas hot cases are cases that end in an arrest.

Many factors affect an investigation, including the time and place at which it occurs. Interviews are fact-finding missions, whereas interrogations, after establishing a baseline, are designed to match acquired information from the interview to a confession. The resistance of a suspect to admission of the crime is affected by many factors, ranging from fear of employment termination to fear of retaliation. Suspect denials are usually centered on survival perspectives, and suspect lies can be ascertained through an observation of the sensory channels, such as eye moments. Interrogations have been criticized on the basis that good confessions are really a product of the collaboration between investigator and criminal. There is always a probability of error, and police organizational traps may sometimes lead to failure in criminal investigations.

Because the burden of proof plays an important role in an investigation, it is up to the investigators to ensure that the investigatory process is both legal and logical. Field investigators gather evidence, and laboratory personnel perform forensic assessments of that evidence. Services provided by forensic personnel (including field investigators) vary, and police are sometimes reluctant to call in forensics.

The *CSI* effect influences criminals, police, and the public, who may be persuaded to believe untruths about forensic science; forensic science cannot always provide evidence to aid in the investigation or conviction of a suspect. The *CSI* effect does have some advantages, however, in that it has made the police focus on enhancing their initiatives to detect and apprehend offenders.

Key Words

Auditory eve movements
Baseline
Burden of proof
Chain of custody
Clearance rates
Cold case
CSI effect
Interrogation
Interview
Investigation
James Bond syndrome
Reid technique
Sensation eye moments
Sensory channels
Visual eye movements

Discussion Questions

1. Define a criminal investigation and explain its purpose.
2. Describe the process of a criminal investigation.
3. Define the chain of custody and articulate its importance to the investigation.
4. Characterize myths associated with investigators and investigations.
5. Describe and compare cold cases and hot cases.
6. Discuss an investigator's control of factors affecting the CSI.
7. Define and compare an interview and an interrogation.
8. Discuss the relevance of a baseline as part of an interrogation.
9. Explain how investigators deal with suspect response, resistance, denials, and lies.
10. Articulate the chief tasks of an interrogator concerning suspect deception.
11. Characterize the strategies used by interrogators.
12. Identify the primary criticism of interrogations.
13. Characterize the role burden of proof plays in the investigation and interrogation practices of investigators. In what way does this responsibility guide investigators' practices as they relate to suspects and defendant searches?
14. Identify strategies designed to improve the criminal investigation process.

15. Explain the association and process of forensics as it relates to criminal investigation.
16. Explain the tasks of a crime scene analyst.
17. Characterize police reluctance to call in crime labs. In what way do you agree or disagree with their reluctance?
18. Describe how the literature supports police reluctance to call in crime labs.
19. Describe the *CSI* effect and characterize its positive and negative effects on the justice practice in general, including investigations.

Notes

1. Mark H. Moore, Robert Trojanowicz, and George L. Kelling, *Crime and Policing* (Washington, DC: National Institute of Justice, 1988).
2. Bureau of Labor Statistics (2007), http://www.bls.gov/oco/ocos160.htm (accessed March 24, 2007).
3. Patricia Cornwell is a contemporary U.S. author. She is widely known for writing a popular series of crime novels featuring the fictional heroine Dr. Kay Scarpetta, a medical examiner. In 2002, Cornwell claimed to have solved the mystery of the Jack the Ripper murders and accused noted artist Walter Sickert of committing the crimes, though her conclusions have been criticized.
4. Bureau of Justice Statistics, *Local Police Departments, 2003* (NCJ 210118) (Washington, DC: U.S. Department of Justice, May 2006), http://www.ojp.usdoj.gov/bjs/pub/pdf/lpd03.pdf (accessed October 16, 2006).
5. Bureau of Justice Statistics, "Law Enforcement Management and Administrative Statistics, 2000: Data for Individual State and Local Agencies with 100 or More Officers" (NCJ203350) (2000), http://www.ojp.usdoj.gov/bjs/pub/pdf/lem00in.pdf (accessed March 12, 2007).
6. Bureau of Labor Statistics (2007).
7. Lawrence F. Travis III and Robert H. Langworthy, *Policing in America: A Balance of Focus* (Upper Saddle River, NJ: Prentice Hall, 2008).
8. Charles R. Swanson, Neil C. Chamelin, and Leonard Territo, *Criminal Investigation*, 8th ed. (Boston: McGraw Hill, 2008).
9. Samuel Walker and Charles M. Katz, *The Police in America: An Introduction*, 6th ed. (Boston: McGraw Hill, 2008).
10. *Crime Scene Investigation: A Reference for Law Enforcement Training: Special Report* (NCJ 200160) (Washington, DC: U.S. Department of Justice, National Institute of Justice, June 2004), http://www.ncjrs.org/ pdffiles1/200160.pdf (accessed April 8, 2008).
11. Personal communication between the writer, the investigator, and a student in New Orleans, March 2007.
12. "Chain of Custody" (2004), http://tpub.com/legalman/80.htm (accessed March 14, 2007).
13. Daniel Glaser, *Profitable Penalties* (Thousand Oaks, CA: Pine Forge, 1997). Also see Marcus Felson, *Crime and Everyday Life*, 3rd ed. (Thousand Oaks, CA: Pine Forge, 2002).
14. Stefan Lovgren, "CSI Effect Is Mixed Blessing for Real Crime Labs," *National Geographic News* (September 23, 2004), http://news.nationalgeographic.com/news/2004/09/0923_040923_csi.html (accessed June 25, 2006).
15. Bureau of Justice Statistics, "Rape and Sexual Assault: 1992–2000" (NCJ 194530) (2002), http://www.ojp.usdoj.gov/bjs/pub/pdf/rsarp00.pdf (accessed March 13, 2007).
16. Council on Sexual Assault (2004), http://www.safefromabuse.com/assault_myths.html (accessed July 14, 2007).
17. Julia Layton, "How Crime Scene Investigation Works" (2006), http://science.howstuffworks.com/csi.htm (accessed August 7, 2006).

18. *Sourcebook of Criminal Justice Statistics* (2003), 457, http://www.albany.edu/sourcebook/pdf/t557.pdf (accessed March 13, 2007).

19. "Estimated Number of Arrests," *Sourcebook of Criminal Justice Statistics* (2003), 344, http://www.albany.edu/sourcebook/pdf/t41.pdf (accessed March 13, 2007).

20. "FBI Hate Crimes Statistics" (2007), http://www.fbi.gov/ucr/hc2006/incidents.html (accessed November 25, 2007).

21. In 1924, Edward Young Clarke, an advertising executive in the state of Louisiana, pleaded guilty in federal court to violating the Mann Act (an anti-prostitution measure enacted in 1910). Source: "FBI Hate Crimes Statistics," http://www.fbi.gov/ (accessed November 25, 2007).

22. "FBI Hate Crimes Statistics."

23. Jack McDevitt, Jack Levin, and Susan Bennett, "Hate Crime Offenders," *Social Issues* 58 (2002): 303–17.

24. Michael Shively and Carrie F. Mulford, "Hate Crime in America: The Debate Continues," *NIJ Journal* 257 (2007), http://www.ojp.usdoj.gov/nij/journals/257/hate-crime.html (accessed November 25, 2007).

25. Bureau of Justice Statistics, "Hate Crimes Reported by Victims and Police" (2007), http://www.ojp.usdoj.gov/bjs/abstract/hcrvp.htm (accessed November 25, 2007).

26. B. J. Karem and C. Napolitano, "In a Room with Madness," *Playboy* (March 2003).

27. Tanya Eiserer, "Dallas Detectives Sick of Worn-out Cars," *The Dallas Morning News* (March 27, 2007), http://www.dallasnews.com/sharedcontent/dws/dn/latestnews/stories/032506dnmetcopcars.407f4a9.html (accessed June 3, 2008).

28. Moore, Trojanowicz, and Kelling, *Crime and Policing.*

29. Swanson, Chamelin, and Territo, *Criminal Investigation*, 178.

30. David E. Zulawski and Douglass E. Wicklander, "Interrogation: Understanding the Process," in *Corrections Perspective*, ed. Dennis J. Stevens (Madison, WI: Coursewise Press, 2002), 3–7.

31. David R. Givens, *Crime Signals: Body Language of Murderers, Terrorists, and Thieves* (New York: St. Martin Press, 2007).

32. Don Rabon and Tanya Chapman, *Persuasive Interviewing* (Durham, NC: Carolina Academic Press, 2007).

33. Don Rabon, *Interviewing and Interrogation* (Durham, NC: Carolina Academic Press, 1992), 10.

34. Jack Haley and Madeleine Richeport-Haley, *The Art of Strategic Therapy* (New York: Brunner-Routledge Press, 2006).

35. Don Rabon, *Investigative Discourse Analysis* (Durham, NC: Carolina Academic Press, 2003), 47.

36. Darko Maver, "Defense Strategies and Techniques of Interrogation: Results of Empirical Research," *Policing in Central and Eastern Europe* (1996), http://www.ncjrs.gov/policing/def331.htm (accessed March 17, 2007).

37. Zulawski and Wicklander, "Interrogation."

38. Charles L. Yeschke, *The Art of Investigative Interviewing* (Boston: Butterworth-Heinesmann, 1997), 25–44.

39. David Wilson, "Crime Lab Investigation," *The Guardian* (September 14, 2006), http://www.guardian.co.uk/comment/story/0,,1871878,00.html (accessed February 8, 2007).

40. Leo Katz, *Bad Acts and Guilty Minds: Conundrums of the Criminal Law* (Chicago: University of Chicago Press, 1987).

41. Michael R. Gottfredson and Travis Hirschi, *A General Theory about Crime* (Stanford, CA: Stanford University Press, 1990), 17.

42. Zulawski and Wicklander, "Interrogation," 4.
43. Gottfredson and Hirschi, *A General Theory about Crime*, 14.
44. Stanton E. Samenow, *Inside the Criminal Mind*, rev. and updated (New York: Crown, 2004), DSV IV.
45. Jeffrey Kluger and Coco Masters, "How to Spot a Liar," *Time* 168, no. 9 (2006): 46–48.
46. Rabon, *Interviewing and Interrogation*, 25.
47. Rabon, *Interviewing and Interrogation*, 133.
48. Rabon, *Interviewing and Interrogation.*
49. Fred E. Inbau, John E. Reid, Joseph P. Buckley, and Brian C. Jane, *Essentials of the Reid Technique: Criminal Interrogation and Confessions*, 4th ed. (Boston: Jones and Bartlett, 2005).
50. Nathan J. Gordon, "Confessions of an Interrogator: Ten Principles That Guide a Successful Interrogation—by Making It Easier for a Suspect to Confess," *Security Management* 46, no. 10 (2002), http://www.truthinjustice.org/get-confession.htm (accessed March 17, 2007).
51. Inbau, Reid, Buckley, and Jane, *Essentials of the Reid Technique*; Rabon, *Interviewing and Interrogation.* This perspective was also confirmed by classroom detectives at the Boston and New Orleans police departments.
52. Rabon, *Interviewing and Interrogation.*
53. Howstuffworks (2007), http://people.howstuffworks.com/police-interrogation1.htm (accessed March 25, 2007).
54. Inbau, Reid, Buckley, and Jane, *Essentials of the Reid Technique*; Gordon, "Confessions of an Interrogator"; Rabon, *Interviewing and Interrogation*; Zulawski and Wicklander, "Interrogation."
55. Frank Schmalleger, "World of the Career Criminal," in *Corrections Perspective*, ed. Dennis J. Stevens (Madison, WI: Coursewise Press, 2002), 50–53.
56. Eric Leberg, *Understanding Child Molesters: Taking Charge* (CA: Sage, 1997); Samenow, *Inside the Criminal Mind.*
57. Samenow, *Inside the Criminal Mind.*
58. Anne M. Coughlin, *Interrogation Stories* (Charlottesville, VA: University of Virginia, March 2007), http://www.law.virginia.edu/pdf/workshops/0607/coughlin.pdf (accessed March 25, 2007).
59. D. Kim Rossmo, "Criminal Investigative Failures: Avoiding the Pitfalls," *FBI Law Enforcement Bulletin* 75, no. 10 (2006): 12–19.
60. Rossmo, "Criminal Investigative Failures."
61. Rossmo, "Criminal Investigative Failures."
62. Rossmo, "Criminal Investigative Failures."
63. Rossmo, "Criminal Investigative Failures."
64. Stanford M. Lyman and Marvin B. Scott, *A Sociology of the Absurd* (New York: General, 1989), 31–34.
65. Lyman and Scott, *A Sociology of the Absurd*, 113.
66. Karl Ask and Par Anders Granhag, "Motivational Sources of Confirmation Bias in Criminal Investigations: The Need for Cognitive Closure," *Journal of Investigative Psychology and Offender Profiling* 2 (2005): 43–63.
67. John E. Hess, *Interviewing and Interrogation for Law Enforcement* (New York: W.H. Anderson Company 1997), 70.
68. Dennis J. Stevens, "Police Training and Management Impact Sexual Assault Conviction Rates in Boston," *Police Journal* 79, no. 2 (2006): 125–54.
69. Peter Greenwood and Joan Petersilia, "The Criminal Investigation Process," in *Policy and Policy Implications* (Washington, DC: U.S. Department of Justice, 1975). Also see Richard Forsyth, "Increase Ethical Conduct," *Law and Order* 51, no. 5 (2003): 101–5.

70. Richard Saferstein, *Criminalistics: An Introduction to Forensic Science*, 9th ed. (Upper Saddle River, NJ: Prentice Hall, 2006).

71. Swanson, Chamelin, and Territo, *Criminal Investigation*, 660.

72. Saferstein, *Criminalistics*, 13.

73. Layton, "How Crime Scene Investigation Works."

74. H. W. "Rus" Ruslander, "The Role of the Crime Scene Investigator" (2006), http://www.criminalistics.us/thumbnails.html (accessed August 7, 2006).

75. Saferstein, *Criminalistics*, 13.

76. Nicholas P. Lovrich, Travis C. Pratt, Michael J. Gaffney, et al., *National Forensic DNA Study Report: Final Report* (NIJ 203970) (Washington, DC: U.S. Department of Justice, February 2004), http://www.ncjrs.gov/pdffiles1/nij/grants/203970.pdf (accessed July 1, 2006).

77. Matthew J. Hickman and Joseph L. Peterson, "Census of Publicly Funded Forensic Crime Laboratories, 50 Largest Crime Labs, 2002." (NCJ 205988) (Washington, DC: U.S. Department of Justice, September 2004), http://www.ojp.usdoj.gov/bjs/pub/ascii/501c102.txt (accessed July 6, 2008).

78. Eric Lambert, Terry Nerbonne, and Phillip L. Watson, "The Forensic Science Needs of Law Enforcement Applicants and Recruits: A Survey of Michigan Law Enforcement Agencies," *Journal of Criminal Justice Education* 14, no. 1 (2003): 67–81.

79. Craig A Leam, Kevin D. Browne, and Ian Stringer, "Sexual Recidivism: A Review of Static, Dynamic and Actuarial Predictors," *Journal of Sexual Aggression* 11, no. 1 (January): 45–69.

80. Based on the disciplinary records of those inmates while incarcerated. *Deadly Speculation: Misleading Texas Capital Juries with False Predictions of Future Dangerousness* (Houston, TX: Texas Defender Service, 2004).

81. Dennis J. Stevens, "Forensic Science, Wrongful Convictions, and American Prosecutor Discretion," *Howard Journal of Criminal Justice* 47, no. 1 (2008): 31–51.

82. Sutham Cheurprakobkit and Gloria T. Pena, "Computer Crime Enforcement in Texas: Funding, Training, and Investigating Problems," *Journal of Police and Criminal Psychology* 18, no. 1 (2003): 24–37.

83. Margaret A. Berger, "Raising the Bar: The Impact of DNA Testing on the Field of Forensics," in *Perspectives on Crime and Justice: 2000–2001 Lecture Series*, ed. Alfred Blumstein, Laurence Steinberg, and Carl C. Bell (Washington, DC: National Institute of Justice, 2002).

84. Francesco Romolo Saverio and Margot Pierre, "Identification of Gunshot Residue: A Critical Review," *Science International* 119, no. 2 (2001): 195–211.

85. Kathleen J. Sternberg, Michael Lamb, and Graham M. Davies, "The Memorandum of Good Practice: Theory versus Application," *Child Abuse and Neglect* 25, no. 5 (2001): 669–81.

86. Stephen M. Cordner, Michael P. Burke, and Malcolm J. Dodd, "Issues in Child Homicides: 11 Cases," *Legal Medicine* 3, no. 2 (2001): 95–103.

87. Cordner, Burke, and Dodd, "Issues in Child Homicides."

88. Craig L. Nelson and David C. Winston, "Detection of Medical Examiner Cases from Review of Cremation Requests," *American Journal of Forensic Medicine & Pathology* 27, no. 2 (2006): 103–5.

89. Lonnie D. Ginsberg, "And the Blood Cried Out: A Prosecutor's Spell-Binding Account of the Power of DNA," *American Journal of Forensic Medicine & Pathology* 18, no. 2 (1997): 218–37.

90. Yasuaki Hagimoto and Hiroki Yamamoto, "Analysis of a Soldered Wire Burnt in a Fire," *Journal of Forensic Sciences* 51 (2006): 87–94.

91. Michael J. Saks and David L. Faigman, "Expert Evidence after Daubert," *Annual Review of Law and Social Science* 1 (2005): 105–30. *Daubert* stands for a trilogy of Supreme Court cases as well as revisions of the Federal Rules of Evidence. Together they represent U.S. law's most recent effort to filter expert evidence offered at trial.

92. Shirley Melnicoe et al., *Police Research Catalog: Police-Related Research Supported by the National Institute of Justice 1969–1981* (Washington, DC: U.S. National Institute of Justice, 1982).
93. Cordner, Burke, and Dodd, "Issues in Child Homicides."
94. Personal communication between the author and a Boston prosecutor who wishes to remain anonymous, December 2005.
95. Donald E. Shelton, "The CSI Effect: Does it Really Exist?" (NIJ 221501) (Washington, DC: U.S. Department of Justice, 2008), http://www.ncjrs.gov/pdffiles1/nij/221501.pdf (accessed April 8, 2008).
96. Kit R. Roane, "The CSI Effect," *U.S. News and World Report,* (2005, April 25), http://www.usnews.com/usnews/culture/articles/050425/25csi.htm (accessed April 8, 2008).
97. Bureau of Justice Statistics, *Fact Sheet: 50 Largest Crime Labs, 2002* (NCJ 205988) (Washington, DC: U.S. Department of Justice, 2004), http://www.ojp.usdoj.gov/bjs/pub/pdf/50lcl02.pdf (accessed June 31, 2008).
98. Lovrich et al., *National Forensic DNA Study Report.*
99. Innocence Project (2006), http://www.innocenceproject.org/press/ (accessed August 8, 2006).
100. Ronald C. Huff, "Wrongful Conviction and Public Policy: The American Society of Criminology 2001 Presidential Address," *Criminology: An Interdisciplinary Journal* 40, no. 1 (2002): 1–18.
101. Richard Willing, "*CSI* Effect Has Juries Wanting More Evidence," *USA Today* (August 5, 2004), http://www.usatoday.com/news/nation/2004-08-05-csi-effect_x.htm (accessed June 25, 2006).
102. Lois A. Tully, "Taking the Initiative: Practitioners Who Perform Frontline Research," *NIJ Journal* 258 (2007), www.ncjrs.gov/pdffiles1/nij/219603d.pdf (accessed April 8, 2008).
103. Willing, "*CSI* Effect Has Juries Wanting More Evidence."
104. Roane, "The CSI Effect," *U.S. News and World Report.*
105. Suzanne Smalley and Ralph Ranalli, "Police to Unveil *CSI*-Style Unit for Evidence," *Boston Globe* (January 22, 2006), http//wwww.boston.com (accessed June 30, 2006). Also see David Bernstein, "Boston Agrees to Pay 3.2 Million to Stephen Cowans," *The Phoenix* (August 6, 2006), 1.
106. Personal communication between a Boston police officer and the author, July 29, 2006.
107. Innocence Project (2001), http://www.innocenceproject.org/case/display_profile.php?id=98 (accessed June 30, 2006).
108. BBC News, "Governor Clears Illinois Death Row," (January 11, 2003), http://www.ccadp.org/news-ryan2003.htm (accessed August 10, 2006). Also see Huff, "Wrongful Conviction and Public Policy."
109. Lovrich et al., *National Forensic DNA Study Report,* 7.
110. National Institute of Justice, *Forensic Examination of Digital Evidence: A Guide for Law Enforcement* (NCJ 199408) (Washington, DC: U.S. Department of Justice, April 2004), http://www.ncjrs.gov/txtfiles1/nij/199408.txt (accessed July 1, 2006).
111. Dennis J. Stevens, "Forensic Science, Wrongful Convictions, and American Prosecutor Discretion," *Howard Journal of Criminal Justice* 47, no. 1 (2008): 31–51.
112. Stevens, "Forensic Science, Wrongful Convictions, and American Prosecutor Discretion."
113. Stevens, "Forensic Science, Wrongful Convictions, and American Prosecutor Discretion."
114. Cheurprakobkit and Pena, "Computer Crime Enforcement in Texas."
115. Tully, "Taking the Initiative."
116. Mark Hansen, "Actions," *American Bar Association Journal* 29, no. 9 (2006): 14–15.

18

Proactive Strategies: Undercover Operations, Stings, Gangs, and Drugs

Nearly all men can stand adversity, but if you want to test a man's character, give him power.

—Abraham Lincoln

LEARNING OBJECTIVES

When you finish reading this chapter, you will be better prepared to:

- Describe strategies to capture repeat offenders
- Characterize federal contributions to local repeat-offender programs
- Describe typical undercover police strategies
- Describe sting strategies employed by local and federal agencies
- Explain the relationship among violence, youth gangs, and drugs
- Provide historical and contemporary accounts of the United States' toughest gangs
- Describe traditional gang suppression models and contemporary gang injunction models
- Explain why methamphetamine is a powerful foe
- Compare supply reduction strategies with demand reduction strategies in terms of their ability to curtail drug abuse
- Characterize new approaches to policing drugs

CASE STUDY: YOU ARE THE POLICE OFFICER

You are a patrol officer at roll call when your sergeant tells you to meet with the detectives in the next room. Small snickers and a few roars roll from the lips of the patrol officers as they file out through the small door to their waiting cruisers. Your partner is paired with a patrol sergeant, implying you will be with the detectives for a while, and you both nod at each other in understanding. Your mind races through all the reports you filed in the past weeks, wondering which report they will criticize. You march hard into the room and launch a strong stare at the detectives, silently expressing the message that you won't allow them to put down your work.

"What is it?" You demand with tight lips, as your eyes scan both detectives.

"You're kind of a young-looking thing," the detective says with a smile, catching you off guard. You shrug.

"We have an immediate need for someone to pose as a criminal for a few weeks at an educational facility. We've been advised by a confidential informant that a threat exists to hack into the school's computer system and destroy hard-to-replace data and possibly vandalize the premises. The chief wants you to pose as a co-ed and make contact with those suspects who we think will commit this criminal act."

You've often thought about officer deception and secrecy, but you never imagined it would be developed in such an informal fashion or for a crime of this nature. "We'll do all the training and ensure your safety providing you follow the rules," said the other.

When you asked which school, the response pushes you into a chair. "Gee, I graduated from that college a few years ago."

The lead detective said, "We know. That's why you're perfect for the job, officer. With you, we don't have to build a cover story and you could be on campus tomorrow without all the fanfare and paperwork."

You learned from your academy training that undercover officers build cases by gathering evidence for prosecutors, but you also learned that the chain of custody was often at risk in such operations. From the sounds of this assignment, there would be little risk or danger.

"Why not," you said in a sing-song way and shrugged again. "Not bad," you thought, "I'll get paid extra for sitting in a classroom talking to college kids. But I'll miss the excitement of patrol and first-response activities, which make me feel like I'm helping others."

"One other thing," the taller detective stated quietly. "There are concerns that the suspects are linked to a domestic terrorist group and it could be dangerous for you if they discover your identity. So no one," he emphasizes with a single finger, "including friends, family, or your partner can know about your assignment."

Your thoughts rushed around, and you wondered how covertly you could carry your service weapon into a classroom. "No weapons," the detective added, "you can't risk discovery and you can't sound or act like the patrol officer who walked in that door," as his finger points to the door.

1. In what way are confidential informants reliable in aiding the police in the detection of criminals?
2. If the police are deceptive in their process so as to identify and apprehend criminals, in what way could that behavior be justified?
3. What evidence is there that this patrol officer's behavior could provide indicators to criminals that she is a police officer?

Introduction

Policing repeat offenders, drugs, and gangs through proactive strategies and undercover deployment of detectives who employ confidential informants and sting initiatives are similar narratives—hence their inclusion in the same chapter. However, whether the description of a police strategy is linked to patrol, service, or investigation, one way to traditionally characterize that strategy is reactive: A crime happens and, assuming it is reported or observed by police, the officers react or respond to it. This chapter is about strategies designed to help reactive responses through proactive efforts, which attempt to sort out, identify, and detain suspects before and during their crime spree. Some of these strategies include undercover investigations, informants, and stings employed by police with an eye toward ensuring safety, especially in public places. The media aid the police in implementing these proactive strategies to control crime.

These strategies cross over into other police operations, resulting in what appears to be duplicative efforts. In the final analysis, these strategies are related. Considering the power an officer individually possesses when performing any of the strategies described in this chapter, the officer's success ultimately reflects the quality of his or her character, as Abraham Lincoln suggests.

Proactive Police Strategies: Television Programs and Amber Alerts

America's Most Wanted and John Walsh

John Walsh of *America's Most Wanted* (*AMW*) is driven by a personal tragedy: His only child was abducted, tortured, and killed. The show that he hosts has, since its inception, aided in the capture of almost 1,000 fugitives from justice.[1] Nielsen reports that *AMW* has been watched by millions of viewers over the past 20 years.

Another result of *AMW*'s influence was the development of Adam Walsh's Law (Adam was John Walsh's son), which strengthened the rules governing the National Sex Offender Registry, so that the U.S. Marshals Service now has the authority to apprehend violators who fail to register. John Walsh says, "In 20 years on *America's Most Wanted*, I've seen law enforcement at its best. But I've also seen horrible mistakes that have cost lives . . . I know those mistakes were made because of lack of training, lack of resources and lack of money."

Cops

Cops is a popular television show that has aired since 1988 and offers an inside view of police services in various communities across the United States. While the show shows police engaged in reactive police practices, its creator, John Langley, implies that *Cops* is a powerful tool in pushing proactive initiatives, because it provides officers with perspectives from other real-life departments and does not use actors or creative writers to alter reality. One officer interviewed about *Cops* reports that, in fact, his knowledge about search and seizures was enhanced by watching *Cops*.

Dateline

Dateline is an NBC magazine-style series that has resulted in the capture of numerous suspects, most notably pedophiles (see its website at http://www.msnbc.com). Often *Dateline* presents series of episodes on the same theme—for example, "To Catch a Predator," "To Catch a Con," and "To Catch an ID Thief." For instance, one series showed Houston police officers posing as "fences" to buy stolen cars, an operation that resulted in numerous arrests. Its most popular series has been "To Catch a Sexual

Predator," in which sexual predators communicate with young girls on the Internet; when the offenders are asked to drive to the underage girl's home, they do so—only to be arrested by local police officers.

Amber Alert

The **Amber Alert** was created in 1996, as a powerful legacy of nine-year-old Amber Hagerman, a bright little girl who was kidnapped while riding her bicycle and brutally murdered in Arlington, Texas. Needless to say, the entire community was shocked. Residents contacted radio stations in the Dallas area and suggested they broadcast special "alerts" over the airwaves so that they could help prevent such incidents in the future. In response to the community's concern for the safety of local children, the Dallas/Fort Worth Association of Radio Managers teamed up with local law enforcement agencies in northern Texas and developed this innovative early-warning system to help find abducted children quickly.[2] Statistics show that, when abducted, a child's greatest enemy is time.

In April 2003, President George W. Bush signed the Amber Alert legislation, turning the initiative into a national program. The Amber Alert system is now mandated across the United States. For instance, on December 1, 2006, an Amber Alert was issued for a 28-day-old Florida boy who was abducted at knifepoint in Fort Myers from the child's mother. The child, Bryan Gomes, has black hair and weighed 12 pounds at the time of his abduction. The suspect is a 28- to 30-year-old female who remains free.

■ Proactive Strategies to Capture Repeat Offenders

Many **proactive police strategies** aim to capture repeat offenders before they reoffend. Why target repeat offenders?

- In a birth cohort study, Wolfgang estimates that a small number of suspects (6 percent of any group of young men) commit an extremely high percentage of all serious crimes.[3]
- Evidence-based findings of Lawrence Sherman and police studies conducted across the country (Chapter 9) reveal that "less than 3 percent of the street addresses and 3 percent of the population in a city produce over half the crime and arrests."[4] For example, one study found that 6 percent of the addresses in Minneapolis accounted for 60 percent of the calls for police service.

Ronald V. Clarke and John E. Eck argue that this phenomenon can be called the 80–20 rule, where in theory 20 percent of some things are responsible for 80 percent of the outcomes.[5] In the real world, seldom is the split 80–20, yet a small percentage of something or some group is involved in a large percentage of an outcome. For instance, an analysis was performed of construction site thefts and burglaries among 55 homebuilders in Jacksonville, Florida. Eleven of the builders (20 percent of the group) experienced 85 percent of all thefts and burglaries at construction sites reported to the Jacksonville Sheriff's Department during the period from January to September 2004 (see Chapter 4 for a profile on this department). One principle underlying crime prevention efforts, according to Clarke and Eck, is the notion that crime is highly concentrated on particular people, places, and things.[6]

Chronic offenders tend to live criminal lifestyles, associate with pro-criminal associates, generally commit a high percentage of crimes, and often escape detection because many of their crimes are never reported.[7] Therefore, concentrating resources upon those individuals who are most likely to commit the most crime can yield the greatest preventive benefits.

Structure of Repeat-Offender Programs

> The rationale of repeat-offender programs (ROP) is a simple one: Identifying, apprehending, and convicting repeat offenders will provide more public safety than waiting for those suspects to reoffend.[8]

Three types of repeat-offender programs are distinguished:[9]

- Programs targeting suspected high-rate offenders for surveillance and arrest
- Special warrant services targeting suspected high-rate offenders who have outstanding warrants or are wanted for probation or parole violations
- Case-enhancement programs that provide prosecutors with full information about the criminal history of high-rate offenders

The Chandler, Arizona, Police Department (APD), for instance, developed a ROP to identify high-risk recidivists and violent persons. Its ROP focuses on post-arrest enhancement of prosecution using police/prosecutor teams to ensure appropriate follow-up related to corrective action within the justice system. The program also seeks to enhance awareness of the officers about repeat offenders, especially those prone to theft and burglary, because the APD believes that repeat offenders manipulate the justice system more than others and commit most of the crime. Therefore, the ROP devotes greater attention to repeat offenders and lobbies to increase their prison time when convicted.

In this ROP, detectives conduct thorough background investigations using targeting criteria to determine a candidate's motivation to commit crime. In particular, they check the following items:

- Current activity: participation in criminal events either as perpetrator or accomplice. Can be supported by pawn records, arrest or booking papers, confidential informant, and other sources.
- Substance abuse: types and quantity of drugs used and whether candidate previously failed drug abuse program or is involved in sale or manufacturing of drugs. Can be supported by admissions, informants, reports, and other sources.
- Lifestyle: living beyond legal means of support; associates heavily involved in crime.
- Failed probation: previously failed to successfully complete probationary period.
- Felony convictions: felony conviction in the last 10 years.
- Past informant: do not use any ROP target as an informant.
- Family background: committing property crimes against family or neighbors.

National Center for the Analysis of Violent Crime

The National Center for the Analysis of Violent Crime (NCAVC), which is supported by the FBI, has as its mission the combination of investigative and operational support functions, research, and training so as to provide assistance, without charge, to federal, state, local, and foreign police agencies investigating unusual or repetitive violent crimes.[10] The NCAVC also provides support through expertise and consultation in nonviolent matters such as national security, corruption, and white-collar crime investigations. Typical cases that NCAVC services include child abduction or mysterious disappearance of children, serial murders, single homicides, serial rapes, extortions, threats, kidnappings, product tampering, arsons and bombings, weapons of mass destruction, public corruption, and domestic and international terrorism.

To accomplish its mission, the NCAVC is organized into three components.

Behavioral Analysis Unit

The Behavioral Analysis Unit (BAU) provides behavioral based investigative and operational support by applying case experience, research, and training to complex and time-sensitive crimes, typically involving acts or threats of violence. The program areas addressed include crimes against children, crimes against adults, communicated threats, corruption, and bombing and arson investigations. The BAU develops "criminal investigative analyses," which include a behavioral and investigative perspective, a review and assessment of the facts of a criminal act, interpretation of offender behavior, and interaction with the victim, as exhibited during the commission of the crime.

Child Abduction and Serial Murder Investigative Resources Center

The Child Abduction and Serial Murder Investigative Resources Center (CASMIRC) was established through legislation that provides the U.S. Attorney General with investigation powers to assist federal, state, and local authorities in matters involving child abductions, mysterious disappearances of children, child homicide, and serial murder across the United States. The overall strategic goal of CASMIRC, as set forth in the legislation, is to reduce the impact of these crimes.

Violent Criminal Apprehension Program

The Violent Criminal Apprehension Program (VICAP) is designed to facilitate cooperation, communication, and coordination between law enforcement agencies and provide support in their efforts to investigate, identify, track, apprehend, and prosecute violent serial offenders. VICAP is a nationwide data information center designed to collect, collate, and analyze crimes of violence—specifically, murder. The following types of cases are a priority of the VICAP:

- Solved or unsolved homicides or attempts, especially those that involve an abduction; are apparently random, motiveless, or sexually oriented; or are known or suspected to be part of a series
- Missing persons, where the circumstances indicate a strong possibility of foul play and the victim is still missing

- Unidentified dead bodies, where the manner of death is known or suspected to be homicide
- Sexual assault cases

The FBI provides software for the VICAP database to local police agencies, free of charge. This program has been embraced by both large and small police agencies nationwide.

Other Repeat-Offender Initiatives

Other programs shepherded and funded by the federal government to aid local police in targeting repeat offenders include community relations initiatives such as the Weed and Seed initiative discussed in Chapter 10. COPLINK can also play a role in identifying repeat offenders. Likewise, **hot spot deployment**, use of checkpoints, and other initiatives driven by the "broken windows" perspective can be considered proactive police strategies. Other methods of proactive initiatives include undercover detective work, confidential informants, and sting operations.

Undercover Operations

Four broad categories of police work are distinguished:[11]

- Overt and nondeceptive (conventional police work)
- Overt and deceptive (tricking a suspect into admission of guilt)
- Covert and nondeceptive (passive surveillance)
- Covert and deceptive (undercover)

> Theoretically, being "undercover" means that an officer "sheds the shield and the uniform" and takes on a different persona other than that of a police officer, yet remains under supervision and regulation of police authority.

Traditional Use of Undercover Operations

Undercover operations have traditionally been employed to detect "victimless" or consensual crimes, such as violations linked to narcotics or sex.[12] Because of the consensual nature of such offenses, "victims" may actually complain to the police about them or, for that matter, provide evidence related to the crime. To prosecute those offenders, undercover officers assumed various roles in victimless crimes. According to one source:[13]

- Approximately 350,000 people are in jail for victimless crimes.
- Some 1.5 million individuals are on parole or probation for victimless crimes.
- Annually, 4 million suspects are arrested for victimless crimes in the United States.
- Nearly $50 billion is spent to punish individuals of victimless crimes.

Can crimes ever be "victimless" from the perspective of the police? Prostitution, for example, aids in the transmission of sexual diseases, financially supports criminal organizations, and does not contribute to the public good or to the tax base of a community, argues Melissa Farley of Prostitution Research and Education.[14] Farley adds that prostitution entails sexual harassment, rape, battering, verbal abuse, domestic violence, a racist practice, a violation of human rights, childhood sexual abuse, a consequence of male domination of women, and a means of maintaining male domination of women. It is difficult to believe prostitution has no victims, Farley argues, given that 78 percent of women who sought help from the Council for Prostitution Alternatives reported being raped an average of 16 times a year by their pimps and 33 times a year by "johns" (customers). Consistent with Farley's thoughts, the Waco (Texas) Police Department (WPD) reports that many prostitutes are carriers of sexually transmitted diseases (STDs) or human immunodeficiency virus (HIV).[15] Given these data, prevention strategies that attempt to control prostitution fit quite well with the police's public safety mission.

Current Use of Undercover Operations

> **Undercover operations** involve officers who pose as criminals and victims in an effort to discover those individuals who hold criminal intent and are willing to violate the law.[16]

Today, undercover operations play a major role in sting initiatives. For example, detectives of the St. Louis (Missouri) County Police, whose jurisdiction includes the city of St. Louis, go undercover as narcotic dealers and drug buyers to infiltrate drug cartels that manufacture and distribute illegal narcotics.[17] Other St. Louis County undercover detectives investigate complaints of prostitution activity, sexual misconduct crimes, and illegal liquor sales.

Undercover work has changed in recent years, expanding in scale and appearing in new covert practices directed at new targets and new crimes such as a computer crimes and hate groups. Few organizations are immune from undercover investigation, including members of organized crime and public officials (as evidenced by Operation Tarnished Badge, discussed later in this chapter).

Whereas uniformed police officers rely or depend on the authority of the uniform to carry out their various assignments (approximately 50 different types), undercover officers aggressively and proactively seek out criminal offenders.[18] Specifically, undercover officers create an opportunity for criminality, and individuals who avail themselves of that opportunity are arrested.

While undercover work might be highly productive, it is arguably the "most problematic area of policing" for the following reasons:[19]

- It involves deception by officers, including lying about identity and intentions.
- Officers befriend criminals.
- Friends and family relationships are weakened, opening the door to inappropriate values and rewards.
- Officers tend to be supervised less often and remain in the field for prolonged periods.
- Undercover work requires secrecy and encourages deviant behavior.

- Most undercover officers are young and inexperienced (so they are difficult to detect—more on this point later in the chapter).
- Most training is conducted by veteran officers in the field rather than in a classroom.

As anticipated, not wearing a uniform can remove expected obstacles to acquiring information. It is part of the same intelligence-gathering function as surveillance, eavesdropping, use of informants, and espionage. It allows an officer to circulate in areas where the police are not ordinarily welcomed. One task of an undercover officer is to "build a case" by gathering (i.e., purchasing drugs, guns, or contraband cigarettes) evidence that can be used as part of a successful prosecution. The undercover officer might seek an arrest warrant and, once served, the officer's identity might be revealed (unless the investigation is ongoing). Often undercover officers convert one or more of their contacts into informants. A typical six-month operation might yield as many as 100 arrest warrants; at other times, depending on the case, a single arrest might be applauded.

Typical Undercover Roles

There are many ways to initiate the undercover role. In less-risky operations, undercover officers may create their own cover stories depending on the type of crime involved (drugs, guns, contraband, gambling, "subversive" groups), but will eventually need more support from the department. There is often a need to create observable altercations between undercover officers and the police to prove criminality or loyalty to the bad guys, and undercover officers pretend to have other identities than their own.

For instance, a typical pattern in a drug operation is for an informant to bring an undercover officer into a group of drug traders in as an acquaintance, business associate, or girlfriend or boyfriend of the confidential informant (CI). As time passes and alliances grow, the CI is removed from the investigation. Another undercover officer might then be brought in as the boyfriend or girlfriend of the first undercover officer, who now claims that he or she is no longer involved with the CI.

One key to a successful undercover operation is to manage CIs so that they are cut out of investigations, thereby reducing any risk to the CI and avoiding any involuntary slips that might identify the undercover officer. The ideal targets of undercover drug or gang initiatives are the "big" dealers or leaders, but most often undercover officers start by going after the "small fry," accumulating suspects, evidence, and more CIs.

Before an undercover investigation begins, commanders and sometimes state attorneys make decisions as to length of time and goals of the operation. Depending on the undercover officer's cover story, false documents and computer records may be developed.

Consequences of Undercover Work

Undercover work puts officers at the greatest risk of corrupting their own integrity. Uncommitted undercover officers can easily lose perspective. Some officers may become emotionally attached to their job and sometimes to the suspects, such that they compromise their own values and try drugs. Danger and temptation can play a role in the late stages of the operation. Undercover officers can become self-conscious after a while, feeling as though they have "officer" written across their foreheads.[20] They may feel insecure and anxious about regular work and continued employment with the department once the mission is closing down. Some become paranoid about danger and arm themselves with secret weapons. Sometimes other forms of self-survival may emerge, such as surveillance of their police supervisors or handlers. Undercover officers who have "gone too native" can become dependent on drugs, alcohol, and sex.

For all these reasons, the process by which undercover officers are selected and supervised is vital to their safety. Most often, undercover officers are processed back into the police ranks through a "debriefing" period.

Selection of Undercover Officers

The best undercover officers are new recruits to policing, assuming their physical appearance fits the assignment. They may have been interviewed and identified for such duty while attending the training academy or while waiting on some civil service eligibility list, but often they finish up a rookie year or two first as a regular officer before becoming part of undercover operations. In some cases, a candidate may be sworn in secretly by the chief the first day on the job. The department may have some special need in an ongoing investigation for someone who fits the "mold" and a new, outside person knows none of the people of interest.

For instance, a department of corrections probation officer trainee (who was also a university student of the author) had the physical appearance (despite her age of 23) and similar expressions of a typical high school student. She was pressed into service as a police undercover officer and enrolled in a high school (some distance from her hometown) where illegal substances were readily available and guarded by a corrupt police officer.[21] The trainee was sworn in by an official with the state troopers, much to her father's delight; he was a retired trooper. After two school semesters, the drug distributors (including a teacher) and the corrupt police officer were identified, apprehended, and charged by the local prosecutor. This undercover officer "marched" with her classmates at graduation and "disappeared" to a far-off college campus. The officer sought permanent employment with the troopers and years later became an investigator.

Fresh recruits from other jurisdictions are often preferred for undercover operations, as are ethnic-looking recruits with foreign language skills.[22] New, inexperienced officers have not taken on the nuances or mindset of a seasoned officer and are considered more difficult to detect.

The use of new faces does present one problem, which is related to the nature of the criminals targeted by undercover operations. Most chronic criminals—and particularly those who trade in narcotics—anticipate attempts to apprehend them through a CI or to "flip them" them into a CI. Many have hidden funds and contacts for bail and legal representation. Many will have established an entrapment defense and made moves toward an immunity deal before it is actually needed. Most successful drug traders rarely make mistakes, and one mistake they make less often than lower-level traders is that they are rarely addicts. Thus to build a successful case that includes undercover testimony with tape recordings or video recordings against a sophisticated criminal would require an undercover officer with superior skills. This contradiction makes the selection process a difficult one. The undercover officer needs to be an unknown who lacks the mindset and body language of an officer, yet possesses the skills and mental motivation of a supercop.

That said, most often it is not the successful criminal who is apprehended, but rather the unsuccessful criminal. It is easy to build a case against these individuals, including documentation of their illegal behavior and winning a conviction. Unsuccessful criminals make mistakes, especially while intoxicated, high, or fatigued. Even at this level, however, the criminal may make some attempt to "verify" an undercover officer's story, often through unconventional methods.

Detection of Undercover Officers

Some successful criminals know a lot about policing and may even be officers themselves. Their attention to detail is probably one reason why they are successful. Nonetheless, how an undercover officer posing as an addict to buy illegal drugs dresses, what kind of vehicle the officer drives, and where and how the officer lives are easy indicators of detection. For instance, the clothing of a drug addict usually doesn't fit right because most addicts are constantly losing or gaining weight. Most undercover officers can't simulate this particular "fit" of clothing: They'll only look sloppy and carry themselves like they have their "street uniform" on. Scraggly beards that look recently grown also are giveaways. In addition, the cars they drive may be too well maintained. A dope addict's car usually has three different brands and styles of tires, a bunch a fast-food and candy wrappers all over the inside, and screaming kids in the back.

Body language is hard to alter, suggesting that it may be easy to detect officers who attempt to portray someone they're not. For instance, if an undercover officer is thinking about putting away the bad guys, his or her eyes may reveal the story of anticipated success. Many officers are too full of positive thoughts, too active, and too curious, so they wear sunglasses. Dope addicts, by contrast, will often stubbornly or masochistically blind themselves by not wearing sunglasses even when they should, and their eyes will look sunken, like they haven't slept in days.

Other giveaways that a person may be an undercover officer include these behaviors:

- Being too sure about the price
- Constantly making phone calls during a deal
- Being overly eager to buy
- Offering sex in exchange for doing business
- Being too familiar or being too unfamiliar with things, places, and people

Other red flags include money and skin flashes, weapons and weapon handling, and too many questions.

Like body language, body odor may also suggest that a person is an undercover officer. The body produces a unique odor that reflects the person's diet, lifestyle, and health. Most junkies live a difficult lifestyle, and often that lifestyle is less centered in eating the right foods and paying particular attention to physiological concerns such as rashes, abrasions (as result of many accidents), and other proper hygiene practices.

Prevention of Undercover Officer Corruption

The Commission on Accreditation for Law Enforcement Agencies (CALEA) requires "written procedures for conducting vice, drug, and organized crime surveillance, undercover, decoy, and rapid operations." These procedures cover the processing of officers supplied with false identification, disguises, and other necessary credentials as needed on the job, and require that the department follow specific supervisor guidelines.

In addition, jurisdictions may provide their own regulations related to undercover operations. For instance, the city of Seattle has an ordinance limiting police investigations into political, religious, and private sexual activities.[23]

Advocates of undercover officers recommend that more power should be provided to undercover officers, while proponents suggest that more power is what police undercover advocates have asked for since recorded history.[24] Another method used to control undercover practices is a checks and balances system of review in a specific time period—for instance, every three months.

Confidential Informants

Confidential informants are vital to most investigations, especially when investigators are building a case against gangs, drug dealers, and terrorists.

A **confidential informant (CI)** is an individual who is caught up in criminal acts of his or her own doing and, once apprehended by police, will exchange information about other criminals to obtain judicial leniency arising from the consequences of his or her own crimes.

The leniency accorded to a CI can include remaining free (not being arrested), facing a lesser criminal charge (charged by a prosecutor), and receiving less or no prison time (judicial sanctions). Sometimes money is the motive of the exchange; at other times a CI is motivated by revenge, egotism, self-preservation, and the desire to play an important role in something.[25] Being a CI can be a scheme to gain recognition or attention, but might also involve someone concerned with justice (but not theirs).

Basis for Liability

Assistance in aiding police to provide public safety cuts two ways.[26] Officers engage in an exchange relationship: information for leniency. They may also compromise their own integrity by opening a window of opportunity to officer corruption and personal frustration, leading to their alienation from the justice system. Then, too, how often might informants manufacture evidence solely to obtain the protection of the police in their quest to commit crime or just to mess with officers? Jerome Skolnick shows that in "Westville," narcotic officers allowed informants to rob other dealers, while burglary detectives allowed their informants to engage in drug trafficking.[27]

Situations in which CI have committed acts producing injury to third persons have not consistently resulted in liability for the contact officer. There is always the possibility that the CI could place others, including police officers and their agencies, at risk, whereby the officers could be held liable for injury if the informant is shown to be acting within the scope of employment, especially if the CI's employment position with the department is not clearly defined in writing. There might also be the claim of a constitutional rights violation by a CI who is acting under the direction of a contact officer, thereby causing a violation of a protected civil right under Title 42, U.S. Code, Section 1983, or even a criminal violation under the federal Criminal Code, Title 18, U.S. Code, Section 241 (conspiracy).

A CI is a usually an individual who favors criminal activities over noncriminal acts and who is more likely to commit a crime of violence against a third party or an innocent person. The legal issues created by a CI's relationship with the police might come into play in this scenario.

There are also moral issues linked to violent acts committed against others when the offender is a CI. "In a sense, I'm allowing it [harm] to happen to an innocent person. I tell him, (the CI), if you're working for me and the department, you can't harm other people. This one CI told me, 'Hell, man, if I don't jack them up, the dudes I run with will think I'm working for the man [officers],'" says a Boston detective. For that reason, CI needs to know the parameters of the exchange relationship.

Informant Policy Details

To counter the risks associated with use of CIs, policy manuals often contain provisions for informant development and informant management. Police agencies that are accredited by the Commission on Accreditation for Law Enforcement Agencies (CALEA), for example, must adhere to a set of requirements related to the security of CI files, criteria for paying CIs, and precautions to be taken when using such informants. Many agencies add their own rules to CALEA's CI requirements. For instance, the Chicago Police Department mandates that CIs must be interviewed by a detective's supervisor, despite detective complaints that such an interview would compromise the CI relationship.

An effective informant policy provides a comprehensive plan for informant coverage, from initial recruitment to termination of the relationship. For example, a typical departmental policy might prohibit the following conduct among officers with CI:

- Socializing or partying with informants or their family members
- Becoming romantically involved with informants
- Buying anything from an informant or selling anything to an informant
- Borrowing money or receiving gifts from informants
- Contacting informants alone
- Paying contacts without another officer as a witness
- Allowing informants to sign anything, such as a receipt for payment, without the entire receipt being completed
- Living with informants or their family members
- Trading healthcare providers, social favors, or lifestyle amenities with informants or their family members

Because officers cannot control CI behavior, one way to avoid or reduce liability for the officer and the department is by creating a written description of the affiliation of the CI with the department and the officers, specifying the nature of the department's relationship to the informant. An informant should be required to sign a "contract" with the department setting out provisions such as expectations on the part of the department and the informant, especially payments and activities.

Payments to Informants

Payments to informants are particularly dangerous for handling officers and departments. They should always be controlled, with receipts being obtained for all monies delivered and expenses paid. Having a second officer or supervisor witness the payment is one recommended method of ensuring verification.

Informants frequently are hungry for money and attempting to "work off" their criminal acts. Officers should take care not to provide too much immunity for criminal acts because an officer has no legal authority to provide it; therefore, the matter must be referred to the appropriate prosecuting agency for an immunity determination.

CIs may be involved in local, state, and federal investigations. For instance, the FBI targets organizations involved in traditional and nontraditional organized crime activities and corrupt public officials who engage in crimes such as murder, arson, loan sharking, extortion, bribery, illegal gambling, and illegal possession of stolen interstate property (see the FBI website at www.fbi.gov). The special agents assigned to these matters use a wide range of investigative techniques, but emphasize the use of CIs and the gathering of evidence through court-authorized electronic surveillance and undercover operations. As a result, CIs play a significant role in high-profile arrests.

Case Study: A CI Speaks from Prison

One CI was willing to talk about her participation with undercover Boston police officers and the prosecutor's investigators from her prison cell at the Massachusetts Correctional Institute in Framingham.[28] Her pseudonym here is Lillian—a glorious flower gone wild. Lillian explains how the informant process worked:

> I had 17 ladies who turned tricks [prostitution] in downtown Boston at the big hotels, especially when the conventions were in town. One of the ladies was my 16-year-old daughter, who I trained in the "trade." It [prostitution and operating a brothel] paid our bills and the bills of the gals who worked for me, many whom I had known since college. It gave my kid and me a lifestyle better than my college degree could provide. We had a condo overlooking Boston Common [a park in the theater district of Boston] and a home at [Cape Cod], complete with a boat dock and a 40-something-foot boat.
>
> I was busted for running a brothel but was told that if I cooperated with the prosecutor's office that I could continue to operate my business, keep my homes, and wouldn't be charged with reckless endangerment of my daughter. I agreed. I gave them what they wanted for almost two years until they pulled the plug—and here I am. I think one of the prosecutor's friends wanted a piece of my business [this last remark could not be documented].

The information that Lillian provided included personal expenditures and gifts (given to Lillian and her employees), and the time, place, and duration of client services. From time to time, Lillian was also asked to obtain specific information from her clients, which she did not elaborate on during the interview.

CIs often provide false information. Thus, as with any information-gathering and covert operation, there is always a need to validate information.[29] Sting operations often consist of undercover officers posing as criminals and informants posing as themselves—criminals.

■ Sting Strategies

Criminals are often apprehended through police deceptions or sting tactics.[30]

> **Stings** are police strategies designed to hasten compliance to laws, in which police use their knowledge of crime to construct circumstances that invite criminals to commit more crime in a recordable way, thereby enhancing the likelihood of identification of criminals who otherwise might be difficult to detect and prosecute.[31]

Local, state, and federal enforcement agencies conduct stings, sometimes independently and sometimes in cooperation with one or more agencies at various governmental levels. Stings may include personnel at all levels of an agency, including patrol officers. Although agency strategies vary, their objectives are typically similar.[32] Stings can be viewed as a proactive strategy because they are designed to identify and apprehend unknown criminal violators.

Local Police Stings

Local stings conducted by police and sheriff agencies include, but are not limited to, fencing (purchase) of stolen goods, alcohol and cigarette stings targeted at youth, covert cameras designed to detect traffic speeders, stings targeted to detect illegal firearm ammunition sales to youths, drug stings (dealers and users), and prostitution (including clients) stings.

Theft stings are typified by anti-fence operations, whereby police pretend to fence stolen property and transact business with criminals. For example, in Birmingham, Alabama, a storefront operation was established to attract and detect auto thieves.[33] Unfortunately, auto thefts increased by 4.46 cars per day during its operation, but then decreased to previous levels after the sting ended.

The Texas Commissioner of Alcoholic Beverages has promoted the use of youth stings.[34] According to Texas officials, such stings proved to be an excellent means of combating problems related to the unlawful purchase and consumption of alcoholic beverages among minors. Youth stings have gained both credibility and community support among many police departments. For example, another alcohol sting employed underage police cadets to purchase packaged beer at retail stores in Denver.[35] Cadets successfully purchased beer 59 percent of the time, and subsequent stings resulted in beer purchases between 32 percent and 26 percent of the time.

Stings also include the use of covert cameras (i.e., cameras that are hidden from view) to identify speeders.[36] For instance, nine automated cameras designed to detect speeders traveling at 10 miles over the posted speed limit produced $12 million in fees annually for Cincinnati.[37]

The Baltimore Police Department's gun sting was launched as part of the Youth Ammunition Initiative in 2002.[38] The initiative addressed gun violence by targeting illegal firearm ammunition sales, and both businesses and public agencies supplied undercover investigators. By interrupting the flow of ammunition, youth violence was reduced.

In Waco, Texas, and Nashville, Tennessee, prostitution stings include female officers posing as prostitutes. Because of the lack of criminal complaints, prostitution is often seen as a victimless crime, so police must initiate investigations of it on their own.[39] A typical prostitution sting in Waco, for instance, may produce 25 arrests among prostitution customers ("johns") per day. After a customer is arrested, his picture graces the Waco Police Department's website. The Metropolitan Police (Metro PD) in Nashville posts the names of johns on its website as well, and its typical sting results in 40 to 45 arrests.[40]

Often several agencies combine resources and information to conduct a sting, and they often use informants as part of this process. For instance, multiple-county drug task forces and state agency participation are characteristic of many efforts, especially in rural areas such as Jonesville, Virginia.[41] One undercover drug sting operation conducted in Pennington Gap resulted in the arrest of six individuals. According to Lee County Sheriff Gary Parsons, to carry out the sting, his office worked with the Lee County Commonwealth's Attorney's Office, the Southwest Virginia Drug Task Force, and the Bureau of Alcohol, Tobacco, and Firearms. Officers worked from the home of a CI to sell a placebo (which was purported to customers to be OxyContin) to the individuals arrested. While some of those charged were buying the "drug" for personal use, others had the intent to distribute their purchases.

A Whiteside County, Illinois, narcotics sting is another example of how agencies work together; in this sting, County Sheriff's Anti-Crime/Narcotics (SAC/Narc) unit included deputies from both the Whiteside County Sheriff's Office and the Illinois State Police.[42] One sting successfully recovered 10 pounds of marijuana, 18 doses of heroin, and 10 Endocet pills; investigators also seized four vehicles. SAC/Narc identified approximately 750 drug county outlets, and members of the unit see stings as being the most productive strategy for detection of substance abuse crimes.

Federal Stings

Federal stings are conducted at the local, national, and international levels by the Drug Enforcement Administration (DEA), the Federal Bureau of Investigation (FBI), the Department of Homeland Security (HS), The Bureau of Alcohol, Tobacco, Firearms, and Explosives (ATF), and the U.S. Immigration and Customs Enforcement (ICE). Highly publicized stings conducted by federal agencies often focus on terrorism and arms smuggling, child pornography and pedophile activities, bank laundering of illegal funds, and cigarette and drug trafficking.

Terrorism and arms smuggling stings are typified by an operation in New Jersey in which a suspect was charged with providing materials to support terrorism through the smuggling of firearms (IGLA-2 surface-to-air missiles) into the United States.[43] In this case, the buyers were FBI agents pretending to be a terrorist cell, the sellers were undercover Russian enforcement agents, and the apprehended broker was a British citizen.[44]

National stings (e.g., Operations Artus, Avalanche, Candyman, Blue Orchid, Hamlet, and Rip Cord) and international stings (e.g., Operations Ore, Snowball, and Twins) have also detected and apprehended both pornographers and pedophiles.[45] For example, Operation Candyman identified an estimated 7,000 subscribers linked to a pornographer's website. Those arrested included daycare workers, clergy members, law enforcement personnel, and military personnel. Operation Rip Cord identified more than 1,500 child pornographer suspects who also solicited child sex through websites in several countries.

Federal agencies and local agencies often work together in an attempt to control drugs through sting operations. For example, The Tri-Net Narcotics Task Force, in cooperation with the DEA's Mobile Enforcement Team (DEA-MET), conducted a three-month investigation and drug sweep of the Carson City, Nevada, area.[46] The Tri-Net Narcotics Task Force consisted of members of members of the DEA-MET, Douglas County Sheriff's Department, Lyon County Sheriff's Department, Carson City Sheriff's Department, Storey County Sheriff's Department, and the Nevada Division of Investigation. The investigation began in the Carson City area with the MET deployment of special agents with Tri-Net officers. MET special agents are tactically trained forces that assist local law enforcement in attacking the violent drug organizations in their neighborhoods. In the end, 35 drug runners were arrested and $220,000 was confiscated. The Carson City operation was funded by the DEA at no cost to Nevada.

What Is a Legal Sting?

Police are allowed to engage in deceit, pretence, and trickery to determine whether an individual has a predisposition to commit a crime, and they can provide an opportunity for an individual to commit that crime, all without being guilty of entrapment.[47] Guidelines for determining entrapment were established in *Sorrells v. United States* (1932), which set the precedent for taking a "subjective" view of entrapment. Police cannot provoke an act by an innocent person.

> Defendants have the right to challenge evidence, and they can invoke the defense that they would not have broken the law if not tricked into doing it (i.e., entrapment); if the court finds that entrapment occurred, the charges may be dismissed.[48]

For example, a Nebraska farmer was indicted for receiving child pornography through the mail after agents spent 26 months offering him pornography. In *Jacobson v. United States* (1992), prosecutors failed to show that a defendant had been "predisposed, independent of the government's acts and beyond a reasonable doubt, to violate the law," Justice Byron White wrote. That is, a suspect's tendency to commit a specific crime must have existed prior to enforcement attention.

The controversy over whether officers were enforcing the law correctly led to stings within the Los Angeles Police Department (LAPD). Ultimately, the LAPD developed and implemented a plan for organizing and executing regular, targeted, and random integrity audit checks, or sting operations, to investigate whether officers were engaging in at-risk behavior, including unlawful stops, searches, seizures (including false arrests), use of excessive force, or violations of the department's codes.[49]

Stings have remained a viable means to enhance police power despite the *Jacobson* ruling. For instance, in *United States v. Jimenez Recio* (2003), the court ruled that the contraband seizure of drugs does not end the probe into conspirators, and that a sting can be launched to trip up criminals.

When Stings Go Wrong

Not everyone is convinced that stings are worth the time and trouble—and expense. In San Diego County, for example, 138 tobacco retailers paid an annual licensing fee to fund police stings and the salary of a code enforcement officer who issued and renewed licenses. Those caught selling tobacco to minors faced penalties ranging from $100 to $500 and the prospect of losing their license. A licensee argued that because only one of 18 merchants sold illegal products to minors, funds to enforce codes should be refunded and tobacco licenses should cost less.

Stings can have a negative effect on public opinion when no law violators are detected. Also, the wrong person is sometimes arrested during a sting. For instance, police erroneously arrested Christi Hernandez in McKinney, Texas, after an undercover agent bought cocaine from a woman with a similar name.[50] Hernandez lost her job over the false accusations. "We apologize to her, and I apologize also," McKinney Police Chief Doug Kowalski said. The apology came with a cash settlement.

A reasonable expectation of privacy may exist even in a sting sex case (i.e., whether a particular place is isolated enough that a reasonable person would see no significant risk of being discovered). The Massachusetts State Police settled a charge on behalf of a gay man who alleged that troopers harassed him at a highway rest stop. The settlement included the implementation of state police guidelines preventing officers from deliberately trudging into wooded areas near rest stops.

Operation Lightning Strike, an FBI sting aimed at trapping NASA personnel and its contractors, included an agent who pretended to have developed a (phony) device. Millions of dollars were spent on luxury hotel suites, gourmet meals, deep-sea fishing trips, and strip clubs in an effort to identify corruption and kickbacks. Agents had few suspects, but the Lightning Strike scenario was similar to dozens of alleged failed stings uncovered by the *Pittsburgh Post-Gazette*.[51]

The Debate Over Sting Tactics

Stings are often justified because of the detrimental effects that certain violations have upon quality-of-life issues. This perception is consistent with quality-of-life police strategies of taking back public places and making arrests even on small violations.[52]

Because stings employ deception to detect and record criminal activities, it raises questions about officers' behavior: When officers testify against offenders in the courtroom "where the stakes—their jobs, the case"—are much higher, will officers attempt to deceive the jury?[53] Other critics argue that stings carry a different burden: The price for more freedom is likely to be less order, and the price for more order is likely to be less freedom.[54]

Despite these doubts, two points about stings are clear:

- Stings and undercover initiatives tend to be covert tactics, and their successful working components tend to be less publicized than their failures.
- Crime has been on the decline for some time, which may in part be due to stings and undercover initiatives commonly practiced by many police agencies.

■ Policing Gangs

One estimate is that 732,000 gang members reside in almost 3,000 police jurisdictions.[55] In fact, 42 percent of those jurisdictions report that gang problems are getting worse in both frequency and intensity. Urban areas used to hold a monopoly on gang violence, but contemporary gang violence has recently spread to suburbia and rural America along with the increase in methamphetamine ("meth") use.[56]

The National Young Gang Center reports that younger gang members are more prevalent in smaller-population jurisdictions, female gang members are more prevalent in the new gang-problem areas, and African American and Hispanic/Latino youth make up a disproportionate share of gangs relative to their numbers in the general population.[57] Gang involvement has been shown to have more long-term effects on females and more severe effects on their children, and crime victimization is greater among female gang members than among their male counterparts.[58]

> A **gang** is "any denotable group of youngsters who (a) are generally perceived as a distinct aggregation by others in their neighborhood; (b) recognize themselves as a denotable group (almost invariably with a group name); and (c) have been involved in a sufficient number of delinquent incidents to call forth a consistent negative response from neighborhood residents and/or law enforcement agencies."[59]

Proactive repeat-offender programs, undercover strategies, CI recruitment, and sting initiatives are all tactics that are used in an attempt to curb both gang violence and drug trafficking and possession.

Gang Violence

Gang members are responsible for a large proportion of all violent offenses committed in recent years in the United States. In a study that examined several gangs across the country, Rochester, New York, gang members (30 percent of the sample) self-reported committing 68 percent of all adolescent violent offenses committed in that city; in Seattle, gang members (15 percent of the sample) self-reported committing 85 percent of adolescent robberies; and in Denver, gang members (14 percent of the sample) self-reported committing 79 percent of all serious violent adolescent offenses.[60] Police have worked hard to control gangs but have not achieved the positive results they anticipated.[61]

Youth Gang Structure

Gang descriptions are often extended to include terrorist gangs, prison gangs, and criminal gangs as in organized crime. In each of these instances, the use of the word "gang" implies a level of structure and organization for criminal conspiracy that is beyond the capacity of most street gangs.[62] "Most street gangs are only loosely structured, with transient leadership and membership, easily transcended codes of loyalty, and informal rather than formal roles for the members."[63] These thoughts are consistent with current studies on organizational structure and leadership among youth gangs. Gangs appear to represent "an adaptive or organic form of organization, featuring diffuse leadership and continuity despite the absence of hierarchy."[64] Most gangs experience considerable organizational change over time—consolidating, emerging, acquiring smaller gangs, reorganizing, and splintering. A typical gang's "generalist" orientation contributes to its ability to adapt to these changes and survive in a volatile environment.

For instance, Los Angeles County encompasses 88 incorporated cities and dozens of other unincorporated places. Crips gangs are firmly established in 24 of those cities. In 1972, there were 8 Crips-affiliate gangs. That number grew to 45 in four years, and by the end of 1999 199 individual Crips chapters were active in Los Angeles County. Growth of the Crips organization in Los Angeles has since stabilized and even declined in areas that are undergoing demographic change, but copycat gangs are on the rise in other parts of the United States.

Gangs and Drugs

Most primary drug distribution systems are managed by drug cartels or syndicates, traditional narcotic operatives, and other organized gangs such as the 18th Streeters, MS-13, and Asian gangs described later. While gang participation, drug trafficking, and violence tend to occur together, they are not a single, comprehensive problem.[65] Where drug-related violence occurs, it mainly stems from drug use and dealing by individual gang members and from gang member involvement in adult criminal drug distribution networks, rather than from drug-trafficking activities of the youth gang as an organized entity.

Some experts warn that "under favorable conditions," gangs can undergo a "natural evolution" from a loosely organized group into a mature form.[66] As gangs become more prevalent, they become more highly organized, taking on the features of more formal organizations.[67] For instance, the most dangerous gangs in the United States, which include the 18th Street gang, MS-13, and Asian gangs, all trace their roots back to youths.

18th Street Gang: Central Los Angeles

The description of loosely structured leadership might apply to the huge city gang such as the 18th Street gang in South-Central Los Angeles (YouTube.com provides videos about 18th Street and other gangs). Yet all of the gang's members appear to have a singular mission: The approximately 25,000 Hispanics and Chicanos in this street gang control portions of the city through violence, extortion, and racketeering, and they have little compassion for outsiders. Once incarcerated, they "jump in" (are initiated) with prison gangs, which actively recruit new members, and defend their prison "turf" with as much violence as they defend "their" streets.

As youths who followed their families to Los Angeles from Mexico, members of this gang were individually targeted by local gangs such as the Crips and were victimized by the police. Protection was the motivator for forming alliances, which eventually led to the 18th Street organization. Factions of the 18th Street gang are spread throughout the San Fernando Valley, San Gabriel Valley, South Bay, and South-Central Westside and East-side Los Angeles, as well as throughout Mexico and Central America.

MS-13

The MS-13 gang has a uniquely international profile, with an estimated 8,000 to 10,000 members in 33 states in the United States, and tens of thousands more members in Central America.[68] It has had a tremendous effect on law enforcement. In the FBI's Washington, D.C., field office, the number of agents dedicated to gang investigations declined by 50 percent after resources were switching to fighting terrorism in the wake of the September 11, 2001, attacks. In early 2007, however, things changed—largely as a result of the MS-13 gang. An MS-13 gang leader was accused of orchestrating a December bus bombing in Honduras that killed 28 people. He was arrested in Texas, and a seven-city sweep by ICE agents subsequently netted more than 100 reputed MS-13 members.[69] Despite the success of this raid, according to Robert Clifford, head of the FBI task force, "No single law enforcement action is really going to deal the type of blow necessary to dismantle the gang."

MS-13 was started in Los Angeles in the 1980s by Salvadorans fleeing the civil war in their native country.[70] Many of the youths grew up surrounded by violence, so MS-13 can be said to have started as a young gang much like the 18th Streeters. Del Hendrixson, in a case study for a gang-outreach program, remembers an MS-13 member who recounted one of his earliest memories: guarding the family's crops at the age of 4, armed with a machete, alone at night.[71] When he and others reached the Los Angeles ghettos, Mexican gangs preyed on them. The newcomers' response was to band together in a *mara,* or "posse," composed of *salvatruchas,* or "street-tough Salvadorans." Over time, the gang's ranks grew, adding former paramilitary operatives with weapons training and a taste for atrocities. MS-13 eventually entered into a variety of rackets, from extortion to drug trafficking. When police finally cracked down and deported planeloads of members, the deportees quickly created MS-13 outposts in El Salvador and neighboring countries such as Honduras and Guatemala.

Asian Gangs

Similar to Italian, Irish, Mexican, Central American, and Jewish newcomers to America, many Asian newcomers struggled to adapt to the strange ways of their new country, and youth gangs emerged in the name of protection of the immigrants.[72] Many of the immigrants faced a cultural struggle after their arrival in the United States. Feelings of frustration and anger that arose during their attempts to adjust to a new country and a new language were intensified by individuals' victimization by existing gangs. Many youth felt alienated by their parents, family, schools, and ethnic communities, too. Their response to these pressures was to form their own youth gangs, which included Chinese (tongs and triads), Vietnamese, Amerasian Vietnamese, Cambodian, and Laotian immigrants.

Some evidence contradicts the hypothesis that cultural struggles for survival were the primary reason these youths joined gangs. For example, in Westminster, California, where 25 percent of the 85,000 residents are Vietnamese refugees who live in an area known as Little Saigon, 48 percent of all Asian delinquency is related to Asian gang involvement.[73] Researchers have learned that two factors predict Vietnamese young gang involvement: attitudes and exposure to gangs in the neighborhood—and not the cultural impacts suggested previously.

Although protection is certainly one motivator for gang formation, grabbing a piece of the American dream is another. Gang membership is a fast track to obtaining amenities, as opposed to the slower, more traditional routes advocated by parents and old-world religions. Some (although not all) Asian youth gangs have also developed connections to, or even may be controlled by, larger and more formal Asian organized crime groups, which themselves manage elaborate prostitution, gambling, smuggling (including illegal aliens), extortion, and other gang activities around the world (i.e., Chinese tongs and triads).[74]

One problem that law enforcement encounters with Asian youth gangs (and MS-13 and other gangs) is that they target their own home neighborhoods for home invasions, rape, and other forms of victimization.[75] Immigrants and newcomers are more likely to keep their valuables hidden somewhere inside their house or business instead of in a bank. Asians, like other newcomers, are less likely to report such crimes to the police owing to both personal embarrassment and belief that the ensuring publicity would bring shame to their community, which goes against the families' social conditioning.[76] Victims and their family members also worry that they may be deported if they report crimes committed against them (which is one reason why sexual exploitation of young children and women is so pervasive among newcomers).

Additionally, like most newcomers, younger Asian immigrants tend to learn the English language quickly. As a result, they are often pressed into service interpreting contracts on homes, autos, and property for their parents, which puts them in key positions of power within the family structure and gives them control over family resources, despite their young age. As a consequence, parents may hold little control over young gang members. Also there is a concern among family members that they may be deported if they report their victimization (one reason sexual exposition among young children and women is so pervasive among newcomers).

Traditional Police Suppression of Gangs

To date, efforts to control gang violence have been largely unsuccessful. Recall from Chapter 4 the discussion of Chicago's antigang loitering order maintenance policy.[77] When the Chicago police arrested gang members who would not disperse, the American Civil Liberties Union (ACLU), acting on behalf of those gang members, challenged and won a civil suit against the city.[78]

Government, schools, social agencies, and the justice system (both local and state) are conspicuous by their absence in the fight against gang violence. At best, many show only sporadic interest in the subject and have failed to develop effective long-term policies and programs. Existing social support mechanisms and strategies for suppressing violence have often been unsuccessful because, for the most part, they have been underfunded, understaffed, and overrated.[79]

The typical police suppression strategy employed to control gangs assumes that most street gangs are criminal associations that must be controlled through efficient gang tracking, identification, and targeted enforcement strategy.[80] This perspective suggests that police employ as much aggression as necessary to remove gang members from the streets, rapidly prosecute them, and send them to prison for long sentences.[81] Typical suppression programs include street sweeps that round up hundreds of suspected gang members, special probation terms for gang members, and strict monitoring of parole such that gang members are subjected to heightened levels of surveillance. They also include more stringent parole revocation rules, prosecution programs that target gang leaders and serious gang offenders, civil procedures that use gang membership as a justification for arrest for conspiracy or unlawful associations, and school-based law enforcement programs that use surveillance and buy–bust operations.[82]

The traditional approach to gang control is loosely based on deterrence theory.[83] That is, police agencies attempt to influence gang members or eliminate gangs entirely by dramatically increasing the certainty, severity, and swiftness of criminal justice sanctions.

Contemporary Police Approaches to Youth Gangs

Strategies useful for social or police agencies begin by partnering with schools, employment programs, criminal justice agencies, religious groups, and grassroots organizations where possible when dealing with gangs; they also incorporate the use of local resources, planning, and collaborative procedures.[84] For example, police agencies may become more culturally competent by training their officers to recognize cultural factors and by actively recruiting bilingual police officers, especially persons from the local community. These ideas have been put into practice by many agencies. Some police agencies have also partnered with outside organizations to enhance their **cultural competency**.

New York City Police Department Gang Initiative

The Vera Institute of Justice works with the New York City Police Department to enhance the efficacy of the police liaison with Arab, African, and emerging Latin American immigrant communities.[85] Police instruction to these groups included topics such as the relationship between the police and the community; the community's crime, safety, and policing needs and concerns; and strategies for improving police–community relations.

Los Angeles Police Gang Initiative

In the summer of 2007, Chief William Bratton of the LAPD held a three-day gang summit with his counterparts from Mexico, Central America, and Canada.[86] According to FBI special agent Steven Tidwell, this forum marked the first time that chiefs of police of various countries had shared intelligence on violent street gangs. This cooperative effort represents a new strategy to track transnational gang bangers.

In the past, what usually happened was that gang bangers were deported from Los Angeles and then started their "postgraduate studies" at home in their country of origin, only to later return to Los Angeles with renewed ideas and greater experience. One target of Bratton, for example, was the members of the MS-13 gang, who are routinely deported to El Salvador. This action by the police is known as **mano dura**—the "hard hand" approach.

During the 2007 summit, the visiting chiefs rode along with patrol officers in gang territories. Chief Bratton launched his Los Angeles top-ten gang leader's list. "A major part of what we're trying to do is to remove the mystique or the veil of secrecy about gangs, and lay them out for what they are," Bratton says. Not everyone is convinced, however: A 17-year-old member of the San Fer gang in Sylmar, California, says Bratton's top ten gang leader list will have the opposite effect upon the gangs. Getting on the list will become a badge of honor, the gang banger says.

Chicago Police Department Gang Initiatives

The mayor and its superintendent of police have emphasized that Chicago's street gangs control the vast majority of narcotics sales in Chicago and are responsible for nearly half of Chicago's murders. According to gang intelligence provided by the Chicago Police Department (CPD), there are 68 active street gangs in Chicago, composed of 600 factions.[87] CPD has documented approximately 68,000 individual street gang members.

CPD has partnered with the DEA to develop new violence-reduction strategies. For instance, during the summer of 2007, more than 3,000 drug customers were arrested as part of Operation Double Play. In this sting operation, CPD arrested drug dealers and replaced them with undercover police officers, who then arrested drug customers, most of whom were from outside the community, and impounded their cars. Gang members who sell drugs are now indicted in both federal and state courts. Also, a Deployment Operations Center now maps crime on a daily basis and quickly deploys officers, including a new 160-member Targeted Response Unit, in an effort to curtail crime.

Controlling Gangs

Gangs and gang problems usually endure despite intensive police hard-hand operations.[88] Police agencies do not generally have the capacity to "eliminate" all gangs, nor do they have the capacity to respond powerfully to all gang-related offenses that occur in their jurisdictions. Pledges to do so, though common, are simply not credible to gang members.

The National Youth Gang Center (NYGC) (2007) argues that an over-reliance on one police strategy or another is unlikely to produce fundamental changes in the scope and severity of a community's gang problem.[89] A balance of prevention, intervention, and suppression strategies and programs is likely to be far more effective. For example, the Gang Resistance Education and Training (GREAT) prevention program (see GREAT's website: www.great-online.org) could serve youth in gang-problem areas, while an intervention team could work with active gang members, and a gang suppression unit could target the most violent gangs and gang members. The comprehensive gang prevention, intervention, and suppression model is a flexible framework that can guide communities in developing and organizing such a continuum of programs and strategies.

Comprehensive Gang Prevention, Intervention, and Suppression Model

In Boston, the comprehensive gang prevention, intervention, and suppression model works in the following way. First, fliers are posted explaining police actions and touting group discussions with gang members to be held in community recreational centers.[90] Gang members on probation, parole, and Department of Youth Services (DYS) supervision are required to attend the meetings. Members of the clergy in the gang territories, for example, persuaded congregation members to attend meetings related to Operation Ceasefire. The clergy voiced their support of police actions, asked youths to stop the violence, and reiterated their offers of services and opportunities. Gang members were encouraged to ask questions. The results of Operation Ceasefire included a 63 percent decline in youth homicides, a 32 percent decrease in shots fired, a 25 percent decrease in gun assaults, and a 44 percent decrease in youth gun assault incidents.[91]

Geographic Information Systems and Gang Control

Orange County, California, uses a geographic information system (GIS) to track and monitor gang activity.[92] The Gang Incident Tracking System (GITS) is intended to help the police make more informed decisions to counter gang activity. Introduced in 1993, GITS was evaluated to determine the validity and reliability of the data collected and whether the goals had been met. Researchers who partnered with the Orange County police and sheriff's association to conduct the study found that GITS made it possible to present a reasonably unbiased and complete picture of gang incidents and to establish a baseline of gang activity. Among the components evaluated were the Orange County Chief's and Sheriff's Association (OCCSA's) definition of gangs, the validity of measures used to count gang incidents, and the process for supervisory review of incident reporting. Of particular concern to community activists was whether the police had overestimated gang crime. The evaluation revealed that, in fact, police tended to underreport crime violence.

GITS demonstrates the usefulness of multijurisdictional efforts to understand and ultimately prevent gang crime. Its output has been used by police to deploy personnel, allocate resources, and evaluate gang intervention activities. In particular, this system has made it possible to use GIS technology to create computerized maps that analyze the spatial and temporal distribution of gang activity.

Gang Restraining Orders

As part of the ongoing battle against gangs, many jurisdictions have devised legal means to curtail their activities—namely, restraining orders against gangs. For instance, the LAPD currently has 29 active injunctions in the city involving 38 gangs. "A gang injunction is a restraining order against a group," advises the LAPD'S website. It is created when a civil suit seeks a court order declaring the gang's public behavior a nuisance and asks for special rules directed toward its activities. Injunctions address the community's gang problem before it reaches the level of felony crime activity consistent with quality-of-life arrests discussed in Chapter 4.

Gang injunctions have a clearly demonstrable positive effect on the neighborhood area covered. In smaller areas, gang nuisance activity has been permanently removed. In larger areas, where gangs have been entrenched for years, the gang's hold on the area can be reduced and monitored by only a small team of police officers. Gang injunctions seem to be one answer in moving police toward gang control.[93] One analysis indicates that, in the first year after the injunctions are imposed, they cause violent crime to fall by 5 to 10 percent.[94]

Tentative Conclusions about Prevention of Gang-Related Crime

The most common response to youth violence has often been punitive enforcement and aggressive crime control: "a short-term, and ultimately, very costly approach which increases social exclusion and may exacerbate it."[95] Finding longer-term solutions has proven more problematic, largely because the roots of gangs go so deep within communities.

An *American Journal of Psychiatry* study shows that children who experience malnutrition exhibit strikingly increased behavioral disorders and aggressive behavior as they grow older.[96] This study, which looked at children between the ages of 9 and 17 years of age, found that children who suffered certain nutritional deficiencies demonstrated a 41 percent increase in aggression at age 8. At age 17, 51 percent of them characterized violent and antisocial behavior. The only difference between more aggressive and less aggressive youths, according to the study, was the diet. When those children become members of existing gangs, the complexion of a gang can change with violence including intimidation and murder. A gang is often conceived and nurtured by an individual who uses it as a vehicle to raise himself or herself to a position of power among his or her peers.[97]

As researchers try to develop innovative studies whose results might alter the direction of violent crime, eventually they learn (as with most criminal justice problems) that there is no single best strategy and no easy solution to youth gang problems.[98]

Policing Drugs

Policing drugs and policing gangs are often one and the same in terms of police strategy. As noted earlier, many gangs are involved in drug distribution.

Drug Use in the General Population

A University of Michigan study reports that drug and alcohol use among youths are on the decline as compared with previous periods.[99] There were an estimated 1.8 million arrests for drug abuse violations (among people of all ages) in the United States in 2005. In the 2004 Survey of Inmates in State and Federal Correctional Facilities, 32 percent of state prisoners and 26 percent of federal prisoners said they had committed their current offense while under the influence of drugs.[100] Among state prisoners, drug offenders (44 percent) and property offenders (39 percent) reported that the highest incidence of drug use was at the time of the offense, and many committed crimes in an effort to obtain a supply of alcohol and drugs. Among federal prisoners, drug offenders (32 percent) and violent offenders (24 percent) were the most likely to report drug use at the time of their crimes.

Based on victims' perceptions, approximately 1.2 million violent crimes occur each year in which victims are certain that the offender had been drinking or used drugs at the time of the offense. An estimated 8 of every 10 new admissions to the department of corrections (i.e., inmate) have a drug or alcohol abuse issue.[101] One concern is that hard-core addicts are not about to change their behavior or their personal goals, no matter what happens to them—including incarceration.[102]

Methamphetamine: A Powerful Foe

State and local police report that methamphetamine trafficking and abuse has become their most pressing illegal drug problem in recent years, surpassing crack cocaine in this regard.[103] Although offenders manufacture a variety of illicit drugs in clandestine labs, methamphetamine accounts for 80 to 90 percent of the clandestine labs' total drug production.

To respond to the problem of clandestine methamphetamine labs, local governments must understand and analyze their particular problem carefully so that they can design a more effective response strategy. By understanding and analyzing local problems, government should have a better understanding of the factors contributing to it. After this analysis, local government should be able to establish a baseline for measuring effectiveness and consider possible responses to address the problem.

Ideally, such a guideline provides a general overview of the problem and of responses to it. It begins by describing the problem and reviewing factors that increase the risks of its realization. It then identifies a series of questions to help local governments analyze their local problem and concludes with a review of responses to the problem and what is known about them from evaluative research and practice.

Response strategies provide a foundation of ideas for addressing the problem. It is critical that responses be tailored to local circumstances and that each response is justified based on reliable analysis.

In 2006, the Office of Drug Control Policy noted that more than 40 states had implemented some type of restriction on retail transactions involving products containing certain chemicals that could be used to make methamphetamine.[104] For instance, Wal-Mart no longer allows customers to purchase Sudafed directly; instead, customers have to ask a pharmacist at the counter for the product and the pharmacist requests purchase from a oversight committee of sorts, which then approves or disapproves a sale. Customers are limited to one transaction for this product per week.

In 2004, there were approximately 17,750 methamphetamine laboratory incident seizures. In 2005, this number fell to 12,500, a decline of more than 30 percent. In 2006, the Combat Methamphetamine Epidemic Act (CMEA) went into effect. Early 2006 data suggest the decline in meth lab incidents is continuing, a trend that is primarily attributable to the enactment of various state laws targeting this drug of abuse. In some states, enactment of the CMEA was followed by a swift and sudden decline in meth lab incidents, sometimes as much as 75 percent or more. In other states, the decline was less dramatic.

The director of the National Drug Control Policy explains the scope of the meth lab problem:

> A meth lab is a hazardous waste site, and cleanup is costly and dangerous. Dismantling meth labs calls for specially trained personnel, hazardous waste equipment, and hazardous waste disposal. According to a 2003 estimate by the El Paso Intelligence Center (EPIC), for every pound of methamphetamine produced in a clandestine lab, about five pounds of toxic waste is left behind. And in some cases law enforcement personnel deal with as much as 10 pounds of leftover chemicals and by-product sludge. At some lab sites, toxic chemicals can enter the water table, causing further damage and cleanup costs.[105]

Drug Enforcement Strategies

Police have traditionally employed two strategies to combat drug trafficking and use: supply reduction and demand reduction.

Supply-Reduction Strategies

The traditional **supply-reduction strategy** (dealing with suppliers) includes four strategies:

- Buy and bust (arrest)
- Disrupting drug syndicates by trading up—arresting lower-level dealers and using them as confidential informants
- Penetrating the drug syndicate through undercover work
- Intense enforcement in a specific part of the jurisdiction

Often officers of a criminal investigative unit tasked with drug enforcement work together with agents from the DEA. They may use confidential informants, cooperating witnesses, and police officers in undercover assignments to penetrate the gang at its most vulnerable point: the lines of distribution. In one case, investigators infiltrated a gang (in this case, the King Gang in Baltimore) and purchased heroin from key members, including King, the leader. Once the investigators built a case, followed by arrests and prosecutions, police officials considered aggressive post-arrest pursuit of the gang's financial assets.

Unfortunately, most gang drug leaders learn how to resist police encroachment and effectively block traditional enforcement methods. Investigators often obtain little information regarding the size and scope of a gang's influence, which makes it difficult to justify an investigation in the first place.

Drug traffickers who belong to gangs may become relentless in their efforts to thwart investigations—tampering with evidence, intimidating witnesses, and accepting long sentences rather than providing information about other members, according to Bureau of Justice Assistance. Sometimes gangs resolve problems with independent operators in their style—which usually means murder.

An example of a supply-reduction strategy was Operation Closed Market, a "violence reduction" policing plan that had been evolving since 1998 in Chicago. A similar program helped New York City cut its homicide toll from 2,245 in 1990 to 580 by 2003. In Chicago, authorities called this approach putting more "officers on the dots"—that is, keeping mobile patrols in places where violent crimes often occur, in a form of hot spot policing. The strategy focused on redeployment of officers into high-crime neighborhoods, and "block sweeps," in which narcotics tactical units, patrol officers, and about 200 temporarily reassigned administrative personnel all converged on "open-air drug markets." Both CPD and NYPD used high-powered, remote-surveillance-camera vans, including nearly 1,000 cameras mounted on light poles; undercover operatives posing as dealers; and squad cars parked on every known corner to "disrupt and discourage" drug commerce. The end result in 2004 of Operation Closed Market in Chicago was an overall decrease in violence: 1,000 fewer shootings and 154 fewer homicides, and 38 "street corner" drug conspiracy cases.[106]

Demand-Reduction Strategies

The second strategy used by police to control drugs is to change the demand for the illegal substance. Such a **demand-reduction strategy** (dealing with users) works through punitive and social worker models, including prevention programs and DARE programs.

At what point should demand-reduction strategies be implemented? By parents? By schools? Studies have looked at these questions, and it appears that parents should initiate their children's education about drugs. In one study, in school districts where extensive antidrug information was spread among a community on billboards, television, and school packets, parents in 12 cities were relatively consistent with their remarks about drug programs linked to their children:

- A typical response from urban parents was that youths learned about drugs from their own parents "because the drug culture is so pervasive in their community."[107] Those parents typically reported that they considered talking to their children about drugs when those children learned about local drug violence incidents. Youths, by contrast, reported that the presence of drug violence was so strong that many urban youths do not expect to live long, and they are often stressed out or fearful about drug violence and death.

- Non-urban parents report that most parents do not talk to their children about drugs because they do not know what to say. These parents say that youths usually obtain their information about the dangers of drugs from their peers at school, and that their children are also constantly stressed about the drug violence and gang violence in schools.

Texas has employed a demand-reduction strategy in an effort to deal with its burgeoning substance abuse problem. In 2000, the total economic cost associated with alcohol and drug abuse in Texas was estimated at $25.9 billion.[108] Seventy-one percent of students in grades 7–12 in the state reported using alcohol, and 26 percent were considered binge drinkers. More than 13,500 Texans died from alcohol and drug disorders, 46 percent of them younger than age 25. In response, Texas developed the following goals and strategies to meet these goals:

Prevention: Stop Use Before It Starts

- Educate the public about substance abuse and promote social norms that discourage illegal and inappropriate use of alcohol, tobacco, and other drugs.
- Target youth with clear messages that no use of alcohol, tobacco, or other drugs is acceptable.
- Provide research-based prevention programs to foster positive, healthy lifestyles among youth, equipping them to reject the use of alcohol, tobacco, and other drugs.

Treatment: Heal Texans Who Are Dependent on Alcohol and Other Drugs

- Through education and training, promote early identification and intervention to slow/halt disease progressions, improve prognosis/outcomes, and reduce related problems.
- Incorporate substance abuse screening and referral throughout the health and human services system to facilitate early access to treatment.
- Integrate evidence-based substance abuse services throughout the criminal justice system to break the cycle of addiction and crime.

Enforcement: Disrupt the Market

- Disrupt and deter the flow of illegal drugs into and through Texas.
- Hold individuals accountable for criminal violations and incarcerate offenders who threaten public safety.
- Use drug courts to reduce drug use and criminal behavior.
- Develop better methods to share information across enforcement agencies.

Integration: Create a Unified Response

- Support education and training for professionals involved in prevention, treatment, and law enforcement efforts, including health and human service workers who interact with substance users.
- Maximize limited resources by investing in activities with the greatest impact.

New Approaches to Policing of Drugs

Policing of drugs is important, no matter whether the police attempt to curb the supply by arresting dealers and manufacturers or alter the demand by breaking the cycle of addiction through conducting probable-cause arrests. Nevertheless, officials across the United States are rethinking their arrestable offense policies as they relate to drug possession.

For instance, Chicago's Mayor Richard Daley applauded a police sergeant when the officer suggested that it might be better to impose fines between $250 and $1,000 for possession of small amounts of marijuana rather than prosecute the cases. Chicago now "tickets" possession of small amounts of marijuana.

Chicago was not the first city to reduce the penalty for possession.[109] In Seattle, voters passed an initiative requiring law enforcement officials to make personal-use marijuana cases their lowest priority. In California and Oregon, possession of a small amount of marijuana is a misdemeanor punishable by a fine of $100 to $500. In Colorado, this offense doesn't even rise to the level of misdemeanor; it's a petty offense with a fine of no more than $100.

Many states are also reducing prison sentences for drug possession—California, for example. In the Midwest, the Cincinnati City Council's efforts to reduce drugs and crime in the city have been hindered by lack of Hamilton County jail space. "We all know that we have a jail capacity problem," said a councilman. "For every person admitted, someone has to be released, so we are not arresting first-time offenders," he said.[110] A new law in Cincinnati made possession of 100 grams of marijuana an offense that carries up to a 30-day jail sentence. Possession of 200 grams carries a maximum sentence of a $1,000 fine and up to six months in jail. "There were 4,100 tickets written last year, which means that potentially 4,100 people could be in the court system," said a councilwoman. She noted that police officers would also have to go to court more often.

A Cincinnati police analysis reported there were 42,063 service calls in the first two months of 2006: 869 of these calls were Part 1 charges—activities ranging from auto theft to murders; 221 were violent crime charges; and 2,237 were drug arrests.[111] The new law will add time to the processing of offenders, taking more of their time away from the streets. Cincinnati is thinking hard about revising its laws and make possession similar to a traffic citation.

Treatment

"Substance abuse is a recognized dynamic risk factor; altering the need can increase the likelihood of changing the criminal behavior and closing the revolving prison door," advises the Iowa Department of Corrections.[112] Treatment for drug abuse is cost-effective as compared to incarceration for the same crime: Costs for treatment typically range from $1,800 to $6,800 per client, compared with incarceration costs of $35,000 to $75,000 per year. In addition, many addicts are employed, so their health insurance may pay for treatment. "Domestic enforcement costs 4 times as much as treatment for a given amount of user reduction, 7 times as much for consumption reduction, and 15 times as much for societal cost reduction."[113]

One program that has proved effective among incarcerated offenders is the Drug Treatment Alternative-to-Prison (DTAP) program, which originated in Brooklyn, New York. DTAP is a prosecution-run, prison-diversion treatment program targeting nonviolent, repeat felony offenders with serious drug addictions.[114] The program has achieved success in reducing drug abuse and criminal recidivism in its target population. There are two key premises behind DTAP: (1) criminal recidivism of addicts can be reduced if the addiction is treated; and (2) legal coercion can be a powerful motivator to get addicts to succeed in treatment.

Criticism of Treatment

Treatment-oriented approaches to substance abuse have drawn criticism. Evidence suggests that treatment does not necessarily reduce drug abuse or recidivism, as might be expected. For instance, one study suggests that the quality of care can affect both the outcomes of abuse and the recidivism rates.[115] Some studies report few benefits through treatment administered to offenders both in the community and while institutionalized, and improved results (e.g., fewer relapses and less recidivism) are visible but only with continued supervision and only among selected participants.[116] Studies that have evaluated treatment versus incarceration have produced indecisive conclusions. For example, the Iowa Department of Corrections reports in its own study that "the data is not compiled in such a way to see changes over time, nor is there a consistent approach used across programs minimizing the data's usefulness for program comparisons."[117] Research implies that not all addicts (and alcoholics), including those who have not been apprehended or jailed, such as methamphetamine addicts in the general population, are amenable toward treatment or education, nor do they always wish to abandon their drug lifestyle.[118]

Shooting the Wolves

Some observers suggest that some individuals should be left alone to fight the fight. According to this perspective, society is better off if we intervene as little as possible and then only when absolutely necessary.[119]

Human society is not exempt from the checks and balances that have long been recorded on the planet.[120] There are few reasons why equilibrium should not work to society's advantage and for its survival, as it does in nature. Ecologists (who study organisms and the environment) speak of a balance between predator and prey that assures neither will die out. A balance or equilibrium exists in nature that governs all living things, including humans.

Using nature as a laboratory, consider the idea of herds of caribou that migrate north across Canada during the summer and south again in the winter. Their steady companions are packs of wolves. While wolves reach a top speed of about 25 miles per hour and can run for short spurts, caribou have a top speed of more than 40 miles per hour and can run for long distances. A healthy caribou has little to fear from wolves that might wait until an individual caribou becomes sick or old. The constant pressure from the wolves makes the caribou alert, and running from the predator keeps the herd focused on the journey and in shape to face the cold and the arduous trek.

Now suppose humans shoot the pack of wolves in an effort to protect the caribou. What happens to the herd? Because wolves no longer kill caribou, it grows in size. Instead of a fast, merciful death, caribou now die a slow death from starvation and disease.

Has humanity provided a favorable situation for the caribou by "protecting" them? Nature would maintain the balance but humans have intervened, disrupting an otherwise healthy exchange. Eventually other wolves from others parts of Canada might rejoin the herd so that the ecological balance is restored.

The lesson learned: The cruelty of the wolf is, in fact, merciful. The quick kill in eliminating the sick, injured, and old plays a vital role in the survival of the herd and its fitness. The caribou needs the wolves as much as the wolves need the caribou. This relationship has an analogy in drug-treatment programs: Does every drug addict really prosper from forced intervention, and can that intervention set them up for a relapse or failure?

RIPPED from the Headlines

Gang Suppression Doesn't Work

Anti-gang legislation and police crackdowns are failing so badly that they are strengthening the criminal organizations and making U.S. cities more dangerous. Mass arrests, stiff prison sentences often served with other gang members, and other strategies that focus on law enforcement rather than intervention actually strengthen gang ties and further marginalize angry young men.

"We're talking about 12- to 15-year-olds whose involvement in gangs is likely to be ephemeral unless they are pulled off the street and put in prison, where they will come out with much stronger gang allegiances," said Judith Greene, co-author of "Gang Wars: The Failure of Enforcement Tactics and the Need for Effective Public Safety Strategies." The report is based on interviews and analysis of hundreds of pages of previously published statistics and reports. And though it is valid and accurate, the ideas raised in it are not new, said Arthur Lurigio, a psychologist and criminal justice professor at Loyola University of Chicago.

"These approaches, although they sound novel, are just old wine in new bottles," he said. "Gang crime and violence in poor urban neighborhoods have been a problem since the latter parts of the nineteenth century."

Lurigio, other academics, and gang intervention workers have echoed elements of the report that found gangs need to be viewed as a symptom of other problems in poor communities, such as violence, teen pregnancy, drug abuse, and unemployment. The report says Los Angeles and Chicago are losing the war on gangs because they focus on law enforcement and are short on intervention. It cites a report this year by civil rights attorney Connie Rice, who was hired by Los Angeles to evaluate its failing anti-gang programs. Her report called for an initiative to provide jobs and recreational programs in impoverished neighborhoods.

Mayor Antonio Villaraigosa and Police Chief William Bratton both commended Rice's report. They unveiled a strategy that focused on targeting the city's worst gangs with arrests and civil injunctions that prohibit known gang members from associating with one another in public. Rice describes the city's policy on arresting the city's estimated 39,000 gang members as "stuck on stupid."

Wes McBride, executive director of the California Gang Investigators Association, dismissed the findings of the report, which he said was written by "thug-huggers." The investigators association is a professional organization for police officers. "Are they saying we can't put a thief in jail, we can't put a murderer in jail, that we should spank them, put a diaper on them, pat them on the bottom, hug them, and let them go?" McBride said. "It's obviously a think tank report, and they didn't leave their ivory tower and spend any time on the streets."

"Gang Wars" also criticizes politicians who overstate the threat of criminal gangs and seek tougher sentences. Greene specifically criticized a bill introduced by Senators Dianne Feinstein, Democrat–California, and Orrin Hatch, Republican–Utah, that would make it illegal to be a member of a criminal gang and would make it easier to prosecute some minors as adults. But Feinstein spokesman Scott Gerber said the bill also calls for spending more than $400 million on gang prevention and intervention programs, which he said would be the largest single investment of its kind.

Source: Andrew Glazer, "Report: Gang Suppression Doesn't Work," *Associated Press* (July 18, 2007), http://www.officer.com/online/article.jsp?siteSection=1&id=36916 (accessed July 18, 2007).

Summary

Police target repeat offenders, gangs, and drugs through various traditional and contemporary police strategies, including uncover detective work, use of confidential informants, sting operations, and gang- and drug-specific programs. Local police, as part of a greater effort to provide proactive policing, have altered their strategies in recognition that most crime is committed by few offenders at few locations.

Undercover strategies occur when an officer "sheds the shield and the uniform" and takes on a persona other than that of a police officer. Undercover operations play a major role in sting initiatives, as officers pose as criminals and victims in an effort to discover those individuals who hold criminal intent and are willing to violate the law. Most often undercover officers employ confidential informants who are criminals and provide information in exchange for judicial leniency.

The police often use their knowledge of crime to construct circumstances or sting operations that invite criminals to commit more crime in a recordable way, thereby enhancing the likelihood of apprehending criminals who otherwise would be difficult to detect and prosecute. Local stings are geared toward local jurisdiction issues such as anti-fence operations, alcohol, drugs, cigarettes, and prostitution. Federal stings tend to focus on drugs, terrorism and arms smuggling, child pornography and pedophile activities, and bank laundering of illegal funds. Often federal and local law enforcement agencies work together to combat a specific target, such as drug dealing by gang members.

Traditional police suppression of gangs has employed strategies based on the assumption that most street gangs are criminal associations that must be controlled through efficient gang tracking, identification, and targeted enforcement strategy. Contemporary police approaches to monitoring gangs entail a combination of prevention, intervention, and suppression strategies: For example, a police department might support GREAT prevention programs, while an intervention team works with active gang members, and a gang suppression unit targets the most violent gangs and gang members. The comprehensive gang prevention, intervention, and suppression model is a flexible framework that can guide development of appropriate strategies.

Policing of drugs has traditionally relied on either supply-reduction strategies (dealing with suppliers) or demand-reduction strategies (dealing with users) through punitive and social worker models, including prevention programs and DARE programs. In recent years, state and local police have faced enormous challenges related to methamphetamine trafficking and abuse; it is now their most pressing illegal drug problem, surpassing crack cocaine. When combating drug-related crime, treatment of addicts is more cost-efficient than incarceration, though the results of institutionalized treatment remain equivocal.

Key Words

Amber Alert
Confidential informant (CI)
Cultural competency
Demand-reduction strategy
Gang
Hot spot deployment
John Walsh (*America's Most Wanted [AMW]*)
Mano dura
Proactive police strategies
Reactive police initiatives
Stings
Supply-reduction strategy
Undercover operations

Discussion Questions

1. Articulate contributions made by popular television dramas to criminal justice.
2. Describe strategies to capture repeat offenders.
3. Characterize federal local participation in repeat-offender programs.
4. Provide the rationale for hot spot deployment strategies.
5. Justify, characterize, and describe typical undercover detective strategies.
6. Describe the role of confidential informants in police initiatives.
7. Characterize the role of and rationale for sting initiatives.
8. Describe sting strategies employed by local and federal police agencies.
9. Characterize a legal sting arrest and explain its rationale.
10. Describe the relationship between violence and youth gangs.
11. Characterize several of America's toughest gangs.
12. Describe traditional suppression strategies of gangs and contemporary injunction models.
13. Articulate contemporary police approaches to young gangs.
14. Explain what makes methamphetamine a powerful foe for police.
15. Compare supply-reduction strategies and demand-reduction strategies for policing drugs.
16. Characterize new approaches to policing drugs.

Notes

1. *America's Most Wanted*, 2008, http://www.amw.com/ (accessed April 10, 2008).
2. Code Amber (2007), http://codeamber.org/ (accessed April 6, 2007).
3. Sam Walker and Charles M. Katz, *The Police in America: An Introduction*, 5th ed. (Boston: McGraw Hill, 2008), 298–299; Marvin E. Wolfgang, Robert Figlio, and Thorsten Selling, *Delinquency in a Birth Cohort* (Chicago: University of Chicago Press, 1972).
4. Lawrence W. Sherman, "Attacking Crime: Police and Crime Control," *Crime and Justice* 15 (1992): 159–230. Also see Lawrence Sherman, "Preventing Homicide through Trial and Error" (1992), http://www.aic.gov.au/publications/proceedings/17/sherman.pdf (accessed June 14, 2007).

5. Ronald V. Clarke and John E. Eck, *Crime Analysis for Problem Solvers in 60 Small Steps* (Center for Problem-Oriented Policing, 2007), http://www.popcenter.org/learning/60Steps/index.cfm?page=Authors (accessed December 2, 2007).
6. Clarke and Eck, *Crime Analysis for Problem Solvers.*
7. Stanton E. Samenow, *Inside the Criminal Mind*, rev. and updated (New York: Crown, 2004), 1–8.
8. Beginning with a group of approximately 10,000 boys born in 1945 who lived in Philadelphia from at least ages 10 through 17, the Center for Studies in Criminology and Criminal Law, University of Pennsylvania, has engaged in a longitudinal analysis of the delinquency of the birth cohort. The first publication, which examined the dynamic flow of delinquency committed by 3,500 of the boys, was published by the University of Chicago Press in 1972. This study was the first in the United States to establish a baseline of delinquency probabilities and to analyze the types of delinquency committed over time, with the recording of the seriousness of each of the 10,000 acts committed by the 3,500 boys. The seriousness scores were derived from the earlier work by T. Selling and M. E. Wolfgang, *The Measurement of Delinquency* (New York: Wiley, 1964). The Center for Studies in Criminology and Criminal Law has continued to follow up. For instance, see Paul Friday, Xin Ren, and Elmar Weitekamp, "Delinquency in a Birth Cohort in Wuchang District China 1973–2000," *ICPSR* (2004), http://webapp.icpsr.umich.edu/cocoon/ICPSR-STUDY/03751.xml (accessed April 6, 2007).
9. Walker and Katz, *The Police in America*, 289.
10. Federal Bureau of Investigation, National Center for the Analysis of Violent Crime (2007), http://www.fbi.gov/hq/isd/cirg/ncavc.htm (accessed April 6, 2007).
11. Gary Marx, *Undercover: Police Surveillance in America* (Berkeley, CA: University of California Press, 1988).
12. Andrew L. Choo, *Abuse of Process and Judicial Stays of Criminal Proceedings* (Oxford UK: Clarendon Press, 1993), 149.
13. Library of Halexandria, "Victimless Crimes" (2007), http://www.halexandria.org/dward267.htm (accessed April 28, 2007).
14. Melissa Farley is cited in Charles Montaldo, "The Oldest Profession Is Hardly Without Victims: About Crime," *New York Times* (2007), http://crime.about.com/od/prostitution/a/prostitution.htm (accessed April 26, 2007).
15. Waco, Texas, Police Department, http://www.waco-texas.com/city_depts/police/policeservices.htm (accessed July 15, 2007).
16. Lawrence W. Travis III and Robert Langworthy, *Policing in America: A Balance of Forces* (Upper Saddle River, NJ: Prentice Hall, 2008), 334.
17. St. Louis Missouri County Police Department, http://www.stlouisco.com/police/dci.html (accessed July 15, 2007).
18. Choo, *Abuse of Process and Judicial Stays.*
19. Walker and Katz, *The Police in America*, 291.
20. Thomas O'Connor (2002), http://faculty.ncwc.edu/TOConnor/205/205lect08a.htm (accessed April 4, 2007).
21. Personal and confidential communication between an undercover officer and the author, July 2002.
22. O'Connor (2002).
23. Walker and Katz, *The Police in America*, 291.
24. Edmond Roy, "New Powers for Undercover Police" (2001), http://www.abc.net.au/worldtoday/stories/s272531.htm (accessed April 25, 2007).
25. Gregory D. Lee, *Conspiracy Investigations* (Upper Saddle River, NJ: Prentice Hall, 2008), 43.

26. Robert T. Thetford, "Informant Liability Issues," Institute for Criminal Justice Education (July 14, 2003), http://www.icje.org/informant_liability_issues.htm (accessed April 13, 2007).

27. Jerome Skolnick, *Justice without Trial* (New York: Macmillan, 1994), 129.

28. Personal conversation between the informant and the author during two 15-week college courses offered through Boston University at MCI Framingham, a women's prison facility close to Boston.

29. Robert M. Shusta, Deena R. Levine, Herbert Z. Wong, Aaron T. Olson, and Phillip R. Harris, (NJ: Prentice Hall, 2008), 322.

30. Dennis J. Stevens, "Sting Tactics," in *Encyclopedia of Police Science*, 3rd ed., ed. Jack R. Greene (New York: Routledge, 2006), 2.

31. John Crank, *Understanding Police Culture*, 2nd ed. (Cincinnati, OH: Anderson, 2004).

32. Walker and Katz, *The Police in America*.

33. Robert Langworthy, "Do Stings Control Crime? An Evaluation of Police Fencing Operations," *Justice Quarterly* 6, no. 1 (1989): 27–45.

34. Rolando Garza. "Minor Sting Guidelines for Law Enforcement," in *Texas Alcoholic Beverage Commission* (Austin, Texas: State of Texas, 2005).

35. D. F. Preusser, A. F. Williams, and H. B. Weinstein, "Policing Underage Alcohol Sales," *Journal of Safety Research* 25 (1994): 127–133.

36. A. H. Reinhardt-Rutland, "Roadside Speed-Cameras: Arguments for Covert Sting," *The Police Journal* 74, no. 4 (2001): 312–315.

37. Mark Sickmiller, "Speeder-Catching Cams To Really Pay Off," *9News* (April 10, 2008), http://groups.yahoo.com/group/seroads/message/4081 (accessed January 16, 2005).

38. N. L. Lewin, J. S. Vernick, P. I. Beilenson, J. S. Mair, M. Lindamood, S. P. Teret, and D. W. Webster, "The Baltimore Youth Ammunition Initiative: A Model Application of Local Public Health Authority in Preventing Gun Violence," *American Journal of Public Health* 95, no. 5 2005: 762–765.

39. Walker and Katz, *The Police in America*, 245.

40. Metropolitan Nashville, Tennessee, Police Department (2004), http://www.police.nashville.org/ (accessed March 14, 2007).

41. Sullivan County Sheriff's Office, "Busted in Bristol" (2006), http://www.sullivan-county.com/id6/meth/lee1.htm (accessed April 27, 2007).

42. Raymond Lewis, "Major Whiteside County, Illinois. Drug Sting in Progress," *Lewis Information Services* (Mount Morris, Illinois) http://camerahacks.10.forumer.com/viewtopic.php?t=1641 (accessed April 10, 2008).

43. Federal Bureau of Investigation, "War on Terrorism: International Undercover Operation Stings Deal for Surface-to-Air Missiles" (August 2003), http://www.fbi.gov/page2/aug03/miss081303.htm (accessed April 10, 2008).

44. "Russian Press Hails Missile Sting," *BBC News* (August 14, 2003), http://news.bbc.co.uk/go/pr/fr/-/1/hi/world/europe/3151195.stm (accessed July 27, 2006).

45. David L. Weiss, "Major Child Pornography. Focus of Social Issues: Pornography Quick Facts" (2003), http://www.citizenlink.org/FOSI/pornography/A000001625.cfm (accessed April 10, 2008).

46. Nevada Division of Investigation (1999), http://www.dmvnv.com/news/99-179.htm (accessed April 25, 2007).

47. Dennis J. Stevens, "Sting Tactics," in *Encyclopedia of Police Science*, 3rd ed., 2 vols., ed. Jack R. Greene (New York: Routledge, 2007).

48. Rachel Clarke, "What Is a Legal Sting?", *BBC News* (September 23, 2003), http://news.bbc.co.uk/2/hi/americas/3148267.stm (accessed June 9, 2007).

49. Consent Decree, "In the United States District Court for the Central District of California v. City of Los Angeles, California, Board of Police Commissioners of the City of Los Angeles, and the Los Angeles Police Department," http://news.findlaw.com/hdocs/docs/lapd/usvlaconsentdec.pdf (accessed April 10, 2008).

50. "Texas Police Apologize for Drug Arrest Mix-up" (May 30, 2007), http://www.officer.com/article/article.jsp?siteSection=1&id=36309 (accessed May 30, 2007).

51. Bill Moushey, "Covering Crime," *Pittsburg Post Gazette* (1999), http://backissues.cjrarchives.org/year/99/2/d_l.asp (accessed July 13, 2007).

52. William Bratton and Peter Knobler, *Turnaround: How America's Top Cop Reversed the Crime Epidemic* (New York: Random House, 1998).

53. Crank, *Understanding Police Culture*, 303.

54. Robert Langworthy and Lawrence Travis, *Policing in America. A Balance of Forces*, 3rd ed. (Upper Saddle River, NJ: Prentice Hall), 2003.

55. Arlen Egley, Jr., and Christina E. Ritz, *Highlights of the 2004 National Youth Gang Survey* (Washington, DC: U.S. Department of Justice, Office of Juvenile Justice and Delinquency Prevention, 2006), http://www.iir.com/nygc/publications/fs200601.pdf (accessed April 8, 2007).

56. Bureau of Justice Statistics (2007), http://www.ojp.usdoj.gov/bjs/ (accessed July 15, 2007).

57. National Young Gang Center, "Young Gangs" (2007), http://www.iir.com/nygc/faq.htm#q3 (accessed April 8, 2007).

58. Katherine Williams, G. David Curry, and Marcia I. Cohen, "Gang Prevention Programs for Female Adolescents: An Evaluation," in *Responding to Gangs: Evaluation and Research*, ed. Winfred L. Reed and Scott H. Decker (Washington, DC: U.S. Department of Justice, National Institute of Justice, 2002), 225–64, http://www.ncjrs.gov/pdffiles1/nij/190351.pdf (accessed April 8, 2007).

59. Malcolm Klein, *Street Gangs and Street Workers* (Englewood Cliffs, NJ: Prentice Hall, 1971), 13.

60. T. P. Thornberry, M. D. Krohn, A. H. Lizotte, C. A. Smith, and K. Tobin, *Gangs and Delinquency in Developmental Perspective* (New York: Cambridge University Press, 2003).

61. Walter B. Miller, *The Growth of Youth Gang Problems in the United States: 1970–1998* (Washington, DC: U.S. Department of Justice, Office of Juvenile Justice and Delinquency Prevention, 2001), http://www.ncjrs.gov/pdffiles1/ojjdp/181868-1.pdf (accessed April 8, 2007).

62. Malcolm W. Klein, *Gang Cop: The Words and Ways of Officer Paco Domingo* (Walnut Creek, CA: Alta Mira, 2004), 57. Also see Malcolm W. Klein, "The Value of Comparisons in Street Gang Research," *Journal of Contemporary Criminal Justice* 21, no. 2 (2005): 135–52.

63. Klein, *Gang Cop*, 59.

64. Winifred L. Reed and Scott H. Decker, *Responding to Gangs: Evaluation and Research* (NCJ 190351) (Washington, DC: Department of Justice, 2002).

65. Anthony A. Braga, *Gun Violence Among Serious Young Offenders: Problem-Oriented Guides for Police: Problem-Specific Guides. Series No. 23. Community Oriented Policing Services* (Washington DC: U.S. Department of Justice, 2004).

66. Klein, *Gang Cop*, 59.

67. Deborah Lamm Weisel, "The Evolution of Street Gangs: An Examination of Form and Variation," in *Responding to Gangs: Evaluation and Research*, ed. W. Reed and S. Decker (Washington, DC: U.S. Department of Justice, National Institute of Justice, 2002), 25–65, http://www.ncjrs.gov/pdffiles1/nij/190351.pdf (accessed April 8, 2007).

68. Arian Campo-Flores, "The Most Dangerous Gang in America," *Newsweek:* National News (2007), http://www.msnbc.msn.com/id/7244879/site/newsweek/ (accessed April 11, 2007).
69. Campo-Flores, "The Most Dangerous Gang in America."
70. "MS-13: The Most Dangerous Gang in America," National Geographic 2006, http://blogs.nationalgeographic.com/channel/blog/2006/01/explorer_gangs.html (accessed April 10, 2008).
71. Bajito Onda Organization (2006), http://www.bajitoonda.org/ (accessed December 1, 2007).
72. C. N. Le, "Asian American Gangs. Asian-Nation: The Landscape of Asian America" (April 11, 2007), http://www.asian-nation.org/gangs.shtml (accessed April 11, 2007).
73. Phelan A. Wyrick, "Vietnamese Youth Gang Involvement," *NCJRS* (2000), http://www.ncjrs.gov/pdffiles1/ojjdp/fs200001.pdf (accessed April 12, 2007).
74. Le, "Asian American Gangs."
75. Ko-lin Chin, *Chinatown Gangs: Extortion, Enterprise, and Ethnicity* (New York: Oxford University Press, 1996), 35–59; Ko-lin Chin, Jeffrey Fagan, and Robert Kelly, "Patterns of Chinese Gang Extortion," *Justice Quarterly* 9, no. 4 (1992): 625–46.
76. Malcolm Klein, *The American Street Gang: Its Nature, Prevalence, and Control* (New York: Oxford University Press, 1995).
77. Chicago Municipal Code §8-4-015 [added June 17, 1992].
78. *City of Chicago v. Jesus Morales, et al.* (1997), In the Supreme Court of the United States, No. 97-1121, http://www.aclu.org/scotus/1998/22520lgl19981001.html (accessed June 24, 2006).
79. Reed and Decker, *Responding to Gangs.*
80. Irving A. Spergel, *The Youth Gang Problem* (New York: Oxford University Press, 1995).
81. Spergel, *The Youth Gang Problem.*
82. Malcolm W. Klein, *The American Street Gang: Its Nature, Prevalence, and Control (Studies in Crime and Public),* (Cambridge UK: Oxford University Press, 1993).
83. Klein, *The American Street Gang* (1993), 61–73.
84. Spergel, *The Youth Gang Problem.*
85. Vera Institute of Justice, "Strengthening Relations between Police and Immigrants" (2006), http://www.vera.org/project/project1_1.asp?section_id=4&project_id=70 (accessed April 10, 2008).
86. Mandalit del Barco, "L.A. Police List Most Dangerous Gangs," National Public Radio (April 11, 2007), http://www.npr.org/templates/story/story.php?storyId=7260563 (accessed April 11, 2007).
87. City of Chicago, "Daley, Cline Seek DEA Help in War on Drugs and Gangs" (December 10, 2003), http://www.cityofchicago.org (accessed May 5, 2007).
88. Klein, *The American Street Gang* (1993).
89. Phelan A. Wyrick and James C. Howell, "Strategic Risk-Based Response to Youth Gangs," *Juvenile Justice* 10, no. 1 (2004): 20–29.
90. David M. Kennedy, "Pulling Levers: Chronic Offenders, High-Crime Settings, and a Theory of Prevention," *Valparaiso University Law Review* 31, no. 2 (1997): 449–84.
91. Anthony A. Braga and David M. Kennedy, "Reducing Gang Violence in Boston," in *Responding to Gangs: Evaluation and Research*, ed. Winfred L. Reed and Scott H. Decker (Washington, DC: U.S. Department of Justice, National Institute of Justice, 2002), 265–88, http://www.ncjrs.gov/pdffiles1/nij/190351.pdf (accessed April 8, 2007).
92. James W. Meeker, Katie J. B. Parsons, and Bryan J. Vila, "Developing a GIS-Based Regional Gang Incident Tracking System," in *Responding to Gangs: Evaluation and Research*, ed. Winfred L. Reed and Scott H. Decker (Washington, DC: U.S. Department

of Justice, National Institute of Justice, 2002), 289–301, http://www.ncjrs.gov/pdffiles1/nij/190351.pdf (accessed April 8, 2007).

93. Jeffrey Grogger, "What We Know about Gang Injunctions," *Criminology & Public Policy* 4, no. 3 (2005): 637–41.

94. Jeffrey Grogger, "The Effects of Civil Gang Injunctions on Reported Violent Crime," *Journal of Law and Economics* 45, no. 1 (2007), http://papers.ssrn.com/sol3/papers.cfm?abstract_id=303498#PaperDownload (accessed April 10, 2008).

95. Margaret Shaw, "Youth and Gun Violence: An Outstanding Case for Prevention," International Centre for the Prevention of Crime (2005), http://www.crime-prevention.com (accessed April 12, 2007).

96. Mike Adams, "Lack of Basic Nutrition Creates Generation of Criminals: Prison System Society," *News Target.com* (2005), http://www.newstarget.com/006194.html (accessed April 12, 2007); Cathy S. Widom and Michael G. Maxfield, "Cycle of Violence," *NCJRS* (2001), http://www.ncjrs.gov/txtfiles1/nij/184894.txt (accessed April 12, 2007).

97. Bureau of Justice Assistance, "Gang Characteristics and Growth" (2005), http://www.ncjrs.gov/html/bja/gang/bja1.html (accessed April 12, 2007).

98. Spergel, *The Youth Gang Problem.*

99. University of Michigan, *Monitoring the Future National Survey Results on Drug Use, 1975–2005, Volume II: College Students and Adults Ages 19–45, 2005* (Washington, DC: U.S. Department of Justice, Bureau of Justice Statistics, October 2006), http://www.ojp.usdoj.gov/bjs/dcf/du.htm (accessed April 26, 2007).

100. Bureau of Justice Statistics, "Drug Use and Dependence, State and Federal Prisoners, 2004" (NCJ 213530; 2006), http://www.ojp.usdoj.gov/bjs/dcf/duc.htm (Retrieved April 25, 2007).

101. Dennis J. Stevens, *Applied Community Corrections: An Applied Approach* (Upper Saddle River, NJ: Prentice Hall, 2006), 114.

102. Stevens, *Applied Community Corrections*, 115.

103. Michael S. Scott and Kelly Scott, *Clandestine Methamphetamine Labs*, 2nd ed. (NCJ 215556) (Washington, DC: U.S. Department of Justice, 2006).

104. Office of Drug Control Policy, "Pushing Back Against Meth: A Progress Report on the Fight Against Methamphetamine in the United States" (NCJ 216528; November 2006), http://www.whitehousedrugpolicy.gov/publications/pdf/pushingback_against_meth.pdf (accessed April 13, 2007).

105. John P. Walters, "Methamphetamine: A National Response," *Police Chief* 72, no. 3 (2005), http://policechiefmagazine.org/magazine/index.cfm?fuseaction=display_arch&article_id=566&issue_id=42005 (accessed December 2, 2007).

106. CPD Director of News Public Affairs, Patrick Camden, cited in Charles Shaw, "The Next American City" (April 2005), http://www.americancity.org/article.php?id_article=123 (accessed May 5, 2007).

107. "Testing the Antidrug Message in 12 Cities" (2002), http://www.mediacampaign.org/ (accessed May 5, 2007).

108. State of Texas, "Toward a Drug Free Texas" (2004), http://www.dshs.state.tx.us/sa/2003 DrugReduction.pdf (accessed May 5, 2007).

109. "Marijuana Police Project" (July 2007), http://www.mpp.org/site/c.glKZLeMQIsG/b.1086497/k.BF78/Home.htm (accessed July 15, 2007).

110. Jennifer Greenup, *QueenCityForum.com Magazine* (2006), http://initonitwantit.blogspot.com/2006_04_01_archive.html (accessed July 15, 2007).

111. Greenup, *QueenCityForum.com.*

112. Iowa Department of Corrections, "Does Prison Substance Abuse Treatment Reduce Recidivism?" (2007), http://www.dom.state.ia.us/planning_performance/audits/07-001-DOC_Substance_Abuse_Report070612.pdf (accessed November 29, 2007).

113. Drugwarfacts.org.

114. Charles J. Hynes, "Prosecution Backs Alternative to Prison for Drug Addicts," *Criminal Justice* 19, no. 2 (2004): 28–33, 36–38.

115. Neil M. Thakur, Rani A. Hoff, Benjamin Druss, and James Catalanotto, "Using Recidivism Rates as a Quality Indicator for Substance Abuse Treatment Programs," *Psychiatric Services* 49 (1998): 1347–1350.

116. Ann Chih Lin, *Reform in the Making: The Implementation of Social Policy in Prison* (Princeton, NJ: Princeton University Press, 2000); Robert Martinson, "What Works? Questions and Answers about Prison Reform," *The Public Interest* 10 (1974): 22–54. Also see "Does Punishment Deter?" *NCPA Policy Backgrounder No. 148* (Dallas, TX: National Center for Policy Analysis, August 17, 1998). The NCPA says, "Relatively little comparable research has materialized to refute Martinson's analysis, although this has not been from want of effort. A possible exception may be a modest superiority for the better-designed interventions in the outcomes of juveniles, and some researchers still believe that appropriate correctional service and treatment can cut recidivism sharply for other criminals, too. Also see Doris Layton MacKenzie, *What Works in Corrections* (Cambridge, UK: Cambridge Studies in Criminology, 2006), 248–54; Serge Brochu, Jacques Bergeron, Michel Landry, Michel Germain, and Pascal Schneeberger, "Impact of Treatment on Criminalized Substance Addicts," *Journal of Addictive Diseases* 21, no. 3 (2002): 23–41.

117. Iowa Department of Corrections, "Does Prison Substance Abuse Treatment Reduce Recidivism?" (Des Moines, IA: Iowa Department of Management Performance Audit Program, 2007), 38.

118. T. Hank Robinson, *Moving Past the Era of Good Intentions: Methamphetamine Treatment Study*, Contract No. 12969-04 (Omaha, NE: State of Nebraska, 2005), 87–89. Also see Dennis J. Stevens and Charles S. Ward, "College Education and Recidivism: Educating Criminals Is Meritorious," *Journal of Correctional Education* 48, no. 3 (1997): 106–13.

119. Edwin Schur, *Radical Non-intervention* (Englewood Cliffs, NJ: Prentice Hall, 1973).

120. Page 391 in Eugene Doleschal, "The Dangers of Criminal Justice Reform," in *The Administration and Management of Criminal Justice Organizations: A Book of Readings*, 3rd ed., ed. Stan Stojkovic, John Klofas, and David Kalinich (Prospect Heights, IL: Waveland, 1999), 378–397.

Third Millennium Changes

Police Subculture, Discretion, and Wrongful Acts

All laws, written and unwritten, have need of interpretation.

—Thomas Hobbes, *Leviathan*, Chapter 26, §146, 21.

LEARNING OBJECTIVES

When you have finished reading this chapter, you will be better prepared to:

- Define police subculture
- Describe the characteristics of a police subculture
- Characterize groupthink
- Describe the most talked-about characteristics of police personality
- Define CYA and its function
- Characterize police discretion
- Articulate methods of limiting police discretion
- Characterize police deviance
- Define police misconduct
- Explain police corruption
- Describe factors leading to wrongful acts
- Characterize methods used to control future wrongful acts

CASE STUDY: YOU ARE THE POLICE OFFICER

When you began your basic training for the job of your dreams after college, you were quick to realize that you were part of a team and that the people who you could count on when things got tough were other cops. That was comforting to know, but some trainees got the impression that outsiders held negative feelings toward the police.

After graduation, when you became a rookie, you decided that your training officer (TO) was a professional officer and you were grateful that she was your senior mentor. She explained the aspects of police discretion: "To arrest or not to arrest should always be centered in the idea of probable cause, but departmental policies have much to say about the process." She explained how the strength of the evidence and the behavior of a suspect might influence your decision to arrest.

Your TO also warned you to be cautious about using force when making an arrest: If you use force, she noted, there is no way you can undo the harm you brought upon another person. "The purpose of force is to further public safety," she emphasized.

You were taught what to do when pursuing a fleeing felon, which in some ways was a different process than you heard from the academy. Even pursing a fleeing felon on the highways required supervisor approval. You never thought that an officer was limited—but then considering that public safety safeguards can be a broader issue than a single incident, those restrictions make sense, you decided.

For instance, your TO explained that often officers lean on a person's status—"You know," she said, "beggars"—rather than on that person's behavior to justify an arrest or the use of force. You learned from your mentor that police discretion is a double-edged sword.

1. Why would officers learn early in their training to depend only on other officers when things got tough?
2. Why was there a difference in what the rookie learned from his TO and the impressions made upon him during his training at the academy?
3. Explain what the officer meant by the thought "police discretion is a double-edged sword."

Introduction

This chapter highlights police subculture, police discretion, and police wrongful acts (deviance, misconduct, and corruption). These three concepts are strongly associated with one another, and their influence can be found to varying degrees in most police departments across the country. For decades, police administrators have tried to control officers through the development of command and control organizations, recruitment processes, training, supervision, rules and regulations, rewards and punishment, specialization, and routine initiatives such as preventive patrol and rapid response to service calls (also see Chapters 6 and 7).[1] For instance, the Christopher Commission (chaired by Warren Christopher) reported in 1991—months after the Rodney King incident—that a significant number of Los Angeles police officers repeatedly used excessive force against the public and persistently ignored written departmental guidelines. The report went on to say that police management was at the heart of the problem because of its failure to provide professional leadership and to deal with aggressive officers.

Aggressive officer conduct in the Los Angeles Police Department (LAPD) changed little as the result of this report, as evidenced by the Rampart scandal. In this series of events, an elite anti-gang unit known as CRASH (Community Resources Against Street Hoodlums) turned itself into a corrupt, money-hungry group of officers, who eventually found themselves on the wrong side of the law.[2] One report criticizing the LAPD implied that the organizational environment at the LAPD "emphasizes control and the exclusion of scrutiny by outsiders."[3] LAPD management had tried to control personnel through a highly stratified, elaborate discipline system that enforced voluminous rules and regulations, but it appeared to be emphasizing overly aggressive policing.[4]

Twelve years later, in May 2007, LAPD Chief William Bratton reports, there was confusion among his commanders as to which supervisor was actually in charge during an immigration demonstration. Street officers made their own decisions and fired rubber bullets and swung batons to disperse the crowd.[5] This ongoing pattern raised a question: What was learned from the earlier incidents?

> The quality of police management and leadership ultimately determines how often officers seek answers and comfort from their peers versus how often officers are guided by senior street officers.[6]

These findings are consistent with practices at police departments across the nation. For example, a study of the Boston Police Department showed that officers were managed through intimidation, which could easily enhance their dependence on one another and lead to them forming strong bonds of brotherhood—that is, a police subculture.[7]

Police Subculture

College students, even though they might be scattered among various campuses across a state or the nation, tend to share similar values and ideas about many things, including music, sexuality, and job expectations. For instance, they share thoughts about "easy professors" and secretly provide new students with the benefits of their experiences. A similar network exists among police personnel. Unlike student cultures, which have little formal power, the power possessed by officers is relatively high. Officers "are seen as monitors of human conduct, standing over and against rather than with their 'client' public and police soon find that this perception interferes with their ability to maintain social relations."[8]

Because humans are social beings who seek friendship, approval, and recognition, they often join religious organizations, form clubs, and enter occupations to meet new friends, develop networks, and attend social events. When anyone starts a new job or a new school, he or she learns the formal rules of the organization from the administration and the informal ways of doing things from the experienced members (other than administrators) of the organization.

Explanation of a Police Subculture

Police officers tend to share similar values and ideas about their jobs, criminals, and expectations of other officers. This sharing can typify what is referred to as a police subculture.[9] Sometimes the values from the at-large culture and the subculture reinforce each other: Christmas is a legal holiday, and it is also a religious historical event. At other times, however, these values are at odds: The drinking age is 21, yet underage drinking is rampant. The at-large culture is similar to a subculture among specific groups of individuals and its tenets—that is, shared ideas, goals, and lifestyles—are learned in much the same way.

A **police subculture** can be described as learned objectives, shared job activities, similar use of nonmaterial and material items, and acceptance of veteran street officers (as opposed to supervisors) as the gatekeepers who educate other officers about their job obligations, responsibilities, and expectations of the job.

The Nature of Police Subculture

Police subculture shapes the hiring, training, and promotion of officers; that is, police subculture makes the officer. Following are characteristics associated with police subculture. These characteristics are listed alphabetically rather than being shown in some order of prominence. Each characteristic is a piece of a subculture, and each crosses over and can influence other characteristics; each can influence subculture participants at various degrees at various times in their careers.

Characteristics of Police Subculture

- Origins in basic training
- Brotherhood of officers
- Cohesion
- Conservatism
- Criminal blindness
- CYA
- Groupthink
- Institutionalized behavior
- Justification of wrongful acts
- Material essentials
- Mentoring by senior patrol officers
- Personality development
- Secrecy
- Thin blue line
- Two worlds perspective
- Use of force

Origins in Basic Training

Rookies quickly learn that their basic academy experience and, in part, their probationary training are simply rites of passage.[10] What they learned in the classroom is seen as irrelevant to police realities on the job; instead, they learn what they need to know on the job from other street officers. Officers believe that they can rely on other officers for assistance and distance themselves from the public and their supervisors. The cadet seeks a sense of security, honesty, and mutual understanding.[11] Rookie officers strive to meet earlier expectations about policing (changing the world), yet now take cues from their new role models about their jobs, their ideals, and their lifestyles. They outwardly appear to accept police practice that is communicated and rewarded by other officers despite the rookie's previous ideals, training, and expectations.[12]

Brotherhood of Officers

As officers move farther away from the public and their supervisors, they move closer to one another. Bonding produces a "we–they mentality" (officers versus constituents) and can be seen in the nonmaterial essentials shared by officers—language, gestures, ideals, goals, and values that are significant only to officers.

Cohesion

Group identification is enhanced through officer beliefs in the constant danger of their job and the need for unwavering dependence among their peers.[13] Officers seek belongingness or a "we-ness" among their peers. Police jobs are no different than other jobs, with one caveat: Officers perceive that they receive little support from the public whom they protect and the supervisors who manage them.[14] There are few individuals outside of policing whom they can talk with about their activities, including spouses; consequently, individual feelings of loneliness and isolation may be neutralized only by communicating with other officers.[15]

Conservatism

According to one police observer, Carl B. Klockars, it is complex and difficult to describe police conservatism as being a specific characteristic.[16]

> Police officer **conservatism** can be defined as defending the status quo—that is, defending traditionalism.

Conservatives are at the political right, yet police conservatives are not as extreme as reactionaries who struggle with the present and want to return or to restore the aspects of the past. What should be remembered is that the police have the task of protecting the status quo. As Klockars puts it, it does not matter whether the issue is the right of the Nazi organization to obtain a permit to march through a Jewish community or the right of an antiabortion group to block the entrance to a clinic. In both cases, police find themselves cast as the defenders of the law (read: traditional ideology) until the law is challenged. Most officers come to accept their conservative role, in part through the following influences:[17]

- Occupational learning that suggests some officers make choices to become the "good guys" while others make choices to be the "bad guys" [18]
- Recognition that people have choices and that good choices produce law-abiding behavior
- Frustration with the courts' leniency[19]
- Frustration with corrections institutions' early-release practices
- Witnessing traumatic victimization

- Threats of officer litigation[20]
- Making suppression of police abuse into a visible priority[21]
- Witnessing the failure of police reform programs
- Witnessing the failure of offender reform programs

Criminal Blindness

Within police departments, there is a focus on the status of individuals as opposed to their behavior. "Want recognition and respect from your coworkers? Then you walk the walk and talk the talk or die alone on the job," says a New Orleans officer. When experienced police officers instruct "students," they describe a "symbolic assailant," a person whose material possessions, attitudes, and behavior, including his or her excuses about bad behavior, fit the "profile" and will cause trouble for officers.[22] These are the individuals to be stopped and frisked, they say; these are the individuals who have fewer constitutional rights than others. "Training officers teach recruits what to look for in terms of potential danger, and officers tell each other stories about things they have seen that mark individuals as dangerous—a type of tattoo, a piece of clothing marking a gang member, or the like." [23]

CYA

An explanation about the relationship between CYA initiatives and their relationship to police subculture would provide a better understanding of officer behavior. If the police subculture is fueled by ineffective police management, then it could be said that CYA is a tool or a device of police discretion—think of CYA as a form of protection against what some officers call "bullshit." [24] It may be for this reason that CYA ("cover your ass") is a popular initiative among police officers, advises John P. Crank.

CYA can be defined as the behavioral response that is least likely to result in disciplinary action or managerial attention.[25]

Organizations attempt to identify and implement protocols and strategies to increase police officer accountability.[26] For instance, one way to enhance constituent quality-of-life experiences is to upgrade the quality of police candidates, so that officers are chosen who will provide quality police services. Agencies may also enhance the cultural competence of officers by hiring more officers of color and women.[27] While this practice has some merit, if the agency fails to address the contradictions between policy and practice, the only thing that is likely to be changed in reality is the intellectual level of the lip-service that officers give to superiors, advises Crank.

One researcher characterizes the tendency of organizational ventures to focus on officers as a preoccupation with the symptoms of a problem rather than with the problem itself.[28] Another writer advises that organizational preoccupation with the officer and not the system represents "the pervasive presence of bullshit," and notes that the "bullshit is as thick in medium-sized departments as it is in large ones." [29] Another expert says that after an evaluation of officer versus commander scenarios, the terms "suspicion" and "betrayal" were appropriate for describing their relationship.[30] Most officers use discretion to distance themselves from their supervisors because it is felt that mimicking the liberal nature of their commanders would be the path to the officer's destruction. Also, part of this perspective suggests that officers pace their work so as not to stand out from others.

How does CYA work? "Do not trust bosses to look out for your interests" or "Always CYA" is the mantra.[31]

Although supervisors were once cadets and worked their way up the bureaucratic ladder, it does not necessarily mean that they were good street officers—perhaps they were just better politicians, according to some officers. This thought is consistent with the Peter Principle. As discussed in Chapter 6, the Peter Principle is a colloquial principle of hierarcheology: "In a hierarchy every employee tends to rise to his level of incompetence."[32]

Groupthink

Groupthink refers to a type of faulty group decision-making process in which groups do not consider all options or alternatives.[33] Instead, they desire unanimity, which may be achieved at the expense of quality decisions because the group has a distorted view of reality. Groupthink occurs when groups are highly cohesive and when individuals are under considerable pressure to make quality decisions. Seven negative outcomes of groupthink are possible:

- Few alternative options (if any) are considered.
- There is little, if any, critical evaluation of others' ideas.
- Alternatives suggested early in the decision-making process are ignored.
- Group members ignore expert opinion and advice.
- Only selective information gathering is performed.
- No contingency plan of action is prepared.
- There is distortion of truth and facts.

Often, rookie officers do not have a chance to develop their own perspective because of groupthink's one–two punch: The first punch—distort the facts—is followed by a right hook—"Here's the real facts, kid." To advance, a rookie either accepts what veteran street officers are "dictating or [finds] another job," says a NYPD veteran.

Institutionalized Behavior

As social relationships form within the department, individual officers typically move closer toward achieving a sense of wellness and a sense of personal wholeness that typically translates into optimal commitment to and performance of the job.[34] When officers are providing police services, their method of delivery tends to become institutionalized, following a standard or a routine. Those standardized techniques often support police officer subcultures to a greater extent than does official policy, because many officers trust one another more than they trust official policy.[35] That is, officers conduct themselves in a certain way to gain approval and respect from their peers because it is "the right thing to do."[36]

Justification of Wrongful Acts

Unfortunately actual officer behavior is often different from what officers say they do—and sometimes it is even illegal.[37] When officers institutionalize their performance to match their subculture's expectation (as described in the previous section), the desired performance could include inappropriate behavior or behavior contrary to official policy.

Material Essentials

Shared material essentials include uniforms, weapons, equipment, technology, and vehicles. When shared material essentials and nonmaterial beliefs become institutionalized, police officers can be said to share an occupational culture or a subculture.

Mentoring by Senior Patrol Officers

Seniority is a key facet of police subculture. Veteran officers can be described as career gatekeepers who instruct, manipulate, or pass along to younger officers what they perceive as appropriate activities and expectations. Gatekeepers maintain a traditional set of attitudes and standardized values. "These are the things a good cop does," says a street officer in Chicago. "If a new guy wants to listen, fine! Because to get along means you learn from the guys who know. If the rookie thinks he doesn't need us old farts, let's see how long it takes him to fry when he's fallen into the fryer."

This method of indoctrination is standard (to varying degrees) among most street officers (and it is found in other occupations, too), and that standard is maintained by gatekeepers through a subculture. The privilege endowed upon senior officers is supported by civil service procedures, unions, and fraternal orders including police associations. The negative side of this practice is that the least experienced officers are assigned the most undesirable jobs such as midnight shifts in the worst part of town. On the positive side, it means that those with the most experience have opportunities to mentor younger officers, thereby enhancing their ability to provide quality services.

Personality Development

Development of specific types of personality is promoted by the police subculture. The Minnesota Multiphasic Personality Inventory (MMPI) and the California Personality Inventory (CPI) tests have produced evidence that traffic officers tend to share certain characteristics: most often, they are "high defended, energetic, dominant, well adjusted, independent, spontaneous, socially flexible, and free from anxiety-related conditions"—results suggesting that "there are personality dimensions that set officers aside from the general public."[38]

Some might think that the officers largely appear to resemble white middle-class males. From a social sciences point of view, however, police can be seen as a homogeneous group, but that group is different from the people whom they serve, who tend to be better educated, receive higher paychecks, and work in a less hazardous environment than the average urban police officer.[39] What can be said is that the characteristics largely demonstrated by most officers resemble the characteristics of the other officers who are within their sphere of influence in the same subculture. This similar personality arises because of the ways that officers see the world and the way they respond to service calls, altercations, stops, and frisks. It is the police subculture that maintains and reinforces the most talked-about personality characteristics among police officers.

In particular, officers are often characterized as cynical, authoritarian, isolationist, and conservative.[40] These characteristics seem to have been around for a while—in fact, they were reported by writers of the Reform Era (discussed in Chapter 2). Nevertheless, these traits remain visible, albeit at different levels and frequencies, among officers today.

A study from 40 years ago reveals that even then officers were considered emotionally maladjusted, authoritarian, impulsive, risk takers, rigid, physically aggressive, lacking in self-confidence, and as having a preference for being unsupervised.[41] This dated study appears to have had regional and methodological flaws, however, and a more recent study shows that most white officers back away from encounters with others. That is, most officers are almost passive in some of their dealings with the public unless a crime is actually in progress. By today's expectations, even when probable cause is present, many officers refrain from confronting a suspect or conducting an arrest for many reasons, including the threat of civil litigation (see Chapter 14).[42] Therefore, it could be argued that any behavior displayed by officers is more a product of the observers' perspective than the reality experienced by officers.

That said, cynical and authoritarian behavior can certainly be learned from the job. Indeed, the job demands that those roles exist to accomplish its mission (or at least officers must provide the appearance that they are cynical and authoritarian).

Cynical and Authoritarian Characteristics Police standards that encourage cynical and authoritative behavior are frustrating in the sense that such behavior may be viewed as unacceptable, yet is often beneficial. Under normal operating conditions, officers often demonstrate cynical and authoritative behavior during a crisis or when involved in a volatile encounter. To control dangerous situations from escalating, cynical and authoritarian perspectives work well toward furthering public safety. It is expected and morally appropriate for officers to take control of a situation and individually to resolve danger.[43]

> **Cynical personalities** can be characterized as pessimistic, suspicious, and sarcastic. **Authoritarian personalities** are characterized by an excessive focus on excessive, rigid belief systems, and the expectation of unquestioned obedience.

Ideally, we want authoritarianism tempered with reason and responsibility. Training, education, and professional supervision can be positive steps toward controlling authoritarian tendencies.[44] Yet, one of the most positive methods of tempering authoritarianism is through establishment of an empowerment process aimed at the community through community relations initiatives.[45] Part of this empowerment process includes a decentralization of the agency—that is, a leveling of police authority to line officers and community members, with an eye toward developing effective problem-solving strategies.[46] Realistically, it might be difficult for many departments to provide an empowered officer under present management models.[47]

Isolationist Characteristics During the Reform Era, police officers became mobile, through the use of automobiles. Those efforts supposedly enhanced professionalism, but they also made officers aware of the emotional threats associated with policing, such as their increased isolation from mainstream society.[48] The isolationism associated with policing might have been even more prominent during the 1960s and 1970s, when the police were referred to as "pigs" and were labeled as symbols of oppression by civil rights protesters. Problem-solving techniques can break down isolationism between police and the community and should help build a "we" instead of "us versus them" mentality.[49]

Secrecy

Officers often have access to other people's property, money, drugs, and personal life experiences, and they often find it easy to manipulate others or influence their behavior (most often in a positive way). Officers routinely have contact with victims of crime, including sex crimes, and how they conduct themselves or abuse those individuals (either intentionally or unintentionally) can depend upon their adherence to the thin blue line (explained below) of secrecy.[50] Knowledge about how to act, what to think, and how to justify police actions (better known as redefining an altercation or circumstance) derives directly from real-world experiences shared by officers through the police subculture.[51]

How often do officers abuse the authority of the job for personal gain or gratification, and how often is that abuse (and other matters) kept secret? In a survey of 925 randomly selected U.S. police officers from 121 departments, researchers found that most officers do not abuse their authority, although the respondents did provide data on different forms of abuse they have observed.[52] While the frequency of that abuse was low, it nonetheless occurs. The fact that the officers who observed the abuse did not report it to their supervisors implies that officers keep a lot of information to themselves about their peers. There are strong indicators that a "blue code of secrecy," also known as the "thin blue line," really exists.

Thin Blue Line

When examining police behavior such as arrest patterns and amounts of force used to make those arrests, the influence of peer pressure cannot be overlooked. For instance, officers are expected to back one another even when their behavior might be inappropriate. To "rat" on another officer will more likely produce disrespect and ostracism than respect and acceptance. Officers see this code as a fact of life. Also, some evidence suggests that when cadets begin basic training, the degree of their ethical orientation linked to idealism and relativism changes; when they reach the streets, their idealism about the world and the things they can fix also changes.[53] It could be argued that the police subculture shapes an officer's ethical perspectives before, during, and after training and once on the job. Occupational conceptions foster isolation and antagonism and provide police with a sense of belonging to a subculture of their own.[54]

Two Worlds Perspective

Police officers live in two worlds: They mirror the police subculture that manipulates them, and they live in the larger world with its cultural nuances at the same time.[55] That is, they live in one world and they work in another. Police subculture is grounded in the everyday interactions with other people, and those interactions take on meaning within the context of the job. Being on the job is different than being at home. When officers allow their job to dominate their realities, home life and relationships become a fantasy. When officers want to cope with their fantasy, they may turn to drugs, alcohol, and corruption (see Chapter 15). Police officers often find themselves having to decide between the expectations of their job and the expectations associated with their off-duty responsibilities.

Use of Force

Although justifiable homicide provisions protect officers from inappropriate prosecution under specific circumstances, police subculture may justify the unnecessary **use of force**. Data gathered from Indianapolis, Indiana, and St. Petersburg, Florida, police departments suggest that the influence of neighborhood members has something to do with the amount of force police exercise during police–suspect encounters.[56] Police officers are significantly more likely to use higher levels of force when suspects encountered live in disadvantaged neighborhoods and neighborhoods with higher homicide rates, net of situational factors (e.g., suspect resistance) and officer-based determinants (e.g., age, education, and training). The effect of the suspect's race is mediated by the neighborhood context. An implication of these findings is that police officers act differently in different neighborhood contexts.

That said, when officers discuss their "day to day" in the locker room or at a weekend get-together, they tend to support and justify the use of unnecessary force, especially when it is used on individuals on the lower economic fringes of society. Because officers routinely work alone or in pairs with little supervision, the only knowledge a department might have about the unnecessary use of force would emanate from a citizen complaint, yet some victims might feel uncomfortable in coming forward.[57]

A former Metropolitan Washington, D.C., police officer and present-day lawyer says, "[I'm]a former U.S. Marine and police officer who has come under gunfire and confronted many fleeing suspects both armed and unarmed. . . . a courageous, physically fit police officer does not behave like a bully or lose control of himself [or herself] when things become tense."[58] Officers who attack constituents can be compared to the members of an unruly mob, who hide behind the front put up by their fellow officers, yet reveal their character as cowards who lack professionalism. "Sadly," the former D.C. officer adds, "in our early tenure as officers, we are instructed on the 'code' of the police subculture. These are norms that are almost always perverse. The first is that if a citizen runs from one of us, we are to beat him severely. Another is that if a citizen physically hurts one of us, we are to hurt that citizen even more before we bring him to the station. And if that citizen has killed an officer, he shouldn't make it to the station alive."

For instance, one study consisting of 558 police officers showed that 75 percent of the officers surveyed from New England to the Gulf of Mexico report they would "rather be tried by twelve than carried by six."[59] That is, officers prefer the prospect of going to court over their behavior to the possibility of leaving an encounter in a coffin, also known as "Shoot first and ask questions later."

What is curious about this finding is that other data from the same sample showed that the officers joined the police because they wanted to help others and put away bad guys. Using force against violators was never their first option. This latter finding is consistent with data provided by the Bureau of Justice Statistics, which shows that among the few officers who did use force during an arrest, pepper spray was used most often.[60] In another large study, most of the officers reported that it is unacceptable to use more force than legally allowable to control someone who physically assaults an officer.[61]

Limiting the Influence of Police Subculture

The police subculture discourages independent thinkers, resulting in few officers who view their responsibilities exclusively in terms of public mandate and official dictum.[62] When peer pressure reinforces a dereliction of duty, in the sense that it encourages officers to make their own rules and act as judge, jury, and executioner, it can lead to the recklessness displayed by Rampart CRASH unit in Los Angeles.

The Community Relations Center[63] advises that police management's response to the police subculture should not be revolutionary, but rather evolutionary. That is, managers should take a stand toward dismantling police subculture's negative influences and should establish an organizational (different from police subculture) culture that begins with a set of desired values. A police subculture's influence can be minimized through both formal means (training) and informal means (here's what you really need to know about your training), values (behavior thought to be appropriate and meritorious), and norms (regulations and laws). First, however, management must limit the influence of the police subculture upon police practice.[64] In fact, larger police agencies will likely need to target several subcultural groups: one at the patrol level, another at the supervisory and middle management level, and one at the administrative level.[65]

Police officials should develop those values by shaping police discretion and challenging the status quo. Sometimes traditional practices or routines are no longer productive. Consider the story of the four monkeys and the cold shower:[66]

> In a conditioning experiment, four monkeys were placed in a room. A tall pole stood in the center of the room, and a bunch of bananas hung suspended at the top of the pole. Upon noticing the fruit, one monkey quickly climbed up the pole and reached to grab the meal, at which time it was hit with a torrent of cold water from an overhead shower. The monkey quickly abandoned its quest and hurried down the pole. After the first monkey's failed attempt, the other three monkeys each climbed the pole in an attempt to retrieve the bananas, and each received a cold shower before completing the mission. After repeated drenchings, the four monkeys gave up on the bananas.
>
> Next, one of the four original monkeys was replaced with a new monkey. When the new arrival discovered the bananas suspended overhead and tried to climb the pole, the three other monkeys quickly reached up and pulled the surprised monkey back down. After being prevented from climbing the pole several times but without ever having the cold shower, the new monkey gave up trying to reach the bananas. One by one, each of the original monkeys was replaced, and each new monkey was taught the same lesson: Don't climb the pole.
>
> None of the new monkeys ever made it to the top of the pole; none even got close enough to receive the cold shower awaiting them at the top. Not one monkey understood why pole climbing was prohibited, but they all respected the well-established precedent. Even when the shower was removed, no monkey tried to climb the pole. No one challenged the status quo.

One implication of this tale is the realization that precedents, when enacted into "policy manuals and training programs, can far outlive the situational context that created them," advise Bennett and Hess.[67] Simply telling officers, "That's the way it's always been done" can do more harm than good. When officers don't know what they don't know and—worse yet—aren't even aware that they don't know, they are kept from empowerment, and their problem-solving efforts toward public safety issues can be seriously compromised, write Bennett and Hess. Encouraging officers to think creatively and responsibly by developing innovative problem-solving skills will challenge the negative aspects of a police subculture mentality and enhance the quality of police services.

Police Officer Discretion

In 1953, U.S. Supreme Court Justice Robert H. Jackson believed that the criminal justice system was in a state of crisis and called for a justice agency study by the American Bar Association.[68] In the study, discretion was found to be used at all levels of the criminal justice system, dispelling the notion that police made arrests simply on the basis of whether a law had been violated. In addition, the study found that low-level decision making was a significant contributor to the crime-control and problem-solving capabilities of police agencies. The realization of the sheer pervasiveness of police officer discretion was stunning, and a hot debate arose about ways to control discretion.

One method of control would be to take advice from Thomas Hobbes: "All laws, written and unwritten, have need of interpretation." That is, officials can write police discretion off the books. Because full enforcement is not an option and because danger is ever present, however, officers must use discretion in enforcing laws, advise Joycelyn M. Pollock and Ronald F. Becker.[69]

There is more to policing than merely enforcing laws, adds George L. Kelling.[70] In Kelling's view, discretion-based policing that reflects the neighborhood's values and the community's sense of justice is more likely to do justice than policing that follows a rule book by rote. Kelling maintains that officers must exercise discretion in their performance on the job.

Nature of Police Discretion

Criminal justice personnel are endowed with discretionary power—that is, the freedom to act on their own—in their jobs.

> **Police discretion** can be characterized as the freedom to act on one's own choices. Although the law legitimizes the police officer's power to intervene, it does not—and cannot—dictate the officer's response in every given situation.[71]

Discretion allows officers freedom to provide quality police services without interference from superiors or regulations in the best interests of the constituent and the community. Its application is often shaped by the police subculture, situational and contextual factors, and procedural guidelines, among other factors.

Factors Affecting Police Discretion

Officers cannot enforce all laws all the time. As a consequence, officers employ discretionary authority in their daily routines, regarding decisions to arrest, stop and question, frisk, write traffic tickets, take a report on a crime, investigate a crime, use physical force, and use deadly force. Rules cannot be developed that will apply for each and every action an officer might possibly consider taking. However, certain factors do influence officers' use of discretion.

Seriousness of the Crime

The seriousness of the crime often dictates whether the police will process an offense formally or informally. For instance, legal factors tend to place more constraints on police officers' discretion in crimes such as homicide and sexual assault, which are perceived as more serious than other offenses. Also, the seriousness or priority of a crime varies with jurisdictions.[72] For example, when their dispatchers receive a 911 emergency call, members of the Boston Police Department might respond to a carjacking before officers respond to an armed robbery, members of the New Orleans Police Department might respond to a home invasion before officers respond to gang violence, and members of the Denver Police Department might respond to a hate crime before officers respond to a home invasion.

Relationship between Victim and Offender

A stranger who commits an offense is more likely to be arrested than a person who commits the same offense but is known to the victim. Officers tend to avoid arrests for minor offenses when the victim and the offender are in a close or intimate relationship with each other, mandatory domestic violence reports aside. Officers are more likely to simply break up an altercation between sisters or friends, whereas they tend to arrest an individual who is suspected of assaulting a stranger.[73] It is not uncommon for police to take sides with a retail proprietor in disputes between businesses and customers, and their willingness to arrest a shoplifter might depend on the proprietor's views on the subject. Mandatory arrest policies are designed to ensure that all felonious assault cases result in an arrest regardless of the relationship between the two parties (see Chapter 10 for more details).

Suspect's Demeanor

Officers employ more severe options, including both reasonable and unjustified force, when confronted with what is perceived as a suspect's disrespectful and antagonistic behavior.[74] One study found that disrespectful suspects report a greater likelihood of police using force.[75] Another study showed, however, that officers escalate the use of force in only 20 percent of altercations among nonresistant suspects.[76] Police action can also invite disrespect from an otherwise cordial suspect and audience.

Department Policies

When policies are inefficient, discretionary behavior is often the most important guide for officers in their dealings with the public.[77] Official attempts to control officer discretion tend to take the form of an ongoing give-and-take relationship.[78] Bureaucratic rules, regulations, and standard operating procedures may be bolstered by extensive use of technology to ensure that departmental policies are followed by officers. A common dilemma faced by police administration (see Chapters 6 and 7) is the tension created when the discretion of officers clashes with the imposition of hierarchical controls.[79]

Extralegal Factors: Race, Gender, and Social Class

Research suggests that extralegal factors such as race, gender, and social class play major roles in police discretion. For example, white officers are more likely to arrest suspects than black officers, yet black suspects are most often arrested by black officers.[80] Departments with relatively more black police officers than white officers are found to have the largest gap in the arrest probabilities for white and black offenders, although whites are more likely to be arrested for assaults than are blacks, regardless of the racial composition of department.[81] Minority status and general situation also influence officer discretion to stop and question suspects.[82] For example, there is evidence that members of poverty-stricken communities are more likely to be stopped, questioned, searched, and arrested than other constituents, regardless of the situation.[83] The "white policing syndrome" (routinely negative white police perspectives of, and treatment of, black constituents in local communities),[84] some say, is readily apparent, although some suggest this syndrome is linked more to social class than being merely color coded.[85] Additionally, it must be acknowledged that professional black officers share the burden of disrespect and feelings of betrayal ("Tom" or "Uncle Tom") from black constituents, particularly in Southern communities where racism (on all sides) is alive and well, making the discretionary stop-and-question process more difficult for officers to pursue.[86]

Fleeing Felon Rule

The fleeing felon rule once gave police broad discretion to use deadly force, but that situation changed when two Memphis police officers fatally shot 15-year-old Edward Garner.[87] In 1985, the Supreme Court ruled in ***Tennessee v. Garner*** that a police officer pursuing an unarmed suspect may use deadly force to prevent the suspect's escape only if the officer has probable cause to believe that the suspect poses a significant threat of death or serious physical injury to the officer or others.

During the Memphis altercation, the officer's flashlight allowed him to see the suspect's face and hands, and the officer was reasonably sure that Garner was unarmed. The officer ordered Garner to halt, but Garner began to climb the fence. Believing that Garner would certainly flee if he made it over the fence, the officer shot him in the back of the head. Garner had $10 and a purse in his possession from a previous burglary. The officers acted according to a Tennessee state statute and official policing department policy authorizing deadly force against a fleeing suspect.

As a result of the *Garner* ruling, police discretion has been limited such that officers can use deadly force only when a reasonable belief exists that imminent danger exists for officers or others. Some jurisdictions have eliminated police discretion related to the use of deadly force altogether, by changing "shoot to kill" directives into "shoot to stop" orders.

Deployment

Perception of higher crime rates in certain areas of a city can lead to greater deployment of officers in those neighborhoods, also called hot spot policing.[88] It is easy to see why officers might interpret this kind of deployment as a silent request from police command to increase arrest rates. Changes in deployment by themselves require justification. For example, officers may be deployed in greater numbers to a certain neighborhood because there is a lot of crime in the area as shown by new-arrest rates.[89] Politics and the media often team up to promote the notion of a crime crisis in urban poor America, or what some call "Ghetto Wars,"[90] to obtain funding from various organizations including the federal government; this funding, in turn, produces heavy callout of patrol and tactical units. Once those units have completed their missions, high arrest rates may be used to justify continuing the callout or deployment.[91] Although an increase in crime can certainly lead to an increase in police presence, an increase in minority presence sometimes has the same results.[92] Management must strategically decide how best to deploy the officers based on need and resources.

Strength of the Evidence

Officers tend to conduct an arrest in situations where the evidence supports a probable cause arrest.[93] When probable cause is weak, an arrest is less likely. In crimes against persons, the strength of the evidence lies with the victim or a witness, and the perceived character of either victim or witness will often influence an officer's arrest decision.[94] Ultimately, officers arrest an estimated 58 percent of individuals suspected of committing felonies and 44 percent of persons suspected of committing misdemeanors.[95]

Civil Liabilities

Officers conduct an arrest more often in cases involving suspects who are less likely to pursue civil litigation against an officer than suspects who are perceived as more likely to litigate (see Chapter 14 for details). The ability of a suspect to effectively use the legal system against the police plays a role in police discretion.[96] How large a role the threat of litigation plays can depend on the seriousness of the crime, the offender's prominence, and the officer's disciplinary history.[97]

Legislation and Statutory Issues

With some arrests, the state's attorney will not prosecute a suspect because of statutory or legislation issue. For example, the Baltimore City State's Attorney has made policy and resource decisions that specific petty crimes, such as littering, will not be criminally prosecuted following an arrest unless the person is a repeat offender.[98] Members of the Baltimore Police Department have been advised to use caution when deciding whether to arrest persons on these charges.

Behavior of Constituents

Police behavior often varies depending on the circumstances and the behavior of constituents; in other words, an officer's behavior is often a product of the individuals whom the officer has encountered in the field. The presence of an audience affects both police and suspects. Often an audience may aggravate the relationship between a suspect and the police or an exchange between an officer and a suspect.[99]

For example, a young man was beaten to death in front of a family center in Baltimore, where children watched the crime unfold.[100] When officers arrived, they began separating children because they were witnesses, and the elderly suspect became tense. Officers learned that a young girl had been raped by the dead victim and her grandfather—the suspect—had tracked him down and killed him. An arrest was conducted, but it was not a simple process because the children knew the raped girl and apparently knew both the dead victim and the grandfather.

Many officers can recount tales of difficult arrests aided—or hindered—by an audience. Sometimes bystanders may believe they're helping the police, while at other times they believe an arrest to be unfair.

Limiting Police Discretion

Officers cannot legally enforce the law at a whim, nor would they have enough time during their shift to do so. They typically face two types of limitations: legal and administrative.

Legal limits include the following considerations:

- U.S. Supreme Court cases (e.g., *Scott v. Harris,* April 2007)[101]
- State court decisions (e.g., the age of consent varies from state to state)
- State law (e.g., mandatory arrest laws in domestic violence calls)

Administrative limits can take the following forms:

- Jurisdictional authority (e.g., cases can raise federal or state jurisdictional issues)
- Supervision (e.g., the LAPD's emphasis on aggressive enforcement)
- Strategy directives (e.g., zero-tolerance policies that dictate an arrest for any violation)

But do such restrictions really govern police officers' actions in the field? Although the task of policing implies that one set of routines is performed, actual enforcement is different because officers do not enforce all the laws, all the time. Officers spend most of their time attending to quality-of-life functions and 15 percent of their time engaged in enforcement initiatives—for example, responding to burglary calls or trying to find stolen cars.[102] Despite all the tactical training, expectations about gunfights, and other myths about the exciting job of the police, real-life policing is often boring and mundane.[103] Some officers might be looking for "action" in all the wrong places, and when they find it, they may respond in such a way that their supervisors take issue with their actions but their peers think "they did the right thing."[104] These contradictions might appear to be isolated incidents, yet most officers (and disciplinary reports) imply that an incredible gap exists between official policy and officer practice.

Consequences of Police Discretion

In some instances, police officers have found themselves being criticized by the public because of the use or misuse of discretionary authority. For instance, where do officers draw the line between police discretion and favoritism (preferential treatment)?

Preferential treatment is especially likely to be given in cases involving other police officers, politicians, and selected others. Consider the events that led to *R. v. Beaudry* (2007 SCC 5). In this case, a police officer was charged with obstructing justice when he failed to gather the necessary evidence to bring criminal charges against another officer whom he had reasonable grounds to believe had been operating a motor vehicle while intoxicated. During the trial, the officer accused of obstruction maintained that he exercised a proper use of police discretion. The trial judge had a different opinion and concluded that the officer had not exercised his discretion appropriately when he deliberately failed to test the offender's alcohol level; instead, the judge ruled, the officer granted preferential treatment to another police officer, which amounted to obstruction of justice. The officer was convicted, and a majority in the Court of Appeals subsequently upheld the conviction. As this case demonstrate, making the wrong call on use of discretionary power may be construed as a failure to carry out one's duty.

In another case, a woman complained to police that she was being beaten by her husband. The officer who responded to the service call decided not to arrest the husband despite probable cause evidence, and subsequent 911 reports indicated that violence against the woman continued. Such a decision—that is, not to arrest—can represent a failure of duty.[105] To counter this problem, some jurisdictions have implemented mandatory arrest policies in domestic violence complaints, which cuts into the discretionary authority once afforded to officers.

One of the most egregious displays of misuse of police discretion occurred when a five-year-old child was handcuffed and driven to the police station from her kindergarten class at St. Petersburg, Florida.[106] It was reported on a popular television program that this kindergarten student acted out in class; when three officers responded, they decided to act, despite learning that her parents were driving to the school.

Some observers think that police discretion can easily be abused and gives rise to outcomes that can be described as conspicuous disparity or a failure to perform.[107] One researcher suggests that the best alternative is for police management to educate officers on how to think about their individual jobs and to talk about this issue with them in an effort to improve their performance.[108]

Administrative Rule Making

Administrative rule making can guide police discretion by creating written departmental rules requiring written reports.[109] For instance, such a policy may include the following elements:[110]

- *Confining discretion.* Rules may be developed that fix the personal boundaries of officers. For instance, management may replace assumed "shoot to stop" orders with a formal defense-of-life standard. This allows officers to understand which directives are applicable in an altercation.
- *Structuring discretion.* A rational system for developing policies should be developed that includes both a method of evaluation and a method of change. In this manner, the process of policy and policy development—once a closely held secret—becomes exposed to everyone, enabling officers to understand what the policy is and how to change it.
- *Checking discretion.* Report writing and report evaluation aid in control. For instance, rules may state that each time an officer discharges his or her weapon, a report must follow, which is then reviewed by supervisors. Reports should be filed and witnessed.[111] In this way, officers and their supervisors may be held consistently accountable for their actions.

There are some limitations to administrative rule making. First, it is impossible to write a rule for every situation; there are always ambiguous situations that arise or situations whereby the police subculture can help an officer redefine (change the scope or nature of an interaction) a situation to fit a more favorable rule. Second, creating more rules encourages officers to learn to deal with rules in negative ways by describing (half-truths) of their activities to fit regulations. Third, complex written rules may create uncertainty for officers because they may not know which rule or even which philosophy to follow.

In an attempt to satisfy police command, officers might make different decisions than expected regardless of the rules or the philosophy of management. For example, in their effort to second-guess police policy, officers might develop informal ways of handling situations involving troublesome people such as the homeless, mentally disturbed, prostitutes, juveniles, and people under the influence of alcohol or drugs. Namely, they might simply dump them, through what is referred to as **police-initiated transjurisdictional transport (PITT)**—that is, by transporting a troublesome person out of the officer's jurisdiction and releasing the person into his or her own recognizance somewhere else. Although the informal policy of transporting troublesome persons to another location (i.e., police "dumping" of problematic citizens) is generally acknowledged by police practitioners, it has rarely been discussed in the policing literature or systematically studied.[112] Dumping is a low-visibility activity that occurs outside the legal and moral norms; regardless of written administrative rules, few would ever know of the transfer.

Wrongful Acts

Substandard officer behavior can be described through several approaches. Employment practices of many departments show that officers are dismissed (involuntarily separated) for substandard performance, which includes deviance, misconduct, and corruption. Of course, many officers leave policing in good standing.[113]

Before examining wrongful acts, it should be acknowledged that while the media overemphasize police corruption, only a small number of officers account for a great proportion of misconduct allegations, citizen complaints, civil suits and judgments, and other indicators of violence or corruption.[114] For instance, the NYPD's Michael Dowd was the subject of 20 or more misconduct allegations in the seven years before he was arrested; his Philadelphia counterpart, John Baird, was the subject of 26 or more complaints before his arrest. Chicago's Joseph Miedzianowski was convicted of hiding a wanted killer from police, shook down drug dealers for cash and narcotics, provided warring gangs with semi-automatic handguns, misused confidential informants, fixed state drug cases, and placed undercover officers' lives in jeopardy by revealing their identities.[115] A veteran of the CPD's gang crime unit, he was nevertheless a major drug dealer in the city. (See the "Ripped from the Headlines" feature for more about Miedzianowski's successful intimidation of federal special agents.)

In this book, "police wrongful acts" is used as a generic description of poor police officer performance, which may be parsed into three parts that are inconsistent with an officer's official authority, organizational authority, values, criminal and civil laws, and standards of ethical conduct.

> **Wrongful acts** are a generic concept that includes three categories: police deviance, police misconduct, and police corruption.

Each category is distinctively different from the others, and each can influence the others in numerous ways at different times in the career of an officer.

Types of Police Wrongful Acts

Some officers engage in deviance by virtue of their positions as police officers. For example, Frank Serpico, an NYPD officer, reported corruption in his department. His allegations led to the formation of the Commission to Investigate Alleged Police Corruption (also known as the Knapp Commission, after its chairperson Whitman Knapp). The Knapp Commission identified two classes of corrupt officers: "grass eaters" and "meat eaters."

- **Grass eaters** are petty corrupt officers who, as a result of peer pressure, accept gratuities and solicit payments from contractors, tow-truck operators, gamblers, and the like, but do not actively pursue corruption payments. Grass eaters imitate the daily deviants they see and investigate, and their wrongful acts are seen as a way to prove their loyalty to the police subculture.
- **Meat eaters** are premeditated corrupt officers who aggressively seek out situations where they can exploit others for financial gain, such as shaking down pimps and drug dealers. They rationalize their own guilt by convincing themselves that the victim deserves the shakedown.

Police Deviance

One definition of police deviance is as wrongful acts characterized as violations of unwritten norms (rules), values, and expectations, but that does not include a victim.

> **Police deviance** includes all kinds of activities that the police engage in but in which they demonstrate unethical deviation from unwritten norms, values, and expectations.

Police deviance can be described as breaking normative police bonds by engaging in favoritism and nepotism. For example, it could involve using police authority to provide unfair breaks to relatives, friends, and other people an officer wants to make friends, such as helping a sister-in-law out of a speeding citation or flirting with a suspect who a patrol officer stopped for speeding. Civil servants such as teachers, mental health workers, and police officers, among others, are held to a higher standard, and their behavior on and off duty are subject to accountability (see Chapter 14).

A deviant act by an officer directed toward a police organization can be described as internal deviance. "[W]hen deviance is directed internally and is intended for a designated target (usually another member of the organization), deviance is guided by the structure of the organization."[116] The right thing to do in this case is to fulfill a set of implied expectations rather than stated (laws) expectations. For instance, officers may choose to exploit officers at the lowest levels of the police hierarchy. These abuses may come in the form of use of authority to force compliance with illegal organizational activities, domination, or sexual exploitation.[117] Officers at the lowest levels of the police hierarchy might direct their deviance toward the organization as a whole. Officers deviate through resistance to domination and exploitation by the police agency.

Police Misconduct

Police misconduct is better known as "behavior unbecoming an officer"; this term describes officer performance that is inconsistent with an officer's official (written) authority, organizational (written) authority, values, and standards of ethical conduct."[118] Most often, these violators can be regarded as grass eaters. Officers may engage in deception, destruction of property, perjury, and false reporting in an attempt to undermine the agency, organizational members, or the criminal justice system.[119] In most cases, these wrongful acts are regulated by status, regulations, and ethical standards.

> **Police misconduct** is the violation of (written) departmental procedures, regulations, or codes and may lead to civil litigation and organizational consequences.

A chief of police, now a circuit court judge, identifies the following acts as areas of police misconduct in which disciplinary actions against officers have been supported by court decisions:

> Abuse of sick leave, failure to adequately enforce traffic laws, lying about drug use of an acquaintance, failure to investigate crimes while off duty, threatening another officer with physical violence, unacceptable job performance, unexcused absences from work, off-duty drunkenness, excessive parking tickets, leaving duty to conduct personal business, off-duty firearms incidents, failure to complete reports, failing to obey a direct order, falsifying overtime records, failure to report misconduct of a fellow officer, failure to inventory confiscated property or evidence, cheating on promotion exams, patronizing a bar while on sick leave, and refusing to take a polygraph.[120]

Largely, these acts violate organizational regulations. At some point, however, they may evolve into actions that are defined as criminal acts or police corruption depending on the seriousness of the violation and the remedies pursued by the department. The Knapp Commission and other sources have identified several types of police wrongful acts that are formally classified as misconduct (and sometimes corruption):[121]

- Taking gratuities
- Bribery
- Shakedowns
- Use of profanity
- Having sex while on duty or while participating in duty-related activities
- Sleeping on duty
- Drinking and abusing drugs while on or off duty
- Perjury
- Misuse of confidential information

Taking Gratuities

A gratuity is the receipt of freebies—free meals, free services, or discounts on products. While these may be considered fringe benefits of the job, they nonetheless violate the police code of ethics because they involve financial rewards or gains that occur by means of the officer's position. Taking a gratuity is considered corruption because it places the officer in a compromising position ("one favor deserves another").

Gratuities (favors) may take any of several forms: mooching, chiseling, and bribery. In mooching, there is an implied favor; chiseling exists when the officer is quite blatant about demanding free services. These smaller gratuitous services and freebies may eventually lead to more serious abuses such as bribery, payoffs, and outright theft.

Bribery

Bribery is a matter of an officer accepting money or other forms of services in exchange for favors. The officer crosses the line when he or she employs extortion (receiving property or money from another through coercion or intimidation or threatening an individual with physical harm); payoffs are payments demanded by an officer in return for police services or lack of them. Police theft is similar to larceny by any other individual. Police skimming can be described as taking evidence from crime scenes or other places officers have access to by the nature of their jobs and using those items for personal gain.

Shakedowns

Shakedowns occur when an officer succeeds in coercing a bribe (i.e., money, services, or an item). In exchange, the officer agrees not to conduct an arrest; for instance, he or she looks the other way when drugs are sold, does not file a complaint, or does not impound property. For example, NYPD officers Dennis Kim and Jerry Svoronos and brothel operators Gina Kim and Geeho Chae were all arrested on bribery charges relating to the protection of a brothel in Flushing, New York.[122] Agents seized approximately $800,000 in cash, believed to be the proceeds of the brothel, from Kim and Chae's vehicle and residence.

Officer shakedowns can take many forms. For example, officers who are drug abusers may confiscate illegal drugs from dealers to feed their own addiction or the addiction of an intimate other or someone they want to be intimate with.

Use of Profanity

Profanity can be described as the use of offensive, obscene, or profane language; The following typology exists:

- Words having religious connotations (e.g., "hell," "goddamn")
- Words indicating excretory functions (e.g., "shit," "piss")
- Words connected with sexual functions (e.g., "fuck," "prick")

Words and phases containing religious connotations are considered the least offensive of the three, whereas profanity related to sexual functions is considered the most offensive. Some argue that use of such language by officers is purposive and does not indicate a loss of control or catharsis, especially given that officers are trained to communicate in a "command voice" so that they can have a "command presence." Sometimes they utilize profanity for the following purposes:[123]

- To gain the attention of citizens who may be less than cooperative
- To discredit someone or something, such as an alibi defense
- To establish a dominant–submissive relationship
- To identify with an in-group, the offender, or police subculture
- To label or degrade an out-group

Of these purposes, the last usage could be of the most concern because it indicates that the officer has made the transition from prejudice to discrimination, especially if racial slurs or epithets are involved. Of course, some officers don't know any better or don't care how they affect others—until their behavior is corrected by a civil suit or effective supervision.

Having Sex while on Duty or while Participating in Duty-Related Activities

Police sometimes give breaks in return for sexual favors from offenders or suspects. For instance, a pornography star claimed that a state trooper who stopped her on a highway let drug charges slide in exchange for oral sex. In this case, Tennessee State Trooper James Randy Moss was suspended from duty. As follow-up to the woman's complaint, investigators interviewed the porn star, and they also found that the trooper had used his patrol laptop to access the porn star's website while on duty.[124]

Sleeping on Duty

Sleeping on duty typically occurs when an officer sleeps in his or her patrol vehicle. For instance, a 19-year veteran of the Cudahy, Wisconsin, police department was found to have "demonstrated a complete lack of investigative initiative" by not questioning a woman who was later sexually assaulted and held captive by a man with whom the police officer allowed the woman to leave.[125] In making her complaint, the woman awoke the police officer while he napped in his squad car behind a skating rink.

Drinking and Abusing Drugs while On or Off Duty

When officers drink and abuse drugs, the combination of this behavior with their training and authority often places others at risk (see Chapter 14). A study that examined career histories of NYPD officers over a 21-year period found a considerable number of officers were dismissed because of their refusal to take part in a police drug-testing program.[126] Other officers were caught possessing or trafficking drugs.

Perjury

Police perjury usually is a means to effect an act of corruption by leaving out certain pertinent pieces of information so as to "fix" a criminal prosecution.[127] The giving of "dropsy" evidence is typical of such perjury, where the officer testifies untruthfully that he or she saw the offender drop some narcotics or contraband. Lies that Miranda warnings have been given, when they haven't, are also typical. Lying in court is called "testilying," reflecting the fact that officers are trained witnesses.[128] Other actors in the system—including supervisors and even judges—are often aware of perjury, according to one source.[129] The rationale behind this type of misconduct is simple: The officer gets credit for a good bust; the supervisor's arrest statistics look good; the prosecutor earns another conviction; the judge can provide words of wisdom to the defendant without endangering his or reelection prospects; the defense lawyer earns his or her fee in dirty money; and the public is thrilled that another criminal is off the streets.[130]

Misuse of Confidential Information

Police leaks can be typically described as efforts to intentionally divulge confidential information that may detail a drug raid or some other police matter. A police leak (with a little twist) is typified by the case (March 7, 2007) referred to as "Tarnished Blue" in Memphis, Tennessee.[131] Deputy Thomas Braswell of the Shelby County Sheriff's Office received $500 from an informant (a federal agent) who posed as a dealer in steroids and cocaine. The informant received information from Braswell about police CIs, including printouts from the Sheriff's Office computer system that listed driver's license data and history, birth certificate data, and Social Security information. Penalties for this charge (Section 1951) include a prison term of up to 20 years and fines of as much as $250,000.

Police Corruption

Police corruption occurs when the officer has criminal intent and commits criminal behavior, either while the officer is off duty, while the officer is on duty, or when he or she uses the job to perform the crime. Off-duty officers may commit crimes that are violations of the criminal code (e.g., off-duty burglaries, domestic assaults, or tax evasion), all of which are certainly crimes and have little to do with the officer's job.[132]

> **Police corruption** includes those forbidden acts that an officer commits outside the job, on the job, and use of the job for personal gain.

For instance, a 10-year Chicago police veteran, Robert Gallegos, was off duty when his son came home and reported that a teenager quarreled with him at school.[133] Gallegos tracked the 15-year-old boy down to an alley near the high school and broke his jaw. Witnesses said that the young man and two friends were approached by Gallegos and another man. Gallegos threw the teenager into a garbage dumpster and punched him about the face. After the teen fell to the ground, the other man kicked him.

On-the-job officers can also commit corruption when they abuse their authority, exercise inappropriate discretion, and employ excessive force. Sometimes it is the job itself that allows the officer to commit the crime. For instance, an officer who steals drugs from an evidence locker and sells those drugs has engaged in police crime because his or her employment status created access to the evidence locker and, therefore, made the crime possible.[134] Often, these violators represent true meat eaters.

The following forms of police corruption were reported by James J. Frye and Robert Cane.

Personal Use of Confiscated Property

Local agencies are protected through the Comprehensive Forfeiture Act of 1984 when they confiscate property from criminals. It is illegal for officers to confiscate drugs, money, and other items and then use them for personal gain or consumption. Such corruption might be intended to further an officer's drug habit or lead to a more comfortable lifestyle, but it negatively influences family members, friends, and coworker circumstances.

Personal Use of Departmental Equipment or Property

This violation can involve equipment, weapons, vehicles, money, and personnel.

Falsifying Evidence

Falsified evidence can include forged evidence or tainted evidence that is used either to convict an innocent person or to guarantee the conviction of a guilty person. Some evidence is forged because the person doing the forensic work finds it easier to fabricate evidence than to perform the actual work involved. For example, Pamela Fish, the star forensic analyst for the Chicago Police Department, lied about the extent of her work; the falsified evidence led to the conviction of many defendants who were eventually exonerated.[135]

False Arrest and Imprisonment

False Arrest A false arrest claim is an intentional tort and requires a showing of (1) an arrest under process of law, (2) without probable cause, and (3) made maliciously as found in *Desmond v. Troncalli Mitsubishi*, 243 Ga. App. 71, 532 S.E.2d 463 (2000).

That is, a false arrest is physically detaining someone without the legal right to do so.[136] (Also see Chapter 14 for more detail.) In this particular case, Victor E. Kappeler defined false arrest as the unlawful seizure and detention of a person by a police officer.[137] In most cases, false arrest claims against police officers consist of those cases where individuals believe that the police did not have probable cause to stop them.[138] Only when the arresting party knowingly with intent detains someone who has not committed a crime, is the false arrest itself a crime. To avoid liability resulting from claims of false arrest, officers can obtain a warrant; securing a search warrant prior to arrest takes the authority of determining probable cause away from police officers, and places the responsibility on judges, argued Kappeler. (See Chapter 14 for more detail.)

False Imprisonment False imprisonment can be defined as the unlawful detention of the person for any length of time, whereby the individual is deprived of his or her personal liberty, argued Harvey S. Gray.[139] The elements of the cause of action are (1) the detention and (2) the unlawfulness of the action as found in *Rbarbe v. Collins*, 219 Ga. App. 63, 463 S.E.2d 922 (1995).

When officers imprison someone illegally, it is a serious matter. For instance, in Riverhead, New York, a bar owner escorted one of his customers into the nearby police department for littering.[140] Chief George Hesse and three officers dragged the tourist into a room and kicked him. The Suffolk County District Attorney said that the police "acted as thugs in police uniforms," and charged the chief and his officers with unlawful imprisonment, reckless endangerment, and hindering prosecution. (See Chapter 14 for more detail.)

Stalking

Stalking is when a person intentionally, and for no legitimate purpose, engages in a course of conduct directed at a specific person, and knows or reasonably should know that such conduct is likely to cause reasonable fear of material harm to the physical health, safety, or property of such person, a member of such person's immediate family, or a third party with whom such person is acquainted. How often do officers stalk others? Not very often. Nevertheless, the chief of Perry and Brady Township (Pennsylvania) was arrested for stalking several young women in his jurisdiction.[141] A 19-year-old woman reported that the chief called her 37 times in a six-day period. She testified that the chief commented on her appearance, telling her she was "beautiful" and "hot." The report on this incident also included interviews with two other women who said the chief had made inappropriate comments to them.

Denial of Due Process Rights

When an officer deprives a constituent of a constitutional or federally protected right during the course of any police activity, it is violation of a fundamental principle governing American society. Outcomes arising from a denial of due process rights may include both psychological and physical harm, including death. Remedies arising for those violations may include criminal or civil sanctions.

A specific example is demonstrated by the case of Philadelphia officer Christopher Rudy, who was on duty but visiting friends and drinking alcohol at a warehouse.[142] There had been a dispute between the warehouse owner and Frank Schmidt, who was accused of stealing items from the warehouse. Rudy, who was friendly with the owner, was at the warehouse when Schmidt telephoned about the dispute. Schmidt said he was afraid of the owner, but Rudy told him to come to the warehouse to talk about the theft. Said Rudy, "I'm an officer; ain't nothing going to happen." When Schmidt arrived, the warehouse gates were locked, he was beaten and threatened, and the owner put a gun to his head. Rudy watched and laughed. Rudy got a 12-day suspension for failing to take police action, inflicting physical abuse, providing false statements, and conduct unbecoming a police officer, but he was eventually returned to active duty.

Brutality

Although brutality includes excessive use of force (discussed later), it merits its own category. Police brutality consists of excessive violence or threats of violence by the police that do not support a legitimate police function.[143] It can include the unnecessary use of physical force, assault, verbal attacks, and threats by officers.

Excessive Use of Force

Police are authorized to use justifiable or legal force in specific circumstances, but excessive force is another matter.

> **Excessive use of force** occurs when the level of force applied during a police function exceeds the level considered justifiable under the circumstances.

When excessive use of force is apparent, police activities come under public scrutiny. The use of force among police officers is a relatively rare event, but studies suggest that when it does occur, it often escalates to the level of excessive force.[144]

Excessive force is distinguished from justifiable homicide,[145] which is defined as follows:

> **Justifiable homicide** is an act committed with the intention to kill or to do grievous bodily injury, under circumstances which the law holds sufficient to exculpate the person who commits it.

Most often the taking of another's life is considered justifiable under the following circumstances:

- When a judge or other magistrate acts in obedience to the law
- When a ministerial officer acts in obedience to a lawful warrant, issued by a competent tribunal
- When a subaltern officer or soldier kills in obedience to the lawful commands of his or her superior
- When the party kills in lawful self-defense

Because police possess the legal authority to use force and to detain suspects, it is sometimes difficult to distinguish appropriate use of force from inappropriate use of force. Although there are many examples of situations where police use of force was excessive, inappropriate, or illegal, the actual use of force by police officers is infrequent.[146] In a study of six U.S. jurisdictions, most adult custody arrests (83 percent) did not involve any use of force, most uses of force (77 percent) involved weaponless tactics, and the most frequent weaponless tactics (49 percent) were grabbing or choking the suspect.[147]

Another study on the use of force (although it is somewhat dated) found that of the 21 percent of the U.S. public older than age 16 who had experienced at least one

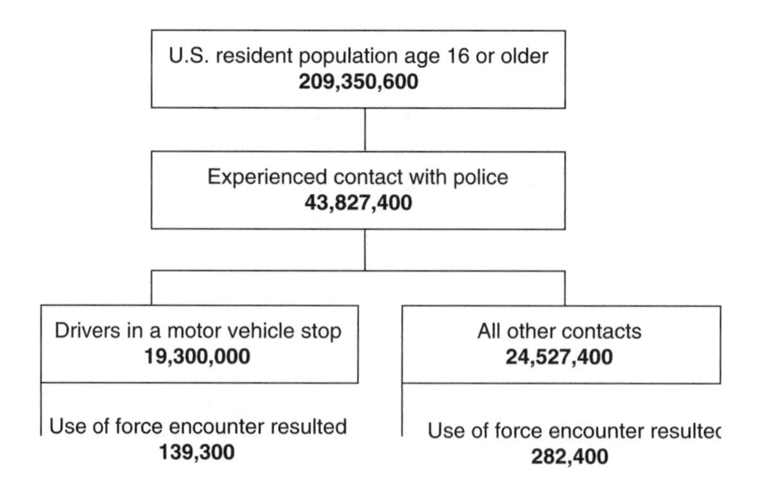

Figure 13–1 Contacts between the Police and the Public
Source: Bureau of Justice Statistics, "Contracts with the Police and the Public" (NCJ 184957; 2001), http://www.ojp.usdoj.gov/bjs/pub/pdf/cpp99.pdf (accessed May 18, 2007).

face-to-face encounter with a police officer (approximately 44 million people), an estimated less than half of 1 percent were threatened with use of force or had force used against them by police (Figure 13–1).[148]

The use of force can lead to other problems, such as fractious race relations. For instance, the National Advisory Commission on Civil Disorders (better known as the Kerner Commission, after its chairperson, Illinois governor Otto Kerner) is perhaps best known for its conclusion that the United States is moving toward two societies: one black, one white—but separate and unequal. The Kerner Commission also looked into the causes of the many urban riots, however, and it concluded that every riot during the 1960s could be traced back to police brutality.[149] Police brutality or alleged police brutality is often closely associated with riots: The more excessive force used by police and the more intense the accusation of police brutality, the greater the riot's intensity.

Perceptions of the sources of riots have changed somewhat in recent years, however. For example, consider the racially charged riot that occurred in Cincinnati following the acquittal of an officer charged in the killing of an unarmed pedestrian, Timothy Thomas. Leaders of the rioters include black politicians and activists, led by the Rev. Damon Lynch, whose slogan was "Fifteen black men"—between February 1995 and April 2001, 15 black males younger than 40 were killed by Cincinnati police or died while in their custody. These African American leaders claimed that Cincinnati's officers were indiscriminately mowing down black citizens.[150]

But were the police actually guilty of the charges thrown in their direction? In this case, some think that Lynch may have profited from racial tension and stirred up the racial hatred against police that set the stage for the riots.[151] Lynch said that the list of 15 police victims showed the "depraved nature of Cincinnati's officers," when, in fact, the depraved nature of the criminals appeared to be the root of the problem. For instance,

Harvey Price, who headed the roster of "victims," axed his girlfriend's 15-year-old daughter to death and then held a SWAT team at bay for four hours with a steak knife, despite being maced and hit with a stun gun.[152] When he lunged at an officer with the knife, the officer shot him. Besides the Thomas shooting, only three of the other 14 cases raised serious questions about officer misjudgment and excessive force. In two of the cases, the "victims" attempted to kill police officers using automobiles, including one incident in which a 12-year-old car thief dragged an officer to his death before succumbing to his own wounds.[153] The notion that race was the controlling element in the 15 Cincinnati deaths does not fit the evidence and the available information.

The Police Foundation's national survey has consistently explored officer views on the abuse of police brutality and authority.[154] (See Table 13–1.) Key findings from this survey include the following reports by officers:

- It is unacceptable to use more force than legally allowable to control someone who physically assaults an officer.
- Extreme cases of police abuse of authority occur infrequently.
- Departments take a "tough stand" on the issue of police abuse.
- At times their fellow officers use more force than necessary when making an arrest.
- It is not unusual for officers to ignore improper conduct by their fellow officers.
- Training and education are effective ways to reduce police abuse.
- A department's chief and first-line supervisors can play important roles in preventing police from abusing their authority.
- Community-oriented policing reduces or has no impact on the potential for police abuse.

The survey has also found that race remains a divisive issue among officers. In particular, black and nonblack officers reported significantly different thoughts about the effect of a constituent's race and socioeconomic status on the likelihood of police abuse of authority. Among the 912 officers who responded to the survey, almost 25 percent agreed that police are not permitted to use as much force as is often necessary in making arrests. Twenty-one percent reported that it is sometimes acceptable to use more force than is legally allowable to control someone who physically assaults an officer.

TABLE 13–1 General Attitudes toward the Use of Force (in Percents)

Question	Strongly Agree	Agree	Disagree	Strongly Disagree
Police are not permitted to use as much force as is often necessary in making arrests.	06	25	61	08
It is sometimes acceptable to use more force than is legally allowable to control someone who physically assaults an officer.	03	21	55	20
Always following the rules is not compatible with getting the job done.	04	39	50	08

Note: All percents rounded

Source: Adapted from David Weisburd and Rosann Greenspan, "Police Attitudes toward Abuse of Authority: Findings from a National Study," National Institute of Justice (Washington, DC: Department of Justice), http://www.ncjrs.gov/txtfiles1/nij/181312.txt (accessed July 22, 2008).

Finally, 39 percent said that always following the rules is not compatible with getting the job done, although 50 percent disagreed with that view.

Factors Leading to Wrongful Acts

Some officers succumb to greed, opportunity, and poor moral judgment when faced with the temptation to commit a wrongful act—similar to behavior of the populations they protect. Selected factors that promote wrongful acts are profiled here.

Police Subculture

Police subculture can be a dangerous rival to public safety and justice. For instance, it took six months for Chicago police commanders to admit that six off-duty officers attacked and assaulted four visiting businessmen, despite videotaped evidence that showed the attack.[155] In a similar case, which was also caught on videotape, another off-duty Chicago officer attacked and assaulted a female bartender. Both videotapes were shown to the CPD's Office of Professional Standards, but the subsequent investigations stalled when officers closed ranks and protected their own.[156] Ultimately, a federal investigation into the crimes motivated the Chicago mayor to request the chief's resignation.

Absence of Standards

When ineffective leaders in a rigid hierarchical police organization fail to develop adequate standards or govern officers' conduct effectively, their perceived weakness can give rise to a code of silence among officers. Such a united front may enhance the police subculture and opens the door to wrongful acts.[157]

Norms That Prevent Reporting of Wrongful Acts

"Ratting out another officer is the second worst thing an officer could do. The first worst thing is letting the officer get away with shit," says a Chicago officer. An officer may ignore misconduct or corruption, which could be seen as equivalent to the wrongful act itself. Often the police subculture deals with the wrongful act internally.

Why not inform authorities about the problematic behavior? According to Herman Goldstein, "If an officer wants to count on another officer when his own life is in danger, he cannot afford to develop a reputation for ratting." [158]

A Nashville officer adds that another way to deal with misconduct of other officers is "not to rank [be promoted] up, because if you do and you know about shit [wrongful acts committed by officers], then you have an obligation. I intend on being a patrol officer until I retire."

Neighborhood Influence

Encounters between police officers and suspects take place in many different types of neighborhoods, and extralegal factors sometimes influence the outcomes of these police–suspect encounters. In particular, residents of economically disadvantaged, predominately African American neighborhoods express lower levels of satisfaction with the police.[159] For example, low-income blacks living in high-crime neighborhoods in Washington, D.C., report more often that the police unfairly use excessive force against their low-income neighbors than against the middle-class blacks and whites from two nearby affluent neighborhoods. Since the behavior of suspects influences the discretion of officers, they are more likely to retaliate when suspects demonstrate disrespectful attitudes and behaviors.

Police sometimes use offensive language, such as profanity and racial slurs, that aggravate already-volatile situations. This finding should be taken with a grain of salt, however, because officers do not report the amount of profanity and disrespectful attitudes often employed by volatile populations. Nevertheless, working on a daily basis with a volatile population could lead to a hardened attitude toward the individuals encountered, or at least that might be the perspective promoted by the police subculture.

Little Supervision and Less Accountability

A study conducted among narcotic officers reveals that ignorance and idealized versions of justice originally played a role in many of these officers' corrupt behavior, which can be seen as following a progression or process.[160] In fact, factors predicting the corruption of a narcotic officer who continually denies due process (arrests without probable cause or reasonable suspicion) and continually applies street justice (such as assaults, theft of product, and extortion) include being originally motivated by the notion of an idealized justice system. Such officers tend to see themselves as the good guys who have to win the war on drugs or America will perish. Narcotic officers, however tend to be supervised far less than other types of officers and are less often held accountable for their behavior—which allows them to slip into a pattern of wrongful acts without drawing notice from their superiors. At the same time, their idealized perspectives are often visible to drug traders, who may manipulate those officers into doing their bidding. Once hooked by the trader, officers may have few choices about right or wrong and committing legal versus illegal acts. Some officers fall victim to criminal influence through manipulation, only to later be kept in line by intimidation.

Hiring Process Inefficiencies

When researchers examined bad officers, their findings were consistent with those reported in Chapters 8 and 15: Hiring practices make a difference.[161] That is, a candidate's prior history is a predictor of future police misconduct and corruption. According to this perspective, bad officers are not made; they saw the benefits of deviance, misconduct, and corruption before they ever became police officers.

Researchers who examined officer wrongful acts at the NYPD argue that prior employment disciplinary and reliability problems or high average rates of annual complaints while employed by the NYPD put the officer at higher risk than other officers for engaging in career-ending police misconduct. In contrast, officers who appear to be committed to the NYPD organization—as evidenced by promotion to supervisory ranks and increased years on the job—are significantly less likely than other officers to engage in career-ending occupational deviance.[162]

Education

Some argue that a college degree might provide an officer with a broader understanding of society and cultural issues and, therefore, reduce prejudice.[163] Other studies suggest that college-educated officers might exercise better judgment in terms of police discretion practices, which can translate to less reliance upon others and the police subculture.[164] A college graduate is seen as being able to provide police services equally as well as an officer without a college degree, but the influence of a police subculture would be less meaningful to him or her and the decisions linked to police discretion would be grounded in better judgment. One study reports that college-educated officers tend to be the focus of fewer constituent complaints and use or threaten to use force less often than other officers.[165] Other researchers suggest that college-educated officers are personally more satisfied with themselves than non-college-educated officers and are more likely to make quality decisions for their families and their departments.[166]

Chronic Criminal Behavior

One view of corrupt police behavior is that it consists of acts committed for personal gain.[167] Another expert sees police misconduct and corruption as a profit-motivated behavior.[168] Both thoughts are consistent with the idea that police misconduct can include both legal and illegal behavior, depending on organizational regulations. Yet both also reinforce the perspective that humans act with intent. Corrupt officers have the intent of committing a crime, which implies that they were corrupt before they ever became officers.[169] (Also see Chapter 8.)

The criminological literature makes no distinction between corrupt violent officers and criminals. Instead, it advises that any chronic criminal behavior can be a product of the following factors:[170]

- Lack of self-control (instant impulse, rather than conscious reasoning)
- Low self-esteem
- Cultural mandates
- Life experiences
- Relationship expectations
- Rage (suggesting little self-control)
- Faulty reasoning; poor problem-solving techniques
- Pure selfishness
- Drugs/alcohol acting as an enabler to commit desired crime
- Apathy

RIPPED from the Headlines

Jury Awards $9.75 Million in Police Corruption Suit

A jury awarded nearly $10 million Thursday to a pair of federal agents who say a corrupt Chicago police officer ruined their lives. The agents claimed the city failed to fire the officer when they blew the whistle on him.

Diane Klipfel and Michael Casali say they have been fighting for 15 years to get someone to believe them. Klipfel, a former ATF agent, and Casali, who is still with the bureau, investigated former Chicago officer Joseph Miedzianowski in 1992. Miedzianowski has since been convicted of corruption and is serving a life sentence in prison.

"We were federal agents. I still am, and I mean this happened to us. You can imagine a citizen trying to make a complaint against a bad officer. What would happen to them?" said Michael Casali, plaintiff.

The couple accused Risley and the department of covering up Miedzianowski's illegal activity. And jurors believed them.

It took seven more years after Klipfel and Casali first made allegations against Miedzianowski before he was eventually charged and convicted by the FBI, not the city.

Diane Klipfel and Michael Casali's attorney said the city would have to pay the entire $9.75 million because Miedzianowski is believed to have no money. She said the couple's attorneys would ask the court to award them fees in the $1 million range.

Source: John Garcia, "Jury Awards $9.75 Million in Police Corruption Suit," *ABC* (March 22, 2007), http://abclocal.go.com/wls/story?section=local&id=5059911 (accessed June 27, 2007).

Summary

Police subculture is a body of knowledge that emerges through a shared application of practical skills to concrete problems encountered through daily routines and in the normal course of activities or a shared lifestyle. Officers live in two worlds: the one they live in when off the job (the larger culture) and the one they work in (the police subculture). Characteristics of this subculture include its origins in basic training, the tendency toward groupthink, secrecy, and shared values and personality traits. For example, the four most talked-about personality characteristics of police officers are cynicism, authoritarianism, isolationism, and conservatism.

Police managers sometimes attempt to limit the influence exerted by police subculture through formal and informal training, instillation of desired values, and development of norms through rules and regulations. Many officers respond to management's attempt to limit the police subculture influence through passive resistance, including CYA ("cover your ass") techniques.

Officer discretion comprises the use of an officer's personal judgment to make decisions about how to enforce the law without outside influence. Factors affecting discretion include the seriousness of the crime and service call priorities, among others. Moves to limit discretion involve both legal and administrative approaches, especially administrative rule-making tasks that are consistent with confirming, structuring, and checking discretion.

Wrongful acts committed by officers include police deviance (unethical deviation from unwritten norms, values, and expectations), police misconduct (violation of written procedures, regulations, and codes), and police corruption (forbidden acts committed outside of the job or on the job, and use of the job for personal gain). Some of the factors that could potentially lead to wrongful acts include the police subculture, the code of silence adhered to by officers, neighborhood influences, and inefficiencies in the police hiring process.

Key Words

Administrative rule making
Authoritarian personality
Conservatism
CYA
Cynical personality
Excessive use of force
Grass eaters
Groupthink
Justifiable homicide
Meat eaters
Police corruption
Police deviance
Police discretion
Police misconduct
Police subculture
Police-initiated transjurisdictional transport (PITT)
Tennessee v. Garner 1985
Use of force
Wrongful acts

Discussion Questions

1. Define police subculture and explain why officers live in two worlds.
2. Describe the major characteristics of police subculture. In what way do two of these characteristics shape police subculture more than the others?
3. Characterize groupthink, including its negative outcomes, and link those outcomes to police officers.
4. Describe the four most talked-about characteristics of a police personality. Which of these four characteristics seem to be the most visible among officers? Provide examples.
5. Explain CYA and its function.
6. Characterize police discretion and provide information about its importance in police work.
7. Describe the primary factors affecting police discretion. In what way is one of these factors the most influential factor linked to officer discretion?
8. Articulate methods of limiting police discretion. In what way is one of these methods better than the others?
9. Define police deviance, and provide some practical examples of it.
10. Define police misconduct, and characterize the various forms it can take.
11. Define police corruption, and describe the various forms it can take. Which of these forms is the most vicious?
12. Describe seven factors leading to wrongful acts. Which of these factors is the most effective in driving poor police performance?
13. Characterize the most appropriate method of controlling future wrongful acts.

Notes

1. George L. Kelling, *Broken Windows and Police Discretion* (NCJ 178259) (Washington, DC: U. S. Department of Justice, 1999), http://www.ncjrs.gov/txtfiles1/nij/178259.txt (accessed June 4, 2007).

2. PBS, "Rampart Scandal Timeline," *Frontline* (n.d.), http://www.pbs.org/wgbh/pages/frontline/shows/lapd/scandal/cron.html (accessed April 24, 2007).
3. Samuel Walker and Charles M. Katz, *The Police in America: An Introduction*, 6th ed. (New York: McGraw Hill, 2008), 159.
4. Erwin Chemerinsky, *An Independent Analysis of the Los Angeles Police Department's Board of Inquiry Report of the Rampart Scandal* [also known as the Christopher Commission] (Los Angeles: Los Angeles Police Protective League, 2000), 5.
5. *Los Angeles Times* (May 27, 2007), 1.
6. John P. Crank, *Understanding Police Culture* (New York: Anderson, 2004), 32–34. Also see Robert Trojanowicz, Victor E. Kappeler, Larry K. Gaines, and Bonnie Bucqueroux, *Community Policing: A Contemporary Perspective*, 2nd ed. (Belmont, CA: Wadsworth, 1998), 267.
7. Dennis J. Stevens, "Police Training and Management Impact Sexual Assault Conviction Rates in Boston," *Police Journal* 79, no. 2 (2006): 125–54.
8. John Kleinig, *The Ethics of Policing* (New York: Cambridge University Press, 1996).
9. William A. Westley, *Violence and the Police* (Cambridge, MA: MIT Press, 1979), 159–60.
10. Page 134 in Robert E. Worden, "The Causes of Police Brutality: Theory and Evidence on Police Use of Force," in *The Police in America: Classic and Contemporary Readings*, ed. Steven G. Brandl and David S. Barlow (Belmont, CA: Wadsworth, 2004), 128–72.
11. O. Mink, A. S. Dietz, and J. Mink, "Changing a Police Culture of Corruption: Implications for the Police Psychologist," paper presented at the 29th Annual Society of Policing and Criminal Psychology Conference, Cleveland, OH, 2000.
12. Dennis J. Stevens, *Applied Community Policing in the 21st Century* (Boston: Allyn and Bacon, 2003), 45.
13. Kleinig, *The Ethics of Policing.*
14. *Sourcebook of Criminal Justice Statistics 2004* (2006), Table 2.12, http://www.albany.edu/sourcebook/pdf/t212.pdf (accessed June 9, 2007). For police supervisor opinion, see Anthony V. Bouza, *Police Unbound: Corruption, Abuse, and Heroism by the Boys in Blue* (Amherst, NY: Prometheus, 2001), 112–32.
15. Patricia A. Kelly, "Stress: The Cop Killer," in *Treating Police Stress,* ed. John M. Madonna, Jr., and Richard E. Kelly (Springfield, IL: Charles C. Thomas, 2002), 33–55.
16. Carl B. Klockars, "The Legacy of Conservative Ideology and Police," in *The Police and Society: Touchstone Readings*, ed. Victor E. Kappeler (Mt. Prospect, IL: Waveland, 1995), 349–57.
17. These ideas are guided by David H. Bayley, 1991; David L. Carter and Louis A. Radelet, *The Police and the Community*, 6th ed. (Upper Saddle River, NJ: Prentice Hall, 1999), 181; Wesley G. Skogan and Susan M. Hartnett, 1997; and Dennis J. Stevens, 2002b, 2001b.
18. Trish Oberweis and Michael Musheno, "Policing Identities: Cop Decision Making and the Constitution of Citizens," *Law & Social Inquiry* 24, no. 4 (1999): 897–923.
19. Richard Kelly, "What Needs to Be Done," in *Treating Police Stress: The Work and the Words of Peer Counselors*, ed. John M. Madonna, Jr., and Richard E. Kelly (Springfield, IL: Charles C. Thomas, 2002), 217–24.
20. Victor E. Kappeler, *Critical Issues in Police Liability*, 4th ed. (Prospect Heights, IL: Waveland, 2006), 4–8.
21. Stephen D. Mastrofski, Michael D. Reisig, and John D. McCluskey, "Police Disrespect toward the Public: An Encounter-Based Analysis," *Criminology* 40, no. 3 (2002): 519–52.
22. Jerome Skolnick, "A Sketch of the Policeman's Working Personality," in *Policing Key Readings*, ed. Tim Newburn (Devon, UK: Willan, 2004), 264–79.
23. Crank, *Understanding Police Culture*, 33.

24. Dennis J. Stevens, *Police Officer Stress: Sources and Solutions* (Upper Saddle River, NJ: Prentice Hall, 2008).
25. Crank, *Understanding Police Culture*, 322.
26. Donna Hale and T. Bricker, "Police Organizational Change: Strategies for Effective Police Management in the 21st Century," in *Visions for Change: Crime and Justice in the 21st Century*, ed. Roslyn Muraskin and Albert R. Roberts (Upper Saddle River, NJ: Prentice Hall, 2005), 390–404.
27. For an excellent discussion on necessary cultural change, see David E. Barlow and Melissa Hickman Barlow, *Police in a Multicultural Society: An American Story* (Prospect Heights, IL: Waveland, 2000), 8–10.
28. Peter K. Manning, "Policing and Reflection," in *Critical Issues in Policing: Contemporary Readings*, 4th ed., Roger G. Dunham and Geoffrey P. Albert (Prospect Heights, IL: Waveland, 2001), 149–58.
29. Crank, *Understanding Police Culture*, 311. Consistent with this thought, also see Elizabeth Reuss-Ianni, *Two Cultures of Policing: Street Cops and Management Cops* (New Brunswick, NJ: Transaction Books, 1983), 16.
30. John Van Maanen, "Kinsmen in Repose: Occupational Perspective of Patrolmen," in *Policing: A View from the Street*, ed. Peter K. Manning and John Van Maanen (Santa Monica, CA: Goodyear, 1978), 115–28.
31. Reuss-Ianni, *Two Cultures of Policing*, 16. Also see Crank, *Understanding Police Culture*, 52–54.
32. Laurence J. Peter, *The Peter Principle* (New York: Morrow, 1972).
33. "Groupthink: Small Group Communication" (n.d.), http://www.abacon.com/commstudies/groups/groupthink.html (accessed July 15 2007).
34. For more detail, see Carter and Radelet, *The Police and the Community*; Mink, Dietz, and Mink, "Changing a Police Culture of Corruption."
35. Crank, *Understanding Police Culture*. Also see George L. Kelling and Catherine M. Coles, *Fixing Broken Windows* (New York: Free Press, 1996), 73–79.
36. Crank, *Understanding Police Culture*.
37. Jon B. Gould and Stephen D. Mastrofski, "Suspect Searches: Assessing Police Behavior under the U.S. Constitution," *Criminology & Public Policy* 3, no. 3 (2004): 315–25. This study examines police conformity to the law by evaluating direct observations of police searches in a medium-sized U.S. city against the applicable constitutional standards. The results paint a disquieting picture, with nearly one third of searches performed unconstitutionally and almost none visible to the courts. The research links police misconduct to the municipality's "war on drugs," but, perhaps surprisingly, the majority of constitutional violations were concentrated among a small number of otherwise model officers who were engaged in community policing.
38. M. A. Surrette, J. M. Ebert, M. A. Willis, and T. M. Smallidge, "Personality of Law Enforcement Officials: A Comparison of Law Enforcement Officials' Personality Profiles Based on Size of Community," *Public Personnel Management* 32, no. 2 (2003): 279–85.
39. Kevin Brewer, *Psychology and Crime* (London: Harcourt Heinemann, 2000), 3. Also see Internal Revenue Service, www.irs.gov (accessed April 11, 2008); Federal Bureau of Investigation, "Officers Assaulted and Officers Killed," www.fbi.gov (accessed April 11, 2008).
40. Carter and Radelet, *The Police and the Community*, 181.

41. Arthur Neiderhoffer, *Behind the Shield* (New York: Anchor Press, 1967). Also see J. Lefkowitz, "The Police and the Criminal Justice System" (n.d.), http://uwf.edu/swright/ (accessed May 27, 2007).

42. Dennis J. Stevens, "Civil Liabilities and Arrest Decisions," in *Policing and the Law*, ed. Jeffery T. Walker (Upper Saddle River, NJ: Prentice Hall, 2002), 53–71. Also see Dennis J. Stevens, "Civil Liability and Selective Enforcement," *Law and Order* 49, no. 5 (2001): 105–8.

43. R. V. Erickson and K. D. Haggerty, *Policing the Risk Society* (Toronto: University of Toronto Press, 1997).

44. Geoffrey P. Alpert and Mark H. Moore, "Measuring Police Performance in the New Paradigm of Policing," in *Critical Issues in Policing*, 3rd ed., ed. Robert Dunham and Geoffrey P. Alpert (Prospect Heights, IL: Waveland, 1997), 265–81.

45. Dennis J. Stevens, "Community Policing and Managerial Techniques: Total Quality Management Techniques," *Police Journal* 74, no. 1 (2001): 26–41.

46. Stevens, "Community Policing and Managerial Techniques."

47. Stevens, "Police Training and Management Impact Sexual Assault Conviction Rates in Boston."

48. Carter and Radelet, *The Police and the Community.*

49. Wesley Skogan, "The Impact of Community Policing on Neighborhood Residents: A Cross-Site Analysis," in *The Challenge of Community Policing: Testing The Promises*, ed. D. Rosenbaum (Thousand Oaks, CA: Sage, 1994), 167–81.

50. Thomas Barker and David L. Carter, *Police Deviance,* 1st ed. (Cincinnati, OH: Anderson, 1994), 191.

51. David Weisburd and Rosann Greenspan, "Police Attitudes toward Abuse of Authority: Findings from a National Study," National Institute of Justice (2000), http://www.ncjrs.gov/txtfiles1/nij/181312.txt (accessed December 5, 2007).

52. Weisburd and Greenspan, "Police Attitudes toward Abuse of Authority."

53. Dennis W. Catline and James R. Maupin, "A Two Cohort Study of the Ethical Orientations of State Police Officers," *Policing: An International Journal of Police Strategies & Management* 27, no. 3 (2004): 289–301.

54. Trojanowicz, Kappeler, Gaines, and Bucqueroux, *Community Policing*, 267.

55. Crank, *Understanding Police Culture.*

56. William Terrill and Michael D. Reisig, "Neighborhood Context and Police Use of Force," *Journal of Research in Crime and Delinquency* 40, no. 3 (2003): 291–321.

57. John S. Dempsey and Linda S. Forst, *Introduction to Policing*, 4th ed. (Belmont, CA: Thomson/Wadsworth, 2008).

58. Christopher Cooper, "Entrenched in Subculture is at Root of Police Brutality and Bias Cases," *CommonDreams.org* (July 21, 2000), http://www.commondreams.org/views/072100-105.htm (accessed May 26, 2007).

59. Stevens, *Police Officer Stress*, 40.

60. Bureau of Justice Statistics, *Law Enforcement Statistics* (Washington, DC: U.S. Department of Justice, Office of Justice Programs, 2000), http://www.ojp.usdoj.gov/bjs/lawenf.htm (accessed July 14, 2004).

61. Weisburd and Greenspan, "Police Attitudes toward Abuse of Authority."
62. Thomas Barker and David L. Carter, *Police Deviance,* 3rd ed. (Cincinnati, OH: Anderson, 2001), 276.
63. Community Relations Service, *Principles of Good Policing: Avoiding Violence between Police and Citizens* (Washington, DC: U. S. Department of Justice, September 2003), http://www.usdoj.gov/crs/pubs/principlesofgoodpolicingfinal092003.htm#74 (accessed June 25, 2007).
64. Stevens, *Applied Community Policing in the 21st Century.*
65. Carl Klockars, "The Modern Sting," in *Thinking about Policing,* 2nd ed, ed. G. Klockars and S. Mastrofski (New York: McGraw Hill, 1991), 258–67.
66. The following example comes from Wayne W. Bennett and Karen M. Hess. *Management and Supervision in Law Enforcement,* 4th ed. (Belmont, CA: Wadsworth, 2007), 500.
67. Bennett and Hess, *Management and Supervision in Law Enforcement,* 500.
68. Kelling, *Broken Windows and Police Discretion.*
69. Joycelyn M. Pollock and Ronald F. Becker, *Ethics Training: Using Officers' Dilemmas* (Washington, DC: Federal Bureau of Investigation, 1996), http://www.fbi.gov/publications/leb/1996/nov964.txt (accessed December 3, 2007).
70. Kelling, *Broken Windows and Police Discretion.*
71. Linda A. Teplin, "Keeping the Peace: Police Discretion and Mentally Ill Persons," *National Institute of Justice Journal* (July 2000), http://www.ncjrs.gov/pdffiles1/jr000244c.pdf (accessed June 25, 2006).
72. Henry M. Wroblewski and Karen M. Hess, *Introduction to Law Enforcement and Criminal Justice,* 8th ed. (Belmont, CA: Thomson-Learning, 2006).
73. Brenda K. Uekert, "Lake County, California, Arrest Policies Project," NCJRS (2003), http://www.ncjrs.gov/pdffiles1/nij/grants/201874.pdf (accessed May 30, 2007).
74. Dempsey and Forst, *Introduction to Policing.*
75. Albert J. Reiss, Jr., "Consequences of Compliance and Deterrence Models of Law Enforcement for the Exercise of Discretion," *Law and Contemporary Problems* 47 (1984): 88–89.
76. William Terrill, "Police Use of Force: A Transactional Approach," *Justice Quarterly* 22, no. 1 (2005): 107–38.
77. Kelling, *Broken Windows and Police Discretion.*
78. Robin Shepard Engel and Robert E. Worden, "Police Officers' Attitudes, Behavior, and Supervisory Influences: An Analysis of Problem Solving," *Criminology* 41, no. 1 (2003): 131–64.
79. Gordon P. Whitaker, "What Is Patrol Work?", in *Policing Perspectives: An Anthology,* ed. Larry K. Gaines and Gary W. Cordner (Los Angeles, CA: Roxbury, 1982), 213–33.
80. Robert A. Brown and James Frank, "Race and Officer Decision Making: Examining Differences in Arrest Outcomes between Black and White Officers," *Justice Quarterly* 23, no. 1 (2006): 96–126.
81. David Eitle, Lisa Stolzenberg, and Stewart J. D'Alessio, "Police Organizational Factors, the Racial Composition of the Police, and the Probability of Arrest," *Justice Quarterly* 22, no. 1 (2005).
82. Geoffrey P. Alpert, James M. MacDonald, and Roger G. Dunham, "Police Suspicion and Discretionary Decision Making during Citizen Stops," *Criminology* 43, no. 2 (2005): 407–34.

83. Samuel Walker, Cassia Spohn, and Miriam Delone, *The Color of Justice: Race, Ethnicity, and Crime in America*, 3rd ed. (Belmont, CA: Thomson-Learning, 2004).
84. Kenneth Bolton, Jr., and Joe R. Feagin, *Black in Blue* (New York: Routledge, 2004).
85. William Julius Wilson, *The Declining Significance of Race: Black and Changing American Institutions* (Chicago: University of Chicago Press, 1980).
86. Bolton and Feagin, *Black in Blue.*
87. Charles R. Swanson, Leonard Territo, and Robert W. Taylor, *Police Administration: Structures, Processes and Behavior*, 6th ed. (Upper Saddle River, NJ: Prentice Hall, 2004), 123–25.
88. Alpert, MacDonald, and Dunham, "Police Suspicion and Discretionary Decision Making."
89. In philosophy, this circular thinking is known as a tautology—use of redundant language that adds no information.
90. Stevens, *Applied Community Policing in the 21st Century.*
91. Bolton and Feagin, *Black in Blue.*
92. Thomas D. Stuckey, "Local Politics and Police Strength," *Justice Quarterly* 22, no. 2 (2005): 139–69.
93. Donald Black, ed., *The Manners and Customs of the Police* (New York: Academic Press, 1980).
94. Robert A. Brown and James Frank, "Police–Citizen Encounters and Field Citations: Do Encounter Characteristics Influence Ticketing?", *Policing: An International Journal of Police Strategies & Management* 28, no. 3 (2005): 435–54.
95. Donald Black, "The Social Organization of Arrest," in *The Manners and Customs of the Police*, ed. Donald Black (New York: Academic Press, 1980), 90–110.
96. Victor E. Kappeler, *Critical Issues in Police Civil Liability* (Mt. Prospect, IL: Waveland, 2001), 7.
97. Stevens, "Civil Liabilities and Arrest Decisions."
98. American Civil Liberties Union, "An Action Plan to Eliminate Improper Arrests" (n.d.), http://www.aclu-md.org/aPress/Remedies.pdf (accessed June 2, 2007).
99. Michael D. Reisig, John D. McCluskey, Stephen D. Mastrofski, and William Terrill, "Suspect Disrespect toward the Police," *Justice Quarterly* 21, no. 2 (2004): 241–68.
100. "Baltimore Crime Blog" (September 16, 2005), http://baltimorecrime.blogspot.com/2005_09_11_archive.html (accessed June 3, 2007).
101. Recently, in *Scott v. Harris*, the U.S. Supreme Court for the first time squarely confronted the Fourth Amendment implications of dangerous, high-speed vehicular chases. The driver in the case, Victor Harris, had failed to pull over when police flashed their lights. Instead, he fled, and police pursued, at faster and faster speeds. When there appeared to be no other way to apprehend Harris, Officer Timothy Scott rammed into the suspect's car, thereby causing a crash that rendered Harris quadriplegic. Harris subsequently sued Scott under Section 1983, the federal statute that allows plaintiffs to seek money damages for violations of their rights under the Constitution and federal statutes. Harris claimed that Scott had used excessive force in violation of the Fourth Amendment. On an interlocutory appeal by the government, the Supreme Court held that the suit should have been dismissed on summary judgment. The Court ruled, by an 8–1 majority, that Scott acted "reasonably" under the circumstances.
102. Peter K. Manning, "The Police: Mandate, Strategies, and Appearances," in *The Police and Society: Touchstone Readings*, ed. Victor E. Kappeler (Mt. Prospect, IL: Waveland, 1995), 97–126.

103. Klockars, "The Legacy of Conservative Ideology and Police."
104. Egon Bittner, "The Police on Skid-Row: A Study of Peace Keeping," in *The Police and Society: Touchstone Readings*, ed. Victor E. Kappeler (Mt. Prospect, IL: Waveland, 1995), 265–91.
105. G. Maggs, "Flexibility and Discretion," *Police Work and Liberal Society* 7 (1992): 517.
106. Anderson Cooper, "Police Handcuff Five-Year-Old," *Anderson Cooper 360 Degrees*, CNN (April 22, 2005).
107. George E. Rush and Sam Torres, eds., "Police Discretion," in *The Encyclopedic Dictionary of Criminology* (Belmont, CA: Wadsworth, 2001).
108. Kelling, *Broken Windows and Police Discretion.*
109. Walker and Katz, *The Police in America*, 372.
110. Kenneth C. Davis, *Police Discretion* (St. Paul, MN: West, 1975).
111. Sam Walker, *The New World of Police Accountability* (Thousand Oaks, CA: Sage, 2008), 41.
112. William R. King and Thomas M. Dunn, "Dumping: Police-Initiated Transjurisdictional Transport of Troublesome Persons," *Police Quarterly* 7, no. 3 (2004): 339–58.
113. James J. Fyfe and Robert Kane, "Bad Cops: A Study of Career-Ending Misconduct among New York City Police Officers" (September 2006), http://www.ncjrs.gov/pdffiles1/nij/grants/215795.pdf (accessed June 6, 2007).
114. Fyfe and Kane, "Bad Cops."
115. Todd Lightly, "Ex-cop a Drug Dealer, Jury Told," *Chicago Tribune* (February 7, 2001), http://www.mapinc.org/drugnews/v01/n231/a09.html (accessed June 27, 2007).
116. Victor E. Kappeler, Richard D. Sluder, and Geoffrey P. Alpert, *Forces of Deviance: Understanding the Dark Side of Policing* (Prospect Heights, IL: Waveland, 1994), 3.
117. Dorothy H. Bracey, "NYPD Integrity" (n.d.), http://web.jjay.cuny.edu/~nbenton/crju709/conf/kimmelman/assess.html (accessed June 4, 2007).
118. Barker and Carter, *Police Deviance*, 3rd ed., 4.
119. Kappeler, Sluder, and Alpert, *Forces of Deviance*, 4.
120. Emory A. Plitt, Jr., "Police Discipline Decisions," *Police Chief* (1983): 95–98.
121. Dennis J. Stevens, "Corruption among Narcotic Officers: A Study of Innocence and Integrity," *Journal of Police and Criminal Psychology* 14, no. 2 (1999): 1–10.
122. "Two New York City Police Officers and the Operators of a Queens Brothel Charged in Bribery Scheme," *New York Jewish Times* (2006), www.nyjtimes.com/Stories/2006/NYPD (accessed April 4, 2007).
123. Thomas O'Connor, "Police Deviance and Ethics" (n.d.), http://faculty.ncwc.edu/TOConnor/205/205lect11.htm (accessed June 4, 2007).
124. Kristina M. Hall, "Porn Star Says Trooper Ignored 'Happy Pills' in Exchange for Sex," *Tennessan.com* (May 23, 2007), http://tennessean.com/apps/pbcs.dll/article?AID=/20070523/NEWS03/70522085/1001 (accessed May 30, 2007).
125. Dan Benson, "Cudahy Police Officer Resigns" (December 2004), http://findarticles.com/p/articles/mi_qn4196/is_20041214/ai_n11004575 (accessed June 4, 2007).
126. Fyfe and Kane, "Bad Cops."
127. O'Connor, "Police Deviance and Ethics."
128. Carloine Gemli, "Staging the Nation: Theatricality in the Law," *Elements* (Spring 2005): 55–58, http://www.bc.edu/research/elements/issues/2005s/elements-spring2005-article6.pdf (accessed June 5, 2007).

129. Alan M. Dershowitz, *Life Is Not a Dramatic Narrative. Law's Stories: Narrative and Rhetoric in the Law* (New Haven, CT: Yale University Press, 1996).
130. Dershowitz, *Life Is Not a Dramatic Narrative.*
131. Federal Bureau of Investigation (March 2007), www.fbi.gov (accessed April 15, 2007).
132. Kappeler, Sluder, and Alpert, *Forces of Deviance.*
133. Andrew L. Wang, "Teen Says Cop Beat Him," *Chicago Tribune* (May 28, 2007), 14.
134. Fyfe and Kane, "Bad Cops."
135. Steve Mills, Flynn McRoberts, and Maurice Possley, "When Labs Falter, Defendants Pay," *Chicago Tribune* (October 2004), http://www.truthinjustice.org/labs-falter.htm (accessed June 4, 2007).
136. The Legal Dictionary, http://legal-dictionary.thefreedictionary.com/false%20arrest (accessed June 9, 2008).
137. Victor E. Kappeler, *Police Civil Liability: Supreme Court Cases and Materials* (Mt. Prospect, IL: Waveland Press, 2001), 21.
138. Carol A. Archbold, *Police Accountability, Risk Management, and Legal Advising* (NY, USA: LFB Scholarly Publishing LLC, 2004), 85.
139. Harvey S. Gray, "Police Liability for False Arrest, False Imprisonment and Malicious Prosecution," http://www.grsmb.com/CM/Resources/Police-Liability.asp (accessed June 9, 2008).
140. Frank Eltman, "4 Officers Face Charges in 05 Tourist Beating," *Chicago Tribune* (March 27, 2007), 5.
141. Liz Shepard, "Area Police Chief Charged with Stalking" (May 10, 2007), http://argus-press.com/articles/2007/05/10/news/news1.txt (accessed June 6, 2007).
142. Mark Bowden, "Major Offenses by Philadelphia Cops Often Bring Minor Punishments," *Philadelphia Inquirer* (November 19, 1995), http://www.hrw.org/reports98/police/uspo111.htm (accessed April 11, 2008). Also see "Shielded from Justice: Police Brutality and Accountability in the United States," http://www.hrw.org/reports98/police/uspo111.htm (accessed April 11, 2008).
143. R. Kania and W. C. Mackey, "Police Violence as a Function of Community Characteristics," *Criminology* 15, no.1 (1977): 27–48; Barker and Carter, *Police Deviance,* 3rd ed., 5.
144. Robert E. Worden and Robin L. Shephard, "Demeanor, Crime, and Police Behavior: A Reexamination of the Police Services Study Data," *Criminology* 34 (1996): 83–105. Also see Weisburd and Greenspan, "Police Attitudes toward Abuse of Authority."
145. Lect Law (n.d.), http://www.lectlaw.com/def/j059.htm (accessed June 3, 2007).
146. National Institute of Justice, "Use of Force by Police: An Overview of National and Local Data" (1999), http://www.ncjrs.gov/pdffiles1/nij/176330-1.pdf (accessed May 19, 2007).
147. Merle Stetser, *The Use of Force in Police Control of Violence: Incidents Resulting in Assaults on Officers* (New York: LFB Scholarly Publishing, 2001).
148. Bureau of Justice Statistics, "Contracts with the Police and the Public" (NCJ 184957; 2001), http://www.ojp.usdoj.gov/bjs/pub/pdf/cpp99.pdf (accessed May 18, 2007).
149. Stephan Thernstrom, Fred Siegel, and Robert Woodson, *The Kerner Commission Report and the Failed Legacy of Liberal Social Policy* (1998).
150. Heather Mac Donald, *Lessons from Cincinnati: A Vivid Guide in How Not to Handle Riots* (Manhattan Institute for Policy Research, July 22, 2001).
151. Jeffrey Gunn, "Cincinnati Police Not as Racist as Columnist Thinks" (April 23, 2001), http://media.www.dailylobo.com/media/storage/paper344/news/2001/04/23/

Opinion/Cincinnati.Police.Not.As.Racist.As.Columnist.Thinks-70976.shtml (accessed June 6, 2007).

152. Mac Donald, *Lessons from Cincinnati.*

153. Gunn, "Cincinnati Police Not as Racist as Columnist Thinks."

154. Weisburd and Greenspan, "Police Attitudes toward Abuse of Authority,"

155. David Heinzmann, "Businessmen Sue Chicago Police Officers in Alleged Bar Beating," *Chicago Tribune* (May 8, 2007), http://criminal.lawyers.com/news-headline/Businessmen-sue-Chicago-police-officers-in-alleged-bar-beating-l:609954285.html (accessed May 19, 2007).

156. Angela Rozas, "Officer in Taped Bar Beating Faces 14 New Charges," *Chicago Tribune* (April 27, 2007), http://www.topix.net/content/trb/13414525350941297303092673767313237549 24 (accessed May 19, 2007).

157. James J. Fyfe, "Good Policing," in *The Administration and Management of Criminal Justice Organizations*, ed. Stan Stojkovic, John Klofas, and David Kalinich (Prospect Heights, IL: Waveland, 1999), 113–35.

158. Herman Goldstein, "Controlling and Reviewing Police–Citizen Contacts," in *Police Deviance*, 3rd ed., ed. Thomas Barker and David L. Carter (Cincinnati, OH: Anderson, 1994), 323–53.

159. Reisig, McCluskey, Mastrofski, and Terrill, "Suspect Disrespect toward the Police."

160. Stevens, "Corruption among Narcotic Officers."

161. Fyfe and Kane, "Bad Cops."

162. Fyfe and Kane, "Bad Cops."

163. Willard M. Oliver, *Community Oriented Policing: A Systematic Approach to Policing* (Upper Saddle River, NJ: Prentice Hall, 2008), 227–29.

164. Thomas Allen Johnson, *The Effects of Higher Education/Military Service on Achievement Levels of Police Academy Cadets* (ERIC: Ed437891), Ph.D. dissertation, Texas Southern University, 1998.

165. William C. Terrill and Stephen D. Mastrofski, "Situational and Officer-Based Determinants of Police Coercion," *Justice Quarterly* 19, no. 2 (2002): 215–48.

166. Dennis J. Stevens, "College Educated Officers: Do They Provide Better Police Service?", *Law and Order* (1999): 37–41.

167. Lawrence W. Sherman, *Scandal and Reform: Controlling Police Corruption* (Berkeley, CA: University of California Press, 1978), 30.

168. Herman Goldstein, *Policing a Free Society* (Cambridge MA: Ballinger, 1977).

169. Fyfe and Kane, "Bad Cops."

170. Michael R. Gottfredson and Travis Hirschi, *A General Theory of Crime* (Stanford, CA: Sanford University Press, 1999); Marcus Felson, *Crime and Everyday Life*, 3rd ed. (Thousand Oaks, CA: Sage, 2002); Curt R. Bartol and Anne M. Bartol, *Criminal Behavior: A Psychological Approach* (Upper Saddle River, NJ: Prentice Hall, 2008); Stanton Samenow, *Before It's Too Late* (New York: Basic, 1998); American Psychological Association, *Diagnostic and Statistical Manual of Mental Disorders*, 4th ed., Text Revision (Washington, DC: Author, 1994).

POLICE

Police Accountability and Civil Liabilities

The ancient Romans had a tradition: Whenever one of their engineers constructed an arch, as the capstone was hoisted into place, the engineer assumed accountability for his work in the most profound way possible: He stood under the arch.

—Michael Armstrong

LEARNING OBJECTIVES

When you finish reading this chapter, you will be better prepared to:

- Characterize the relationship between police and justice
- Explain the purpose of police accountability
- Describe the rotten apple theory
- Characterize internal and external methods of police accountability
- Describe the exclusionary rule and the U.S. Supreme Court rulings related to it
- Explain the differences between state liability suits and federal violations
- Characterize the effects of civil liability suits
- Identify methods for changing civil liability risks

CASE STUDY: YOU ARE THE POLICE OFFICER

You've been on the job for a week. You wanted to change the world, but now you see that being a good cop means doing the right thing, and doing the right thing usually means making decisions that are legally and morally within the meeting the rule of law. It was so easy in college when your class discussed police behavior, because you just offered simple solutions. Your classmates knew you were going to the police academy after graduation and figured that you must know what you were talking about.

Your professor lectured time and again that police agencies in a democracy are accountable to the public whom they serve, but you insisted that the public would give up its rights in the name of public safety. "Like with terrorists, if you have nothing to hide, then everybody would allow a police search of their personal possessions, right?"

You remember the time your instructor asked something like, "How would you control the officers who took advantage of their authority in the name of changing the world?" Your response was a simple statement: "They should be punished."

Now you're beginning to see that the best way "to do the right thing" is to think carefully through the each situation before making decisions and to be personally accountable for your actions. You must rely on your training and the professional standards to guide you in changing the world, you decided. You thought that the few bad officers ("rotten apples in the barrel") could be easily removed and the other officers would continue to do the right things. Now you've learned that every 911 call gives you a chance to make many different decisions. After all, you are a police officer. Maybe close supervision, internal affairs, and civil liabilities can help remind other cops what the right thing to do is—something you already know.

1. Why should officers respond legally and morally when providing police services?
2. What did the officer mean by saying that it is necessary to rely on your training and the professional standards to change the world?
3. What is it that this officer already knew about doing the right thing?

Introduction

Complaints against police officers can act as a barometer that measures the community's satisfaction with the service provided by the police.[1] The number and type of complaints filed against officers can indicate whether problems exist, but most often officers work hard to accommodate the public and professional managers can use compliments as indicators for improved training, detection of problematic personnel, identification of areas of potential legal liability, and officer risk (i.e., risk management initiatives). Recent challenges confronting police organizations and their personnel include issues related to police accountability, which can lead to litigation against officers, their training officers (TOs), and their supervisors, as well as the police organization itself (see Chapter 8). There are three primary reasons why police officers must be held accountable for their actions:

- Due-process revolution: All individuals encountered by local police should not have to demand their constitutional rights.
- Facing conflicts of interest and opportunities to commit crime is inevitable in policing.
- Limited local police financial and personnel resources often set limits on equipment and technology, investigative initiatives and forensic analysis, personnel training and expertise, and support staff including supervision.

As noted in Chapter 7, most employees work up to 33 percent of their capabilities, 10 percent are high achievers, and 10 percent are low achievers, but those low achievers tend to cause 90 percent of the problems in most organizations. If this information is applicable to police officers, perhaps weeding out police candidates who are potentially problematic should take precedent during the hiring process. It cannot be overemphasized that the best predictor of future bad cops is their history prior to becoming officers. If departments devoted more effort to hiring ethical and moral candidates in the first place, then perhaps personal accountability would be less of a problem among police organizations (see Chapter 8).[2] Maybe hiring committees should stand under the "capstone" (as suggested by Michael Armstrong) of police corruption charges when the candidates they hired are criminally indicted.

Disturbing Accounts of Police Officers

How should police administrators respond to the following events?

- A police officer in Los Angeles was caught on video tape applying a choke hold to a handcuffed 16-year-old boy; the officer was arrested for suspicion of assault.[3]
- Four Chicago police officers broke into the home of a Chicago fire fighter, beat him, and accused him of being a drug dealer.[4] When the fire fighter filed a complaint with the Chicago Police Department, the police internal affairs investigator threatened the fire fighter with a loss of his job and jail if he pursued the complaint.
- A state trooper in Illinois stopped a vehicle and forced the young occupants to strip naked.[5] Other drivers came forward when the complaint was publicized. The trooper pleaded guilty to two counts of official misconduct; was sentenced to 30 months of probation and 200 hours of public service; was ordered to pay $1,000 to a crime stoppers' organizations and to seek sex-offender health treatment.
- A federal grand jury in Los Angeles indicted six gang members, including three police officers, on civil rights, narcotics, and weapons charges for allegedly participating in an organized enterprise that "invaded" private homes as though the participants were conducting legitimate police operations, but in reality were staging home-invasion robberies to steal drugs, money, and weapons.[6]
- Seven officers in New Orleans engaged in what can be called a chaotic post–Hurricane Katrina shooting spree that left two dead, including a mentally disabled man.[7] The district attorney said that the rules governing the use of lethal force could not be suspended during a state of emergency. Six days after the storm, police commandeered a truck and responded to what they said was "an officer down" incident during an armed attack on the bridge. Later it was learned that no officers had been shot. Seven officers were eventually indicted, some for first-degree murder. Although a police investigative unit cleared the officers of wrongdoing, widespread public criticism of what has come to be known as the Danziger Bridge incident symbolized the lawlessness and disorder of the days that followed the hurricane.

The expectations of the public place police officers in a vulnerable position regarding their legal obligations and expose them to charges of misconduct.[8] Personal accountability through internal methods (e.g., early warning, internal affairs) or external methods (e.g., oversight, litigation) can be an expensive proposition for police agencies, governments, and—ultimately—the public. When government assumes the responsibility to provide services or to protect constituents, people injured by inadequate performance deserve compensation; innocent parties who suffer injury should have an avenue for redress. That includes injuries that result from a department's "failure to train," as characterized in Chapter 8.

Americans demand that the police judiciously conduct their activities within the parameters of constitutional guarantees. Police officers, as insisted upon by the public and supported through court mandates, cannot function as if they were an occupational army; this statement might bother some officers, yet it is centered in urban realities that many of us do not have to encounter for various reasons.[9] It is also expected that police performance will not appear to suggest that some constituents are an "exception" to the rule of law regardless of their status.

With all those ideas in mind (and probably many others unintentionally neglected here), the centerpiece of police work becomes clear: Police are expected to support social order, but police performance toward social order is not an end in itself. Constitutional rights take priority over police performance.

> Policing is a means to justice and to the sanctity of individual liberty.[10]

Through law, order, and public safety, government gains legitimate power over the population it governs. Herein lays the focus of this chapter—namely, police accountability and civil liabilities. The issue here is the use of the coercive powers of government, under the color of law and in the name of good, which really explains the ugliest portions of human history (as opposed to democracy).[11] The question asked is this: Who polices the police?[12]

One response to this question comes from Los Angeles Police Chief William J. Bratton: "[T]he reality is, no police department is immune from bad cops. I have no tolerance for intentional misconduct and will deal with it forcefully and aggressively. Supervision, safeguards and civilian oversight are used to monitor employees and ensure quality police service"[13] Police officers and their departments must be accountable, primarily through their own means or internal methods of police accountability, but also through externally initiated means owing to factors outside the police organization.

Police Accountability

Definition

> **Police accountability** means that police officers and their organizations are accountable for their behavior and their recklessness.[14]

Inherent in this definition is the notion that government agencies are accountable to the public, at least within the American democratic (there are other forms of democracy which may not include public accountability) society.[15] Agents of the government are expected to serve and protect the due process specified by law for each individual encountered and to treat each person in a respectful manner. In particular, officers are expected not to use more force than necessary nor to exhibit bias against any group of individuals.[16] Accountability also means that police organizations maintain written policies and procedures to effectively ensure that their personnel meet the high standards associated with true professionalism. Thus police accountability also refers to the idea that police organizations develop policies prohibiting and investigating bias and unlawful use of force, and that they have procedures for receiving and investigating complaints about alleged misconduct.

Purpose of Police Accountability

In America, police organizations and other government agencies are held to a higher standard than their counterparts in many other nations. Accordingly, the legal and moral expectation is that officers will not violate the rule of law. In addition, accountability is a means to reinforce the notion that the function of the police is to enforce the law as opposed to make the law. "Police compliance with the law is one of the most important aspects of a democratic society," argue Wesley G. Skogan and Tracey L. Meares.[17] One function of the police is to enforce the rule of law fairly and to follow the rules that circumscribe police power.

> One purpose of police accountability is to legally and morally ensure that the police conform to the rule of law.

Recall that the framers of the Constitution had in mind that no single government agency should possess unlimited power. Rather, power is pooled through a system of checks and balances that is intended to protect the public from arbitrary government intervention.[18]

■ Theoretical Frameworks for Understanding Police Misconduct

Rotten Apple Theory

The **rotten apple theory** offers one explanation of why police corruption occurs. This perspective, which is often advocated by police chiefs and policymakers, holds that the officers involved in police misconduct perform that behavior on their own without the knowledge or support of their supervisors or coworkers.[19] Corrupt officers are seen to be only a few bad officers among many good officers—that is, "rotten" apples in the barrel. Once those officers are neutralized or removed from service, according to this theory, the problem is resolved and the remainder of the officers continue to behave in an appropriate way. In effect, management blames the individual officer for the problem of misconduct.

Some police organizations have established elaborate accountability mechanisms in an attempt to control officers, and one of those methods is to blame the officer for his or her misconduct.[20] The rotten apple theory justifies this perspective by stating that the individual officer is responsible. In this way, it offers the "simplest and easiest-to-fix interpretation of police misconduct."[21]

Unfortunately, focusing solely on individual responsibility can fail to account for ethical problems that officers face in relation to their commitments toward the task of policing.[22] Officers often encounter problems that undermine their best efforts to be ethical; in many cases, those ethical dilemmas go unrecognized. For instance, drug enforcement activities in which officers face administrative pressures to apprehend drug traders often push officers to use confidential informants (CIs), most of whom are of highly suspect character in the first place.[23] Yet officers do not possess the legal authority to allow CIs to commit crimes (see Chapter 12). This standard of operation inverts the hierarchy of sanctions, in that CI trade information for leniency. At the same time, officers are often rewarded for high arrest rates—which creates a conflict of interest for them in dealing with CIs.

One example of how the rotten apple theory can be used to explain officer misconduct is provided by Officer Antoinette Frank, who has been described as "cold as ice" by fellow officers at the New Orleans Police Department. She stood in the cramped kitchen of the Kim Anh Restaurant on the night of March 3, 1995, a 9-mm pistol clutched in her hand.[24] Kneeling on the dirty floor at Frank's feet were 17-year-old Cuong Vu and his 24-year-old sister, Ha. Cuong was an altar boy at St. Brigid Catholic Church. He played high school football and wanted to be a priest. His sister wanted to become a nun. Both worked long hours at their parents' restaurant. Frank fired nine bullets into their bodies. Then she turned her anger on her former partner, Officer Ronnie Williams, who was working security in the restaurant. Williams was found in uniform, face down behind the bar in a pool of blood. He had been shot twice in the head and once in the back. Frank robbed the restaurant and was later captured when she responded to an "officer down" call from dispatch.

Other examples supporting the rotten apple perspective includes the Miami River police officers who became involved with drugs and corruption in the 1980s. They were considered rotten apples, because there was little evidence of systematic corruption throughout the department.[25] In New York City, the Mollen Commission concluded that corruption in the NYPD was limited to a few officers. Likewise, recent cases of police misconduct in Los Angeles (Rampart's CRASH gang unit) and in Chicago appear to be the activities of a few "rotten apple" officers.

How else might widespread police corruption be explained? For example, the Knapp Commission found that corruption in the NYPD was "widespread . . . with a strikingly standardized pattern."[26] Throughout the 1980s and 1990s, the Metropolitan Washington, D.C., Police suffered through a host of scandals involving corruption, mismanagement, deception, political cover-ups, and abuse of authority.[27] By contrast, other urban police agencies—such as those in Arlington, Texas; Alexandria, Virginia; and Orlando, Florida—experienced few, if any, scandals, patterns of corruptive behavior, or significant citizen complaints over a 10-year period (1997–2007) after new chiefs were hired by those departments. For example, Arlington's Chief Theron Bowman's attitude is "simply to do the job better than anyone else," which includes selective hiring and providing cutting-edge training and equipment (such as "dragon skin" armor).[28] "Keeping police safe is also my job," says Bowman. (Check out the Bowman's foreword written for this textbook.) These examples suggest that maintaining an effective police organizational structure might neutralize a "rotten apple" situation through the use of professional police management tactics in areas such as hiring, training, and performance evaluation practices. It could be argued that the more professional the chief and his or her district commanders, the more professional the officers.

Structural Explanations of Corruption

Structural explanations of police misconduct have also been put forth.[29] Misconduct can and does occur in every profession, of course. However, officers face numerous opportunities to engage in misconduct because corruption is inherent in the job, especially given that officers tend to behave in ways that would be considered criminal among ordinary professions, such as the use of the force and detainment. Officers are involved with violators on a daily basis, suggesting that the opportunity to engage in deviance is ever present. At other times, officers may be manipulated into committing deviant acts by criminals, corrupt officers, supervisors, and court personnel including prosecutors.

For the most part, policing is low-visibility work, meaning that the activity of officers is often isolated and discrete.[30] To gain the acknowledgment of police supervisors, however, officers typically have to have an outstanding arrest record, recommendations from high command, and employment in a specialized but respected unit such as a homicide, sex crimes, narcotics, or tactical unit. At the same time, employment in these units can enhance the opportunities among officers to engage in corrupt behavior because of the isolation and the "element" (variety of criminals encountered). For that reason, among others, internal methods of police accountability are a priority among police managers.

Methods of Monitoring Police Behavior

There are three methods of monitoring police behavior:

- Self-monitoring
- Internal monitoring
- External monitoring

Self-Monitoring

The most efficient method of monitoring police behavior is self-monitoring by the officers themselves and by other officers. For this method of monitoring to be effective the social values and positive behavioral patterns of officers must be in place prior to their employment.

For instance, one officer had been mistreated by police as a youth.[31] Years later, the officer says, he treats all people that he comes across with respect, even if he knows an individual is a drug dealer. He is kind and courteous to them, even if the courtesy is not returned. For example, rather than approaching a person he suspects is dealing drugs and coldly ordering that individual to lean against a wall for a "pat-down," the officer takes the time to talk to the person first and then, almost nonchalantly, does the pat-down in the middle of a conversation. The officers hopes that this approach will relax the person, or at least allow him or her to see that the pat-down is nothing "personal"; it's just part of the officer's job.

Another officer spent a great deal of time trying to get the children on his beat (a particularly troubled area with only low-income residents) to work hard in school and respect the law. He also worked diligently with his partner to secure the respect of the residents and prided himself on his hard-won good relations with the residents, who had earlier been distrustful of the police.

Most often, accountability is conducted through police initiatives administered internally (supervision, close supervision, performance evaluations, early warning systems, early intervention systems, internal affairs units, and professional standards).

Internal Monitoring of Police Accountability

Accountability is enforced through internal monitoring of police behavior. The challenge is to convince officers to make use of and benefit from routine supervision rather than resisting and resenting it and, perhaps, for active supervisors to become facilitators rather than authoritative figures.[32] Officer behavior can be monitored and changed through the span of control, which assumes that any supervisor can effectively supervise only a limited number of police (see Chapter 10). Ideally, sergeants will provide close supervision to the officers they manage, who should number fewer than 3 to 10 officers (though in some departments sergeants supervise from 20 to 30 officers).[33] When fewer officers are included in the span of control, it would be more efficient to write reports, performance evaluations, monitor routine behavior, and make the necessary recommendations for training.

How much does the supervisor's attitude influence the officer's behavior? In one study, more than 80 percent of the officers surveyed agreed that a patrol officer would be more likely to use force unnecessarily and, the study implied, to engage in corrupt police behavior if he or she observed a supervisor using unnecessary force.[34]

Internal Monitoring Methods

Internal methods of officer accountability include supervision, close supervision, performance evaluations, early warning and early intervention systems, internal affairs units, and professional standards.

Supervision **Supervision** relates to the method of managing officers. A Police Foundation study found that quality routine supervision is the best way to prevent police abuse.[35] Among the officers who participated in this study, first-line supervision was found to be the best method of controlling officer abuse of power.

Close Supervision **Close supervision** suggests that supervisors be present at service calls, during arrests, and at crime scenes. Supervisors should ride along on calls to observe officers in action and to learn any danger signs displayed by their charges.[36] If an officer who is involved in a disproportionate number of motor vehicle crashes receives no remedial driving training, for example, the supervisor may have missed an early warning signal (discussed later in this section). If an officer transports a handcuffed, compliant, misdemeanor suspect to headquarters at dangerous speeds with lights and sirens activated, the supervisor must address this danger sign. If an officer's suspects always arrive at headquarters bleeding, the supervisor should recognize another danger sign. A supervisor who has not counseled an officer who repeatedly commits any of the deadly errors is missing an early warning sign of future misconduct, advises Joseph Petrocelli.

Performance Evaluations Performance evaluations are standardized assessments that are conducted yearly or more often and documented in writing by sergeants and middle police managers. Such evaluations are intended to better control police conduct, to enhance police practices, and to aid in promotions and disciplinary issues. Also, most performance evaluations provide numerical ratings (e.g., ranging from 1 to 5) of officer achievement of goals. Most officers evaluated receive a 4, which says little about the quality and diversity of the work efforts performed.

Early Warning/Intervention Systems **Early warning systems (EWS) and early intervention systems** are recent initiatives related to police accountability. The objective of both strategies is to identify negative behavioral patterns among police officers before the behavior of those officers becomes problematic or leads to outright corruption, unnecessary use of force, or other forms of illegal or immoral performance.[37] For example, the results of professional performance evaluations may be used to help detect potential problem officers. In one study involving the background files and academy records of nearly 2,000 officers within the Philadelphia Police Department (PPD), researchers were able to reasonably predict characteristics associated with future disciplinary officer problems.[38] Some of those characteristics included citizen complaints, internal investigations, and departmental discipline. The attitudes of officers regarding police work, their department, and inappropriate police conduct were also explored through the use of a survey administered to almost 4,000 patrol officers within the PPD. Researchers were able to then use the survey results to predict future problematic behavior among police officers. The strongest indicator of problematic officer behavior was frequency of departmental discipline, followed by physical abuse complaints, internal investigations, and off-duty incidents. Other indicators of future problematic behavior included youthful age, having traffic offenses, and having prior contact (before employment) with the criminal justice system.

Performance indicators that hint at potentially problematic performance (as compared to other officers in the same jurisdiction) can also include these early warning signs:

- Repetitive use of force
- Extraordinary number of traffic stops
- Usually high number of violent arrests
- Constituent complaints
- Numerous civil suits
- Frequent vehicle and felony pursuits that end in tragedy
- Numerous "resisting arrest" charges brought against suspects
- Frequent sick leaves
- Complaints from other officers

The presence of any of these markers, if it exceeds the average for most officers in the same jurisdiction, could suggest that an officer chooses violence as his or her best strategy as opposed to other forms of conduct. Most officers do not turn to violence or the threat of violence as their first choice when confronted with a suspect, as mentioned in Chapters 4, 10, and 12.[39]

In some cases, early warning systems and early intervention systems may be managed by outside independent agencies as opposed to internal personnel. Also, many firms market digital evaluation equipment such as policepath.one, whose software includes prerecorded entries from which evaluators make selections; the system then spits out a sophisticated assessment concerning the performance of an officer. It is strongly recommended that the results of such early warning signals and systems be used as guides in providing appropriate counseling and training for officers who exhibit any of the previously mentioned danger signals, argue Samuel Walker, Geoffrey P. Alpert, and Dennis J. Kenney.

It goes without saying that Pre-employment evaluations (see Chapters 8 and 15) are another type of early warning system that can take some of the guesswork out of the hiring process by identifying potentially problematic candidates before they ever become full-fledged officers. Departments have a duty to take reasonable precautions in hiring and retaining officers who are psychologically disturbed. The wife of one police officer was shot by her husband, who then killed himself;[40] the wife then sued the City of New York for negligence in allowing her husband to carry a gun. In this case, the court held that, to avoid liability, a department has to show that it has taken reasonable precautions in hiring and/or retaining officers who are psychologically disturbed. The wife was awarded $500,000 in compensatory and punitive damages.

Other means of monitoring police officers for early warning signs of corruption are more intrusive. When a police officer returns home after a shift, he or she places the police radio on its home charger, which the department makes officers buy from a police department merchant. Radios can be turned on from a remote, and some conversations in the home can be heard and recorded. Private security companies can be contracted through the district attorney's office to place certain or at random officers under surveillance. For instance, police officers and their families may be monitored in hopes of catching them in an illegal activity. Officers' cell phones are sometimes tracked through GTD (global tracking devices), and conversations can be easily monitored. Mortgage offices are checked repeatedly for officers purchasing property. Conversely, high-ranking officers reside in different jurisdictions so that patrol officers cannot monitor their movements as easily.

Internal Affairs Units and Professional Standards According to Tim Dees, most sworn officers became officers for the right reasons, yet even good officers make mistakes—and then there are officers who just shouldn't be officers.[41] Recognizing that mistakes and corruption are always possible, many departments employ a specialized police unit whose responsibility is to investigate police misconduct. Members of the **internal affairs (IA)** unit or office of professional standards (OPS), also called quality assurance officers or policy compliance officers, react to constituent complaints or employ proactive initiatives in response to unverified evidence of officer misconduct. Often IA or OPS personnel are directly under the command of the police chief or commissioner or some other designated official.

Professional standards that may be used to define misconduct may, for example, include those outlined in accreditation standards such as those established by the Commission on Accreditation for Law Enforcement Agencies (CALEA), as discussed in Chapter 7. In the Kingsport (Tennessee) Police Department, Chief James Keesling reports that use of OPS personnel for monitoring compliance with CALEA accreditation standards lowered the number of constituent complaints by approximately 60 percent in his jurisdiction.[42]

Most states have established statutes that mandate a department's response to the officer disciplinary process. For example, Florida Statute 112.532 is commonly known as the Police Officer's Bill of Rights. This statute, combined with the officer's labor agreement and departmental rules and policies, defines the manner in which an internal investigation will proceed. Officers cannot be harassed, intimidated, or threatened by supervisors, nor can they be subjected to unreasonable periods of interrogation. Officers have the right to legal counsel or union representation during an investigation. They also have the right to seek civil action in cases where accusations were deliberately false or vindictive.

Computer-Based Programs for Internal Affairs Tracking

In today's policing, a computer solution is available for most police work—and IA is no different in this respect.[43] Several applications have been developed that can aid IA investigations, but each involves "creating a relational database of personnel and complaint information, and then 'mining' that database to identify relationships that might not otherwise be immediately evident."[44] There are four major software products available:

- BlueOrder and RMAAT from Delta Systems Design and Liekar System Solutions, respectively
- IA Trak from the Institute of Police Technology and Management (IPTM) at the University of Florida
- IAPRO from CI Technologies (http://www.iaprofessional.com)
- Administrative Investigations Management (AIM) from On Target Performance Systems

For example, the IA Trak software was developed by IPTM (http://www.iptm.org) to meet the standards taught in internal affairs courses at the University of Florida as well as standards set forth by CALEA. IA Trak allows agencies to keep track of personnel's conduct and performance issues, and it tracks internal affairs investigations as well as issues related to use of force, pursuits, personnel accidents, suspect processing, and prisoner injuries, among other things. Alerts can be customized by the police agency as part of an early warning system when a particular employee reaches a minimum threshold, possibly averting potential problems. Users can also generate charts, graphs, and statistical analysis as necessary.

Another software application available from Segreant (intentionally misspelled) Software is more focused on early warning and intervention with at-risk personnel, rather than on investigations management. Segreant (http://www.segreant.com) can track incidents ranging from pursuits to uses of force to calls that have a special impact on certain individuals.[45]

The Internal Affairs Process

In the Boston Police Department (BPD), the Internal Affairs Department (IAD) follows a strict process when investigating potential officer misconduct.[46] After researching a complaint against an officer, the IAD investigator prepares a report and submits it with his or her recommendation to the IAD team leader. After further review, the report is forwarded through the chain of command to the chief of the Bureau of Internal Investigations (BII). After the BII chief reviews and accepts the report, both

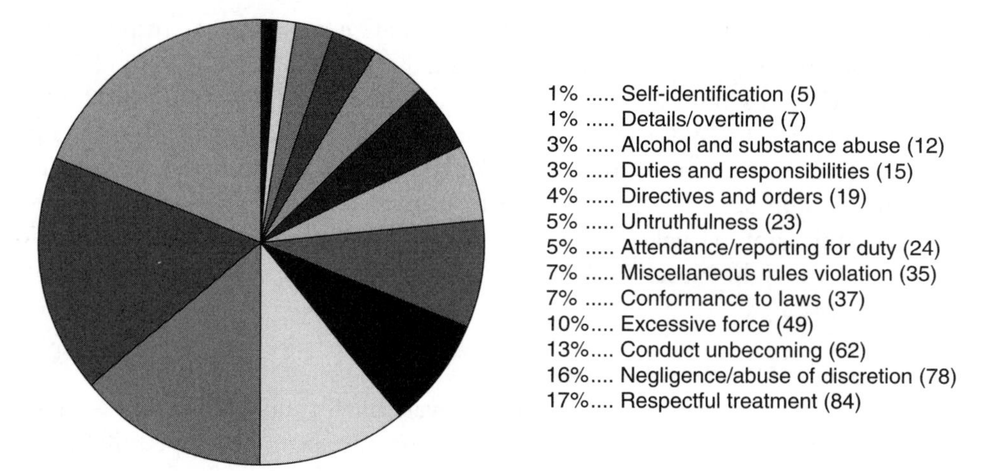

Figure 14–1 Allegations against Boston Police Department Personnel, 2005
Source: Adapted from Boston Police Department, "Annual Report: City of Boston" (2006), http://www.cityofboston.gov/police/pdfs/Annu_Rep_05.pdf (accessed June 16, 2007).

the report and a recommendation are forwarded to the legal advisor for the BPD, and ultimately to the police commissioner.

Within one case, there could be multiple allegations with varied dispositions and recommendations. Outcomes could include the following:

- *Sustained:* Sufficient evidence supports the complainant's allegations and personnel are subject to disciplinary action. This finding may reflect a need for some action.
- *Not Sustained:* The investigation failed to prove or disprove the allegations. This is the weakest finding, as it reflects the investigator's inability to prove or disprove the allegations.
- *Unfounded:* The investigation reveals that the action complained of did not occur.
- *Exonerated:* The investigation reveals that the action complained of did occur, but it was reasonable, proper, and legal. Nevertheless, this finding may reflect a need for training or a change/creation of a policy.

If a citizen is not satisfied with the investigative process, he or she may make an appeal to the Community Appeals Board.

In 2005, there were 467 allegations against BPD personnel (see Figure 14–1), of which "respectful treatment followed by negligence/abuse of discretion" led the list. After these allegations went through the investigatory process, 244 cases were sustained, 73 cases were not sustained, 91 cases were unfounded, 8 cases were exonerated, and 51 cases are pending (see Figure 14–2).

Some Statistics about Constituent Complaints

In 2004, an estimated 30,000 civil action suits were filed against the police across the United States.[47] The percentage of officially sustained complaints in 2002 ranged from 6 percent among county police departments to 12 percent among sheriffs' offices.[48] The rate was 6.6 complaints per 100 full-time sworn officers. About 8 percent of the complaints were officially sustained—

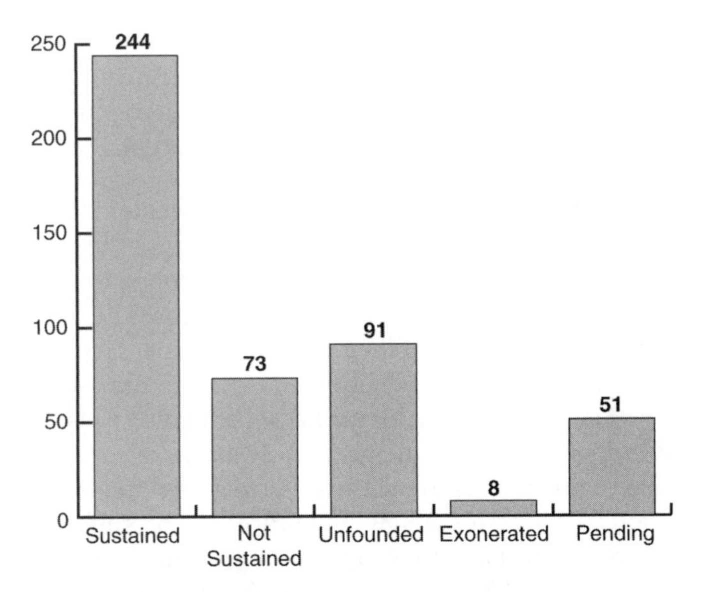

Figure 14–2 Dispositions of Individual Allegations against Boston Police Department Personnel, 2005
Source: Adapted from Boston Police Department, "Annual Report: City of Boston" (2006), http://www.cityofboston.gov/police/pdfs/Annu_Rep_05.pdf (accessed June 16, 2007).

that is, there was sufficient evidence to justify disciplinary action against the officer or officers; 34 percent were not sustained; 25 percent were unfounded, meaning the complaint was not supported by facts or the alleged incident did not occur; 23 percent ended in exonerations because the police actions were deemed lawful and proper; and 9 percent ended in other dispositions, such as complaint withdrawal.

Criticism of Internal Discipline Practices

Supervision, close supervision, performance evaluations, early warning systems, and internal affairs units and professional standards have short comings.

Criticism: Supervision: One study argues that the effects of active supervision on officer compliance are ambiguous.[49] It's unclear whether active supervisors have an influence on street officers or not. The presence of supervisors at calls for service may force patrol officers to follow agency policy and the "wishes of the complainant."[50] This tactic could produce resentment of the presence of supervisors and even their reluctance to call supervisors to any altercation.

Criticism: Close Supervision: One study of close supervision asked three questions:[51]

- How frequently do patrol officers engage in searches?
- How often do their searches meet constitutional standards?
- What explains the proclivity to search unconstitutionally?

The results of this study "[paint] a disquieting picture, with nearly one-third of searches (115) performed unconstitutionally and almost none visible to the courts."[52] The officers who participated in the study worked in a department that is ranked in the upper 20 percent nationwide, and the observers included an appellate judge, a former federal prosecutor, a government attorney, and the authors of the study (Gould and Mastrofski). Thus, even while professional observers were present, police officers conducted unprofessional and unconstitutional searches. For example, Gould and Mastrofski offer the following description about an unconstitutional "egregious" search:

> A black male in his late twenties, while riding a bike, was told to pull over by two officers from their patrol cruiser. The biker refused and the officers called for backup. When another vehicle was seen, the biker stopped and an officer told him the police received a report that he was selling drugs. The rider denied it several times while one officer searched the rider's knapsack. Another officer padded the rider's pockets down. Nothing. "I bet you are hiding them under your balls. If you have drugs under your balls, I am going to f— your balls up," one of the officers said. "You sure are nervous. I wonder why you are so nervous," another officer commented, and told the biker to get behind the police car door, and pull his pants down to his ankles. The officer put on rubber gloves, and felt around the biker's private parts for drugs. Finding nothing, the officer said, "I bet you are holding them in the crack of your ass. You better not have them up your ass." And with that the biker turned, bent over, and spread his cheeks. The officer put his hands in the biker's rectum, but found no drugs. As the four officers walked back to their cars, one commented, "I know he had some drugs."

The illegal searches discovered by these researchers were highly concentrated in a few officers: 16 percent of the officers in the sample accounted for 24 (70 percent) of the illegal searches. Furthermore, the officer who single-handedly accounted for the most illegal searches was also one of the most articulate, low-key officers observed.

If officers were closely supervised, would unconstitutional searches and inappropriate officer behavior be reduced? Hanniman reports that it is unlikely that routine supervision of officers will alter officer behavior.[53] Kelling describes the more recent contention that officers are "pushing the Fourth Amendment" to the verge of or beyond what is legally permissible.[54] Nonetheless, Gould and Mastrofski clarify that officers in the previously mentioned study were ordered by their supervisors to crack down on drug pushers, supporting the theory that officers, when pushed toward making more arrests, can violate human rights in the name of "good."

Standardized evaluation reports have also come in for criticism as means to monitor officers for corruption can be seriously flawed, says Sam Walker and Charles M. Katz. For instance, most performance evaluations do not necessarily address the work officers do. Reports often lack clarity, and they tend to suffer from the "halo" effect (ratings in one area reflect the ratings in other areas). At an extreme, the literature implies that good officers will cover for bad officers, suggesting that some sergeants will write good performance evaluations for their subordinates even knowing that those officers might be corrupt.[55]

Consider the case of Sergeant Timothy White, a 16-year veteran of the Massachusetts State Police, who was convicted on charges of stealing 27 pounds of cocaine, marijuana, and ecstasy. White was once Massachusetts' most respected state troopers with excellent performance evaluations; he even served as media spokesman. The trooper's life unraveled when he was charged with assaulting his wife Maura, selling drugs, and using his position on the Narcotics Inspection Unit to pilfer 13 kilos of cocaine from evidence storage. Prior to that, prosecutors alleged that White, his wife, and fellow partiers engaged in coke-fueled kinky sex, partner swapping, and bisexuality. Had there been early signs of corruption? If so, they were never reflected in his performance evaluations.

Likewise, early warning systems may fail to catch police misconduct for several reasons. First, only a small proportion of officers in a department are responsible for a large proportion of the problems (i.e., 10 percent of the low achievers are responsible for 90 percent of the problems), and they may slip through the cracks of the early warning system. Second, software is expensive and time-consuming to use; it requires almost daily input from competent personnel if it is to aid in investigative pursuits or, for that matter, become an efficient early warning system for the department.

A study of 558 patrol officers employed by departments in Massachusetts, North Carolina, and Florida, reported that almost one half of the officers said that the disciplinary action taken against officers was "always" stressful.[56] These results may imply that most patrol officers have little confidence in the process initiated by IA units, OPS, or other police officials investigating police misconduct. Furthermore, budgets are often inadequate to staff internal affairs units, an idea supported by the revelations about the LAPD following the Rampart scandal. In that case, the LAPD reported that its IA unit could investigate only 10 percent of all constituent complaints.[57]

Other police departments have also struggled with IA demands. When West Palm Beach (Florida) Police Department (WPBPD) officials met in executive session in 2006, Captain Mark Anderson noted that the goal of his IA unit was to streamline management reporting systems related to IA functions.[58] Some of the specific goals included a revision of paperwork flow, control of officers linked to the early warning system, and for Anderson to personally meet, in a one-to-one setting, with every uniformed police sergeant and provide handouts on use of force and other departmental recommendations. An aide of WPBPD Chief Delsa R. Bush indicates that the captain has been unable to fulfill IA unit objectives because of a lack of personnel due to shortfalls in payroll budgeting for the unit.

Frank Colaprete advises that the IA function is a necessary evil within departments.[59] However, he suggests that it can be the most complex and sensitive investigation conducted by a police organization: "Insurmountable barriers of attitudes, politics, and egos block the road."[60] Also, an IA investigation must be fair: "balancing the rights of the complainant should be equal to and not overshadow the rights of the accused officer." Despite this mandate, officers under investigation are not necessarily treated as though they are "innocent until proven guilty," and often the rights of officers are breached in favor of a complainant.

Differences in Complaints

It is difficult to measure complaints—and hence to determine the scope of internal monitoring investigations—because of differences in measuring periods and differences of department officer size and populations. The Atlanta Police Department (APD) reportedly investigated 56 complaints that alleged the use of unauthorized force in one year.[61] Compare this low rate to that for the San Francisco Police Department (SFPD), which had a slightly larger force (approximately 2,000 sworn officers at SFPD versus 1,500 sworn officers at APD) at the time of the Atlanta complaints. The SFPD's Commission's Office of Citizen Complaints receives 1,000 complaints each year, with approximately half of the complainants alleging unnecessary force or unauthorized action by the police. By comparison, the San Jose (California) Police Department, which had approximately 1,200 sworn officers during the time period measured, received 198 unnecessary force complaints. In other words, the San Jose force was 20 percent smaller than the Atlanta force but, if the APD official tally is to be believed, it receives three times as many unnecessary force complaints.

The Portland (Oregon) Police Bureau (PPB) reports that its (sworn 1,015) officers use more force, more frequently than their peers in Minneapolis, San Diego, San Jose, and Seattle.[62] PPB reports almost 38,000 arrests per year;[63] in 4,600 cases, officers used force to control the arrest, where use of force is defined as by the PPB as physical (handcuffs, control holds, take-downs, pressure points, and hobbles), Taser, less-lethal munitions, blunt impact (with hands or feet), and pepper spray. These cases result in between 140 and 200 complaints each year, of which 0.2 percent (approximately 35) are sustained.[64]

Why might these discrepancies arise? Two explanations have been suggested:

- Civil liabilities violations vary from jurisdiction to jurisdiction.
- Complaints might represent a different process in each department.

When police managers do not respond as well to police corruption as expected by their civilian bosses, the outcomes can resemble what occurred in Chicago in 2007. In that city, Mayor Richard M. Daley ordered that all investigative reports about police officer misconduct be sent directly to his office instead of to the police superintendent's office.[65] Under the mayor's plan, the Office of Professional Standards of Chicago would become a separate city department with subpoena powers under the direct control of the mayor. "We must assure every Chicagoan that we are doing everything possible to prevent abuse by police," says Mayor Daley.

These thoughts are congruent with findings by the Bureau of Justice Statistics, which has suggested that "caution must be exercised when interpreting complaint data inasmuch as volumes, rates and dispositions may well vary by agency characteristics, such as the size and type of the agency as well as the policies and procedures related to the handling of complaints."[66] They also raise another question: Would more cases be sustained or exonerated if constituents were in charge? That is, how effective are external methods at ensuring police accountability?

External Monitoring of Police Accountability

External methods of police accountability can take the form of constituent oversight committees and the courts, which include rulings in civil suits. When police practices give rise to unlawful and disrespectful police behavior, police officer performance can be mandated in an effort to move the offending department toward compliance.

External Monitoring Methods

Constituent Oversight Committees **Constituent oversight committees** consist of members of the community who are charged with performing external reviews of police department actions and advising/directing police policy and behavior. In communities across the country, residents participate to some degree in overseeing their local police agencies.[67] The most active oversight boards investigate allegations of police misconduct and recommend actions to the chief or sheriff. In many jurisdictions (except federal agencies), oversight committees review complaints, budgets, hiring and promotion procedures, deployment initiatives, service call priorities, grievance procedures, use of force limits, pursuit of felons, sex offender registries, training, and general work policies. Other boards review the findings of internal police investigations and recommend that officials approve or reject the findings. In still other committees, an auditor investigates the process by which the police or sheriff's department accepts or investigates complaints and reports to the department and the public on the thoroughness and fairness of the process.[68] Not all oversight committees are extensively used, however, nor do they necessarily possess **subpoena power** (legal authority to demand evidence).

The use of constituent oversight committees has grown in recent years. A Bureau of Justice Statistics study of county, municipal, and sheriff's departments found that as many as 21 percent of those agencies had established a civilian complaint review board and that an estimated 10 percent of those boards had subpoena power within their jurisdiction.[69] Although oversight committees are common among many police agencies, their missions, responsibilities, and authority vary depending on the department and the interests of the committee members. There is no single model of oversight. Nevertheless, most oversight systems can be classified as one of four types:[70]

- Type 1: Citizens investigate allegations of police misconduct and recommend findings to the chief or sheriff (examples: Berkeley, Flint, Minneapolis, San Francisco). These oversight systems are the most expensive largely because professional investigators must be hired to conduct the investigations; lay citizens do not have either the expertise or the time to do so.
- Type 2: Police officers investigate allegations and develop findings; members of the committee review and recommend that the chief or sheriff approve or reject the findings (examples: Orange County, Rochester, St. Paul). These systems tend to be inexpensive because volunteers typically conduct the reviews.
- Type 3: Complainants may appeal findings established by the police or sheriff's department to members, who review them and then recommend their own findings to the chief or sheriff (example: Portland, OR). This type of oversight system can be inexpensive because it relies on volunteers.
- Type 4: An auditor investigates the process by which the police or sheriff's department accepts and investigates complaints and reports on the thoroughness and fairness of the process to the department and the public (example: Tucson, AZ). These systems tend to fall in the middle of the price range. On one hand, as in type 1 systems, only a paid professional has the expertise and time to conduct a proper audit. On the other hand, typically only one person needs to be hired because the auditing process is less time-consuming than conducting full-scale investigations of citizen complaints.

Oversight committees often combine several of these features, and they serve jurisdictions in different ways. For instance, the city of Santa Rosa, California, added a tax resource to city government spending, approved by constituents.[71] The plan provided funding for a variety of specific programs benefiting the community, including additional fire and police department personnel and equipment, four interim fire stations in various locations throughout the city, and gang prevention and youth programs run by the Recreation and Parks department. A constituent oversight committee was appointed to ensure that all revenues received would be spent only on permissible uses, which were defined in the ordinance establishing the special tax. For example, the committee reviewed police salaries and approved an increase in payroll and the hiring of two field and evidence technicians.

Oversight committees are often part of life at colleges and universities. University campus oversight committees evaluate criminal activity and deviant behavior. For instance, at the University of Texas at Austin, Professor M. Michael Sharlot chaired the first Police Oversight Committee for the university.[72] The 12-member committee is composed of faculty, staff, and students, and is the principal institutional channel of communication between members of the university community and the University of Texas at Austin Police Department (UTPD).

According to Samuel Walker, arguments in favor of oversight committees include the existence of widespread police corruption, made possible because police organizations have "consistently failed to investigate allegations of misconduct and to discipline guilty officers."[73] This perspective assumes that the police hierarchal organization cannot or will not appropriately conduct an investigation in case of corruption or complaints against police made by constituents. David H. Bayley echoes this thought, noting that jurisdiction after jurisdiction has been forced to share responsibility for maintaining appropriate levels of discipline with newly created oversight committees because of the enormous amount of corrupt behavior reported within police organizations.[74] As a result of this trend, which has been aided by the courts, the police have lost their monopoly on determining whether officers are behaving legally and morally, says Bayley. In part, this "due-process revolution," as some call it, is associated with increased deployment of the courts to handle police corruption.

Other arguments in favor of oversight committees include the opportunities they provide for constituents to voice their opinion, the way they encourage constituents to be responsible to their community, and their ability to enhance police compliance. In addition, public oversight often creates solidarity between police and community and adds to the effectiveness of community relations initiatives.

Oversight committees are not immune to criticism, however. They probably experience higher levels of sustained complaints or complaints assigned a continuing status, which would result in less disciplinary action for officers who were factually guilty of misconduct. Consequently, the deterrence effect from the latter cases is probably minimal.

Other arguments against the use of oversight committees include the following points:[75]

- Police misconduct and abuse of force are not necessarily a serious problem in all departments.
- Most officer misconduct is dealt with through command personnel, where meaningful punishment can be distributed as a deterrent for other officers.
- Individuals who are not trained investigators will experience difficulties in conducting a thorough and fair investigation.
- Police internal review procedures can perform equally well, if not better, in investigating complaints and administering justice than can oversight committees.
- Constituents are generally more lenient with guilty officers than are their fellow officers.
- Oversight can deter officers from effective police work.

- Oversight can provide less satisfaction to complainants.
- Oversight undermines police professionalism by intruding on the professional autonomy of the police.
- Constituents tend to be unfamiliar with ongoing investigations and may aid criminal activities through ambition or ignorance.

Although many arguments both for and against oversight committee participation in police matters lack empirical support, feelings about these systems run deep among both officers and the public. The reality is that policing—like law, medicine, social work, and higher education—requires particular individuals with particular skills at certain times in "highly specific situations."[76]

The courts, and especially the U.S. Supreme Court, have influenced police accountability through their decisions in numerous court cases. Some of these rulings continue to be hotly debated. This chapter summarizes eight key cases:

- *Weeks v. United States*, 1914
- *Miranda v. Arizona*, 1966
- *Mapp v. Ohio*, 1967
- *Katz v. United States*, 1968
- *Terry v. Ohio*, 1968
- *New York v. Quarles*, 1984
- *Nix v. Williams*, 1984
- *United States v. Leon*, 1984

In the case of *Weeks v. United States*, 232 U.S. 383 (1914), police officers entered the residence of Fremont Weeks without a warrant and seized personal materials that were later used to convict Weeks of transporting lottery tickets.[77] Weeks took action against the police and wanted his materials returned. The Court held that both the seizure and the government's refusal to return his personal property violated Weeks' constitutional rights.[78] (See Chapter 10 for more details.)

The insistence on freedom from "unreasonable" search and seizures is a fundamental right associated with the notion that "Every man's house is his castle"—an idea expressed by the colonists as a result of their experiences with the English Crown.[79] A product of *Weeks v. United States* was the first application of the *exclusionary rule* (see Chapter 10 for a definition and details); however, *Boyd v. United States* (1886) paved the way for the exclusionary rule's use because it made the fruits of police misconduct inadmissible in court, argues Jeffery T. Walker.[80] The **exclusionary rule** suggests that evidence secured by illegal means and in bad faith cannot be introduced at a criminal trial.[81]

There are lawful warrantless searches that lead to lawful arrests, and these incidents rarely provoke disputes over constitutional issues. However, some want the exclusionary rule abolished because it limits police and prosecutorial potential to gather evidence, indict suspects, and convict defendants, according to Nancy E. Marion and Willard M. Oliver.[82] For instance, some police officers believe that the exclusionary rule limits the police's ability to provide public safety: "[If] there is any legal barrier I want eliminated, it's the freggin' exclusionary rule," said a Boston officer in 2007, 93 years after *Weeks*.[83] Prior to *Weeks*, the rules and procedure governing use of evidence in the U.S. judicial system reminded faithful to the idea that the process by which evidence was obtained had very little to do with the permissibility of its use in court. Originally, *Weeks* applied to federal prosecution cases, states Jeffrey Walker,[84] until *Mapp v. Ohio* (1961) and *Katz v. United States* (1967).

Exclusionary advocates believe the exclusionary rule and Fourth Amendment safeguards play an important role in the separation-of-powers principles.[85] "When agents of the executive branch (the police) disregard the terms of search warrants, or attempt to bypass the warrant-issuing process altogether, the judicial branch can and should respond by 'checking' such misbehavior," explains Timothy Lynch of the Cato Institute. An implication suggested by exclusionary rule proponents is that the law makers cannot be law enforcers in a democratic nation, and vice versa.

In the case of *Miranda v. Arizona*, 384 U.S. 435 (1966), Ernesto Miranda was arrested and interrogated at the Phoenix Police Department after being identified by the victim of a rape–kidnapping.[86] After two hours of questioning, the police had obtained a signed written confession of guilt from Miranda, who never went beyond ninth grade at school. At his trial, his written confession was admitted into evidence and Miranda was found guilty of kidnapping and rape. The Supreme Court, however, held that prosecutors could not use statements stemming from custodial interrogation of the defendant unless those statements demonstrated the use of procedural safeguards "effective to secure the privilege against self-incrimination."[87]

Whether Miranda had been told that anything he said could be used against him was unclear. But the police admitted—and this admission was to prove fatal for the prosecution—that neither before nor during the questioning had Miranda been advised of his right to consult with an attorney before answering any questions or that he could have an attorney present during the interrogation. Specifically, the Court stated that a confession could be admissible under the Fifth Amendment self-incrimination clause and the Sixth Amendment right to an attorney clause only if the suspect had been made aware of his or her rights and had then waived those rights. The Court through *Miranda* attempts to aid police. Once *Miranda* warnings are provided, the burden shifts to the defendant to prove that a custodial interrogation confession was coerced. *Miranda* symbolized the legal system's determination to treat even the lowliest and most despicable criminal suspect with dignity and respect.[88]

Mapp v. Ohio, 367 U.S. 643 (1961), supported Fourth Amendment rights and the exclusionary rule linked to unreasonable searches and seizures and evidence gathered from those searches (see Chapter 10). Fourth Amendment issues were not used in criminal prosecutions in either state or federal courts until *Mapp v. Ohio*.[89] In the *Mapp* case, the Cleveland Ohio Police received information from an informant that Dolree Mapp was harboring a fugitive. The police broke down the door and entered Mapp's home, producing a phony warrant as justification to search her home. In the basement, officers discovered a trunk filled with obscene materials. The police arrested Mapp, and the prosecutor used the contents of the trunk to convict Mapp of violating an Ohio law that prohibited the possession of obscene materials. Mapp said she was holding the trunk for a friend and didn't know what was in it. The warrant was never found.

After Mapp's conviction, her lawyer argued that the search was unlawful and, therefore, that the evidence was obtained illegally and should have been excluded at her trial. The U.S. Supreme Court ruled in favor of Mapp, declaring that "all evidence obtained by searches and seizures in violation of the Constitution is, by [the

Fourth Amendment], inadmissible in a state court."[90] The Supreme Court clarified that the exclusionary rule (see above and Chapter 10) also applies to states, which means that states cannot use evidence gained by illegal means to convict anyone. The idea of the "fruit of a poisonous tree" was not addressed until *Wong Sun v. United States* (1963). However, in *Nix v. Williams* (1984), the Court ruled that tainted evidence may be admissible in court if officers would have discovered it anyway.

Katz v. United States, 389 U.S. 347 (1967), clarified suspects' Fourth Amendment rights, particularly those rights linked to an expectation of privacy. In its ruling, the Supreme Court stated that "Wherever a man may be [a hotel room or a telephone booth], he is entitled to know that he will remain free from unreasonable searches and seizures."[91] That is, individuals are protected from unreasonable searches and seizures, which in this case included a telephone booth; thus wiretaps by authorities without a warrant are illegal.[92] In this case, Charles Katz used a public pay phone booth in Los Angeles to place bets in Miami and Boston. Federal Bureau of Investigation (FBI) agents attached an electronic eavesdropping device to the phone and recorded Katz's conversations. As a result of those recordings, Katz was convicted of illegal gambling. His conviction was overturned because the Court said that "the Fourth Amendment protects people, not just places."

A year after the *Katz* ruling, the Supreme Court ruled that the Fourth Amendment prohibition on unreasonable searches and seizures is not violated when a police officer stops a suspect and searches the individual without probable cause. In *Terry v. Ohio*, 392 U.S. 1 (1968), a police officer had "no probable cause to arrest (or stop) Terry for anything, but he (the officer) had observed circumstances that would reasonably lead an experienced, prudent policeman to suspect that Terry was about to engage in burglary or robbery," the Court stated.[93] The Court also concluded that the revolver seized from Terry was properly admitted into evidence because the officer "had reasonable grounds to believe" that the suspect "was armed and dangerous, and it was necessary for the protection of himself and others to take swift measures to discover the true facts and neutralize the threat of harm if it materialized." In *Terry*, the exclusionary rule was limited based on an officer's reasonable suspicion that a suspect can be a danger. Officers make many "*Terry* stops" while on duty. Because of the important interest in protecting the safety of police officers and the public, the Supreme Court held that police officers have the authority to stop someone and do a quick surface search of that person's outer clothing for weapons if the officer has a reasonable suspicion that the individual has, is about to, or is actually committing a crime.[94]

New York v. Quarles, 467 U.S. 649 (1984), established the public safety "exception" to the rule proscribed in its historic decision in *Miranda v. Arizona*.[95] In this case, after receiving the description of Quarles, an alleged assailant, a police officer entered a supermarket, spotted him, and ordered him to stop. Quarles stopped and was frisked by the officer. Upon detecting an empty shoulder holster, the officer asked Quarles where his gun was. Quarles responded. The officer then formally arrested Quarles and read him his Miranda rights. The question in this case was whether the court should suppress Quarles's statement about the gun and the gun itself because the officer had failed at the time to read Quarles his Miranda rights. The U.S. Supreme Court held that this evidence was admissible because there is a "public safety" exception to the requirement that officers issue Miranda warnings to suspects. Because the police officer's request for the location of the gun was prompted by an immediate interest in assuring that it did not injure an innocent bystander or fall into the hands of a potential accomplice to Quarles, his failure to read the Miranda warning did not violate the suspect's constitutional rights.

Nix v. Williams, 467 U.S. 431 (1984), also dealt with admissibility of evidence. Williams was arrested for the murder of a 10-year-old girl whose body he disposed of along a gravel road. State law enforcement officials engaged in a massive search for the child's body. During the search, after responding to an officer's appeal for assistance, Williams made statements to the police (without an attorney present) that helped lead the searchers to the child's body. The defendant's Miranda rights were read to him only after his arrest. The question in this case was whether evidence resulting in an arrest should be excluded from trial because it was improperly obtained. The Supreme Court ruled that it could be admitted. In this decision, it relied on the "inevitable discovery doctrine," as it held that the exclusionary rule did not apply to the child's body as evidence since it was clear that the volunteer search teams would have discovered the body even without Williams's statements.

United States v. Leon, 468 U.S. 897 (1984), further refined the exclusionary rule, which requires that evidence seized illegally by police be excluded from criminal trials. Leon was the target of police surveillance based on an anonymous informant's tip. The police applied to a judge for a search warrant of Leon's home based on the evidence from their surveillance. A judge issued the warrant and the police recovered large quantities of illegal drugs. Leon was indicted for violating federal drug laws. A judge concluded that the affidavit for the search warrant was insufficient; it did not establish the probable cause necessary to issue the warrant. Thus the evidence obtained under the warrant could not be introduced at Leon's trial. But is there a "good faith" exception to the exclusionary rule? According to the U.S. Supreme Court, there is such an exception. The justices held that evidence seized on the basis of a mistakenly issued search warrant could be introduced at trial. The exclusionary rule, argued the majority, is not a right, but rather a remedy justified by its ability to deter illegal police conduct. In *Leon*, the costs of the exclusionary rule outweighed the benefits. The exclusionary rule is costly to society: Guilty defendants go unpunished and people lose respect for the law. The benefits of the exclusionary rule are uncertain: The rule cannot deter police in a case like *Leon*, where they act in good faith on a warrant issued by a judge.

A recent U.S. Supreme Court decision that will change police practice deals with no-knock searches. In 2006, the Court ruled that drugs or other evidence seized can be used in a trial even if police failed to knock and announce their presence. This decision marks a major shift in its rulings on illegal searches by police. This case clearly undercuts a nearly century-old rule that says evidence found during an unlawful search cannot be used. It also offers a sign that the Supreme Court might broaden the power of the police in the future.[96] The thinking of the Court is that tossing out evidence acquired in violation of the knock-and-announce rule—albeit with a valid warrant—could mean "releasing dangerous criminals."

Not all of the justices concurred with this ruling. "It represents a significant departure from the court's precedents," Supreme Court Justice Stephen Breyer wrote in dissent, joined by Justices John Paul Stevens, David Souter, and Ruth Bader Ginsburg. "It weakens, perhaps destroys, much of the practical value of the Constitution's knock-and-announce protection," he added.

The knock-and-announce rule is based on a centuries-old idea of home privacy and the Fourth Amendment protection against unreasonable searches. It requires police to knock, announce themselves and wait a "reasonable" time, which justices noted can be about 20 seconds, before entering. The practice shields occupants from surprise and property damage.

> Federal judges and magistrates may lawfully and constitutionally issue **"no-knock" warrants** where circumstances justify a no-knock entry, and federal law enforcement officers may lawfully apply for such warrants under such circumstances. Although officers need not take affirmative steps to make an independent re-verification of the circumstances already recognized by a magistrate in issuing a no-knock warrant, such a warrant does not entitle officers to disregard reliable information clearly negating the existence of exigent circumstances when they actually receive such information before execution of the warrant.[97]

U.S. District Court decisions often proceed to the U.S. Supreme Court for evaluation. In 2008, a significant case to follow linked to changing police practice is the federal judge's ruling in August 2006 stating that the government's **warrantless wiretapping** program is unconstitutional. A U.S. District Court judge struck down the National Security Agency's program, which the judge said violates the rights to free speech and privacy. "We must first note that the Office of the Chief Executive has itself been created, with its powers, by the Constitution. There are no hereditary kings in America and no power not created by the Constitution."[98]

Harris v Scott, 2007 [81]: In Chapter 9, you read about Victor Harris, who was left a quadriplegic after Deputy Thomas Scott rammed his car off the road.[99] In *Harris v. Scott* (2007), Justice Antonin Scalia wrote for the majority, "The car chase that [Harris] initiated in this case posed substantial and immediate risk of serious physical injury to others. Scott's attempt to terminate the chase by forcing [Harris] off the road was reasonable." According to this decision, police are not subject to litigation when a clear and present danger exists and can be documented (e.g., with video cameras).

Despite the Supreme Court rulings, there are significant limitations on the role of the courts associated with police accountability, for the following reasons:[100]

- The courts cannot supervise an officer's day-to-day performance.
- It is next to impossible to ensure that all officers follow the court's decision.
- Most police work does not end in an arrest and, when it does, it may never come to the attention of the court.
- Officers may or may not be informed of the court's decision.
- Court-imposed rules can encourage evasion or lying by officers.
- Many suspects waive their rights after an arrest.
- Accepting a plea bargain typically eliminates suspect's constitutional guarantees.

When the court or the police fail to protect the rights of constituents, those constituents have the right to litigate through civil liabilities or to bring suit against an officer and his or her organization, which includes officials and training instructors.

Federal Investigations Federal agencies may also investigate local police corruption. For example, federal agents raided the New Haven (Connecticut) Police Department (NHPD) headquarters and charged the head of NHPD's narcotic division with stealing thousands of dollars planted by the FBI during sting operations.[101] Federal authorities also arrested a police narcotic detective and three members of a family of bail bondsmen who bribed police to catch criminals.

In an ongoing investigation of police corruption, the FBI stashed $27,500 in the trunk of a car parked in the Long Wharf area of Hartford, Connecticut. Authorities said two Hartford detectives were told the car belonged to a drug dealer, and the two detectives searched it without a warrant and split the money.

Compliance of police officers with regulations and laws can expand and strengthen the notion of accountability.[102] Accountability works best in the following circumstances:

- Organizations are directly accountable to those whom they serve—their constituents.
- Performance expectations, based on what the constituent values, are clear.
- Performance can be monitored regularly by both police and constituents.
- Police have the freedom to make the changes necessary to improve.
- Constituents have the ability to make performance matter.

Civil Liabilities

There are two ways to litigate against an officer: a state court lawsuit filed as a tort law claim, and a federal lawsuit filed under Title 42 of the United States Code.

State (Tort) Liability Law

Victims of police abuse can litigate for civil damages under state law. In general, these tort lawsuits involve allegations of false arrest, false imprisonment, malicious prosecution, assault, battery, or wrongful death.[103] These suits can be settled by money awards. The standard of proof in such cases is the preponderance of evidence—a standard much lower than that required to convict someone in a criminal case. Unlike criminal cases, liability cases are tried in civil court. It is common for an individual officer to be the target of a single accusation; in such litigation, the officer's personal assets are on the line.[104]

There are three types of torts under state law. Evidence rules, precedent, and judicial discretion play a role in determining which type of tort law will be applied:

- *Strict Liability.* If the injury or damage is severe, and it is reasonably certain that the harm could have been foreseen, then the law dispenses with the need to prove intent or mental state. The only issue is whether the officer or department should pay the money award; because officers don't usually have any money, the department almost always pays. (Examples: reckless operation of vehicle; excessive SWAT tactics)
- *Intentional Tort.* An officer's intent must be proven, using a "foresee-ability" test—that is, whether the officer knowingly engaged in behavior that was substantially certain to bring about injury. (Examples: wrongful death, assault, false arrest, false imprisonment)

- *Negligence.* In negligence cases, intent or mental state does not matter. What does matter is whether some inadvertent act or failure to act created an unreasonable risk to another member of society. Most states have three levels of negligence: (1) slight or mere (absence of foresight); (2) gross (reckless disregard); and (3) criminal. (Examples: speeding resulting in traffic accident, not responding to a 911 call)

Other types of state tort liability are associated with specific types of lawsuits:

- *Wrongful Death.* Wrongful death may be alleged when an officer thinks a suspect is reaching for a weapon, he or she shoots the suspect, and no weapon can be found on or near the suspect. Merely alleging that a suspect appeared to be reaching for a weapon is no defense.
- *Assault and Battery.* A police assault would occur if an interrogator threatens to beat a suspect senseless unless he or she confesses. Battery is (paradoxically) defined more loosely as any offensive contact without consent.
- *False Arrest.* False arrest involves the unlawful restraint of a person's liberty without his or her consent. For example, using the caged area of a patrol car as a holding area, several officers surrounding a suspect, or ordering someone to remain in an investigator's office could can be interpreted as false arrest.
- *False Imprisonment.* This tort is different from false arrest in that an officer may have had probable cause to conduct an arrest, but later the officer violates certain pretrial rights, such as access to a judge or bondsman.

Federal (Title 42 of USC 1983) Law

Most civil actions against police officers for misconduct are filed in federal court as a violation of Title 42 U.S.C. § 1983. Such a suit is referred to as a civil rights claim and is essentially a charge that an individual has had his or her U.S. Constitutional rights violated. States cannot be sued in a civil rights claim, but municipalities and sheriffs can be sued in the following circumstances:

- They are acting under color of state law (a legal phrase meaning that a person is claiming or implying the acts he or she is committing are related to and legitimized by his or her role as an agent of governmental power—for instance, an off-duty police officer detaining an individual).
- They are violating a specific Amendment right in the Constitution.

The Civil Rights Act of 1871 reads as follows:

> Every person who under color of any statute, ordinance, regulation, custom, or usage, of any State or Territory or the District of Columbia, subjects, or causes to be subjected, any citizen of the United States or other person within the jurisdiction thereof to the deprivation of any rights, privileges, or immunities secured by the Constitution and laws, shall be liable to the party injured in an action at law, Suit in equity, or other proper proceeding for redress, except that in any action brought against a judicial officer for an act or omission taken in such officer's judicial capacity, injunctive relief shall not be granted unless a declaratory decree was violated or declaratory relief was unavailable. For the purposes of this section, any Act of Congress applicable exclusively to the District of Columbia shall be considered to be a statute of the District of Columbia.

For most of its existence, the Civil Rights Act of 1871 had very little effect on police performance. The legal community did not think the statute served as a check on state officials, let alone police officers, and did not often litigate under the statute. But that changed in *Monroe v. Page,* 365 US 167 (1961), when the Supreme Court articulated three underlying mandates of the statue:[105]

- "To override certain kinds of state laws"
- To provide "a remedy where state law was inadequate"
- To provide "a federal remedy where the state remedy, though adequate in theory, was not available in practice"

Page opened the door for renewed interest in Title 42, USC 1983. Today the statute stands as one of the most powerful means by which federal courts can protect those whose rights are deprived. Title 42, USC 1983 is most often used to sue officers and other state officials who allegedly deprived a plaintiff of his or her constitutional rights within the criminal justice system. Most often, violations of the First, Fourth, and Fourteenth Amendments are alleged.

Advantages and Disadvantages of Civil Liability Suits

The justice system often awards substantial monies to resolve liability cases, but these lawsuits have both advantages and disadvantages for constituents. Advantages of suits include the following points:[106]

- Such suits are subject to lower burden of proof standards than those required in a criminal case.
- Victims can personally initiate a civil action.
- Direct compensation to the victim often occurs.
- Police organizations may attempt to develop policy consistent with suit standards.
- Litigation can be an impetus for better training and more responsible practices.

Disadvantages of civil liability suits include the following issues:

- Litigation is costly and time-consuming.
- Officers ultimately may be judgment-proof and protected by sovereign immunity.
- Judicially imposed barriers limit the value of remedies under Title 42, USC 1983.
- Successful judgments do not necessarily hold officers accountable for their actions.
- Successful judgments do not necessarily result in changes in police practice.
- Federal agents are protected through a doctrine of immunity, whereas local officers are often fair game.
- Officers may engage in fewer proactive police activities as a way to insulate themselves from litigation.

Differences in Civil Litigation between Female and Male Officers

"Studies have shown that male officers are much more likely to use excessive force and engage in misconduct than female officers," reports Chief Penny Eileen Harrington.[107] One conclusion that can arise from those studies is that gender and civil liabilities are significantly related variables. That is, constituents bring civil suits against male officers more often than female officers.

In a study conducted by the National Center for Women and Policing (NCWP) in cooperation with City of Los Angeles, different findings and conclusions arose.[108] Female officers were involved in excessive force lawsuits at substantially lower rates than their male counterparts, and no female officers were named as defendants in cases involving sexual assault, sexual abuse, molestation, or domestic violence. Of the $66.3 million in judgments or out-of-court settlements paid by the City of Los Angeles for which the gender of the officers could be determined, $63.4 million (96 percent of total payouts) was attributable to male officers' misconduct. In contrast, $2.8 million (4 percent) was attributable to female officers' misconduct. Of concern, male officers disproportionately accounted for the excessive force lawsuit payouts involving killings and assault and battery: Male officer payouts for killings exceeded female officer payouts by a ratio of 43:1, and male officer payouts for assault and battery exceeded female officer payouts by a ratio of 32:1.

In all, Los Angeles incurred $10.4 million in judgments and settlements in cases involving police officers sued for sexual assault, sexual abuse, molestation, and domestic violence, according to the NCWP.

Implications of Litigation Linked to Violence

There are substantial costs when police practice incorporates unconstitutional behavior. For example, unconstitutional police behavior affects both the rights of individuals and the legitimacy of policing. In fact, the financial costs may be dwarfed by the human costs of such misconduct. For instance, in the NCWP study mentioned in the previous section, researchers noted that domestic violence represents the single largest category of calls made to police departments, and tens of thousands of women are physically assaulted each year in the city. Given these facts, failure by police to control violence not only has particularly high consequences for women in the community, but also negatively affects the credibility of and community trust in the police.

Effects of the Threat of Civil Liability Suits among Police Officers

When officers are overly worried about civil liability suits, they may engage in fewer proactive police activities. In particular, they may shy away from making an arrest (including a probable cause arrest), using force (including returning fire), and conducting legal searches and initiating encounters with suspects, seeing this kind of "disengagement" as a way to insulate themselves from personal accountability.[109] Some scholars might argue that data sets show that attitudes toward civil liability among officers are weak and inconsistent predictors of their behavior.[110] Others argue that "many police officers do not conduct probable cause arrests when both offenders and evidence suggest they should."[111]

One implication of this failure to engage in proactive policing is that reported crime rates and actual crime rates are less strongly correlated than might otherwise be expected. In a study of 658 officers from 21 police agencies spread across the United States, when the data were pooled and examined, the distinctive patterns that emerged lent support to the idea that the threat of litigation among officers influenced arrest decisions. In other words, the potential for litigation plays a significant role in officer discretion to arrest or not to arrest a suspect.

Specifically, when officers were asked about their experiences with arrests today (the day they received the questionnaire) as compared to the previous three and five years of their experiences, six out of 10 participants reported that they conducted fewer arrests, even though they confronted more suspects today then they had in the earlier periods. When the officers were asked about arrest obstacles, their rank-ordered responses were as follows:

- Paperwork
- Training/skills
- Warrants/citations
- Minor crime
- Civil liabilities
- Supervisor/judge/prosecutor
- Lack of staff/backup
- Laziness/no incentive
- Community policing/change

When the participants were asked follow-up questions about the training/skills item, more than one third of the officers said civil liability training was needed. "How to keep my butt out of court as a defendant," one officer wrote on a survey. The responses of the participants were strongly associated with a lack of training, especially training linked to CYA strategies (see Chapter 13 for more detail).

Consistent with their responses were their indicators of arrest. That is, when the officers rank-ordered which arrests they most frequently made, the findings were as follows:

- Adult white male DWI
- Adult white female DWI
- Altercations with immigrants
- Domestic violence (probably because of mandates)
- Adult professional male drug possession

The offenders who were least likely to be arrested had committed the following crimes:

- Youths shoplifting
- Female youth smoking
- Female youth smoking a joint
- Elderly shoplifters
- Officer engaged in domestic violence

Different officers in different parts of the country would probably develop a different list of their most likely and least likely to arrest. Nevertheless, this pattern suggests that officers arrest individuals who are least likely to file a lawsuit against them. Should these data be consistent at any level, the implication would be that the reported lower crime rates experienced across jurisdictions in the United States might be inconsistent with reality.

This idea also has some disturbing implications in terms of police misconduct, suggesting that corrupt officers might actually prey on the most vulnerable (i.e., least likely to file lawsuits) constituents. In Cedar Town, Georgia, for example, an officer pleaded guilty in federal court of stopping Hispanics and stealing money from them.[112] Officer Douglas Damiano admitted that he stopped the motorists under the pretext of a traffic stop, then asked for wallets and searched vehicles looking for money. The rogue cop said he chose his victims based on his perception that the illegal immigrants were less likely to report him to authorities.

Changing Civil Liability Rates

You have already decided that hiring moral and capable officers in the first place is a good practice. But sometimes it doesn't matter. Americans have a penchant for litigation, and they particularly enjoy suing cops—even good officers.[113]

How can police mitigate this risk? In one study, researchers searched the official records of 151 police departments in Michigan over a 15-year period (1985–1999) and evaluated 11,273 claims of civil liabilities against officers, jailers, and agencies.[114] Most often, it was found that current trends in police liability can be reduced through managerial attention to policy and procedures, a prudent focus on training endeavors, and implementation of risk management strategies (a good risk management system is a continuous process of analysis and communication). Loosely explained, police management and professional training can make the difference in legal exposure among officers to civil liability litigation.

RIPPED from the Headlines

Policing the Cops: Time for a Police Complaint Database

We alert cities to the danger of possible terrorist attack. We alert people to the presence of sex offenders in their neighborhood. Now, as the most recent police brutality case in the Golden State so clearly demonstrates, it's time to alert people to the complaint and discipline history of their local police officers, so they can better gauge the risks in their communities. Americans need a Police Complaint Database.

Does it surprise us that Officer Jeremy Morse of Inglewood, California, had a history of police brutality before he punched a helpless, handcuffed juvenile in the mouth? . . . As the crime rate plummeted in the 1990s, the number of law enforcement personnel who were sent to prison multiplied five times.

In response to several different police brutality scandals, such as the Abner Louima scandal in New York City and the Rampart scandal in Los Angeles, communities around the country have been challenging their city councils to give civilian review boards the power to investigate citizen complaints against police officers.

Typically, police departments guard the files of police officers and legal access is severely limited. So mayors, police chiefs, and members of the police commissions have typically balked, saying they investigate their own just fine.

A comprehensive database—organized by local civilian review boards to keep track of all citizen complaints against police officers—would be a helpful community tool to identify a pattern of excessive force by police officers. And the list should be made public via the Internet.

Source: Adapted from Joe Loya, "Policing the Cops: Time for a Police Complaint Database," *Pacific New Service* (February 8, 2002).

Summary

Policing is a means to justice and to upholding the sanctity of individual liberty. Police accountability means that police officers and their organizations are accountable for their behavior and their recklessness when it runs counter to the mission of policing. The purpose of police accountability is to legally and morally ensure that officers and departments conform to the protections guaranteed by the Constitution.

The most effective method of policing police is through self-monitoring, which means that the values of officers must be in place prior to their hiring by the police department. Internal monitoring methods (i.e., within the police department), such as early warning systems and internal affairs investigation, and external monitoring methods (i.e., outside of the department) such as oversight committees, lawsuits, and federal investigations, are also remedies toward controlling police accountability.

The rotten apple theory is one explanation used by police chiefs and policymakers in the face of police corruption. This perspective holds that the officers involved in police misconduct perform that behavior on their own without the knowledge or support of their supervisors or coworkers. Although misconduct occurs in every profession, officers often have more opportunities to engage in misconduct for two reasons:

(1) the behaviors of the public whom they encounter most often and (2) their ability to legally use force and detainment powers. Ideally, professional police management tactics in the areas of hiring, training, and performance evaluation should eliminate the "bad apples," who account for only a small proportion of all officers. The best indicators that officers might be characterized as problematic are frequency of departmental discipline, followed by physical abuse complaints, internal investigations, and off-duty incidents.

Supervision could potentially control police conduct, but close supervision can produce resentment because of the presence of supervisors and even patrol officer reluctance to call supervisors to an altercation. Of course, officers have been observed by their supervisors and continue to perform in an unconstitutional manner.

Several U.S. Supreme Court rulings have limited officer performance. At the same time, other rulings (such as those dealing with no-knock policies and pursuits) have provided greater legal protection for officers. At lower court levels, people who feel they have been victims of police abuse can litigate for civil damages under state and federal laws.

Unconstitutional police behavior affects both the rights of individuals and the legitimacy of policing. At the same time, the threat of civil liability plays a significant role in officer discretion to arrest a suspect. In general, police liability can be reduced through managerial attention to policy and procedures, a prudent focus on training endeavors, and implementation of risk management strategies.

Key Words

Civil liability
Close supervision
Constituent oversight committees
Early intervention systems
Early warning systems (EWS)
Exclusionary rule
External methods of police accountability
Internal affairs (IA)
Internal methods of officer accountability
"No-knock" warrants
Police accountability
Rotten apple theory
State (tort) liability law
Subpoena power
Supervision
Warrantless wiretapping

Discussion Questions

1. Characterize the relationship between police and justice. In what way do you agree with this perspective?
2. Articulate the importance of sound values before police employment.
3. Describe police accountability and explain its purpose.
4. Describe the rotten apple theory. In what way can police departments weed out rotten apples?
5. Characterize internal methods of police accountability. Of those discussed in this chapter, which monitoring strategy seems the most reliable and why?
6. Describe the criticism of internal disciplinary practices. In your viewpoint, which criticism seems too far-fetched? Why?
7. Characterize external methods of police accountability. Of those discussed in this chapter, which monitoring strategy seems the most reliable and why?

8. Identify the most influential rulings from U.S. Supreme Court cases that have changed police practice. In your opinion, which ruling seems least fair to police practice? To constituents?
9. Describe the criticism of external monitoring methods of police accountability.
10. Explain the differences between state liability suits and federal violations filed under Title 42, U.S.C. § 1983. What are the limitations of state courts and what are the advantages of federal courts from the perspective of the public?
11. Describe the differences between female and male officer civil liability suits.
12. Characterize the effects of civil liability suits.
13. Describe the role that civil liability can play in arrest decisions by officers. In what way do these findings make sense to you?
14. Identify methods of changing civil liability rates.

Notes

1. Richard R. Johnson, "Citizen Complaints: What the Police Should Know," *FBI Law Enforcement Bulletin* (1988), http://findarticles.com/p/articles/mi_m2194/is_12_67/ai_53590195 (accessed April 19, 2008).
2. James J. Fyfe and Robert Kane, "Bad Cops: A Study of Career-Ending Misconduct among New York City Police Officers" (September 2006), http://www.ncjrs.gov/pdffiles1/nij/grants/215795.pdf (accessed June 6, 2007).
3. Policeone.com, http://www.policeone.com/ (accessed July 17, 2007).
4. David Heinzmann and Carlos Sadovi, "Suit Tells of 4 Cops in Night of Terror," *Chicago Tribune* (November 1, 2006), 22.
5. "State Trooper Must Pay $1M for Making Couple Strip: Jeremy Dozier Was Accused of Making Couples Run Around Naked in Cook, Lake Counties" (July 12, 2007), http://cbs2chicago.com/topstories/Jeremy.Dozier.Illinois.2.338245.html (accessed April 19, 2008).
6. Jim Kouri, "Corrupt Cops Lead Crime Gang," *Review America* (March, 10, 2006), http://www.renewamerica.us/columns/kouri/060310 (accessed January 31, 2007).
7. Miguel Bustillo, "7 Officers Indicted in Post-Katrina Shootings," *Los Angeles Times* (December 29, 2006), http://www.usatoday.com/news/nation/2006-12-28-police-charged_x.htm (accessed April 19, 2008).
8. Victor E. Kappeler, *Critical Issues in Police Civil Liability*, 3rd ed. (Prospect Heights, IL: Waveland, 2007).
9. Jeffrey Reiman, *The Rich Get Richer and the Poor Get Prison* (Boston: Allyn and Bacon, 2007).
10. Willard M. Oliver, *Community-Oriented Policing: A Systemic Approach to Policing* (Upper Saddle River, NJ: Prentice Hall, 2008), 5.
11. Walter Williams, "Foundation Works for Individual Liberty, Free Market," *Deseret News* (Salt Lake City), May 17, 2006.
12. Carl Klockars, "The Dirty Harry Problem," *Annals of the American Academy of Political and Social Science* 452 (1980): 52.
13. Kouri, "Corrupt Cops Lead Crime Gang."
14. Kappeler, *Critical Issues in Police Civil Liability.*
15. David H. Bayley and James Garofalo, "The Management of Violence by Police Patrol Officers," *Criminology* 27 (1989): 1–25.

16. "Citizen Oversight of Police: Oversight of Police Accountability" (2005), http://www.policeaccountability.org/citzoversight.htm (accessed January 31, 2007).

17. Wesley G. Skogan and Tracey L. Meares, "Lawful Policing," *Annals of the American Academy of Political and Social Science* 593, no. 1 (2004): 66–83.

18. "US Constitution" (September 17, 1787), http://www.usconstitution.net/ (accessed December 11, 2007).

19. Michael D. White, *Current Issues and Controversies in Policing* (Boston: Allyn and Bacon, 2007).

20. John P. Crank, *Understanding Police Culture* (Cincinnati, OH: Anderson, 2004).

21. White, *Current Issues and Controversies in Policing*, 242.

22. Crank, *Understanding Police Culture*, 39, 279.

23. Peter K. Manning and Lawrence J. Redlinger, "Invitational Edges of Corruption: Some Consequences of Narcotic Law Enforcement" (1978), http://www.ncjrs.gov/App/Publications/abstract.aspx?ID=55529 (accessed April 19, 2008).

24. Chuck Hustmyre (March 4, 1995), http://www.crimelibrary.com/gangsters_outlaws/cops_others/antoinette_frank/index.html (accessed March 4, 2007).

25. White, *Current Issues and Controversies in Policing.*

26. White, *Current Issues and Controversies in Policing.*

27. White, *Current Issues and Controversies in Policing.*

28. Arlington, Texas, Police Department (2007), http://www.arlingtonpd.org/index.asp?nextpg=chief/statement2005.asp (accessed March 4, 2007).

29. White, *Current Issues and Controversies in Policing*, 243.

30. Samuel Walker and Charles M. Katz, *The Police in America: An Introduction*, 6th ed. (Boston: McGraw Hill, 2008).

31. Jon B. Gould and Stephen D. Mastrofski, "Suspect Searches: Assessing Police Behavior under the U.S. Constitution," *Criminology & Public Policy* 3, no. 3 (2004): 315–62, http://www.blackwell-synergy.com/doi/abs/10.1111/j.1745-9133.2004.tb00046.x (accessed February 3, 2007).

32. Wayne Hanniman, "Active Supervision of the Patrol Function," *Canadian Review of Policing* (2005): 1.

33. Special Counsel Merrick Bobb, *9th Semiannual Report* (Los Angeles: The Special Counsel, 1998), 22–23.

34. Hanniman, "Active Supervision of the Patrol Function."

35. David Weisburd, Rosann Greenspan, Edwin E. Hamilton, Kellie A. Bryant, and Hubert Williams, *The Abuse of Police Authority: A National Study of Police Officers' Attitudes* (Washington, DC: Police Foundations, 2001).

36. Joseph Petrocelli, "Preventing the 10 Deadly Errors," *FBI Law Enforcement Bulletin* (November 2006), http://www.fbi.gov/publications/leb/2006/november2006/november06.htm#page10 (accessed February 3, 2007).

37. Jack R. Greene, Alex R. Piquero, Matthew J. Hickman, and Brian A. Lawton, "Police Integrity and Accountability in Philadelphia: Predicting and Assessing Police Misconduct," *NCJRS* (2004), http://www.ncjrs.gov/App/Publications/abstract.aspx?ID=207823 (accessed February 3, 2007).

38. Samuel Walker, Geoffrey P. Alpert, and Dennis J. Kenney, "Early Warning Systems: Responding to the Problem Police Office," *NCJRS* (2001), http://www.ncjrs.gov/txtfiles1/nij/188565.txt (accessed February 3, 2007).

39. Bureau of Justice Statistics, "Police Use of Force: Collection of National Data" (1998), http://www.ojp.usdoj.gov/bjs/pub/pdf/puof.pdf (accessed February 12, 2007).

40. For more details, see *Bonsignore v. City of New York*, 683 F.2d 635 (2d Cir. 1982).
41. Tim Dees, "IA Management Software," *Law and Order* 51, no. 5 (2003): 88–95.
42. CALEA, http://www.calea.org/ (accessed February 6, 2007).
43. Dees, "IA Management Software."
44. Dees, "IA Management Software," 89.
45. Dees, "IA Management Software."
46. Boston Police Department, "Annual Report: City of Boston" (2006), http://www.cityofboston.gov/police/pdfs/2005AnnualReport.pdf (accessed June 16, 2007).
47. I. Silver, *Police Civil Liability* (New York: Matthew Bender Publishers, 2004).
48. Bureau of Justice Statistics, "Citizens Complained" (July 2006), http://www.ojp.usdoj.gov/bjs/pub/press/ccpufpr.htm (accessed July 18, 2007).
49. Hanniman, "Active Supervision of the Patrol Function."
50. Hanniman, "Active Supervision of the Patrol Function."
51. Gould and Mastrofski, "Suspect Searches."
52. Gould and Mastrofski, "Suspect Searches,"
53. Hanniman, "Active Supervision of the Patrol Function."
54. George L. Kelling, *Broken Windows and Police Discretion* (NCJ 178259; 1999), http://www.popcenter.org/ (access February 3, 2007).
55. Antonio Merlo, "Introduction to Economic Models of Crime," *International Economic Review* 45, no. 3 (2004): 677–79.
56. Dennis J. Stevens, *Police Officer Stress: Sources and Solutions* (Upper Saddle River, NJ: Prentice Hall, 2008).
57. Los Angeles Police Commission, *Report of the Rampart Independent Review Panel* (Los Angeles: Author, 2000), 93–96.
58. West Palm Beach, Florida, Police Department, "Department Goals and Objectives" (2006), http://www.wpb.org/police/policies/downloads/-I-03DEPARTMENTALGOALSANDOBJECTIVES.pdf (accessed February 7, 2007).
59. Frank Colaprete, "The Necessary Evil," *Law and Order* 51, no. 5 (2003): 96–100.
60. Colaprete, "The Necessary Evil," 100.
61. Kathy Scruggs, "Angry Harvard Changing Policies," *Atlanta Journal-Constitution*, January 11, 1996, http://www.hrw.org/reports98/police/uspo41.htm (accessed April 19, 2008).
62. Portland (Oregon) Police Bureau, http://www.portlandonline.com/shared/cfm/image.cfm?id=154384 (accessed July 18, 2007).
63. Christopher Stone, *Civilian Oversight of the Police in Democratic Societies* (Vera Institute of Justice, 2002), http://www.vera.org/publication_pdf/179_325.pdf (accessed July 16, 2007).
64. Portland (Oregon) Police Department, "Complaints," http://www.portlandonline.com/police/pbnotify.cfm?action=ViewContent&content_id=223 (accessed July 18, 2007).
65. Deanna Bellandi, "Chicago Mayor Seeks Police Misconduct Oversight" (May 4, 2007), http://www.officer.com/article/article.jsp?siteSection=5&id=35997 (accessed June 2, 2007).
66. Bureau of Justice Statistics, "Citizens Complained."
67. Peter Finn, "Citizen Review of Police: Approaches and Implementation" (March 2001), http://www.ncjrs.gov/txtfiles1/nij/184430.txt (accessed February 14, 2007).
68. Finn, "Citizen Review of Police."
69. Bureau of Justice Statistics, "Law Enforcement Management and Administrative Statistics, 2000: Data for Individual State and Local Agencies with 100 or More Officers" (NCJ 203350; 2000). http://www.ojp.usdoj.gov/bjs/pub/pdf/lem00in.pdf (accessed March 12, 2007).

70. Finn, "Citizen Review of Police."

71. City of Santa Rosa, "Measure O Citizen Oversight Committee" (September 7, 2005), http://ci.santa-rosa.ca.us/as/pdf/admin/pdf/MeasO_Annual%20Report0405.pdf (accessed December 6, 2006).

72. University of Texas at Austin, "Faulkner Appoints UT Police Department Advisory Committee" (July 28, 2003), http://www.utexas.edu/opa/news/03newsreleases/nr_200307/nr_utpd030728.html (accessed December 6, 2006).

73. Samuel Walker, *Police Accountability: The Role of Citizen Oversight* (Belmont, CA: Wadsworth/Thomson, 2001), 55.

74. David H. Bayley, *Police for the Future* (New York: Oxford University Press, 1994).

75. Guided by Samuel Walker, *Police Accountability.*

76. Bayley, *Police for the Future*, 65.

77. U.S. Supreme Court Media: The Oyez Project, *Weeks v. United States,* 232 U.S. 383 (1914), http://www.oyez.org/cases/1901-1939/1913/1913_461/ (accessed June 4, 2008).

78. "The Constitution of the United States of America: Forth Amendment," 197, http://www.gpoaccess.gov/constitution/html/amdt4.html (accessed June 4, 2008).

79. FindLaw Constitutional Center, "U.S. Constitution, Fourth Amendment: Annotations," http://supreme.lp.findlaw.com/constitution/amendment04/01.html (accessed June 4, 2008).

80. Jeffrey T. Walker, "Law of the State and the State of the Law: The Relationship between Police and Law," in *Policing and the Law*, ed. Jeffrey T. Walker (Upper Saddle River, NJ: Pearson Prentice Hall, 2002), 1–24.

81. Legal Explanations.Com, http://www.legal-explanations.com/definitions/exclusionary-rule.htm (accessed June 4, 2008).

82. Nancy E. Marion and Willard M. Oliver, *The Public Policy of Crime and Criminal Justice* (Upper Saddle River, NJ: Pearson Prentice Hall, 2006), 384.

83. Personal communication between the writer and a Boston police officer in confidence, summer 2007.

84. Walker, "Law of the State," 4.

85. Timothy Lynch, "Policy Analysis: In Defense of the Exclusionary Rule," Cato Institute's Center for Constitutional Studies, http://www.cato.org/pubs/pas/pa-319es.html (accessed June 4, 2008).

86. U.S. Supreme Court Media: The Oyez Project, *Miranda v. Arizona,* http://www.oyez.org/cases/1960-1969/1965/1965_759/ (accessed June 4, 2008).

87. U.S. Supreme Court Media: The Oyez Project, *Miranda v. Arizona,* http://www.oyez.org/cases/1960-1969/1965/1965_759/ (accessed June 4, 2008).

88. Answers.com, http://www.answers.com/topic/miranda-v-arizona?cat=biz-fin (accessed June 4, 2008).

89. Walker, "Law of the State," 4.

90. U.S. Supreme Court Media: The Oyez Project, *Mapp v. United States* (1961), http://www.oyez.org/cases/1960-1969/1960/1960_236/ (accessed June 4, 2008).

91. Supreme Justice.com, *Katz v. United States,* http://supreme.justia.com/us/389/347/case.html (June 4, 2008).

92. Findlaw.com, "U.S. Supreme Court: *Terry v. Ohio*. 382 U.S. 1968," http://caselaw.lp.findlaw.com/scripts/getcase.pl?court=US&vol=389&invol=347 (accessed June 4, 2008).

93. Findlaw.com, "U.S. Supreme Court: *Terry v. Ohio*."

94. Walker, "Law of the State," 11.

95. Steven Andrew Drizin, "Fifth Amendment: Will the Public Safest Exception Swallow the Exclusionary Rule?", *Journal of Criminal Law & Criminology* 75, no. 3 (1984): 692–715.

96. "Justices Allow No-Knock Searches," *USA Today* (June 15, 2006), http://www.usatoday.com/news/washington/judicial/2006-06-15-scotus-knocking_x.htm (accessed February 15, 2007).

97. Patrick Plilbin, Deputy Assistant Attorney General, Office of Legal Counsel, "Authority of Federal Judges and Magistrates to Issue 'No Knock' Warrants" (Washington, DC: U.S. Department of Justice, June 12, 2002), http://www.usdoj.gov/olc/noknock.htm (accessed April 19, 2008).

98. Anthony D. Romero, "ACLU Wins NSA Challenge in Federal Court: Email to ACLU Members" (August 17, 2006). Romero was the Executive Director of the American Civil Liberties Union (ACLU) in 2006–2007.

99. John Hill, "High Speed Police Pursuits: Dangers, Dynamics, and Risk Reduction," *FBI Law Enforcement Bulletin* (July 2002), http://findarticles.com/p/articles/mi_m2194/is_7_71/ai_89973554 (accessed April 19, 2008).

100. Walker and Katz, *The Police in America*, 221.

101. "FBI Investigation" (n.d.), http://www.courant.com/news/local/hc-13194812.apds.m0274.bc-ct—polimar13,0,5964885.story?coll=hc-headlines-home (accessed July 17, 2007).

102. Peter Hutchinson, "Compliance: Getting 'Em to Understand That Carrying Out the Garbage Just Isn't What it Used to Be!" (2005), http://www.psgrp.com/resources/compliancegetting.html (accessed December 3, 2006).

103. "How to Get Even: Civil Remedies Under State Law. Boston: Massachusetts Police Brutality" (2005), http://www.massbrutality.org/sueing0.html (accessed April 19, 2008).

104. Thomas O'Connor, "Civil Liabilities" (2004), http://faculty.ncwc.edu/TOConnor/205/205lect12.htm (accessed March 4, 2007).

105. The Civil Rights Act of 1871, also known as the Ku Klux Klan Act of 1871, now codified and known as Title 42, USC 1983, is an important federal statute. It was originally enacted a few years after the Civil War. One of the main reasons behind its passage was to protect Southern blacks from the Ku Klux Klan by providing a civil remedy for abuses then being committed in the South. The statute has been subjected to only minor changes since then, but has been the subject of voluminous interpretation by courts. See Wikipedia for more detail: http://en.wikipedia.org/wiki/Civil_Rights_Act_of_1871.

106. Kappeler, *Critical Issues in Police Civil Liability*.

107. Samuel Walker and Dawn Irlbeck, "Driving While Female: A National Problem in Police Misconduct," http://www.pennyharrington.com/drivingfemale.htm (accessed April 19, 2008).

108. Katherine Spillar, Penny Harrington, and Michelle Woods, "Gender Differences in the Cost of Police Brutality and Misconduct" (September 2000), http://www.womenandpolicing.org/ExcessiveForce.asp?id=4516 (accessed July 15, 2007).

109. Kappeler, *Critical Issues in Police Civil Liability*.

110. Kenneth J. Novak, Brad W. Smith, and James Frank, "Strange Bedfellows: Civil Liability and Aggressive Policing," *Policing: An International Journal of Police Strategies and Management* 26, no. 2 (2003): 352–68.

111. Michael F. Masterson and Dennis J. Stevens, "The Value of Measuring Community Policing Performance in Madison, Wisconsin," In *Policing and Community Partnerships*, ed. Dennis J. Stevens (Upper Saddle River, NJ: Prentice Hall, 2002), 77–92.

112. Bill Torpy, "Ex-Ga. Officer Admits Robbing Drivers during Traffic Stops," *Atlanta Journal-Constitution* (March 13, 2007), http://www.policeone.com/officer-misconduct-internal-affairs/articles/1228641/ (accessed March 23, 2007).

113. Kappeler, *Critical Issues in Police Civil Liability*.

114. Darrell L. Ross and Madhava R. Bodapati, "A Risk Management Analysis of the Claims, Litigation, and Losses of Michigan Law Enforcement Agencies: 1985–1999," *Policing: An International Journal of Police Strategies and Management* 29, no. 1 (2006): 38–57.

HOSKINS
LOS ANGELES
POLICE

Police Officer Stress

It is totally reprehensible that the cops we expect to protect us, come to our aid, and respond to our needs when victimized should be allowed to have the worst fatigue and sleep conditions of any profession in our society.

—William C. Dement, M.D.[1]

▸ ▸ LEARNING OBJECTIVES

When you finish reading this chapter, you will be better prepared to:

- Define stress
- Characterize the scope of stress's effects
- Identify the general sources of stress among most police officers
- Define critical incidents and explain their relationship to critical responses
- Explain the relevance of debriefing
- Describe and identify general work stressors
- Characterize individual remedies to control stress
- Characterize stress treatment and options involving person-centered programs
- Identify obstacles associated with person-centered programs
- Articulate primary issues linked to the medication of stress

▸ ▸ CASE STUDY: YOU ARE THE POLICE OFFICER

You've been a police officer for seven years, and you have two years to go until you're 30. "That's pretty good," your wife has told you more than once. "And you spend your free time with your family and you're a college grad, too. People look up to you. I look up to you."

Your wife's father retired from the department, and the two of you often talk about the "profession." He doesn't think that policing has a legitimate claim to a professional status, because "back in the day" the officers did everything. Today, officers are trained and retrained in specialized tasks. You could hear your father-in-law say, "In the day, we never needed SWAT, each of us had weapons, and we dealt with the criminals they way they should be dealt with—not by following this crap about due process. After a gunfight, we'd have a few beers and call it a night."

It was usually then that your mother-in-law rained down with something like, "That was the problem, Hank, you boys never called it a night. You'd drink until the sun came up." The retired officer shrugged it off and smiled. "All the boys bellied up to the bar. That's where we mapped out our strategy for the next day."

His wife chimed back, "That's just it! Your strategy was that you had a drinking problem and not once would any of your supervisors help you. Your drinking almost cost our marriage." Your father-in-law would be quiet for a while and the family would continue with whatever was being celebrated—one of your kids' birthday or a holiday of some sorts.

One time, the old man pushed back: "Honey, all my supervisors were drinking with us. If one of us had a problem, no one would listen. It was a sign of weakness not to drink with the boys and a bigger sign of weakness to complain about it."

"See that's the thing, pops," you said. "We've been trained to deal with the normal wear and tear of the job, and our supervisors encourage us to discuss and resolve issues including things that interfere with good police service every week in our classroom." You knew you'd be sorry for that comment because the next words out of his mouth were damaging, "Are you saying something about us being fools 'back in the day?' I didn't do such a bad job in raising my daughter, now did I?"

1. In what way is modern American policing a professional occupation compared to earlier generations of police officers?
2. If police officers and their supervisors drank frequently together in previous police generations, in what way would the police subculture justify wrongful acts?
3. How does the training of current officers differ from previous officer training or perspectives concerning police officer stress and related issues?

Introduction

This chapter is about the sources of police officer stress and solutions to control it, both individually and professionally. Complicating this issue, stressed officers often avoid seeking help because it would be construed by coworkers (and reinforced through the police subculture) as indicating that the officer is weak, untrustworthy, and unlikely to back up an officer during a critical altercation.[2] Other complications arise because of police organizations' reactions to such requests for help. Typically, management blames the officer for his or her stressed condition, which in turn reduces the officer's chances of promotion (or recovery).[3]

In a desperate attempt to avoid embarrassment, potential disciplinary action, or lawsuits, many officers resolve stressful feelings privately, even covertly.[4] For officers to admit that they feel suicidal or have domestic problems is considered to be close to admitting a loss of control[5]—"And being a cop is all about control."[6]

What you need to know about stress from the beginning is that officers are not cut from the same cloth:

Officers are affected differently by events and circumstances that produce stress. Most officers accept the responsibility and consequences of their stress and deal with it in numerous ways.

To think about stress, it is helpful to think about an analogy from construction. Building materials are often tested to discover their breaking points. A steel bar, for instance, is subjected to increasing pressure to discover its weaknesses or the stress point at which the bar will fracture or eventually break. Each building material has a different stress resistance. Similarly, each officer reacts differently even when confronted by similar events and circumstances.

Definition of Stress

There are many definitions of stress, depending on who you ask: Medical doctors articulate physiological (organism functions) images, psychologists focus on cognitive phenomena, and sociologists associate stress descriptions with environmental icons. A reasonable definition for our purposes follows:

Stress is the normal wear and tear that our bodies and minds experience as we react to physical, psychological, and environmental changes in our daily lives.[7]

Stress is the term used to define the body's automatic physiologic reaction to circumstances that require behavioral adjustment.[8] When a person is confronted by a threat—physical or emotional, real or imagined—the hypothalamus causes the sympathetic nervous system to release epinephrine and norepinephrine (also known as adrenaline and noradrenaline, respectively) and other related hormones.[9] When released into the body, these messengers cause a person to enter a state of arousal (metabolism, heart rate, blood pressure, breathing rate, and muscle tension all increase). This "fight-or-flight response" was first described by Walter B. Cannon,[10] though many practitioners would suggest that another response in this situation might be "freezing" or ignoring a stressor.[11] Thus "stress is a physical, chemical, or emotional factor that causes bodily or mental tension resulting from factors that tend to alter an existing equilibrium."[12] It takes us out of our comfort zone and forces us to fight, flee, freeze, or ignore a circumstance or event. At the same time, stress can propel us to change and adapt to our surroundings physically, mentally, and socially.

The biological concept of homeostasis helps put this idea into perspective. Homeostasis (a balance) is the process that keeps all bodily functions, such as breathing and blood circulation, working together harmoniously.[13] In short, the body and mind automatically seek a balance or comfort within the physical and social environments. To see how homeostasis works, run in place for a few minutes and sit down. The running stresses your body, temporarily putting it out of balance. After you sit for a while, regardless of what you want to happen, physically your body returns to normal, or at least the way it was before you starting running. As this example makes clear, stress is an inevitable part of life.

In general terms, stress is the nonspecific response of the body to any demand. Just staying alive creates a demand on the body, as life requires maintaining energy and balance. We're hungry, we eat; we want new clothes, we work; we're lonely, we meet people; we feel unsafe, we seek protection. Although stress is a fundamental part of being alive, it affects each person differently, according to Hans Selye, the person who coined the term "stress."[14] The key to understanding the scope of stress is to understand the relationship between the external (stimuli) stressors encountered and the person's internal responses to those stressors or events. Sometimes those responses are involuntary and cannot be avoided. For instance, when a person is nervous about taking an examination at school, his or her palms may become moistened.

Why do people exhibit different reactions to stress? Because a human body contains its own pharmacy, manufacturing chemical substances as needed to maintain homeostasis, the response of a body to stimuli is generally programmed in such a way associated with those chemicals. Also each of us is different: culturally (the values, norms, and beliefs we hold), socially (where we are within our community and who we are with), and mentally (predispositions, patience, learning skills, goals, knowledge). Therefore, stress and the way we respond to stimuli are necessary parts of life.

Responses to Stress

Of course, the exact response to stress depends on the individual. You should be getting the sense that stress is all around us and can be both detrimental and beneficial to ordinary human development. That is, both bad stress (distress) and good stress (eustress) affect every person on this planet, almost every minute of their existence. Selye notes that, without stress, there could be no life. Just as distress produces disease, so it is plausible that eustress can promote wellness.

Distress: Bad Stress

Distress, which refers to the "bad" type of stress (i.e., the opposite of eustress), arises when excessive adaptive demands are placed on a person.[15] This occurs when demands are so great that they lead to bodily and mental damage with or without the person's consent or contribution. Distress is damaging, excessive, or pathogenic (disease-producing) stress, and its consequences in relation to the health of officers and their constituents are staggering.

Consequences of Distressed Officers

The consequences of negative untreated (individually or professionally) stress can be chilling. Stressed officers are 30 percent more likely to experience health problems than other personnel, 3 times more likely to abuse their spouses, 5 times more likely to abuse alcoholism, 5 times more likely to have somatization (multiple, recurrent), 6 times more likely to experience anxiety, and 10 times more likely to be depressed—yet police officers are the least likely of all occupational group members to seek help.[16]

Stress doesn't go away simply because an officer ignores its individual effects. Instead, it builds up over time. The literature shows that police officers have a high burnout rate, caused in part by their habit of ignoring stress, a lack of coping with family problems, occupational stress, low status, low public trust, and likelihood to become involved in a civil litigation (for which the risk is higher for police officers than for members of any other occupation).[17]

Several factors contribute to police stress, including characteristics of the individual officer. Some researchers argue that "critical incidents alone do not cause most law enforcement officers undue stress; neither do cumulative stressors, such as organizational and job factors, nor personal stressors, such as physical and psychological elements. Instead, the confluence of all of these different factors does."[18] For example, cumulative stress can contribute to high rates of gastrointestinal disorders, high blood pressure, and coronary heart disease in the police community.

Types of Traumatic Stress Disorders

Five types of traumatic stress disorders are distinguished:[19]

- Acute stress disorder (ASD): **post-traumatic stress disorder (PTSD)**, consisting of overwhelming fear and revulsion
- Conversion reaction: hysteria, the development of physical symptoms such as blindness or paralysis in response to stress
- Counter-disaster syndrome: excessive excitement and inappropriate overinvolvement leads to many accidents
- Peacekeepers' acute stress syndrome: rage, delusion, and frustration in response to atrocities
- Stockholm syndrome: identification by the victim with the perpetrators of violence

Acute stress disorder can develop within one month after an individual experiences or sees an event that involved a threat or actual death, serious injury, or another kind of physical violation to the individual or others, and responded to this event with strong feelings of fear, helplessness, or horror.[20] This diagnosis was established to identify those individuals who would eventually develop PTSD. ASD was brought to light as it became clear that for a short period, people might exhibit PTSD-like symptoms immediately after a trauma.

PTSD is characterized as an extreme reaction to stress revolving around core symptoms experienced after a life-threatening event, which include reexperiencing the trauma in the form of nightmares and intrusive thoughts and avoiding reminders of the event, among other reactions. It is defined by the American Psychological Association (APA) in the *Diagnostic and Statistical Manual of Mental Disorders* (*DSM-IV*) as follows:

> The essential feature of post-traumatic stress disorder is the development of characteristic symptoms following exposure to an extreme stressor involving direct personal experience of an event that involves actual or threatened death or serious injury, or other threat to one's physical integrity; or witnessing an event that involves death, injury, or a threat to the physical integrity of another person; or learning about unexpected or violent death, serious harm, or threat of death or injury experienced by a family member or other close associate. The person's response to the event must involve intense fear, helplessness, or horror. . . . persistent reexperiencing . . . persistent avoidance of stimuli associated . . . persistent symptoms of increase arousal . . . [that] must cause clinically significant distress or impairment in social, occupational, or other important areas of functioning.[21]

The APA introduced PTSD as a diagnosable condition in 1980. Most discussions of these extreme reactions to stress revolve around a set of core symptoms that is experienced after a life-threatening event:

- Reexperiencing the trauma in the form of nightmares and intrusive thoughts
- Avoiding reminders of the event
- Experiencing numbing to the point of not having loving feelings, increased arousal in the form of exaggerated startle response, hyper-vigilantism, and sleeping difficulties

Recent research suggests that certain factors predict the likelihood of someone experiencing PTSD and its symptoms.[22] Most predictive of PTSD is a dissociative experience during or in the immediate aftermath of the traumatic event and high levels of emotion during or shortly after the traumatic event.[23] Other predictors of PTSD include individual prior trauma, psychological adjustment before the trauma, and a family history of mental illness. Prior exposure to a similar event probably is the most difficult factor to understand because it is counterintuitive.[24] In most other aspects of life, experience helps. Unfortunately, many police officers discover that repeated exposure to certain events can have seriously detrimental effects. It is almost as if repeated events reinforce the original traumatic event. The frequency, duration, and intensity of stressors also determine the person's proclivity to develop PTSD. Likewise, stress reactions vary among individuals because perceptions of situations differ and reactions are subjective.

The debilitating symptoms of PTSD are not the worst things that can happen. Sadly, among police officers, job-related stress frequently contributes to the ultimate maladaptive response to stress: suicide.[25] Furthermore, there is a significant association between held-in anger and the presence of a circulatory disorder in individuals with PTSD.[26]

It goes without saying that not every police officer who experiences one or more traumatic event becomes a candidate for PTSD. Traumatic event experiences can lead to highly trained personnel who enhance their resilience (see Chapter 9 and discussed later in this chapter) capabilities as a result of coping with those experiences, and these officers continue to provide quality police services (and continue to make sound lifestyle choices about themselves, their family members, and their friends).[27]

Nevertheless, individuals who are prone to stress disorders and fail to seek treatment—either internally in the police organization, from outside independent providers, or personal strategies—can lead to development of a full-blown traumatic stress disorder.

Eustress: Positive Stress

> **Eustress** can be defined as a pleasant or curative stress that can affect individuals positively when they are participating at competitive events, taking examinations for a course, or working to excel so as to obtain a promotion.[28]

At appropriate levels, stress can increase both efficiency and performance. For instance, Officer David Murphy of Newbury, Massachusetts, says that stress pushes him to work harder to achieve higher goals and provide quality police service.[29] In today's high-powered competitive police environment, developing the appropriate stress response (fight, flight, or freeze) is essential to success. Yes, freezing or doing nothing is sometimes part of the problem—but experienced officers eventually learn that timing has a lot to do with achieving positives outcomes during altercations. Stress is essential for the productivity and well-being of personnel, too.

Sources of Stress

> One key to understanding stress is understanding the relationship between external stressors and internal reactions to those stressors.

The source of stress may be referred to as an external stressor. External stressors can include the feelings we have about circumstances or events (stimuli) depending on our individual perceptions arising out of uncertainty, lack of control, pressure, and an unsafe or unprofessional social environment, to name a few sources. Almost every aspect of police work can produce these feelings—but then individuals may be confronted with additional pressure as a result of stress stemming from their daily lives and family, gender, working in a rural area, personality, and job-related issues.[30]

Daily Living and Family

Stress agents associated with daily living include changing relationships, unfulfilled expectations about relationships, inconsistencies between lifestyle and values (e.g., unfulfilled living arrangements), money problems (e.g., credit-card debt), loss of self-esteem (e.g., falling behind professionally or in training), and fatigue or illness (e.g., as a result of poor diet, lack of sleep, or lack of exercise).

Several studies reveal that many officers see their families as less of a stressor than might be expected, whereas many families see the officer's job as a source of stress for them in their daily routines.[31] Spouses of officers say that rotating shift work and overtime, concern over the spouse's cynicism, the need to feel in control in the home, and an inability or unwillingness to express feelings are among the biggest problems they face with their officer-spouses. According to marriage therapists, the willingness of both the husband and the wife to own up to their responsibility for needing to change is usually the best predictor of success when it comes to managing family-related stress.

Gender

Female officers share similar general work stressors as male officers, but significant differences emerge in female experiences with stressors associated with the treatment they receive from male officers, supervisors, the courts, and the public.[32] These stressors may become turning points leading to voluntary resignation among female police officers owing to perceptions of a stagnated police career, an intense experience that brought accumulated frustration to a head, a lack of a sense of fulfillment on the job, family considerations, the conduct of coworkers toward the female officer, particular department policy or policies, and new employment opportunities.

In a study among 133 female officers, their responses to questions about job stress and work stressors were largely similar to the answers given by male officers. Significant differences emerged in their answers about stress derived from other police officers, their supervisors, the courts, and the public, however. (See Chapter 9 for more details on on-the-job issues faced by female officers.)

In particular, the dual role that female officers seek may be a source of stress. As mentioned previously, female officers want the respect that comes from being sworn officers as well as the respect that our society traditionally extends to women by virtue of their gender—and often that does not necessarily happen. For instance, a female officer explains it this way:

> We have pressures the men never thought of. There was the expectation by the men, as soon as I came on the job, that I would automatically screw up. And then there's also the other problem—the expectation that because you wear the uniform, you are somehow no longer a woman. You cannot win. I keep telling people that by putting on the uniform, I didn't have a sex change operation. But few seem to believe [me].[33]

Female Officers and Stress from Supervisors

Female officers are often frustrated by some of the actions taken by their supervisors, who frequently think of them as incapable of responding to at-risk encounters (see Chapter 9). For example, in my own experiences while riding in a police cruiser with an Aiken, South Carolina, police officer in the late afternoon, the dispatcher advised that an armed robbery was in progress at a convenience store. The officer advised her estimated time of arrival (ETA) as 3 minutes. As her police cruiser made the corner and she had the location in sight, her sergeant advised her to "back off." "But sir, I am almost on the parking lot," she responded. "Break it off now, that's an order," the supervisor said over the radio.

The following day I asked why the officer was pulled from the call. Was it because the sergeant realized that she had nonsworn personnel in the cruiser, I wondered? "No, that didn't matter. I often get bumped from those calls," the former U.S. Army MP said.

Female Officers and Stress from the Community and Other Officers

Female officers also encounter resistance from constituents at times. One female officer, upon investigating a neighborhood dispute in the countryside, was told by the caller that he would not speak with her—that she should "go home and send a real cop."[34] When the officer refused, the man dialed 911 but was told to deal with the officer; he ignored her.

The other officers on the force can also create stress for female officers. One officer experienced constant harassment from a male officer who kept telling her she was not up to the job. On the advice of another male officer, she finally dropped her belt at the stationhouse and told him, "Okay! Let's go." They engaged in a tussle before the sergeant separated them.[35]

Female Officers: Tough and Honorable Cops

Many studies emphasize conformity to male stereotypes by female officers but fail to provide a representative sampling of police tasks and situations, while overemphasizing the violent and dangerous aspects of policing.[36] Many female officers take less aggressive—yet what they feel are no less effective—approaches to stressful work situations than male officers typically adopt.

However, women officers can also be great warriors. Consider the Fayetteville, North Carolina, officer who took a call about a family fist fight in the park. Upon arriving, the five-foot-tall, slender officer (from Hawaii) asked the men what was going on while still sitting in her cruiser.[37] She was waiting for backup. One of the men told her that since he was an Army Ranger (Fort Bragg is adjacent to "Fayettenam," as it is called), he didn't need her help. The officer responded that he was right and that things would go better if they could wait a bit, but talk. With that, the man moved toward the officer's vehicle in a highly aggressive manner. The officer exited (advising her passenger—the author of this book—to remain in the vehicle) and locked the doors of the police cruiser. In a quick motion, she slammed her aggressor to the ground, and pushed down on his esophagus with her boot, holding him steady on the ground while she advised him of his Miranda rights.

As for being macho, Officer Moira Smith was the first officer to report to the World Trade Center when the first aircraft rammed the tower on September 11, 2001. Officer Smith was killed while attempting to rescue trapped victims.[38]

Deputy (Orange County, Florida) Sheriff Jennifer Fulford exhibited a similar type of heroism.[39] As Isola Allen was preparing to drive her son to school one morning, three men pulled into her driveway and forced Isola from her van and into the house, leaving her eight-year-old son and two-year-old twin daughters in the back seat. The boy called 911 from the van with his mother's cell phone. When the police arrived, the men sent Isola outside to tell police that everything was okay, but instead she screamed to police, "My babies, my babies," pointing toward the van in the garage. Deputy Fulford immediately went to the garage to protect Isola's children. A gun battle ensued between her and the three men. When it was over, Deputy Fulford was bleeding on the floor of a garage; she had been shot 10 times. "I wasn't going to die there, and I heard the voices of the children in the van," she said. Despite her injuries, she summoned up enough courage to stop the three men with the aid of other officers who arrived on the scene. The children were unharmed.

Other Sources of Stress for Female Officers

Depending on individual personality and experience, some female officers may be offended and intimidated by degrading language than others. Other female officers find exchanges of insults to be a way to use humor to relieve stress. Other sources of stress for female officers result from the following issues:[40]

- Lack of acceptance by the predominantly white, male police force and subsequent denial of needed information, alliances, protection, and sponsorship from supervisors and colleagues
- Lack of role models and mentors
- Pressure to prove oneself to colleagues and the public
- Exclusion from informal channels of support
- Lack of influence on the decision-making process

Many female officers imply that there is a "good old boy" system in place, which locks them out in terms of obtaining choice assignments and promotions. However, even among male officers who are not part of the "good old boy" system are overlooked for choice assignments and promotions.

Rural Policing and Stress

Stress in small-town and rural policing is a very serious issue and one that has been inadequately addressed in both research and training. Police officers and sheriff's deputies in these settings face many of the same types of stressors as their urban counterparts, plus unique stressors related to security, social factors, working conditions, and inactivity, explains Willard M. Oliver.[41] Evidence suggests that officer safety and security needs to be addressed more explicitly not only for its own sake, but also to help alleviate the high levels of stress experienced by many officers in small-town and rural environments.

Personality

General personality characteristics of police officers are summarized in Chapter 13. For the purpose of discussing stress, psychologists often divide individuals into two types: type A personality and type B personality.[42]

The **type A personality** is an aggressive, hyperactive "driver" who tends to be a workaholic, is highly active, and is likely to experience high stress levels more often than the **type B personality**, who tends to be laid back, is reserved, and experiences lower stress levels than the type A personality.

There are two cardinal features of type A personality: "time urgency or time impatience, and free-floating (all pervasive and ever-present) hostility."[43] For instance, Arthur—a type A personality—talks really quickly, at the rate of 140 words per minute or more. His voice is grating, harsh, irritating, excessively loud, and just generally unpleasant. His posture is tense with abrupt jerky movement. Every few minutes, he raises his eyebrows in a tic-like fashion. Likewise, every few minutes, he raises or pulls back one or both shoulders in a tic-like fashion.[44] His free-floating hostility includes a constant apprehension about future disasters (which is not a symptom of an anxiety disorder or depressive disorder). Such a type A individual tends to utilize a pessimistic, explanatory style more often than an optimistic, explanatory style like the type B personality, who tends to be more relaxed than the type A personality.

Type A personalities are generally hard workers who are often preoccupied with schedules and the speed of their performance. By comparison, type B personalities may be more creative, imaginative, and philosophical.

Personality type is determined through a modified version of the Jenkins Activity Survey. Visit the following website to take the survey: http://www.psych.uncc.edu/pagoolka/TypeA-B-intro.html.

The Jenkins Activity Survey was originally formulated to detect behaviors that led to heart attacks. The test consists of 30 multiple-choice items, and possible scores range from 35 to 380. Type A is associated with a high score on this survey, while type B is associated with a low score.

Job-Related Sources of Stress

At work, officers experience three sources of stress:

- The job itself—participation in critical incidents, delivering general police services, and a "lack of fit" of an individual with the police job
- Police subculture (as discussed in Chapter 13)
- Police organization, which includes management styles (e.g., hierarchical top-down directives based on intimation rather than values; discussed in Chapter 7), and autocratic management styles (e.g., managers who seek little input from personnel, a general failure to keep personnel informed, and a disregard of objective processes to help personnel perform their tasks)

Thus the job-related stressors that affect officers include both critical incidents and general work stressors.

Critical Incident Stressors

Critical incidents are part of a typical police officer's job description, job training, obligation, and work role.[45] They are the subject of both legal and moral mandates for sworn personnel.[46] Some officers encounter critical encounters frequently during their career, whereas others—even officers in the same jurisdiction—may experience few, if any, critical incidents.

> A **critical incident** is any situation beyond the realm of a police officer's usual experience that overwhelms his or her sense of vulnerability and leads to a lack of control over the situation.[47]

Characteristics of Critical Incidents

Critical incident stressors that produce stress among officers are "event" specific and have the following characteristics:[48]

- Sudden and unexpected
- Disruptive to an officer's sense of control
- Disruptive to an officer's beliefs, values, and basic assumptions about the world in which the officer lives and works, and the people in it
- Potentially damaging or a life threat
- Leading to an emotional trigger or physical loss

Not all of these characteristics will be present in every critical incident. The degree to which an officer is ultimately affected by an incident depends on five factors that are related to the incident itself (first five factors in the following list), plus three factors that are related to things in existence prior to the incident (last three factors):[49]

- The actual event
- Its intensity
- Its duration
- Its level of unexpectedness
- Primary victimization (recipient of injury/violence)[50]
- Secondary victimization (witnessing injury/violence)[51]
- The mental health of the officer[52]
- The previous experiences of the officer[53]

Often, after both primary and secondary victimization experiences, it is not uncommon for an officer to assume responsibility and to internalize feelings of guilt, which typically exacerbates his or her vulnerability toward a traumatic stress reaction. Those reactions can be either acute (occurring during the actual event) or delayed (occurring minutes, hours, months, or even years after the event). Delayed reactions may include a change in sleep patterns, disorientation, self-doubt, unresolved anger, irritability, fear, anxiety, agitation, flashbacks, and increased use of drugs and alcohol.[54] Also, the traumatic event can be persistently reexperienced in at least one or more of the following ways:[55]

- Recurrent images
- Thoughts
- Dreams
- Illusions
- Flashbacks
- Episodes
- Reliving the experience

Types of Critical Incidents

During a critical incident, an officer is placed in sudden jeopardy, such that a serious threat is made to his or her existence or well-being, or to the well-being of another person.[56] Most often, these incidents consist of high-risk officer–civilian contacts in which an officer reasonably believes he or she might be legally justified in using deadly force, regardless of whether such force is actually used or averted.[57]

Researchers have identified 13 types of critical incident experiences, which are listed here in order of their propensity to produce officer stress:[58]

- Another officer killed or assaulted in an altercation with a suspect
- Barricaded subjects
- Apprehension of emotionally disturbed offenders
- Harming or killing an innocent person
- Harming or killing another officer
- Hate groups and terrorist altercations
- Serving high-risk warrants
- Hostage takers
- Hot or fresh pursuit of a suspect

- Killing a criminal
- Protecting VIPs
- Sniper incidents
- Use of excessive force during an altercation

Crisis Responses

Critical incidents can also be characterized as any situation faced by emergency service personnel that cause those individuals to experience unusually strong emotional reactions that have the potential to interfere with their performance.[59]

> A **crisis response** is often confused with a critical incident. A critical incident is the stressor *event* that initiates the crisis *response*.[60]

Researchers distinguish three primary crisis responses:[61]

- Disruption of psychological homeostasis (the tendency of the body and the mind to gravitate toward a state of equilibrium or balance)[62]
- Failure of the individual's usual coping mechanisms (e.g., sports, hobbies, reading, playing with your children, praying)
- Human distress or dysfunction (usually demonstrated as compulsive behavior of sort)[63]

Because each officer responds differently to events—including critical incidents—the quality of resilience (a perspective described in Chapter 9) possessed by an officer might well play a role in determining the extent to which an officer responds negatively to an event or circumstance. At the same time, it is through critical incident experiences that officers can enhance their intuitive policing styles.

Intuitive Policing

In **intuitive policing**, experienced officers observe behaviors exhibited by criminals that send out danger signals to the officer, moving the officer toward a reaction intended to further the objectives of the police.[64] Intuitive policing represents a decision-making process that officers use frequently, but that is difficult to explain to individuals who are unfamiliar with the concept. Officers learn intuitive policing strategies through experience—and especially critical incident experiences. Danger is something that can be sensed or predicted through observation rooted in experience.[65]

Intuitive policing has been the subject of much criticism, with opponents of such policing implying that officers tend to overreact to perceived danger signals. In the current era of pervasive violence, perhaps better training might help temper unnecessary overreactions while still enabling an officer to maintain adequate vigilance.

Debriefing

Treating stress that results from critical incident experiences can enhance the officer's self-esteem, encourage nonabusive behavior toward an officer's own children and life partner, discourage substance abuse, and prevent suicide. Many basic initiatives related to crisis intervention have been put forth in an effort to mitigate critical incident stress. The most effective crisis intervention remedy is immediate professional intervention, which strives to stabilize an officer and moving him or her toward independent functioning. This method is sometimes called debriefing.[66]

> **Debriefing** is a type of short-term psychological intervention used to help officers who experience temporary extreme emotions to recognize, correct, and cope with them as a result of a critical incident.

The debriefing process is governed by three primary principles:

- Ventilation and abreaction
- Social support
- Adaptive coping

The utility of critical incident stress management has been demonstrated in many crises and disasters around the world. It allows emergency mental health practitioners to tailor the intervention response to individual or organizational needs and is emerging as the international standard of care for police officers and other victims associated with disasters. Strategies used to guide officers through critical incident experiences include the development and use of a tactical plan to adapt to the many circumstances encountered, the communication of coordinated actions with other officers, and the avoidance of independent action or separation of partners during felony pursuits.

One criticism of debriefing is that it does not address long-term conditions—but then that is not one of its objectives. Officer resilience—gained through experience with critical incidents, for example—may play a more important role in determining long-term outcomes. In fact, research suggests that the factors that contribute to the vulnerability of police officers to traumatic stressors are similar to the factors that actually enhance resilience or intuitive policing. See the "Ripped from the Headlines" feature in this chapter which focuses on police officer experiences with critical incident debriefing in Tennessee and emphasizes this somewhat contradictory link.

■ General Work Stressors

General work stressors can be characterized as experiences that arise during a patrol officer's daily work routine. For instance, an estimated 7 of every 10 sworn officers respond most often to 911 emergency calls and provide services associated with patrol responsibilities.[67] The general work descriptions that follow pertain to typical patrol officers and were characterized in Chapter 9 as part of patrol officer general work experiences; all of these factors may contribute to the everyday stress experienced by officers:[68]

- Conflict with regulations
- Domestic violence stops
- Media and publicity
- Losing control
- Child beaten or abused calls

- Excessive paperwork
- Public disrespect
- Another officer reported injured
- Lack of recognition
- Poor supervisor support
- Disrespect of the courts
- Shift work
- Death notification
- Poor fringe benefits
- Accidents in the patrol vehicle

Treating Stress

Individual Remedies to Control Stress

Control and prevention of the negative stress (distress) response are crucial to the health of street officers, to the health of communities, and ultimately to security of the nation as a whole. Distress is more than just a mental process; it produces both psychological and physiological effects. For that reason, coping with, controlling, and preventing negative stressors should be a priority for police. As mentioned earlier, all too often stressed officers avoid seeking help because it might lead to ostracism by coworkers.[69]

Seeking help is never easy. Choosing to be healthy is the best weapon against the negative influences of stress and against living a life of discomfort regardless of the occupation of an individual.[70] Unfortunately, it is difficult to pursue a healthy lifestyle when an individual is unaware of poor lifestyle influences, wishes to ignore the danger signs, or is unwilling to commit to a healthy lifestyle. Conversely, arriving at the realization that changes are necessary in one's life does not automatically mean that an officer will change.

Nonetheless, reading this chapter and other books might be steps toward positive action. Some practical works include the sources listed here:

- Herbert Benson and Eileen M. Stuart, *The Wellness Book* (New York: Carol, 1992).
- Stephen R. Covey, *Seven Habits of Highly Effective People*[71] (New York: Simon & Schuster, 1990); *Principle Centered Leadership* (New York: Summit Books, 1991); and *First Things First* (New York: Simon & Schuster, 1995).
- Katherine W. Ellison, *Stress and the Police Officer*, 2nd ed. (Springfield, IL: Charles C. Thomas, 2004).
- John M. Madonna, Jr., and Richard E. Kelly, *Treating Police Stress: The Work and the Words of Peer Counselors* (Springfield, IL: Charles C. Thomas, 2002).
- Dennis J. Stevens, *Police Officer Stress: Sources and Solutions* (Upper Saddle River, NJ: Prentice Hall, 2008).
- Hans Toche, *Police Officer Stress* (Washington, DC: American Psychological Association, 2001).

Being a police officer and delivering quality police services is easy when an officer is in good physical and mental health. The bottom line: Regardless of how treatment is accomplished (individually or professionally), it is always up to the individual to resolve his or her personal problems. Officers can choose to initiate action on their own to reduce the negative influences of stress by employing various coping methods.

> **Coping** is as an attempt to deal with and overcome problems and difficulties.[72]

Coping is a way to manage stress. Popular methods of managing stress include deep breathing, muscle relaxation, meditation/prayer, positive thinking, self-talk, and mental imagery, for example.[73] Officers can choose one of these techniques or use them in combination—whatever is most productive in promoting recovery and control. No matter which method is selected, it must be consistently practiced—for instance, every day for 20 minutes or every other day for 40 minutes—to see results.[74] Because each person is different, a method or schedule that might work well for one officer might not work for another officer.

When an officer finds a stress management technique that works well for him or her, it should be practiced often to increase its effectiveness. Most important, officers can refine these coping skills to the point that they can employ them as stressful situations occur.[75] Using these techniques provides hope in dealing with most stressful situations and even preventing them from becoming destructive to an officer's overall health.[76]

Stress management techniques also must be used in the context of practicing a healthy lifestyle, including regular exercise, wholesome nutrition and diet, spiritual renewal, and enriching social interactions.[77] Officer choices determine the health of the body, mind, spirit, and social interactions. If any one of these four areas weakens, other areas can weaken as well, leading to total collapse. Conversely, if officers choose to practice a healthy lifestyle in these areas in some way or another, then they choose to take care of their bodies.

Treating Stress through Departmental Person-Centered Stress Providers

Selecting a Stress Program Provider

Many police agencies know that choosing an external provider to treat stress is the most economical way to provide professional services to officers and their families. Many options are available, including private psychologists or other mental health practitioners (individual or group practices). Employee assistance programs (EAPs) are already in place in some agencies and program planners may be amenable to aiding another agency, or other similarly qualified practitioners may be available, particularly in college towns where faculty members could also be qualified to operate programs.[78] An administrator may choose a provider informally, based on personal knowledge or recommendations, or more formally by requesting competitive proposals from multiple practitioners. Although many departments decide to design their own stress program, there are many specialized services available for police officers across the county.

Interventions designed to reduce occupational stress can be categorized according to their focus, content, method, and duration. In jobs that already involve a high degree of decision-making latitude (which is certainly characteristic of policing), **cognitive-**

behavioral intervention (CBI) tends to be more effective than the other intervention types, at least among those studied by researchers.[79]

> Cognitive-behavioral intervention is an action-oriented form of psychosocial therapy that assumes that maladaptive, or faulty, thinking patterns cause maladaptive behavior and "negative" emotions. (Maladaptive behavior is behavior that is counterproductive or interferes with everyday living.) The treatment focuses on changing an officer's thoughts (cognitive patterns) in an effort to change his or her behavior and emotional state.[80]

CBI has the flexibility of being administered in places other than a clinical environment. It is commonly practiced in various forms by treatment organizations serving the criminal justice community.

Specialized Services for Police Officers

A number of outside treatment organizations serve only police personnel.[81] For example, the On-Site Academy in Gardner, Massachusetts, is a nonprofit agency for training and treating emergency services personnel involved in traumatic incidents. Crossroads, in Delmar, New York, provides complete outpatient treatment services for police officers and their families, addressing alcoholism, critical incident stress, anger management, and relationship problems. These providers often contend that officers are more likely to use services that are sensitive to the job-related concerns of police personnel. Many of these agencies are contracted with departments across the United States.

■ Person-Centered Stress Programs

Basic Options

The basic options associated with **person-centered interventions**, also known as EAPs, include external agencies (i.e., outside the department), an in-home agency within a police department, or a hybrid-type strategy consisting of a combination of in-house and external approaches.[82]

External Programs

This section describes five variations on external person-centered stress programs. A psychologist serves police officers in 5 of the 11 agencies within Stanislaus County, California, which is located east of San Francisco. The Modesto Police Department is the largest agency served, with 215 officers; other agencies have as few as 15 officers. The psychologist provides counseling services to officers and their families; most often officers visit her office.

The Counseling International Team, located in San Bernardino, California, is a private psychology practice that has been providing counseling to police officers and fire fighters in more than 80 public safety agencies throughout the state of California since 1983.[83] Seven full-time clinicians and five part-time counselors work at the program's office; the company also refers some cases to five independent mental health professionals who live in the jurisdictions where the client police and fire departments are located. The Counseling International Team has a separate written contract with each agency and bills on a fee-for-service basis for individual counseling, critical incident debriefing, and peer supporter training.

A police psychologist acts as the health resources coordinator for the Palo Alto, California, Police Department. She works as a contract employee and maintains an office at the police station for meetings with clients. This psychologist provides training and counseling for the department's 100 sworn officers. The department also hired an organizational consultant to respond to the department's organizational sources of stress.

The U.S. Postal Inspection Service operates a self-referred counseling program for postal inspectors in the 12 states that make up its Western Region. Contracts are in place with police psychologists chosen from the region. The psychologists bill the Postal Inspection Service for treatment provided to inspectors. A police psychologist, who is not an employee of the Postal Inspection Service, serves as coordinator of the program, putting inspectors who need services in touch with a contracted service provider.

Psychological Services is an outside provider for counseling services for the Tulsa, Oklahoma, police and fire departments. A critical incident response team consisting of peer supporters trained by Psychological Services talks with, refers, and helps train other officers to deal with critical incidents.

In-House Programs

This section describes five variations on in-house person-centered stress programs.

In Philadelphia, negative publicity resulting from eight officer suicides in five years—three of them in 1994 alone—prompted the police department to create an in-house stress manager's position in 1995. Among other duties, the stress manager examines departmental policies and procedures and recommends ways to make them less stressful.

The Michigan State Police Behavioral Science Section trains experienced and new sergeants in techniques to manage critical incident stress among officers. After a critical incident, support comes from the chief down through the command staff to the field officer. Command-level staff also offers assurance and support to family members—including helping with paperwork, providing telephone numbers for follow-up assistance, and spending time with them. Word of the command staff's concern typically spreads through the department grapevine to every officer on the force, instantly improving morale and alleviating stress.

Like most in-house police agencies, the Massachusetts State Police Stress Unit has its roots in the 1970s, when authorities realized they needed to help state troopers and Boston police officers who had an alcohol problem.[84] Psychology entered the world of policing through training centers and the hiring process in the 1980s, and eventually led to the establishment of stress units in the 1990s. The stress units in most police departments and prisons first engaged chaplains as psychologists, then peers, then volunteers, and eventually paid personnel to aid troopers and other officers in achieving better mental health. Today sworn troopers with appropriate university credentials provide stress counseling to officers in Boston, and the stress unit has its own facility, career paths, and chain of command. Other departments across the United States have followed a similar path toward establishing professionalized units and are at different points along this evolutionary path.

The Adams County, Colorado, Sheriff's Department's stress program consists of an in-house peer support program that was initially coordinated and now is also supervised by a contracted psychologist. The psychologist and a peer support team coordinator developed guidelines for and selected members of the peer support team. The psychologist trains the peers as well as other officers, and he meets individually with each team member to review his or her support contacts.

The Rochester, New York, Police Department's Stress Management Unit is housed in the Professional Development Section. An in-house mental health professional provides counseling services, coordinates a small group of peer supporters, and conducts stress training for officers and their family members. The department also contracts with the University of Rochester's Department of Psychiatry for additional mental health services as well as assistance with training program design and clinical reviews.

Hybrid Programs

Most program practitioners utilizing hybrid programs claim to offer the advantages of both the internal and external options, with few of their shortcomings—though the latter point is debatable. Unless they are well coordinated, hybrid programs may run the risk of confusing clients about how the program operates and fostering conflicts between internal and external program staff. This section describes six variations on hybrid person-centered stress programs.

The director of the Erie County Law Enforcement EAP in New York originally served as director of the EAP for all county employees. Now her services are specialized to Buffalo police officers and deputies. The director is still a county employee, serving the sheriff's department on an in-house basis, but is an outside provider or subcontractor for the Buffalo Police Department (an arrangement that required approval from state authorities).

The Rhode Island Centurion Program is operated by a licensed clinical social worker (who is also a sworn active-duty reserve officer with the Coventry Police Department), his wife (a licensed counselor and a sworn active reserve officer), and a network of peer supporters from various law enforcement and correctional agencies. The director of the program is the sole contracted provider of stress or EAP services to eight police agencies, many of them small; the director also furnishes bimonthly stress training or EAP-related services to 10 other police agencies every other month to support these departments' own in-house stress prevention efforts. Contracts are usually with the department's management, union, or both. The Centurion Program acts as an "affiliate" for other departments that request services on certain occasions, such as following critical incidents. The director serves his own department (consisting of 65 sworn officers) as the in-house stress program director, providing direct counseling services to about six officers per year and delivering training and oversight of the department's peer police officer.

The federal Drug Enforcement Administration (DEA) has a five-year contract with an outside provider to coordinate EAP services to DEA employees nationwide. The agency also has a full-time in-house administrator who directs the program from DEA headquarters and supervises the contracted services, which are provided by a combination of contract support unit personnel and a subcontracted area clinician network consisting of practitioners across the country. The DEA trains and certifies agents as trauma team members to respond to critical incidents; those trained aid local police officers with trauma critical incident tactics.

The Bureau of Alcohol, Tobacco, and Firearms employs a private contractor to coordinate professional stress-related counseling services. It operates three peer programs, specializing in critical incidents, substance abuse, and sexual assault. The peer support programs have a group format, aiding personnel and their family members in talking to other government personnel with similar situations. The groups are administered out of the ombudsman's office at the agency's headquarters in Washington, D.C., while the contracted EAP services are supervised by the Office of Personnel.

The Fraternal Order of Police (FOP) has established stress services in several of its lodges across the United States. Through the Pennsylvania FOP Officer Assistance Program (OAP), different lodges throughout the state designate a lodge liaison officer who educates members about the program and calls in a critical incident debriefing team when necessary. The program also offers confidential access to professional counselors for members and their families.

Many police officers can easily access the FOP website, given that this organization represents an estimated 310,000 police officers in the United States.[85] Many officers are familiar with the organization's legal representation during police officer litigation. Some chapters or lodges also have stress management components, such as the New Jersey State Lodge.[86] Its program, which is called COP 2 COP, consists of a free and confidential 24-hour telephone helpline. It is available exclusively to police officers and their families, and is intended to help them deal with personal or job-related stress and behavioral healthcare issues.

Obstacles Associated with Person-Centered Stress Programs

Six obstacles are associated with person-centered stress programs:

- Multimodal stress interventions
- Lack of encouragement to seek treatment
- Perceptions of treatment are a sign of weakness
- "Shedding the uniform"
- Unrealistic view of the job
- Organizational contributors to officer stress

Multimodal Stress Interventions

Stress intervention programs should be multimodal in nature—that is, they should take a holistic approach. Psychologists tend to employ the functional aspects of person-centered techniques,[87] usually from a strict operant perspective or a voluntary action perspective.[88] In essence, they treat stress as though it derives from a single source. According to Gary Kaufman, manager of the psychological services at the Michigan State Police, stress intervention must be multimodal because controlling stress cannot be "a zero-sum game." A thorough process dealing with the multiple causes of behavior requires knowledge of an officer's behavior obtained from multiple theoretical perspectives. For every cause there is always more than a single effect, Kaufman might argue: In a sense, a stressed officer is a product of himself or herself, the job, the community, supervisors, rules, and the individuals she or he encounters.[89]

Lack of Encouragement to Seek Treatment

Treatment of stress-related problems is not encouraged by the public, officials, or the police subculture. Officials believe that officers involved with counselors—and especially stress counselors—should be on "desk duty." The public is fascinated by the police profession, but officer stress is not considered a "glamorous" aspect of the job. Also, for every depiction in the media of real-life police stress, hundreds of complicated gunfights and police brutality stories can be found in films and television; this distortion of reality does a disservice to real-world officers, who are only human (unlike their fictional counterparts).

Although officer stress certainly affects individual officers, it is also an issue for everyone around the affected officers, including their family members and other members of "front-line" professions, who need to fully understand and in a sense "inoculate" themselves against stress as best they can. Police counseling, police peer counseling, and even critical incident stress management (CISM) and critical incident stress debriefing (CISD)[90] are not encouraged or employed often enough, despite their well-known efficacy. Police organizations tend to isolate stressed police officers rather than provide opportunities for treatment.

Perceptions of Treatment are a Sign of Weakness

Seeking professional help or showing emotion during debriefing after a critical event is sometimes seen as a weakness by police personnel.[91] From the perspective of the officer, stress may be ignored for several reasons:[92]

- Initial naiveté and high level of enthusiasm
- Difficulty in measuring performance
- Lack of organizational support
- Poor distribution of resources
- Necessity of dealing with the public and police administration as adversaries

Most officers began their careers with a great deal of optimism and high levels of enthusiasm.[93] The greater the initial naiveté when hired, the more disillusioned an officer may become once in the field. Statistics about crime, response time, and arrest rates reveal very little about the real performance of officers and their ability to reach their personal goals.

According to one officer, the public's perceptions of police officers are usually buried in stereotypes: chain smoking, perched in a stakeout car filled with fast-food debris, making late-night phone calls to former spouses, going on infamous doughnut runs, having affairs, contemplating "eating your gun," and patronizing the ultimate "cop bars." What is dangerous about these characterizations is that all of them, to some extent, "glamorize avoiding coping head on with the underlying causes of the stress."[94]

Female officers tend to seek help more often than male officers do. Despite the extra sources of stress they experience (as explained earlier in this chapter), female officers tend to respond directly to serious stressors more often than male officers for the following reasons:[95]

- Women are willing to talk about their feelings.
- Women reject competitiveness with others.

- Women make a conscious effort to reduce stress through actions such as taking time off from work.
- Women seek emotional support from family members and friends.

In one study, male officers reported more family-related problems than did female officers.[96] For instance, one of five male officers reported that work-related stress had often spilled over into his home life, but only 1 of 20 female officers agreed with that notion. Notably, half of the male officers and 46 percent of the female officers reported difficulties in balancing job and family responsibilities. One implication of this finding is that stress results in their personal lives are similar for both female officers and male officers.

"Shedding the Uniform"

Shedding the uniform is hard. Some officers have learned to separate their personal experiences from their professional experiences, but often it is not that easy to draw a line between work and non-work attitudes when interacting with the public.[97] This phenomenon might be easier to explain if officers had similar views about their jobs, but research has shown that officers have divergent views about the ends and means of what their functions really are when they have their uniforms on.[98]

An example of church participation might clarify the "shedding the uniform" difficulty: The congregation knows the individual is an officer, but the officer pretends he's not. The same thing happens at parties. The minute the officer steps in the door, everyone changes their behavior, even though an officer and his or her spouse were invited. "Cops would like to be one of the guys, but they can't," says Richard Walsh, from the Massachusetts District Police in Boston. "They go to a party, and if they're introduced as a cop, the joint which would have normally got passed around does not, the talk about the 'good buy' on a CB radio that normally would have happened does not, and the cop is not able to surrender his job."[99]

Not being able to separate professional from personal experiences and expectations can result in one of the most commonly shared and most critical stress-inducing syndromes—the figurative inability of an officer to shed the uniform.[100] "I get paranoid about it sometimes," says an officer from Salem, Massachusetts. "I went into a bar to have a drink off-duty, and I saw someone there I arrested for drunk driving. So I left being afraid to let him see me drink. Look, I also smoke pot sometimes and maybe that's bad, but I do it. I feel bad because I'm supposed to set an example and I feel there's a terrible double standard when I arrest someone on a pot charge."[101]

Officers can easily empathize with some victims and aggressors because of the realization that being a victim or an aggressor could happen to anyone under certain circumstances. Each time an officer encounters a victim or an aggressor whom the officer can identify with, his or her positive spirits erode.[102] After some time, the mind builds a wall to protect itself from experiencing more pain. When this happens, some officers can act out their feelings by displaying cold, unfeeling behavior or by becoming cynical toward others while on duty and toward their own family members when at home. Education and peer group discussions can initiate action to guard against cynical behavior.[103]

Unrealistic View of the Job

The most common method of preventing stress is to train officers to recognize its sources and signs, and to develop individual strategies for coping with stress.[104] Often it comes down to an officer saying "no." Officers, like members of most other occupations, should take stock of their role as officers and redefine their original illusions about the job.[105] At the top of the list is the eventual reality that an officer is not going to change the world.[106]

Organizational Contributors to Officer Stress

The primary stressor for many officers is linked to organizational factors, rather than solely contained within the individual officer. That is the police organization is the most influential factor that promotes stress among officers and often those factors remain outside an officer's control. When an officer is treated through person-centered strategies and once 'cured,' (assuming the organization has not changed) the officer continues to work or returns to an unchanged police organization.[107] In such circumstances, person-centered interventions can do more harm than good: The "treated" officer once again experiences organizational or environmental stimuli similar to those that prompted the initial treatment, and those stimuli continue to produce the stressors experienced prior to treatment. The result is a vicious circle, in which stress is followed by a respite from stress, only to have stress emerge once again once the police officer reenters the organizational culture.

If the organizational or environmental stressors remain unchanged, an officer can be set up for failure by being coaxed into believing that he or she is, indeed, the problem.[108] Many officers blame themselves for their predicament and see themselves as operating in a vacuum. To feel alone and betrayed by the system could be a normal reaction at times (though if he or she acts on those feelings, the officer's career will undergo a dramatic change). Most officers see themselves and their performance as unique and independent of any influences.

■ Experiences of Stress Counselors

Stress counselors employed by the Massachusetts State Police also treat municipal officers from Boston and other jurisdictions. When they were asked about their difficult critical incident cases, their responses proved quite revealing about the types of stresses encountered by officers.

One counselor's most difficult case involved another officer with an alcohol abuse problem; nothing worked for him.[109] The counselor worked with this officer for six or seven years of the officer's 15-year career on the force. The pair tried everything to turn around his abuse problem, including Alcoholics Anonymous and counseling. The officer was eventually forced to resign because of alcoholism. The counselor describes his experiences after arriving to the scene of a suicide:

> When I got there, I didn't really know what to expect to see. I remember being taken upstairs by one of the detectives. I knew there was a suicide. Yet somewhere in my mind, I didn't really believe that it had taken place, and as I entered the apartment, we went through the apartment, came into the kitchen area, and

> observed this man who had taken his life with a revolver. I guess it's similar to going through a dream. I didn't really believe what I was seeing, and I went through the motions, and I spent a few minutes in the apartment. Later, I was involved with the family. It was a very unsettling situation. . . . I guess after that I always questioned myself as to what I could have done to have avoided this, the disaster that really took place. I questioned myself for quite a while. I guess that there was nothing [more] that I could've done. But, you know, we've seen some really bad ones make it back. So you always had a little thing in your mind of "Gee, what else could I have done, or what else could somebody have done to turn this gentleman's [officer's] life around?"[110]

Another counselor also noted that his experiences with death and dying cases among law enforcement officers were the most difficult for him. However, he adds another dimension to our understanding of what constitutes a "difficult" critical incident case. This Massachusetts State Trooper, whose job description is that of a stress counselor, says that his stress has always been highest when having to fight with the administration to obtain resources to aid law enforcement officers resolve their stress-related problems:

> Knowing that help is both the moral and cost-effective thing to do, and having to take energy away from treatment, to chisel away at some cement-block, self-serving, egocentric, self-righteous thinking really raises my blood pressure. The analogy holds with policing in general . . . you don't mind fighting with the bad guys, you expect them to try to kill you. . . . it pisses you off when you have to fight with the supposed good guys to get what you need to get the job done.[111]

In Chapter 1, it was suggested that most often, police agencies tend to blame the officer for his or her stressed conditions. One way to interpret the preceding message is that agencies tend not to cooperate with units designed to aid stressed officers, while the police subculture adds its influence to the mix, encouraging officers to avoid treatment.

■ Treating the Organization

Evidence continues to mount that the risks of the streets are less traumatic than some organizational factors that officers face—especially policies developed by nonpolice professionals, top-down bureaucratic practices, and policy initiatives linked to strategies developed as a law enforcement response to the war on crime, drugs, and terrorism.[112] Curing officers is one initiative, but "cured" officers must continue in their work environment that stressed them out in the first place. When organizations attempt to cure themselves, they face many challenges:

- Agencies must change the way they deliver police services and the way they manage their personnel to meet the new challenges of the twenty-first century.
- The support of policymakers and police executives should not be limited to treating individual officer stress; rather, preventive (organizational changes) strategies should take priority in controlling stress.
- Police organizations should utilize a public health or disease prevention model (i.e., educate, treat, and prevent) as part of their stress prevention efforts.

- Recovery and rehabilitation are not for everyone.
- Police agencies need to be healthy before they can treat the community's illnesses and injuries.
- The police organization must support and encourage police officers to practice the primary characteristics consistently found among professional personnel.
- Organizations that employ professionals must encourage their personnel to behave in a socially responsible manner, and their personnel should enjoy considerable freedom and responsibility for their sacrifice.
- Policing should have a legitimate claim to a professional status.
- A reasonable control initiative for preventing stress and moving toward a professionalism model in the department is to hire worthy candidates, to appropriately train candidates prior to their becoming officers, and to train them after they become officers to recognize signs and sources of stress and to develop individual coping strategies.
- A serious pre-employment screening process should be established to alert hiring committees as to the risk potential associated with certain types of personalities.
- The characteristics of "good" officers should become standardized, and those characteristics should be used as a benchmark to identify candidates who have the potential to become professional officers and leaders.

Stress reduction might be difficult to achieve immediately because of a series of complex issues related to policing efforts across the United States, which both include internal and external influences. (Chapter 4 outlines reasons why police organizations should change.) Nonetheless, we have learned that politicians, organizational structures, and federal and local intrusions are a greater source of obstructive stress for municipal and rural police officers, investigators, and administrators as opposed to the actual responsibilities and practices of the job or the families of police personnel.[113] This thought is consistent with the findings of many experts who have explored links between police organizational structures, professionalism (or the lack of it), police reform, and recent changes in police performance.[114] Police organizations should provide the leadership that encourages movement toward becoming a professionalized department. Enhancing the professionalism of officers would represent a giant step toward quality police services.

Thinking about Professionalism

Police command should consider supporting and encouraging police officers to practice the primary characteristics consistent with professional personnel. A profession is an occupation requiring extensive education or specialized training.[115] A profession displays three characteristics to some degree or another: responsibility, expertise, and corporateness.[116]

Responsibility is best represented by practicing experts, who work in a social context and perform a service, such as the promotion of health, education, or justice, that is essential to the functioning of society. The client of every profession is the public at large, either individually or collectively. A profession is a moral unit promoting specific values and ideals that guide its members in their interactions with others. This guide can be a set of unwritten rules or norms transmitted through the professional educational system, or it can be codified into written canons of professional ethics.[117]

Professions consist of experts who possess specialized knowledge and skill in a significant field of human endeavor. This expertise can only be acquired through a prolonged combination of education, training, and experience. It is the basis of objective standards of professional competence—that is, the objective is objectivity.

Professional experts share a common sense of organic unity and consciousness of themselves as a person and as a group apart from other occupations—a characteristic known as corporateness. One origin of this collective feeling or set of experiences can be found in the education, training, and experience necessary for professional competence, the common bond of work, and the sharing of a unique social responsibility.

What Is a Profession?

Organizations that employ professionals should encourage their personnel to behave in a socially responsible manner, and their personnel should enjoy considerable freedom and responsibility for their sacrifice. When we think of a profession, generally there are six elements to consider:[118]

- Full-time stable occupation serving a continuum of societal needs (e.g., mental health personnel)
- Lifelong occupation held by practitioners who identify themselves personally with their area of expertise and the subculture of other similar experts (e.g., Broadway performers)
- Organized as a collaborative group that guides boundaries, performance, and recruitment and training (e.g., The APA's decision-making process consists of practicing psychologists, researchers, educators, and staff members engaged in a dialogue without huge concerns of social rank.)
- Formal and informal education, training, and specialized training (e.g., teachers, whose education includes academic certification, internships, and practice semesters at schools)
- Service orientation in which the primary loyalty is to standards of competence and needs of clients as opposed to profit and organizational efficiency (e.g., prosecutors who seek justice rather than profit)
- Individual and collective autonomy because practitioners have proven their high ethical standards and trustworthiness (e.g., college instructors and university professors within academic limits and degree of academic freedom)

A profession holds a monopoly over the occupation and works in its field to further human progress. Autonomy or discretion of the professional is conditional and ultimately depends on continuous social approval. Without constant self-policing and task success, a profession can narrow its own freedom and destroy public trust as rapidly as it gained its relative autonomy. Empowered professional personnel make informed decisions about their own authority, standards, and autonomy. Table 15–1 summarizes other attributes associated with professionalism.[119]

TABLE 15–1 List of Professional Attributes

- Full-time occupation
- Client-centered
- Service ideal
- Caters to human needs (not wants)
- Efficiency moving towards client-needs mission
- Collaboration between participants of all ranks to enhance profession
- Self-policing
- Ethics
- Pecuniary profit is not primary objective
- Esoteric language (job specific words and phases)
- Professional associations, symbols, artifacts
- Autonomy/Discretion/Judgment

Is Policing a Profession?

Policing should have a legitimate claim to a professional status, given its stance on the three essential characteristics for a profession.

In terms of *responsibility*, police officers are practitioners who have clear expertise in their line of work, and they perform a social service essential to the individual and the collective functioning of society. Policing is a moral, ethical unit that maintains specific formal and informal values and promotes "greater good" ideals to its members and the public.

Concerning *expertise*, policing is characterized by formal and informal norms and policies that are disseminated through a professional training and educational system. The profession is self-policing in terms of dealing with the indiscretions of its members, usually through senior officers trained to specifically deal with those indiscretions. Officers are entrusted with greater authority than their counterparts in most occupations because they have both a professional and bureaucratic mandate to enforce those standards.

Finally, *corporateness* is seen in the unity of street officers, who are bound by a cooperative sense of union based on shared expectations, lifestyles, and obligations. Overall, a police department looks to ethics and integrity as touchstones for judgment and service and, therefore, inherently shares elements of a profession. Some might even talk about a police culture to help make this point.

How many matches for the previously cited criteria can you find that would suggest policing is a profession? How closely does this perspective represent the performance of most officers on the street? In a hierarchical bureaucratic organizational structure, where orders come from the top down and reports come from the bottom up, the choice is easy—either change the definition of a profession or change the way police departments are managed.

RIPPED from the Headlines

Departments Increasingly Look to Help Officers Cope

One moment the man who had just fought Millington, Tennessee, police officers lay handcuffed on the ground, ranting. Then he stopped. Police Sergeant Troy Walls said paramedics had just arrived to treat an officer whose knee was injured in the fight, but Walls directed them to provide care instead to the now-silent man.

Malcolm J. Carruthers, the subdued man later found to have suffered from heart disease and mental illness, died that day in March. Walls said he and the other three officers involved will never forget.

"Do I hate it for his family?" Walls asked in a recent interview. "Yes, my God, they have to live with it in one way every day.

"We have to live with it in another way every day. Does his name ever leave our heads? No, he's with us, somebody we barely knew is a part of our lives as long as we live," he said.

The public seldom hears from officers thrust into what are known as "critical incidents"—traumatic deaths and injuries, at times even caused by decisions an officer makes.

Traditionally, cops rarely talked about the emotional toll that the incidents take on their own lives. Like macho Western movie actor John Wayne, they clammed up and kept going.

Talking about it would make them appear weak. Higher-than-average rates of suicide, alcoholism, and divorce, and shorter-than-average life expectancies followed.

"I know of one particular officer that I can tell you without a doubt that he died over the stress of a traumatic incident that he was involved in, because he just grieved himself to death," said Shelby County Sheriff's Office Inspector Mark Dunbar.

But in recent years, a process known as "critical incident stress debriefing" is providing law enforcement and corrections officers, as well as other first responders, with an outlet to vent pent-up emotions.

Shelby County Sheriff Mark Luttrell, 60, who has fostered the process for deputies and jailers since becoming sheriff in 2002, described it as "cops taking care of cops." "I can give you cases, upon cases, upon cases of officers that basically say that it saved their lives," Dunbar said.

Within days of a critical incident, officers involved are gathered together with volunteer mental health professionals and other officers who have experienced similar trauma. The privacy of what is said in the meetings is protected by state law.

After Tennessee State Trooper Calvin Jenks, 24, was shot to death in Tipton County in January, about 150 people, including troopers, deputies, and officers, gathered in small groups at Hope Presbyterian Church, said Luttrell.

"At the end of the day there were a lot of tears on the floor, a lot of wet eyes, and people left there a little bit emotionally drained but certainly emotionally more stable than they'd been leading up to that," Luttrell said.

Trained officers from Shelby County and elsewhere in Tennessee journeyed to the Mississippi Gulf Coast to debrief officers there after Hurricane Katrina in 2005.

"Ninety percent of the day we deal with the worst part of society, but on good days that other percentage that we get to see, we help a lot of people."

The day Carruthers fell silent was the second critical incident for Walls since he joined the department as a reserve officer in 1993. In October 2004, Walls fatally shot a man, Mark Chumley, 34, on Harrold Cove after Chumley refused to drop a shotgun. "At Christmas time, you're thinking about, did he have kids? It's not something that happens and it's over," Walls said.

Source: Kevin McKenzie, "Departments Increasingly Look to Help Officers Cope," *Police.one* (July 19, 2007), http://www.officer.com/online/article.jsp?siteSection=1&id=36942 (accessed July 19, 2007).

Summary

Stress is defined as the normal wear and tear on our bodies and minds that is experienced as we react to physical, psychological, and environmental changes in our daily lives. It is a necessity for stress to be present in our daily lives. Distress (bad stress) can result in posttraumatic stress and death, whereas eustress (positive stress) helps in the normal processing of lives. Sources of stress for police officers include daily life and family, gender, working in a rural department, personality, and job-related sources.

A critical incident is any situation beyond the realm of an individual's usual experience that overwhelms his or her sense of vulnerability and a lack of control over the situation. Mandatory debriefing sessions among officers exposed to critical incidents is recommended to allay the stress generated by such events, even though critical incident experiences can help officers develop an intuitive policing understanding of police work.

Treating stress individually focuses on helping officers develop coping skills. Treating stress through departmental person-centered stress providers is a huge undertaking. This kind of employee assistance program may incorporate the services of peers, volunteers, or professional providers as part of an external, in-house, or hybrid program.

The stress that arises among officers from working on the streets is often less serious than the stress associated with the organizational factors they face. Sources of organizational stress include policies developed by nonpolice professionals, top-down bureaucratic practices, and policy initiatives linked to strategies developed as a law enforcement response to the war on crime, drugs, and terrorism.

Key Words

Cognitive-behavioral intervention (CBI)
Coping
Crisis response
Critical incident
Critical incident stressors
Debriefing
Distress
Eustress
General work stressors
Intuitive policing
Person-centered intervention
Post-traumatic stress disorder
Primary victimization
Secondary victimization
Stress
Type A personality
Type B personality

Discussion Questions

1. Define stress, and explain in what ways it is positive and negative.
2. Characterize the key to understanding the scope of stress. In what way do you agree with these characterizations? In what way do you disagree with them?
3. Characterize bad stress and explain its consequences for officers at home, at work, and in terms of their mental and physical health.
4. Identify and explain the general sources of stress among most police officers.
5. Define critical incidents, and explain their relationship to critical responses.
6. Identify advantages and consequences of critical incidents for officers.
7. Describe intuitive policing. In what way do you agree with this perspective? In what way do you disagree with it?
8. Explain the relevance of debriefing after a critical incident and its advantages for officers.
9. Describe and identify general work stressors.
10. Characterize individual remedies to control stress. In what way do you agree or disagree with the recommendations presented in this chapter?
11. Characterize stress treatment through person-centered programs.
12. Identify the basic options available in terms of person-centered stress management programs. Which method do you think has the most to offer officers? Why?
13. Identify the obstacles associated with person-centered stress management programs.
14. Articulate the primary issues linked to the medication of stress.

Notes

1. William C. Dement, Stanford University Center of Excellence for the Diagnosis and Treatment of Sleep Disorders, 4th Annual FOCUS Conference on Respiratory Care and Sleep Medicine, Baltimore, MD, April 23, 2004.
2. John P. Crank, *Understanding Police Culture*, 2nd ed. (Cincinnati: Anderson, 2004), 276–78; Dennis J. Stevens, *Police Officer Stress: Sources and Solutions* (Upper Saddle River, NJ: Prentice Hall, 2008), 2.
3. Katherine W. Ellison, *Stress and the Police Officer*, 2nd ed. (Springfield, IL: Charles C. Thomas, 2004), 62.
4. Page 34 in Patricia A. Kelly, "Stress: The Cop Killer," in *Treating Police Stress: The Work and the Words of Peer Counselors*, ed. John M. Madonna, Jr., and Richard E. Kelly (Springfield, IL: Charles C. Thomas, 2002), 33–54.
5. Sandra D. Terhune-Bickler, "Too Close for Comfort: Negotiating with Fellow Officers," *Federal Law Enforcement Bulletin* 73, no. 4 (2004), http://www.fbi.gov/publications/leb/2004/apr2004/april04leb.htm#page_2 (accessed May 30, 2005).
6. Captain Robert Dunford, Commander of Boston Law Academy and Superintendent of Uniform Division, Boston Police Department, Forum at University of Massachusetts–Boston, 2003.
7. Stevens, *Police Officer Stress*, 30. Also see Dennis J. Stevens, "Police Officer Stress and Occupational Stressors," in *Policing and Strategies*, ed. Heith Copes (Upper Saddle River, NJ: Prentice Hall, 2005), 1–24; Dennis J. Stevens, "Origins of Police Officers' Stress Before and After 9/11," *Police Journal* 77, no. 2 (2004): 145–74; Dennis J. Stevens, "Police Officer Stress," *Law and Order* (1999): 77–81.

8. Herbert Benson and Eileen M. Stuart, *The Wellness Book: The Comprehensive Guide to Maintaining Health and Treating Stress Related Illness* (New York: Fireside, 1992).
9. Herbert Benson, "The Mind Body Medical Institute" (2005), http://www.mbmi.org/pages/mbb_s1.asp (accessed March 4, 2006).
10. Walter B. Cannon, *Bodily Changes in Pain, Hunger, Fear, and Rage—An Account of Recent Researches into the Function of Emotional Excitement: 1927* (NY: Cannon Press, 2007), 4–8.
11. Dennis J. Stevens, "Police Officer Stress," in *Encyclopedia of Psychology and Law* (Los Angeles: Sage, 2007), Vol II, 1025.
12. Page 183 in J. F. Volpe, "A Guide to Effective Stress Management," *Law and Order* (October 2000): 183–88.
13. Wayne W. Bennett and Karen M. Hess, *Management and Supervision in Law Enforcement* (Belmont, CA: Wadsworth, 2008), 403.
14. Brain J. Gorman, "Attitude Therapy for Stress Disorders" (2003), http://www.stressdoctor.com/index.htm (accessed April 24, 2005).
15. "Psychology Glossary," http://www.alleydog.com/ (accessed March 4, 2006).
16. Robin Gershon, *Public Health Implication of Law Enforcement Stress* [video presentation], U.S. Department of Justice, National Institute of Justice, Washington, DC, March 23, 1999.
17. David L. Carter and Lewis A. Radelet, *The Police and the Community* (Upper Saddle River, NJ: Prentice Hall, 1999); Mark L. Dantzker and M. A. Surrette, "The Perceived Levels of Job Satisfaction among Police Officers: A Descriptive Review," *Journal of Police and Criminal Psychology* 11, no. 2 (1996): 7–12.
18. Donald C. Sheehan and Vincent B. Van Hasselt, "Identifying Law Enforcement Stress Reactions Early," *FBI Law Bulletin* 72, no. 9 (2003), http://www.fbi.gov/publications/leb/2003/sept2003/sept03leb.htm#page_13 (accessed July 21, 2004).
19. Douglas Paton, "Critical Incidents and Police Officer Stress," in *Policing and Stress*, ed. Heith Copes (Upper Saddle River, NJ: Prentice Hall, 2005), 25–40; John H. Pearn, "The Victor as Victim: Stress Syndromes of Operational Service" (2001), http://www.defence.gov.au/dpe/dhs/infocentre/publications/journals/NoIDs/ADFHealthNov99/ADFHealthNov99_1_1_30-32.pdf (accessed May 24, 2005). Also see *Diagnostic and Statistical Manual of Mental Disorders*, 4th ed. (Washington, DC: American Psychiatric Association, 1994), 431–33. Note: APA does not discuss counter-disaster syndrome.
20. "Acute Stress Disorder," *Psychology Today* (n.d.), http://cms.psychologytoday.com/conditions/acutestress.html (accessed May 9, 2005).
21. *Diagnostic and Statistical Manual of Mental Disorders*, 424.
22. Sheehan and Van Hasselt, "Identifying Law Enforcement Stress Reactions Early."
23. E. J. Ozer, S. R. Best, T. L. Lipsey, and D. S. Weiss, "Predictors of Post-traumatic Stress Disorder and Symptoms in Adults: A Meta-analysis," *Psychological Bulletin* 129, no. 1 (2003): 52–73.
24. Nancy Jo Dunn, Elisia Yanasak, Jeanne Schillaci, et al., "Personality Disorders in Veterans with Posttraumatic Stress Disorder and Depression," *Journal of Traumatic Stress* 17, no. 1 (2004), 75–82.
25. Donald C. Sheehan and Janet I. Warren, eds., *Suicide and Law Enforcement* (Washington, DC: U.S. Department of Justice, Federal Bureau of Investigation, 2002), 205–09.
26. Paige Ouimette, Ruth Cronkite, Annabel Prins, and Rudolf H. Moos, "Posttraumatic Stress Disorder, Anger and Hostility, and Physical Health Status," *Journal of Nervous & Mental Disease* 192, no. 8 (2004): 563–66.
27. Samuel Walker and Charles M. Katz, *The Police in America: An Introduction*, 6th ed. (Boston, McGraw Hill, 2008), 185–86. Also see Hans Toch, *Stress in Policing* (Washington, DC, American Psychological Association, 2002).

28. Benson, "The Mind Body Medical Institute."

29. Officer David Murphy, Newbury, Massachusetts, police department, as reported to assistant researcher Kerri Mahoney, April 26, 2005.

30. Bennett and Hess, *Management and Supervision in Law Enforcement*, 405.

31. Kelly, "Stress."

32. Jennifer M. Brown and Elizabeth A. Campbell, "Sources of Occupational Stress in the Police," *Work and Stress* 4 (1990): 305–18.

33. Kelly, "Stress," 40.

34. Merry Morash and Robin N. Haarr, "Gender, Workplace Problems, and Stress in Policing," *Justice Quarterly* 12 (1995): 113–40. Also see Robin Haarr and Merry Morash, "Police Coping with Stress: The Importance of Emotions, Gender, and Minority Status," in *Policing and Stress*, ed. Heith Copes (Upper Saddle River, NJ: Prentice Hall, 2005), 158–77.

35. Stevens, *Police Officer Stress*, 187.

36. Merry Morash and Jack R. Greene, "Women on Patrol: A Critique of Contemporary Wisdom," *Evaluation Review* 10 (1986): 230–55.

37. Officer Marie Durrant, Fayetteville Police Department. The author was riding with this officer; she was a university student of the author.

38. "Officer Down Memorial Page," http://www.odmp.org/officer.php?oid=15818 (accessed May 17, 2007).

39. Larry Smith, "Meet our 2005 Police Officer of the Year," *Parade Magazine* (September 18, 2005): 4.

40. P. Morrison, "Female Officers Unwelcome—But Doing Well," *Los Angeles Times* (July 12, 1991). Additional literature on stress for female, gay, and ethnic officers provided by the National Center on Women and Policing, 8105 West Third St., Suite 1, Los Angeles, CA 90048, (213) 651–0495, http://www.womenandpolicing.org/publications.asp (accessed April 20, 2008). Also see C. Fletcher, *Breaking and Entering: Women Cops Talk about Life in the Ultimate Men's Club* (New York: Harper Collins, 1995).

41. Willard M. Oliver, "The Four Stress Factors Unique to Rural Patrol Revisited," *Police Chief* 71, no. 11 (2007): 44–47.

42. For information on how to change a type A personality, see Carol F. Lankton, "Transforming a Type A Personality" (November 11, 1998), http://goinside.com/98/11/typea1.html (accessed May 3, 2005).

43. Vijai P. Sharma, *Characteristics of "Type A" Personality* (Mind Publications, 2004), http://www.mindpub.com/art207.htm (accessed May 3, 2005).

44. Sharma, *Characteristics of "Type A" Personality.*

45. Michael S. McCampbell, "Field Training for Police Officers," in *Critical Issues in Policing: Contemporary Readings*, 4th ed., ed. Roger G. Dunham and Geoffrey P. Alpert (Prospect Heights, IL: Waveland, 2001), 107–16.

46. David H. Bayley and Egon Bittner, "Learning the Skills of Policing," in *Critical Issues in Policing: Contemporary Readings*, 4th ed., ed. Roger G. Dunham and Geoffrey P. Alpert (Prospects Heights, IL: Waveland, 2001), 82–106. Also see McCampbell, "Field Training for Police Officers."

47. Roger Soloman, *Police wives.org* (2004), http://www.policewives.org/modules.php?name=News&new_topic=2 (accessed May 22, 2005).

48. Daniel Goldfarb and Gary S. Aumiller, "Critical Issues: Stress Reductions," *The Heavy Badge* (2005), http://www.heavybadge.com/cisd.htm (accessed May 28, 2005).

49. Roger J. Dodson, "Critical Incident Stress" (NCJ 176330; 2000), http://www.ncjrs.org/txtfiles1/nij/176330.txt (accessed May 23, 2005).

50. M. MacLeod and Douglas Paton, "Police Officers and Violent Crime: Social Psychological Perspectives on Impact and Recovery," in *Police Trauma: Psychological Aftermath of Civilian Combat*, ed. J. M. Violanti and Douglas Paton (Springfield, IL: Charles C. Thomas, 1999), 25–36.

51. MacLeod and Paton, "Police Officers and Violent Crime."

52. Added by the author based on discussions with officers involved in critical incidents.

53. Added by the author based on discussions with officers involved in critical incidents.

54. Gershon, *Public Health Implication of Law Enforcement Stress.*

55. *Diagnostic and Statistical Manual of Mental Disorders*, 432.

56. Han Toch, *Stress in Policing* (Washington, DC: American Psychological Association, 2001), 180.

57. W. A. Geller, "Officer Restraint in the Use of Deadly Force: The Next Frontier in Police Shooting Research," *Journal of Police Science and Administration* 10, no. 2 (1985): 151–77.

58. R. C. Davis, *SWAT Plots: A Practical Training Manual for Tactical Units* (Washington DC: U.S. Government Printing Office, 1998). Also see Stevens, *Police Officer Stress*, 132–33.

59. J. T. Mitchell, "When Disaster Strikes: The Critical Incident Debriefing Process," *Journal of Emergency Medical Services* 8 (1998): 36–39.

60. George S. Everly, Jr., "Five Principles of Crisis Intervention: Reducing the Risk of Premature Crisis Intervention" (2000), http://www.icisf.org/articles/Acrobat%20 Documents/TerrorismIncident/5princip.pdf (accessed June 1, 2005).

61. Everly, "Five Principles of Crisis Intervention."

62. *The Psychological Dictionary*, http://allpsych.com/dictionary/h.html (accessed June 2, 2005).

63. This interpretation is based on the author's observations.

64. Anthony J. Pinizzotto, Edward F. Davis, and Charles E. Miller, "Intuitive Policing: Emotional/Rational Decision in Law Enforcement," *FBI Law Enforcement Bulletin* (February 2004), http://www.au.af.mil/au/awc/awcgate/fbi/intuitive.pdf (accessed December 19, 2007).

65. Samuel Walker and Charles M. Katz, *The Police in America: An Introduction*, 5th ed. (New York: McGraw Hill, 2008).

66. Toch, *Stress in Policing.*

67. Bureau of Justice Statistics, "Full Time Officers Assigned to Calls in Local Police Departments," *Sourcebook of Criminal Justice Statistics, 2002* (2002), Table 1.32, http://www.albany.edu/sourcebook/pdf/t132.pdf (accessed June 7, 2005).

68. Stevens, *Police Officer Stress*, 30

69. Crank, *Understanding Police Culture*, 276–78.

70. Benson and Stuart, *The Wellness Book.*

71. This particular book is used at the Federal Law Enforcement in Glencoe, Georgia, to aid law enforcement personnel in dealing with many of the issues that confront them.

72. *Dictionary.com* (2005), http://dictionary.reference.com/search?q=cope (accessed July 25, 2005).

73. Edward A. Charlesworth and Ronald G. Nathan, *Stress Management: A Comprehensive Guide to Wellness* (New York: Ballantine, 1984).

74. Benson and Stuart, *The Wellness Book.*

75. Patricia Carrington, *How to Relax* (New York: Warner Audio Publishing, 1985).

76. Joseph A. Harpold and Samuel L. Feemster, "The Negative Influences of Police Stress," *FBI Law Enforcement Bulletin* (September 2002), http://articles.findarticles.com/p/articles/mi_m2194/is_9_71/ai_92285044 (accessed June 22, 2004).

77. Harpold and Feemster, "The Negative Influences of Police Stress."

78. Philip Trapasso, "Traumatic Incident Reaction in Law Enforcement," in *Treating Police Stress*, ed. John M. Madonna, Jr., and Richard E. Kelly (Springfield, IL: Charles C. Thomas, 2002), 55–68.

79. Jac J. van der Klink, Roland W. Blonk, Aart H. Schene, and Frank J. H. van Diik, "The Benefits of Intervention for Work-Related Stress," *American Journal of Public Health* 91, no. 2 (2001): 270–76.

80. "Health A to Z" (2007), http://www.healthatoz.com/ (accessed December 14, 2007).

81. Julie Esselman-Tomz and Peter Finn, "Developing a Law Enforcement Stress Program for Officers and Their Families," *National Institute of Justice* (1997), http://ncjrs.org/txtfiles/163175.txt (accessed June 18, 2004).

82. Ellen M. Scrivner, *Controlling Police Use of Excessive Force* (NCJ 150063) (National Institute Justice, 1994), http://www.ncjrs.org/txtfiles/ppsyc.txt (accessed June 22, 2004).

83. "Counseling International," http://www.thecounselingteam.com/agencies.html (accessed March 13, 2006).

84. John M. Madonna, Jr., and Richard E. Kelly, *Treating Police Officers: The Work and the Words of Peer Counselors* (Springfield, IL: Charles C. Thomas, 2002), 7–23.

85. Fraternal Order of Police, http://www.grandlodgefop.org/ (accessed February 26, 2006).

86. New Jersey State Lodge, Fraternal Order of Police, http://www.njfop.org/cop2cop.html (accessed February 26, 2006).

87. In learning theory, an action or other unit of behavior that does not appear to have a stimulus.

88. G. Dunlop, L. Kern, M. dePerczel, et al., "Functional Analysis of Classroom Variables for Students with Emotional and Behavioral Disorders," *Behavioral Disorders*, 18 (1993): 275–91; V. M. Durand, *Severe Behavior Problems: A Functional Communication Training Approach* (New York: Guilford, 1990). Also see B. A. Iwata, T. R. Vollmer, J. R. Zarcone, and T. A. Rodgers, "Treatment Classification and Selection Based on Behavioral Function," in *Behavior Analysis and Treatment*, ed. R. Van Houten and S. Axelrod (New York: Plenum, 1993), 101–25.

89. M. M. Bandura and C. Goldman, "Expanding the Contextual Analysis of Clinical Problems," *Cognitive and Behavioral Practice* 2 (1995): 119–41; J. D. Cone, "Issues in Functional Analysis in Behavioral Assessment," *Behaviour Research and Therapy* 35 (1997): 259–75; J. A. Miller, M. Tansy, and T. L. Hughes, "Functional Behavioral Assessment: The Link between Problem Behavior and Effective Intervention in Schools," *Current Issues in Education* 1, no. 5 (1998), http://cie.ed.asu.edu/volume1/number5/ (accessed June 17, 2004).

90. Renee B. Meador, "Model Policy and Procedures for Critical Incident Stress Management/Critical Incident Stress Debriefing for Law Enforcement Internal Programs: Virginia Critical Incident Stress Management, Law Enforcement Critical Incident Stress Management" (2004), http://www.geocities.com/~halbrown/cism_cisd_procedures.html (accessed June 18, 2004).

91. Hal Brown, "Am I Stressed Out?" (2004), http://www.geocities.com/~halbrown/index1.html (accessed June 19, 2004). Brown is a police clinician who has treated officers for stress for most of his professional life.

92. Guided in part by Dennis L. Conroy and Karen M. Hess, *Officers at Risk: How to Identify and Cope with Stress* (Placerville, CA: Custom Publishing, 1992), 210–11.

93. Ellison, *Stress and the Police Officer*, 6.

94. Brown, "Am I Stressed Out?"

95. Morash and Haarr, "Gender, Workplace Problems, and Stress in Policing." This paper focuses on the connection between workplace problems with stress for women and for men working in police departments. Field research was used to identify the problems that women experience in police departments, and quantitative measures were developed to measure these problems in a survey of women and men in 25 departments. Although women and men experience many of the same work-related problems, and although such problems account for a high proportion of workplace stress in both groups, the gendered nature of police organizations causes unique stressors for women. Overall, however, women do not report higher levels of stress than men.

96. Toch, *Stress in Policing*, 106.

97. William Terrill, Eugene A. Paoline, and Peter Manning, "Police Culture and Coercion," *Criminology* 41, no. 4 (2003), 1003–34.

98. Robert E. Worden, "Police Officers' Belief Systems: A Framework for Analysis," *American Journal of Police* 14, no. 1 (1995): 49–81.

99. Kelly, "Stress," 47.

100. Kelly, "Stress," 47.

101. Kelly, "Stress," 47.

102. Gershon, *Public Health Implication of Law Enforcement Stress.*

103. Eugene R. D. Deisinger, "Executive Summary of the Law Enforcement Assistance and Development (LEAD) Program: Reduction of Familial and Organizational Stress in Law Enforcement" (NCJ 192276; 2002), http://www.ncjrs.org/pdffiles1/nij/grants/192276.pdf (accessed June 23, 2004).

104. Meador, "Model Policy and Procedures."

105. Hal Brown, "Depression" (2004), http://www.geocities.com/~halbrown/depression_041002.html (accessed June 19, 2004).

106. Eugene A. Paoline, *Rethinking Police Culture: Officers' Occupational Attitudes* (New York: LFB, 2001).

107. James J. Messina and Constance M. Messina, "Tools for Personal Growth" (2006), http://www.coping.org/growth/beliefs.htm#What (accessed April 4, 2006).

108. Thomas Griggs, Thomas Caves, and Edward S. Johnson, "Reaching Out to North Carolina's Law Enforcement Community" (NCJ 188874; 2001), http://www.ncjrs.org/pdffiles1/nij/grants/188874.pdf (accessed June 24, 2004).

109. Page 85 in John M. Madonna, Jr., "The Tough Moments and the Good Ones," in *Treating Police Stress*, ed. John M. Madonna, Jr., and Richard E. Kelly (Springfield, IL: Charles C. Thomas, 2002), 83–108.

110. Madonna, "The Tough Moments and the Good Ones," 85.

111. Madonna, "The Tough Moments and the Good Ones," 89. The counselor talking is Richard Kelly of the Massachusetts State Police Stress Unit.

112. Ellison, *Stress and the Police Officer*, 6; Madonna and Kelly, *Treating Police Officers*; Stevens, *Police Officer Stress*, 47; Toch, *Stress in Policing*, 61.

113. Toch, *Stress in Policing*. Also see Stevens, *Police Officer Stress*, 382.

114. David L. Carter, *The Police and the Community* (Upper Saddle River, NJ: Prentice Hall, 2001). Also see Hans Toch, *Police Stress* (Washington, DC: American Psychological Association, 2001).

115. *MSN Encarta—Dictionary*, http://encarta.msn.com/encnet/features/dictionary/DictionaryResults.aspx?search=profession (accessed July 2, 2004).

116. Samuel P. Huntington, *Soldier and the State: The Theory and Politics of Civil–Military Relations* (New York: Belknap, September 1981).

117. Huntington, *Soldier and the State*, was used as a guide. The interpretation of his work is that of this author.

118. Allan R. Millett is a Professor of Military History at the Ohio State University and the author of many military books.

119. Carter and Radelet, *The Police and the Community*, 133–16, was used as a guide.

Index

Page numbers followed by *f* or *t* denote figures or tables respectively.

B

E

F

J

K

N

O

P

Q

S

T

U

Y

Z

Photo Credits

Chapter 1

Page 2 © Christopher Penler/ShutterStock, Inc.; Page 6 © Fara Spence/ShutterStock, Inc.; Page 9 © Photos.com; Page 19 © Carlos Villalon/Reuters/Landov; Page 21 © Paul Warner/AP Photos

Chapter 2

Page 34, Page 46 Courtesy of Greater Manchester Police Museum and Archives; Page 49 Courtesy of Library of Congress, Prints & Photographs Division, [reproduction number LC-USZ62-2582]; Page 53 Courtesy of National Archives

Chapter 3

Page 66 © Eva Madrazo/ShutterStock, Inc.; Page 74 © Isaak/ShutterStock, Inc.; Page 76 © Martine Oger/ShutterStock, Inc.; Page 82 © Jeff Greenberg/Alamy Images

Chapter 4

Page 98 © Jim Parkin/ShutterStock, Inc.; Page 107 © Jack Dagley Photography/ShutterStock, Inc.

Chapter 5

Page 134 © Dennis Van Tine/Landov; Page 141 Courtesy of SJPD; Page 147 Courtesy of Harris County Constable Precinct 4

Chapter 6

Page 176 © Cyril Hou/ShutterStock, Inc.

Chapter 7

Page 208 © Mark C. Ide; Page 214 Courtesy of Arlington, Texas Police Department; Page 225 Courtesy of Greenbelt Maryland Police Department

Chapter 8

Page 250 © John Birdsall/age fotostock; Page 271, Page 273 © Mark C. Ide

Chapter 9

Page 292 © UpperCut Images/age fotostock; Page 298 © Mark C. Ide; Page 309 © Jack Dagley Photography/ShutterStock, Inc.; Page 317 Courtesy of Tucson Police Department; Page 321 Courtesy of Des Moines Police Department; Page 323 © Lawrence Roberg/ShutterStock, Inc.; Page 323 © Igor Karon/ShutterStock, Inc.; Page 324 Courtesy of the Fort Lauderdale Police Department

Chapter 10

Page 334 © Sergey I/ShutterStock, Inc.; Page 340, Page 342 © Mark C. Ide; Page 352 © Reuters/Landov; Page 359 © Chinatopix/AP Photos

Chapter 11

Page 376 © Mark C. Ide

Chapter 12

Page 414 © Luis Galdamez/Reuters/Landov

Chapter 13

Page 456 © Photos.com; Page 477 © Brian Bates, *Handout*/Reuters/Landov; Page 477 © Shannon Stapleton/Reuters/Landov

Chapter 14

Page 500 © Corbis/age fotostock; Page 519 © Pilar Echevarria/ShutterStock, Inc.; Page 525 © Crystalcraig/Dreamstime.com

Chapter 15

Page 538 © Gene Blevins/Reuters/Landov; Page 547 © Kenneth Lambert/AP Photos; Page 547 © Mark C. Ide; Page 550 © Yoon S. Byun, *Boston Globe*/Landov